a

MINERAL REFERENCE MANUAL

Ernest H. Nickel
Monte C. Nichols

VNR VAN NOSTRAND REINHOLD
New York

Library of Congress Catalog Card Number 90-21739

ISBN 0-442-00344-7

Printed in the United States of America.

Van Nostrand Reinhold
115 Fifth Avenue
New York, New York 10003

Chapman and Hall
2-6 Boundary Row
London, SE1 8HN, England

Thomas Nelson Australia
102 Dodds Street
South Melbourne 3205
Victoria, Australia

Nelson Canada
1120 Birchmount Road
Scarborough, Ontario MIK 5G4, Canada

16 15 14 13 12 11 10 9 8 7 6 5 4 3 2 1

Library of Congress Cataloging-in-Publication Data

Nichols, Monte C., 1938–
Mineral reference manual / Ernest H. Nickel, Monte C. Nichols.
p. cm.
Includes bibliographical references.
ISBN 0-442-00344-7
1. Minerals. I. Nickel, Ernest H., 1925- . II. Title.
QE372.2.N53 1991
549—dc20

90-21739
CIP

Introduction

This manual represents an alphabetical listing of all valid mineral species and includes the name, formula, current status, crystal system, appearance, hardness, measured and calculated density, type locality, mineral classification, a reference to the origin of the name, an indication of related species, and selected literature references for each species including one for the crystal structure determination when available. This compilation utilizes open literature references available to the authors through May, 1990.

The data herein are a subset of a comprehensive computerized database compiled by the authors and currently marketed by Aleph Enterprises under the name "MINERAL — a Mineral Reference Book for the IBM PC"; information on this database can be obtained from Aleph Enterprises, PO Box 213, Livermore, California, 94551, USA.

Because of space limitations, only a subset of the information from the MINERAL database is provided in this manual. Systematic searches (cross referencing) based on chemical composition, crystallographic parameters, etc. are much more efficiently performed using a computer rather than a book and are not provided for here.

The first item for each entry in the book is the name. Mineral names follow the recommendations of the Commission on New Minerals and Mineral Names (CNMMN) of the International Mineralogical Association (IMA). Following the mineral name is the chemical formula. For the chemical formulas, we have made use of information from crystal structure determinations when available, tempered by rules for the nomenclature of inorganic compounds as put forth by the International Union of Pure and Applied Chemistry (IUPAC). The status of each mineral is indicated by a symbol in a square box immediately after the chemical formula. The following symbols are used: [A] — approved by the CNMMN [1466]; [D] — discredited by the CNMMN [248]; [G] — "Grandfather" status, ie. minerals introduced before the formation of the CNMMN and generally accepted by the mineralogical community [1784]; [P] — polytypes not specifically given approved species status by the CNMMN [63]; and [Q] — questionable status, generally reserved for those species not specifically discredited by the CNMMN but which haven't been accepted either [232]. The numbers in square brackets represent the number of minerals having the designated status.

The next entry indicates the crystal system for the mineral. The abbreviations used include: Cub. — cubic; Tet. — tetragonal (quadratic); Hex. — hexagonal; Rhom. — rhombohedral; Orth. — orthorhombic; Mon. — monoclinic; Tric. — triclinic (anorthic); Amor. — amorphous. Following the crystal system is a brief description of the appearance of the mineral.

This is followed by the type locality, which may be incomplete or lacking for some minerals. We have generally followed the example of Chambers World Gazetteer in that we have used the anglicized version of the country name but have tried to retain local names for subdivisions within a given country. For some local names the anglicized version of a given name is included in parentheses although most local names lack this designation. For example we use Sardegna for Sardinia, Peimonte for Piedmont, Toscana for Tuscany, Roma for Rome, etc. In rare cases we have retained the English version of the name, Mt. Vesuvius, for example. Directional indications appear in parentheses and indicate the portion of a geographical subdivision in which the locality is to be found, Arizona (SW) indicates that the locality is to be found in southwestern Arizona. When used with a town or city this designation indicates the direction in which the locality is to be found, Tucson (SW) indicates that the locality is to be found southwest of Tucson.

Next comes the hardness (Mohs), and the measured and calculated densities (D_m and D_c respectively).

The multiplicity of approaches to the designation of mineral groups has resulted in no small confusion (disagreement), and has caused us to avoid using mineral group names in this edition of the book. We look forward to a consensus on this issue by the mineralogical community. In place of groups we have chosen to use a classification based on a combination of a Roman numeral, an alphabetic character and an Arabic numeral. This designation constitutes the classification symbol based on the system in *Mineralogische Tabellen* by Hugo Strunz (a summary is shown in Appendix A). The Roman numeral and alphabetical character divide the minerals into broad groups based mainly on chemistry, with the silicate subdivision based on silicate polymerization. The Arabic numeral specifies the group to which a particular mineral belongs; a group is defined mainly on the basis of crystallographic similarities, but minerals with similar chemical compositions are sometimes included in the group even though they do not exhibit crystallographic affinities with other members of the class.

The next entry gives the name of the author of the mineral and the date of its introduction into the literature. In some cases where a mineral has been redescribed or the name has been changed from the original, this entry will reflect the person and time when this was done. This is followed by a reference to what we have tried to make the latest, or most authoritative, literature report. We have attempted to include at least one reference for each species. If there is a second reference following the first, it will be a reference to the determination of the crystal structure. The abbreviations used in the references are given in Appendix B.

The final entry, when present, gives the names of similar or related minerals and is preceded by the designation "See also".

There are two other entry formats included in the body of the text. One consists of minerals which have been discredited by the CNMMN since its inception and includes the discredited name, the reason it was discredited, a reference entry, and the originator of the mineral name. The other type of entry is for names which, in the past, were considered mineral names but are now generally accepted to represent groups of minerals, mica for example. The mineral name is present in a non-bold font for both of these categories as contrasted to the valid species for which the name is in a bold-faced font. Including these two formats, there are more than 3800 mineral entries represented in this book.

An abridged synonymy is included as Appendix C. The names included consist mainly of more "recent" synonyms, especially for minerals whose names may have been changed in the recent past. It is designed to be consulted when a reader fails to find a given name in the main text.

On the page following this introduction we include an abbreviated listing of elements and elemental groupings along with the number of occurrences of each in the chemical formulas of this book and the MINERAL database mentioned above.

A number of crystal drawings are also included at various places throughout the book. While many of them have been patterned after drawings in the mineralogical literature, they have all been completely redrawn using the SHAPE program available from Eric Dowty, 521 Hidden Valley Road, Kingsport, Tennessee, 37663, USA.

The authors appreciate the advice and kind assistance of the following people in the preparation of this manuscript: Chr. W. Bauditz, Vejby, Denmark; Pierre Perroud, Genève, Switzerland; Jiri Just, Perth, Australia; Richard Thomssen, Carson City, Nevada, USA; Peter Embrey, London, England; Jessie Hardman, Long Beach, California, USA; and Barbara Nichols, Livermore, California, USA. All of these colleagues made helpful comments and substantive contributions, especially concerning type localities. We also acknowledge John Sampson White at the Smithsonian and Carl Francis at Harvard for their assistance in making type locality source materials available to us. Naturally any remaining errors or omissions are ours.

We welcome readers' suggestions or corrections for all areas of this book, especially for type locality information where our goal is to assemble a definitive list prior to the next edition of this manual. Any correspondance concerning this book can be directed to us at the address below.

Ernest H. Nickel
Monte C. Nichols

Aleph Enterprises
PO Box 213
Livermore California
94551 USA

Elemental Table

"Element"	Symbol	No.	"Element"	Symbol	No.
Aluminum	Al	787	Nickel	Ni	162
Amide	NH_2	3	Niobium	Nb	109
Ammonium	NH_4	41	Nitrate	NO_3	17
Antimony	Sb	233	Osmium	Os	12
Antimony–oxygen	SbO_x	60	Oxonium	H_3O	15
Arsenate	AsO_4	221	Oxygen	O	1337
Arsenic	As	446	Palladium	Pd	91
Arsenite	AsO_3	21	Phosphate	PO_4	422
Barium	Ba	154	Phosphorus	P	429
Beryllium	Be	78	Phosphorus–oxygen	PO_x	424
Bismuth	Bi	217	Platinum	Pt	50
Bismuth–oxygen	BiO_x	28	Potassium	K	302
Boron	B	179	Praseodymium	Pr	1
Boron–oxygen	BO_x	179	Rhenium	Re	4
Bromine	Br	9	Rhodium	Rh	19
Cadmium	Cd	9	Ruthenium	Ru	15
Calcium	Ca	935	Samarium	Sm	3
Carbon	C	55	Scandium	Sc	6
Carbonate	CO_3	243	Selenium	Se	101
Cerium	Ce	128	Selenium–oxygen	SeO_x	15
Cesium	Cs	9	Silicon	Si	1018
Chlorine	Cl	232	Silicon–oxygen	SiO_x	992
Chromium	Cr	81	Silver	Ag	177
Chromium–oxygen	CrO_x	47	Sodium	Na	595
Cobalt	Co	107	Strontium	Sr	75
Copper	Cu	530	Sulfate	SO_4	353
Dysprosium	Dy	3	Sulfur	S	909
Erbium	Er	2	Sulfur–oxygen	SO_x	357
Fluorine	F	246	Tantalum	Ta	66
Gadolinium	Gd	2	Tellurium	Te	167
Gallium	Ga	6	Tellurium–oxygen	TeO_x	50
Germanium	Ge	23	Thallium	Tl	35
Germanium–oxygen	GeO_x	11	Thorium	Th	38
Gold	Au	42	Tin	Sn	105
Hydroxyl	OH	1362	Titanium	Ti	233
Indium	In	7	Titanium–oxygen	TiO_x	50
Iodine	I	14	Tungstate	WO_4	12
Iodate	IO_3	8	Tungstite	WO_3	3
Iridium	Ir	23	Tungsten	W	29
Iron	Fe	997	Tungsten–oxygen	WO_x	25
Lanthanum	La	62	Uranium	U	214
Lead	Pb	441	Uranium–oxygen	UO_x	202
Lithium	Li	69	Uranyl	UO_2	163
Magnesium	Mg	599	Vanadium	V	137
Manganese	Mn	448	Vanadium–oxygen	VO_x	127
Mercury	Hg	70	Ytterbium	Yb	5
Molybdenum	Mo	37	Yttrium	Y	88
Molybdenum–oxygen	MoO_x	25	Zinc	Zn	199
Neodymium	Nd	33	Zirconium	Zr	68

Abbreviated listing of name, symbol(s), and the number of occurences of each element or elemental grouping in the MINERAL database as of May 1990.

A

Abelsonite, $NiC_{31}H_{32}N_4$, [A] Tric. Semi-metallic/adamantine pink-purple, reddish-brown. Green River Formation, Uintah Co., Utah, USA. H= < 3, D_m= 1.4, D_c= 1.45, IXD 01. Milton et al., 1978. Am. Min. 63 (1978), 930.

Abernathyite, $K(UO_2)AsO_4 \cdot 3H_2O$, [G] Tet. Vitreous yellow. Fuemrole mine, Temple Mt., Emery Co., Utah, USA. H= 2-3, D_m=?, D_c= 3.572, VIID 20b. Thompson et al., 1956. Am. Min. 41 (1956), 82. Am. Min. 49 (1964), 1578.

Abhurite, $Sn_3OCl_2(OH)_2$, [A] Rhom. Transparent colorless. Jiddah, Red Sea, Sharm Abhur, Saudi Arabia. H=?, D_m= 4.29, D_c= 4.34, IIIC 15. Matzko et al., 1985. Can. Min. 23 (1985), 233.

Acanthite, Ag_2S, [G] Mon. Metallic black; grey in reflected light. H= 2-2.5, D_m= 7.2, D_c= 7.255, IIA 03. Kenngott, 1855. Can. Min. 12 (1974), 365. Struct. Repts. 32A (1967), 123.

Acetamide, CH_3CONH_2, [A] Hex. Translucent colorless/grey. L'vov-Volȳnskiĭ Basin, USSR. H= 1-1.5, D_m= 1.17, D_c= 1.15, IXA 03. Srebrodol'skiĭ, 1975. Am. Min. 61 (1976), 338 (Abst.).

Achavalite, $FeSe$, [Q] Hex. Metallic. Cacheuta, Argentina. H=?, D_m= 6.53, D_c= 6.716, IIB 09a. Olsacher, 1939. N.Jb.Min.Mh. (1972), 276.

Achrematite, [D] A mixture of mimetite and wulfenite. Am. Min. 62 (1977), 170. Mallet, 1875.

Actinolite, $Ca_2(Mg,Fe)_5Si_8O_{22}(OH)_2$, [G] Mon. Vitreous green, greyish-green. H= 5-6, D_m= 3.27, D_c= 3.22, VIIID 05b. Kirwan, 1794. DHZ (1963), v.2, 249. Zts. Krist. 133 (1971), 273. See also Tremolite, Ferro-actinolite.

Acuminite, $SrAlF_4(OH) \cdot H_2O$, [A] Mon. Transparent colorless. Ivigtut, Greenland (SW). H= 3.5, D_m= 3.29, D_c= 3.305, IIIC 09. Pauly & Petersen, 1987. Am. Min. 73(1988), 1492 (Abst.).

Adamite, $Zn_2AsO_4(OH)$, [G] Orth. Vitreous yellow, brownish-yellow, green, white, etc. Chañarcillo, Chile. H= 3.5, D_m= 4.434, D_c= 4.45, VIIB 04. Friedel, 1866. Acta Cryst. B34 (1978), 715. Am. Min. 61 (1976), 979. See also Eveite, Olivenite, Paradamite.

Adelite, $CaMgAsO_4OH$, [G] Orth. Resinous colorless, grey, bluish-grey, yellow, light green. Långban mine, Filipstad (near), Värmland, Sweden. H= 5, D_m= 3.73, D_c= 3.74, VIIB 11b. Sjögren, 1891. JCPDS 24-208.

Admontite, $Mg_2B_{12}O_{20} \cdot 15H_2O$, [A] Mon. Translucent colorless. Schildmaur, Admont (near), Austria. H= 2-3, D_m= 1.82, D_c= 1.875, Vc 02. Walenta, 1979. Am. Min. 65(1980), 205 (Abst.). See also Mcallisterite.

Aegirine, $NaFeSi_2O_6$, [G] Mon. Vitreous/resinous dark green, greenish-black. Rundemyr, Eker, Kongsberg (near), Buskerud, Norway. H= 6, D_m= 3.5, D_c= 3.577, VIIID 01a. Berzelius, 1835. DHZ, 2nd ed.(1978), v.2A, 482. MSA Spec. Paper 2 (1969), 31.

Aenigmatite, $Na_2Fe_5TiSi_6O_{20}$, [G] Tric. Vitreous black. Ilímaussaq intrusion, Naujakasik, Tunugdliarfik, Greenland (S). H= 5.5-6, D_m= 3.8, D_c= 3.724, VIIID 07. Breithaupt, 1865. Am. Min. 59 (1974), 820. Am. Min. 56 (1971), 427.

Aërinite, $Ca_4(Al,Fe,Mg)_{10}Si_{12}O_{36}(OH)_{12}CO_3 \cdot 12H_2O$, [A] Mon. Translucent blue, blue-green. Caserras, Aragon, Spain. H= 3, D_m= 2.48, D_c= 2.47, VIIID 01b. von Lasaulx, 1876. Am. Min. 73(1988), 1498 (Abst.).

Aerugite, $Ni_{8.5}As_3O_{16}$, [G] Rhom. Translucent blue-green. Terres mine, St. Ausdell, Cornwall, England and/or Johanngeorgenstadt, Sachsen (Saxony), Germany. H=?, D_m= 5.85, D_c= 5.772, VIIB 06. Adam, 1869. Min. Mag. 35 (1965), 72. Acta Cryst. B45 (1989), 201.

Aeschynite-(Ce), (Ce, Ca, Fe, Th)(Ti, Nb)$_2$(O, OH)$_6$, G Orth. Sub-metallic resinous black, brown, yellow. Miask, Ilmen Mts., USSR. H= 5-6, D_m= 4.96, D_c= 6.04, IVD 11. Berzelius, 1828. Dana, 7th ed.(1944), v.1, 793. Struct. Repts. 27 (1962), 535. See also Rynersonite, Vigezzite, Aeschynite-(Y), Niobo-aeschynite-(Ce).

Aeschynite-(Nd), (Nd, Ce, Ca)(Ti, Nb)$_2$(O, OH)$_6$, Q Syst=? Adamantine dark/light brown. Bayan Obo, Inner Mongolia, China. H= 5-6, D_m= 4.8, D_c=?, IVD 11. Zhang & Tao, 1982. Am. Min. 69(1984), 565 (Abst.).

Aeschynite-(Y), (Y, Ca, Fe, Th)(Ti, Nb)$_2$(O, OH)$_6$, A Orth. Sub-metallic resinous black, brown, yellow. Urstad, Hitterö, Norway (S). H= 5-6, D_m= 4.95, D_c=?, IVD 11. Levinson, 1966. Strunz (1970), 208. See also Aeschynite-(Ce), Tantalaeschynite-(Y).

Afghanite, (Na, Ca, K)$_8$(Si, Al)$_{12}$O$_{24}$(Cl, SO$_4$)$_3$•nH$_2$O, A Hex. Transparent bluish. Lapis-lazuli mine, Sar-e-Sang, Badakshan, Afghanistan. H= 5.5-6, D_m= 2.55, D_c= 2.65, VIIIF 05. Bariand et al., 1968. Can. Min. 17 (1979), 47.

Afwillite, Ca$_3$(SiO$_3$)$_2$(OH)$_2$•2H$_2$O, G Mon. Vitreous colorless, white. Dutoitspan mine, Kimberley, South Africa. H= 4.5, D_m= 2.630, D_c= 2.646, VIIIA 10. Parry & Wright, 1925. Min. Jour. 14 (1989), 279. Acta Cryst. B32 (1976), 475.

Agardite-(La), (Cu, Ca)$_6$La(AsO$_4$)$_3$(OH)$_6$•3H$_2$O, A Hex. Translucent colorless, yellowish-green, bluish-green. Red Cloud Fluorite mine, Lincoln Co., New Mexico, USA. H=?, D_m=?, D_c=?, VIID 17. Fehr & Hochleitner, 1984. Am. Min. 70(1985), 871 (Abst.).

Agardite-(Y), Cu$_6$(Y, Ca)(AsO$_4$)$_3$(OH)$_6$•3H$_2$O, A Hex. Translucent blue-green. Bou Skour mine, Jbel Sarhro, Morocco. H=?, D_m= 3.72, D_c= 3.61, VIID 17. Dietrich et al., 1969. N.Jb.Min.Mh. (1983), 385. Acta Cryst. C41 (1985), 161.

Agrellite, NaCa$_2$Si$_4$O$_{10}$F, A Tric. Pearly white, greyish, greenish-white. Kipawa River, Timiskaming Co., Québec, Canada. H= 5.5, D_m= 2.902, D_c= 2.887, VIIID 18. Gittins et al., 1976. Can. Min. 14 (1976), 120. Am. Min. 64 (1979), 563.

Agrinierite, (K$_2$, Ca, Sr)(UO$_2$)$_3$O$_4$•4H$_2$O, A Orth. Translucent orange. Margnac mine, Compreignac, Haute Vienne, France. H=?, D_m= 5.7, D_c= 5.62, IVF 12. Cesbron et al., 1972. Min. Mag. 38 (1972), 781.

Aguilarite, Ag$_4$SeS, G Orth. Metallic black; white in reflected light. San Carlos mine, Guanajuato, Mexico. H= 2.5, D_m= 7.586, D_c= 7.562, IIA 03. Genth, 1891. Can. Min. 12 (1974), 365.

Aheylite, (Fe, Zn)Al$_6$(PO$_4$)$_4$(OH)$_8$•4H$_2$O, A Tric. Translucent pale blue-green. Mira Flores mine, Huanuni dist., Oruro, Bolivia. H=?, D_m= 2.85, D_c= 2.89, VIID 08. Foord & Taggart, 1986. IMA 1986, Abst. p. 102.

Ahlfeldite, NiSeO$_3$•2H$_2$O, G Mon. Vitreous brownish-pink, red. Colquechaca, Potosí, Bolivia. H= 2-2.5, D_m= 3.37, D_c= 3.51, VIG 01. Herzenberg, 1935. Can. Min. 12 (1974), 304. See also Cobaltomenite.

Aikinite, CuPbBiS$_3$, G Orth. Metallic grey; creamy-white in reflected light. Cornwall, England. H= 2-2.5, D_m= 7.07, D_c= 7.25, IID 05a. Chapman, 1843. Am. Min. 74 (1989), 250. Acta Cryst. B27 (1971), 1245.

Ajoite, (K, Na)Cu$_7$AlSi$_9$O$_{24}$(OH)$_6$•3H$_2$O, G Tric. Translucent bluish-green. New Cornelia mine, Ajo, Pima Co., Arizona, USA. H=?, D_m= 2.96, D_c= 2.951, VIIIF 19. Schaller & Vlisidis, 1958. Am. Min. 66 (1981), 201.

Akaganéite, β–FeO(OH, Cl), G Tet. Brown. Akagané mine, Iwate, Japan. H=?, D_m= 3.0, D_c= 3.6, IVF 04. Nambu, 1961. Min. Mag. 33 (1962), 270. Acta Cryst. A35 (1979), 197. See also Feroxyhyte, Goethite, Lepidocrocite.

Akatoreite, Mn$_9$(Si, Al)$_{10}$O$_{23}$(OH)$_9$, A Tric. Vitreous orange-brown. Akatore Creek, Eastern Otago, Dunedin, South Island, New Zealand. H= 6, D_m= 3.48, D_c= 3.47, VIIIE 10b. Read & Reay, 1971. Am. Min. 56 (1971), 416.

Akdalaite, $(Al_2O_3)_4 \cdot H_2O$, [A] Hex. Vitreous/porcelanous white. Solvech fluorite deposit, Karaganda, Kazakhstan, USSR. H= 7.2, D_m= 3.68, D_c= 3.673, IVF 04. Shpanov et al., 1970. Am. Min. 56(1971), 635 (Abst.).

Åkermanite, $Ca_2MgSi_2O_7$, [G] Tet. Translucent colorless, greyish-green, brown. H= 5-6, D_m= 2.944, D_c= 2.944, VIIIB 02a. Vogt, 1884. DHZ, 2nd. ed.(1986), v.1B, 285. N.Jb.Min.Mh. (1981), 1. See also Gehlenite.

Akhtenskite, ϵ–MnO_2, [A] Hex. Dark grey, black. Akhtenskiĭ deposit, Ural Mts. (S), USSR. H=?, D_m= 4.00, D_c= 4.78, IVD 03a. Visser, 1979. Am. Min. 68(1983), 473 (Abst.).

Akrochordite, $(Mn,Mg)_5(AsO_4)_2(OH)_4 \cdot 4H_2O$, [G] Mon. Translucent red-brown. Långban mine, Filipstad (near), Värmland, Sweden. H= 3.5, D_m= 3.2, D_c= 3.26, VIID 02. Flink, 1922. Am. Min. 53(1968), 1779 (Abst.). Am. Min. 74 (1989), 256.

Aksaite, $MgB_6O_7(OH)_6 \cdot 2H_2O$, [G] Orth. Translucent colorless, light grey. Ak-Sai, Kazakhstan, USSR. H= 2.5, D_m= 1.99, D_c= 1.975, Vc 11. Blazko et al., 1962. Am. Min. 48(1963), 209 (Abst.). Am. Min. 56 (1971), 1553.

Aktashite, $Cu_6Hg_3As_4S_{12}$, [G] Rhom. Metallic white. Aktash mercury deposit, Gornyĭ Altaĭ, USSR. H= 4.5, D_m=?, D_c= 5.453, IID 01c. Vasil'ev, 1968. Am. Min. 58(1973), 562 (Abst.). Min. Abstr. 81-2423. See also Nowackiite, Gruzdevite.

Alabandite, α–MnS, [G] Cub. Sub-metallic black; grey in reflected light. Alabanda, Caria, Turkey. H= 3.5-4, D_m= 4.0, D_c= 4.056, IIB 11. Beudant, 1832. N.Jb.Min.Abh. 144 (1982), 107. See also Niningerite.

Alacranite, As_8S_9, [A] Mon. Adamantine orange. Uson caldera, Kamchatka, USSR. H= 1.5, D_m= 3.43, D_c= 3.43, IIB 19. Popova et al., 1986. Am. Min. 73 (1988), 189 (Abst.).

Alamosite, $PbSiO_3$, [G] Mon. Translucent colorless, white. Alamos, Sonora, Mexico. H= 4.5, D_m= 6.5, D_c= 6.30, VIIID 17. Palache & Merwin, 1909. Dana/Ford (1932), 567. Zts. Krist. 126 (1968), 98.

Alazanite, [D] Probably marcasite. Am. Min. 60(1975), 161 (Abst.). Ivanitskii et al., 1973.

Albite, $NaAlSi_3O_8$, [G] Tric. Vitreous/pearly white, bluish, grey, pink, etc. H= 6-6.5, D_m= 2.63, D_c= 2.623, VIIIF 03. Gahn & Berzelius, 1815. DHZ (1963), v.4, 94. Am. Min. 75 (1990), 135.

Albrechtschraufite, $Ca_4Mg(UO_2)_2(CO_3)_6F_2 \cdot 17H_2O$, [A] Tric. Vitreous yellow-green. Jáchymov (St. Joachimsthal), Západočeský kraj, Čechy (Bohemia), Czechoslovakia. H= 2-3, D_m= 2.6, D_c= 2.67, VbD 04. Mereiter, 1984. Acta Cryst. A40 (1984), C-247.

Albrittonite, [D] An artificial substance. Am. Min. 67 (1982), 156. Crook & Marcotty, 1978.

Aldermanite, $(Mg,Ca)_5Al_{12}(PO_4)_8(OH)_{22} \cdot 32H_2O$, [A] Orth. Pearly colorless. Moculta quarry, Angaston (near), South Australia, Australia. H= 2, D_m=?, D_c= 2.15, VIID 28. Harrowfield et al., 1981. Min. Mag. 44 (1981), 59.

Aldzhanite, [D] Inadequate data. Min. Mag. 43 (1980), 1055. Avrova et al., 1968.

Aleksite, $PbBi_2Te_2S_2$, [A] Hex. Metallic pale grey in reflected light. Alekseev mine, Sutamskii region, Stanovoi Range, USSR. H= 2.3, D_m=?, D_c= 7.80, IID 09e. Lipovetskii et al., 1978. Am. Min. 64(1979), 652 (Abst.).

Alforsite, $Ba_5(PO_4)_3Cl$, [A] Hex. Transparent colorless. Big Creek, Fresno Co., California, USA. H=?, D_m=?, D_c= 4.83, VIIB 16. Newberry et al., 1981. Am. Min. 66 (1981), 1050.

Algodonite, Cu_6As, [G] Hex. Metallic grey/white. Algodones, Coquimbo, Chile. H= 4, D_m= 8.38, D_c= 8.72, IIG 01. Field, 1857. Dana, 7th ed.(1944), v.1, 171.

Aliettite, $Ca_{0.2}Mg_6(Si,Al)_8O_{20}(OH)_4 \cdot 4H_2O$, [G] Syst=? Monte Chiaro, Taro Valley, Italy. H=?, D_m=?, D_c=?, VIIIE 08c. Veniale & van der Marel, 1969. Can. Min. 19 (1981), 651.

Allactite, $Mn_7(AsO_4)_2(OH)_8$, [G] Mon. Vitreous/greasy dark/light purplish-red, brownish-red. Moss mine, Nordmark, Sweden. H= 4.5, D_m= 3.83, D_c= 3.94, VIIB 10a. Sjögren, 1884. Dana, 7th ed.(1951), v.2, 785. Am. Min. 53 (1968), 733.

Allanite-(Ce), $Ca(Ce,La)(Al,Fe)_3(SiO_4)_3(OH)$, [G] Mon. Sub-metallic/pitchy/resinous brown, black. Qeqerssuatsiaq, Aluk, Greenland (E). H= 5.5-6, D_m= 4.0, D_c= 3.99, VIIIB 15. Thomson, 1810. Dana, 6th ed. (1892), 522. Am. Min. 56 (1971), 447.

Allanite-(La), $(La,Ca)_2(Al,Fe)_3(SiO_4)_3(OH)$, [A] Mon. Karelia (N), USSR. H=?, D_m=?, D_c=?, VIIIB 15. Levinson, 1966. Am. Min. 51 (1966), 152.

Allanite-(Y), $Ca(Y,La,Ce)(Al,Fe)_3(SiO_4)_3(OH)$, [A] Mon. H=?, D_m=?, D_c=?, VIIIB 15. Semenov & Barinskii, 1958. Am. Min. 51 (1966), 152.

Allargentum, ϵ–$Ag_{1-x}Sb_x$, [G] Hex. Metallic white. Cadesky Vein, Hi-Ho mine, Cobalt, Ontario, Canada. H= 3.5, D_m= 10.0, D_c= 10.12, IIG 02. Ramdohr, 1950. Can. Min. 10 (1970), 163.

Alleghanyite, $Mn_5(SiO_4)_2(OH)_2$, [G] Mon. Translucent bright/greyish pink. Bold Hill, Alleghany Co., North Carolina, USA. H= 5.5, D_m= 4.0, D_c= 4.199, VIIIA 16. Ross & Kerr, 1932. Dana/Ford (1932), 600. Am. Min. 70 (1985), 182.

Alloclasite, $(Co,Fe)AsS$, [G] Mon. Metallic white. Oraviţa (Oravicza), Banat, Romania. H=?, D_m= 5.95, D_c= 5.997, IIC 08. Tschermak, 1866. Can. Min. 11 (1971), 150. Can. Min. 14 (1976), 561. See also Glaucodot.

Allophane, Al_2O_3,SiO_2,H_2O, [G] Amor. Vitreous/sub-resinous pale blue, green, brown, colorless. Gräfenthal, Saalfeld (near), Thüringen, Germany. H= 3, D_m= 1.9, D_c=?, VIIIE 08d. Stromeyer, 1816. Am. Min. 61 (1976), 379. Min. Abstr. 76-3277.

Alluaudite, $(Na,Ca)_2(Mn,Mg,Fe)Fe_2(PO_4)_3$, [A] Mon. Sub-translucent dirty yellow, brownish-yellow, green. Vilate quarry, Chanteloube, Haute-Vienne, France. H= 5-5.5, D_m= 3.45, D_c= 3.62, VIIA 05b. Damour, 1848. Min. Mag. 43 (1979), 227. Am. Min. 56 (1971), 1955. See also Hagendorfite, Ferroalluaudite, Maghagendorfite.

Almandine, $Fe_3Al_2(SiO_4)_3$, [G] Cub. Vitreous/resinous red, brown, black. Alabanda, Caria, Turkey. H= 6.5-7.5, D_m= 4.318, D_c= 4.29, VIIIA 06a. Agricola, 1546. DHZ, 2nd ed.(1982), v.1A, 468. Am. Min. 56 (1971), 791. See also Pyrope, Spessartine.

Almbosite, [D] Inadequate data. Am. Min. 66(1981), 878 (Abst.). Ramdohr & Cevales, 1980.

Almeraite, $KNaMgCl_4 \cdot H_2O$, [Q] Syst=? Translucent reddish. Suria, Barcelona, Spain. H=?, D_m=?, D_c=?, IIIB 07. Tomas & Foleh, 1914. Dana, 7th ed.(1951), v.2, 94.

Alstonite, $BaCa(CO_3)_2$, [G] Tric. Vitreous colorless, white, greyish, cream, pink. Alston Moor, Cumberland, England. H= 4-4.5, D_m= 3.707, D_c= 3.69, VbA 04. Breithaupt, 1841. Lithos 8 (1975), 199. See also Barytocalcite, Paralstonite.

Altaite, $PbTe$, [G] Cub. Metallic white. Altai Mts., Mongolia. H= 3, D_m= 8.15, D_c= 8.31, IIB 11. Haidinger, 1845. Powd. Diff. 2 (1987), 230. See also Clausthalite, Galena.

Althausite, $Mg_2PO_4(OH,F,O)$, [A] Orth. Vitreous grey. Modum, Norway. H= 3.5, D_m= 2.97, D_c= 2.91, VIIB 03c. Raade & Tysseland, 1975. Am. Min. 61(1976), 502 (Abst.). Am. Min. 65 (1980), 488. See also Holtedahlite.

Althupite, $AlTh(UO_2)_7(PO_4)_4O_2(OH)_5 \cdot 15H_2O$, [A] Tric. Transparent yellow. Kobokobo, Kivu, Zaïre. H=?, D_m= 3.9, D_c= 3.98, VIID 21. Piret & Deliens, 1987. Am. Min. 73(1988), 189 (Abst.).

Aluminite, $Al_2SO_4(OH)_4 \cdot 7H_2O$, [G] Mon. Earthy white. Garden of the Paedogogium, Halle, Germany. H= 1-2, D_m= 1.7, D_c= 1.794, VID 03. Haberle, 1807. Zts. Krist. 151 (1980), 141. Acta Cryst. B34 (1978), 2407. See also Meta-aluminite.

Aluminium, Al, [A] Cub. Metallic white. Tsepochechnyĭ intrusive, Siberia, USSR. H= 2.0-2.9, D_m= 2.707, D_c= 2.710, IA 01a. Oleĭnikov et al., 1978. Am. Min.65(1980), 205 (Abst.).

Alumino-magnesio-hornblende, $Ca_2Mg_4Al(Si_7Al)O_{22}(OH)_2$, [A] Mon. Translucent green. H= 5-6, D_m= 3.0, D_c= 3.23, VIIID 05b. Leake et al., 1978. Am. Min. 63 (1978), 1023. MSA Spec. Paper 2 (1969), 117. See also Ferrohornblende, Magnesio-hornblende.

Alumino-taramite, $Na_2CaFe_3Al_2(Si_6Al_2)O_{22}(OH)_2$, [A] Mon. H=?, D_m=?, D_c=?, VIIID 05c. Leake et al., 1978. Am. Min. 63 (1978), 1023. See also Magnesiotaramite, Taramite.

Alumino-barroisite, $NaCaMg_3Al_2(Si_7Al)O_{22}(OH)_2$, [A] Mon. H= 5-6, D_m=?, D_c= 2.94, VIIID 05c. Leake et al., 1978. Am. Min. 63 (1978), 1023. See also Ferrobarroisite, Barroisite.

Aluminobetafite, [D] Inadequate data. Min. Mag. 36 (1967), 133. Kawai, 1963.

Aluminocopiapite, $(Al,Mg)Fe_4(SO_4)_6(OH,O)_2 \cdot 20H_2O$, [G] Tric. Earthy yellow. Temple Rock, Utah, USA, and, Island Mt., Trinity Co., California, USA. H=?, D_m=?, D_c= 2.07, VID 04b. Berry, 1947. Can. Min. 23 (1985), 53.

Alumino-ferro-hornblende, $Ca_2Fe_4Al(Si_7Al)O_{22}(OH)_2$, [A] Mon. Translucent green. H= 5-6, D_m= 3.2, D_c= 3.23, VIIID 05b. Leake et al., 1978. Am. Min. 63 (1978), 1023. MSA Spec. Paper (1969), 117. See also Magnesiohornblende.

Alumino-katophorite, $Na_2Ca(Fe,Mg)_4Al(Si_7Al)O_{22}(OH)_2$, [A] Mon. H=?, D_m=?, D_c=?, VIIID 05c. Leake et al., 1978. Am. Min. 63 (1978), 1023. See also Magnesio-aluminokatophorite, Ferrikatophorite.

Alumino-tschermakite, $Ca_2Mg_2Al_2(Si_6Al_2)O_{22}(OH)_2$, [A] Mon. Translucent green. H= 5-6, D_m= 3.13, D_c= 2.88, VIIID 05b. Leake et al., 1978. Am. Min. 63 (1978), 1023. See also Tschermakite, Ferrotschermakite.

Alumino-winchite, $NaCaMg_4AlSi_8O_{22}(OH)_2$, [A] Mon. H=?, D_m=?, D_c=?, VIIID 05c. Leake et al., 1978. Am. Min. 63 (1978), 1023. See also Winchite, Ferrowinchite.

Alumobritholite, [D] Unnecessary name for aluminian britholite. Min. Mag. 36 (1967), 133. Kudrina et al., 1961.

Alumocobaltomelane, [D] Mixture. Min. Mag. 33 (1962), 261. Ginzberg & Rukavishnikova 1951.

Alumoferroascharite, [D] A mixture of szaibelyite and hydrotalcite. Am. Min. 49 (1964), 1501 (Abst.). Serdyuchenko, 1956.

Alumohydrocalcite, $CaAl_2(CO_3)_2(OH)_4 \cdot 3H_2O$, [Q] Tric. Chalky white, pale blue, violet. Khakassy dist., Siberia, USSR. H= 2.5, D_m= 2.231, D_c=?, VbD 02. Bilibin, 1926. Dana, 7th ed.(1951), v.2, 280. See also Para-alumohydrocalcite.

Alumopharmacosiderite, $KAl_4(AsO_4)_3(OH)_4 \cdot 6.5H_2O$, [A] Cub. Translucent white. Guanaco (NE), Chile. H=?, D_m=?, D_c= 2.676, VIID 14b. Schmetzer et al., 1981. Am. Min. 66(1981), 1099 (Abst.). See also Pharmacosiderite, Sodium pharmacosiderite.

Alumotantite, $AlTaO_4$, [A] Orth. Adamantine colorless. Kola Peninsula, USSR. H= 7.5-8, D_m=?, D_c= 5.623, IVD 14a. Voloshin et al., 1981. Am. Min. 67(1982), 413 (Abst.).

Alumotungstite, $(W,Al)(O,OH)_3$(?), [Q] Cub. Kramat Pulai mine, Kinta, Perak, Malaysia. H=?, D_m=?, D_c=?, IVF 15. Davis & Smith, 1971. Min. Rec. 12 (1981), 81.

Alunite, $KAl_3(SO_4)_2(OH)_6$, [G] Rhom. Vitreous white, greyish, yellowish, reddish. Tolfa, Roma (near), Italy. H= 3.5-4, D_m= 2.69, D_c= 2.839, VIB 03a. Beudant, 1824. Am. Min. 65 (1980), 953. Acta Cryst. 18 (1965), 249.

Alunogen, $Al_2(SO_4)_3 \cdot 17H_2O$, [G] Tric. Vitreous/silky colorless. H= 1.5-2, D_m= 1.77, D_c= 1.791, VIC 04. Beudant, 1832. Bull. Min. 96 (1973), 385. Am. Min. 61 (1976), 311. See also Meta-alunogen.

Alvanite, $Al_6(VO_4)_2(OH)_{12} \cdot 5H_2O$, [Q] Syst=? Vitreous/pearly light bluish-green, bluish-black. Kara Tau, Kazakhstan, USSR. H= 3-3.5, D_m= 2.41, D_c=?, VIIC 24. Ankinovich, 1959. Am. Min. 44(1959), 1325 (Abst.).

Amakinite, $(Fe,Mg)(OH)_2$, [G] Rhom. Translucent pale green, yellow green. Lucky Eastern pipe, Yakutiya, USSR. H= 3.5-4, D_m= 2.98, D_c= 2.74, IVF 03a. Kozlov & Levshov, 1962. Am. Min. 47(1962), 1218 (Abst.).

Amarantite, $Fe_2O(SO_4)_2 \cdot 7H_2O$, [G] Tric. Vitreous red, brownish-red, orange-red. Sierra Gorda, Caracoles, Chile (?). H= 2.5, D_m= 2.2, D_c= 2.14, VID 02. Frenzel, 1888. Dana, 7th ed.(1951), v.2, 611. Zts. Krist. 127 (1968), 261.

Amarillite, $NaFe(SO_4)_2 \cdot 6H_2O$, [G] Mon. Vitreous/adamantine pale yellow (greenish tint). Tierra Amarilla, Copiapó (near), Chile. H= 2.5-3, D_m= 2.19, D_c=?, VIC 07. Ungemach, 1933. Dana, 7th ed., vol. 2, 468.

Amber, [C, H, O], [G] Amor. Greasy yellow, brown, yellowish-white. H= 2-2.5, D_m= 1.0, D_c=?, IXC 01.

Amblygonite, $(Li,Na)AlPO_4(F,OH)$, [G] Tric. Vitreous/greasy white, yellowish, pink, greenish, etc. Montebras, Soumans, Creuse, France. H= 5.5-6, D_m= 3.11, D_c= 3.21, VIIB 02. Breithaupt, 1817. Min. Mag. 37 (1969), 414. Acta Cryst. 12 (1959), 988. See also Montebrasite, Natramblygonite.

Ameghinite, $NaB_3O_3(OH)_4$, [A] Mon. Vitreous colorless. Tincalayu Borax deposit, Salar Del Hombre Muerto, Salta, Argentina. H= 2.5, D_m= 2.030, D_c= 2.037, Vc 05. Aristarain & Hurlbut, 1967. Am. Min. 52 (1967), 935. Am. Min. 60 (1975), 879.

Ameletite, [D] A mixture of nepheline and other minerals. Min. Mag. 36 (1967), 438. Marshall, 1929.

Amesite-2H, $Mg_2Al(SiAl)O_5(OH)_4$, [G] Hex. Translucent pale green. Chester, Hampden Co., Massachusetts, USA. H=?, D_m= 2.78, D_c= 2.70, VIIIE 10b. Shepard, 1867. Am. Min. 66 (1981), 185. Acta Cryst. 9 (1956), 487. See also Amesite-6R.

Amesite-6R, $Mg_2Al(SiAl)O_5(OH)_4$, [P] Hex. Translucent pale green. Chester, Hampden Co., Massachusetts, USA. H=?, D_m=?, D_c= 2.70, VIIIE 10b. Shepard, 1867. Can. Min. 13 (1975), 227. Acta Cryst. 15 (1962), 510. See also Amesite-2H.

Amicite, $K_2Na_2(Al_4Si_4)O_{16} \cdot 5H_2O$, [A] Mon. Translucent colorless. Höwenegg, Hegau, Germany. H=?, D_m= 2.06, D_c= 2.146, VIIIF 14. Alberti et al., 1979. Am. Min. 65(1980), 808 (Abst.). Acta Cryst. B35 (1979), 2866.

Aminoffite, $Ca_3Be_2Si_3O_{10}(OH)_2$, [G] Tet. Vitreous colorless. Långban mine, Filipstad (near), Värmland, Sweden. H= 5.5, D_m= 2.94, D_c= 2.86, VIIIE 22. Hurlbut, 1937. Am. Min. 23(1938), 293 (Abst.). Struct. Repts. 32A (1967), 466.

Ammonioalunite, $NH_4Al_3(SO_4)_2(OH)_6$, [A] Rhom. Vitreous greyish-white. The Geysers, Sonoma Co., California, USA. H=?, D_m= 2.4, D_c= 2.58, VIB 03a. Altaner et al., 1988. Am. Min. 73 (1988), 145.

Ammonioborite, $(NH_4)_3B_{15}O_{20}(OH)_8 \cdot 4H_2O$, [G] Mon. Translucent white. Larderello, Val di Cecina, Piza, Toscana, Italy. H=?, D_m= 1.765, D_c= 1.759, Vc 08. Schaller, 1931. Am. Min. 44 (1959), 1150. Science 171 (1971), 377.

Ammoniojarosite, $NH_4Fe_3(SO_4)_2(OH)_6$, [G] Rhom. Dull/waxy/earthy light yellow. Kaibab Fault, Utah, USA. H=?, D_m=?, D_c= 2.936, VIB 03a. Shannon, 1927. Can. Min. 20 (1982), 91.

Ammonioleucite, $(NH_4,K)(AlSi_2)O_6$, [A] Tet. Resinous/vitreous white. Fujioka, Tatarazawa, Sanbagawa, Gunma, Japan. H=?, D_m= 2.29, D_c= 2.24, VIIIF 11. Hori et al., 1986. Am. Min. 71 (1986), 1022. See also Leucite.

Ammonium hydromica, $(NH_4)Al_2(Si_3Al)O_{10}(H_2O, OH)_2$, [Q] Mon. Kapka, Vihorlat Mts., Slovakia (E), Czechoslovakia. H=?, D_m=?, D_c=?, VIIIE 07a. Kozac et al., 1977. Min. Abstr. 78-4853.

Amphibole, A group name for double-chain silicates with the general formula $(Ca, Na, K)_{0-1}(Ca, Fe, Li, Mg, Mn)_2(Al, Fe, Mg, Mn, Cr, Ti)_5(Si, Al)_8O_{22}(OH, F, Cl)_2$.

Amstallite, $CaAl(Si, Al)_4O_8(OH)_4.(H_2O, Cl)$, [A] Mon. Transparent colorless. Amstall, Austria. H= 4, D_m= 2.40, D_c= 2.380, VIIIE 30. Quint, 1987. Am. Min. 73(1988), 1492 (Abst.). N. Jb. Min. Mh. (1987), 253.

Analcime (cubic), $Na(AlSi_2)O_6 \cdot H_2O$, [G] Cub. Vitreous colorless, white, grey, pink, greenish, yellowish. Cyclops Island, Italy. H= 5-5.5, D_m= 2.27, D_c= 2.26, VIIIF 11. Häuy, 1797. Zts. Krist. 184 (1988), 63. Zts. Krist. 135 (1972), 240. See also Pollucite.

Analcime (monoclinic), $Na(AlSi_2)O_6 \cdot H_2O$, [P] Mon. H=?, D_m= 2.29, D_c= 2.307, VIIIF 11. Zts. Krist. 184 (1988), 63. See also Pollucite, Analcime.

Anandite-2M$_1$, $(Ba, K)(Fe, Mg)_3(Si, Al, Fe)_4O_{10}(S, OH)_2$, [A] Mon. Lustrous black. Wilagedera prospect, North Western Province, Sri Lanka. H= 3-4, D_m= 3.94, D_c= 3.97, VIIIE 06. Pattiaratchi et al., 1967. Min. Mag. 36 (1968), 871. See also Kinoshitalite, Anandite-2Or.

Anandite-2Or, $(Ba, K)(Fe, Mg)_3(Si, Al, Fe)_4O_{10}(S, OH)_2$, [P] Orth. Lustrous black. Wilagedera prospect, North Western Province, Sri Lanka. H= 3-4, D_m= 3.94, D_c= 4.22, VIIIE 06. Filut et al., 1985. Min. Mag. 36 (1968), 871. Am. Min. 70 (1985), 1298. See also Anandite-2M1.

Anapaite, $Ca_2Fe(PO_4)_2 \cdot 4H_2O$, [G] Tric. Vitreous green, greenish-white. Anapa, Taman Peninsula, Russia, USSR. H= 3.5, D_m= 2.816, D_c= 2.811, VIIC 06. Sachs, 1902. Dana, 7th ed.(1951), v.2, 731. Bull. Min. 102 (1979), 314.

Anarakite, [D] Unnecessary name for zincian paratacamite. Min. Mag. 43 (1980), 1055. Adib & Ottemann, 1972.

Anatase, TiO_2, [G] Tet. Adamantine brown, green, grey, black, etc. France. H= 5.5-6, D_m= 3.90, D_c= 3.895, IVD 02. Haüy, 1801. Dana, 7th ed.(1944), v.1, 583. Zts. Krist. 136 (1972), 273. See also Rutile, Brookite.

Ancylite-(Ce), $(Ce, La, Sr, Ca)_2(CO_3)_2(OH, H_2O)$, [G] Orth. Vitreous/greasy yellow, yellowish-brown, grey. Narsarsuk, Greenland (S). H= 4-4.5, D_m= 4.1, D_c= 4.15, VbD 03. Flink, 1900. Dana, 7th ed.(1951), v.2, 291. Am. Min. 60 (1975), 280. See also Calcio-ancylite-(Ce), Gysinite-(Nd).

Andalusite, Al_2SiO_5, [G] Orth. Translucent pink, white, red, grey, yellow, violet, etc. Andalusia, Spain. H= 6.5-7.5, D_m= 3.14, D_c= 3.144, VIIIA 14. Delamétherie, 1789. Zts. Krist. 115 (1961), 269. Zts. Krist. 115 (1961), 314. See also Kyanite, Sillimanite, Kanonaite.

Andersonite, $Na_2Ca(UO_2)(CO_3)_3 \cdot 6H_2O$, [G] Rhom. Translucent green. Hillside mine, Bagdad, Yavapai Co., Arizona, USA. H=?, D_m= 2.8, D_c= 2.86, VbD 04. Axelrod et al., 1951. N.Jb.Min.Mh. (1987), 488. Acta Cryst. B37 (1981), 1946.

Andesine, $(Na, Ca)(Si, Al)_4O_8$, [G] Tric. Sub-vitreous/pearly white, grey, greenish, yellowish, etc. Marmato, Andes Mts., Bolivia. H= 5-6, D_m= 2.68, D_c= 2.68, VIIIF 03. Abich, 1841. DHZ, v.4 (1963), 94. Zts. Krist. 178 (1987), 207. See also Plagioclase.

Andorite, $AgPbSb_3S_6$, [G] Orth. Metallic grey; white in reflected light. Oruro, Bolivia. H= 3.5, D_m= 5.35, D_c= 5.413, IID 09b. Krenner, 1892. N.Jb.Min Abh. 147 (1983), 47. Struct. Repts. 49A (1982), 7. See also Ramdohrite, Uchucchacuaite.

Andradite, $Ca_3Fe_2(SiO_4)_3$, [G] Cub. Vitreous/resinous greenish-yellow, green, brown, black, etc. H= 6.5-7.5, D_m= 3.75, D_c= 3.85, VIIIA 06a. Dana, 1868. DHZ, 2nd ed.(1982), v.1B, 468. Am. Min. 56 (1971), 791. See also Grossular, Schorlomite.

Andremeyerite, $BaFe(Fe,Mn,Mg)Si_2O_7$, [A] Mon. Translucent pale green. Mt. Nyiragongo, Zaïre. H= 5.5, D_m= 4.15, D_c= 4.28, VIIIB 10. Sahama et al., 1973. Am. Min. 59(1974), 381 (Abst.). Am. Min. 73 (1988), 608.

Andrewsite, $(Cu,Fe)Fe_3(PO_4)_3(OH)_2$, [G] Orth. Silky dark green, bluish-green. West Phoenix mine, Liskeard, Cornwall, England. H= 4, D_m= 3.475, D_c= 2.941, VIIB 07. Maskelyne, 1871. Strunz (1970), 319.

Anduoite, $(Ru,Os)As_2$, [A] Orth. Metallic pinkish-white. Unspecified locality, Tibet, China. H= 6.5-7, D_m=?, D_c= 8.69, IIC 07. Yu & Chou, 1979. CIM SV23 (1981), 83. See also Omeiite.

Angelellite, $Fe_4O_3(AsO_4)_2$, [G] Tric. Adamantine/sub-metallic blackish-brown. Cerro Pululus, Argentina. H= 5.5, D_m= 4.9, D_c= 4.762, VIIB 06. Ramdohr et al., 1959. Am. Min. 44(1959), 1322 (Abst.). Struct. Repts. 44A (1978), 265.

Anglesite, $PbSO_4$, [G] Orth. Adamantine/resinous white, grey, yellow, green, blue. Isle of Anglesey, Wales. H= 2.5-3, D_m= 6.38, D_c= 6.321, VIA 07. Beudant, 1832. Dana, 7th ed.(1951), v.2, 420. Am. Min. 63 (1978), 506.

Anhydrite, $CaSO_4$, [G] Orth. Vitreous/greasy colorless, bluish, violet. Hall, Innsbruck (near), Tyrol, Austria. H= 3.5, D_m= 2.98, D_c= 2.960, VIA 07. Werner, 1804. Dana, 7th ed.(1951), v.2, 424. Can. Min. 13 (1975), 289.

Anhydrokainite, $KMgSO_4Cl$, [Q] Syst=? Germany (N). H=?, D_m=?, D_c=?, VIB 02. Janecke, 1913. Dana, 7th ed.(1951), v.2, 596.

Anilite, Cu_7S_4, [A] Orth. Metallic bluish-grey. Ani mine, Akita, Japan. H= 3, D_m=?, D_c= 5.68, IIA 01a. Morimoto et al., 1969. Am. Min. 54 (1969), 1256. Acta Cryst. B26 (1970), 915.

Ankerite, $Ca(Fe,Mg,Mn)(CO_3)_2$, [G] Rhom. Vitreous grey, yellowish-brown, brown, etc. H= 3.5-4, D_m= 3.02, D_c= 3.16, VbA 03a. Haidinger, 1825. TMPM 24 (1977), 279. Am. Min. 74 (1989), 1159. See also Dolomite, Kutnohorite.

Annabergite, $Ni_3(AsO_4)_2 \cdot 8H_2O$, [G] Mon. Adamantine/earthy green. Annaberg, Sachsen (Saxony), Germany. H= 1.5-2.5, D_m= 3.07, D_c= 3.22, VIIC 10. Brooke & Miller, 1852. NBS Monogr. 19 (1982), 60. See also Erythrite.

Annite, $KFe_3(Si_3Al)O_{10}(OH,F)_2$, [G] Mon. Adamantine/vitreous black. Cape Anne, Massachusetts, USA. H= 3, D_m= 3.17, D_c= 2.77, VIIIE 05b. Dana, 1868. Jour. Petrol. 3 (1962), 82. Am. Min. 58 (1973), 889.

Anorthite, $CaAl_2Si_2O_8$, [G] Tric. Translucent colorless, white, greyish, reddish, etc. Monte Somma, Mt. Vesuvius, Napoli, Campania, Italy. H= 6-6.5, D_m= 2.76, D_c= 2.758, VIIIF 03. Rose, 1823. DHZ (1963), v.4, 94. Acta Cryst. 15 (1962), 1005. See also Plagioclase.

Anorthoclase, $(Na,K)AlSi_3O_8$, [G] Tric. Vitreous/pearly white, yellow, red, green, etc. Pantelleria Island, Italy. H= 6-6.5, D_m= 2.59, D_c= 2.60, VIIIF 03. Rosenbusch, 1885. Dana, 6th ed. (1892), 324. Am. Min. 67 (1982), 975.

Anosovite, [D] An artificial product; some compositions = armalcolite. Am. Min. 73 (1988), 1377. Tagirov, 1951.

Antarcticite, $CaCl_2 \cdot 6H_2O$, [A] Rhom. Don Juan Pond, Victoria Land, Antarctica. H=?, D_m= 1.71, D_c= 1.71, IIIC 10a. Torii & Ossaka, 1965. Science 149 (1965), 975.

Anthoinite, $AlWO_3(OH)_3$(?), [G] Tric. Powdery white. Misobo Mt., Kalima dist., Zaïre. H=?, D_m= 4.8, D_c= 4.84, VIF 03. Varlamoff, 1947. Min. Rec. 12 (1981), 81.

Anthonyite, $Cu(OH,Cl)_2 \cdot 3H_2O$, [G] Mon. Translucent lavender. Centennial mine, Calumet, Houghton Co., Michigan, USA. H= 2, D_m=?, D_c=?, IVF 19. Williams, 1963. Am. Min. 48 (1963), 614.

Anthophyllite, $(Mg,Fe)_7Si_8O_{22}(OH)_2$, [G] Orth. Vitreous brownish-grey, yellowish-brown, green. H= 5.5-6, D_m= 3.1, D_c= 3.09, VIIID 06. Schumacher,

1801. Can. Min. 21 (1983), 173. Zts. Krist. 188 (1989), 237. See also Magnesio-anthophyllite, Ferro-anthophyllite.

Antigorite, $(Mg,Fe)_3Si_2O_5(OH)_4$, G Mon. Green, green-blue, white. Val Antigorio, Novara, Piemonte, Italy. H= 2.5-3.5, D_m= 2.6, D_c= 2.52, VIIIE 10b. Schweizer, 1840. Min. Journ. 12 (1985), 299. Rev. Min. 19 (1988), 91. See also Clinochrysotile, Lizardite, Orthochrysotile, Parachrysotile.

Antimonpearceite, $(Ag,Cu)_{16}(Sb,As)_2S_{11}$, Q Mon. Submetallic black. Sonora and Guanajuato, Mexico. H=?, D_m= 6.34, D_c= 6.32, IID 11. Frondel, 1963. Can. Min. 8 (1965), 172. See also Arsenpolybasite.

Antimony, Sb, G Rhom. Metallic white. H= 3-3.5, D_m= 6.7, D_c= 6.688, IB 01. Dana, 7th ed.(1944), v.1, 132. Acta Cryst. 16 (1963), 451.

Antlerite, $Cu_3SO_4(OH)_4$, G Orth. Vitreous green. Antler mine, Mohave Co., Arizona, USA. H= 3.5, D_m= 3.88, D_c= 3.945, VIB 01. Hillebrand, 1889. Dana, 7th ed.(1951), v.2, 544. Can. Min. 27 (1989), 205.

Anyuiite, $AuPb_2$, A Tet. Metallic grey; silvery-grey in reflected light. Bolshoĭ Anyuĭ River basin, USSR (NE). H= 3.2, D_m=?, D_c= 13.49, IA 13. Razin & Sidorenko, 1989. Min. Zhurn. 11(4) (1989), 88.

Apachite, $Cu_9Si_{10}O_{29} \cdot 11H_2O$, A Mon. Translucent blue. Christmas mine, Christmas, Gila Co., Arizona, USA. H= 2, D_m= 2.80, D_c= 3.37, VIIID 04. Cesbron & Williams, 1980. Min. Mag. 43 (1980), 639.

Apatite, $Ca_5(PO_4)_3(F,OH,Cl)$, G Hex. Vitreous, various colors. H= 5, D_m= 3.2, D_c= 3.20, VIIB 16. Werner, 1786. Dana, 7th ed.(1951), v.2, 879. Struct. Repts. 38A (1972), 308. See also Chlorapatite, Fluorapatite, Hydroxylapatite.

Aphthitalite, $(K,Na)_3Na(SO_4)_2$, G Rhom. Vitreous/resinous white, colorless, grey, blue, greenish. Mt. Vesuvius, Napoli, Campania, Italy. H= 3, D_m= 2.656, D_c= 2.70, VIA 05. Shepard, 1835. Dana, 7th ed.(1951), v.2, 400. Acta Cryst. B36 (1980), 919.

Apjohnite, $MnAl_2(SO_4)_4 \cdot 22H_2O$, G Mon. Silky colorless, white, rose, pale green, yellow. Lourenço Marques (Delagoa) Bay, Maputo, Mozambique. H= 1.5, D_m= 1.81, D_c= 1.836, VIC 06. Glocker, 1847. Dana, 7th ed.(1951), v.2, 527. Min. Mag. 40 (1976), 599.

Aplowite, $(Co,Mn,Ni)SO_4 \cdot 4H_2O$, A Mon. Vitreous pink. Magnet Cove Barium Corp. mine, Walton (S), Hants Co., Nova Scotia, Canada. H= 2.5, D_m= 1.97, D_c= 2.006, VIC 02. Jambor & Boyle, 1965. Can. Min. 8 (1965), 166.

Apuanite, $Fe_5Sb_4O_{12}S$, A Tet. Metallic black. Buca della Vena mine, Stazzema, Alpe Apuane (Apennines Alps), Toscana, Italy. H= 4, D_m= 5.12, D_c= 5.32, VIIA 15. Mellini et al., 1979. Am. Min. 64 (1979), 1230. Am. Min. 64 (1979), 1235.

Aragonite, $CaCO_3$, G Orth. Vitreous colorless white, grey, yellowish, blue, etc. Aragon, Spain. H= 3.5-4, D_m= 2.947, D_c= 2.930, VbA 04. Werner, 1796. Am. Min. 50 (1965), 1489. Am. Min. 56 (1971), 758. See also Calcite, Vaterite.

Aramayoite, $Ag(Sb,Bi)S_2$, G Tric. Metallic black; greyish-white in reflected light. Animas mine, Chocoya, Potosí, Bolivia. H= 2.5, D_m= 5.60, D_c= 5.69, IID 07. Spencer, 1926. Am. Min. 36 (1951), 436. Zts. Krist. 139 (1974), 54.

Aravaipaite, $Pb_3AlF_9 \cdot H_2O$, A Tric. Vitreous/pearly colorless. Grand Reef mine, Laurel Canyon, Graham Co., Arizona, USA. H= 2, D_m=?, D_c= 6.37, IIIB 12. Kampf et al., 1989. Am. Min. 74 (1989), 927.

Arcanite, K_2SO_4, G Orth. Translucent colorless, white. Santa Ana Tin mine, Trabuco Canyon, Orange Co., California, USA. H=?, D_m= 2.663, D_c= 2.667, VIA 05. Haidinger, 1845. Dana, 7th ed.(1951), v.2, 399. Acta Cryst. B28 (1972), 2845. See also Mascagnite.

Archerite, $H_2(K,NH_4)PO_4$, A Tet. Translucent buff, colorless. Petrogale cave, Madura, Western Australia, Australia. H=?, D_m=?, D_c= 2.34, VIIA 14. Bridge, 1977. Min. Mag. 41 (1977), 33. See also Biphosphammite.

Arctite, $Na_5BaCa_7(PO_4)_6F_3$, [A] Rhom. Vitreous colorless. Khibina massif, Vuonnemi river, Kola Peninsula, USSR. H= 5, D_m= 3.13, D_c= 3.19, VIIB 18. Khomyakov et al., 1981. Am. Min. 67(1982), 621 (Abst.). Min. Abstr. 89M/4118.

Arcubisite, Ag_6CuBiS_4, [A] Syst=? Metallic grey in reflected light. Ivigtut, Greenland (SW). H=?, D_m=?, D_c=?, IID 11. Karup-Møller, 1976. Am. Min. 63(1978), 424 (Abst.).

Ardaite, $(Pb,Fe)_{10}Sb_6S_{17}Cl_4$, [A] Mon. Metallic greenish-gray in reflected light. Madyarovo deposit, Bulgaria. H=?, D_m=?, D_c= 6.11, IIF 03. Breskovska et al., 1982. Can. Min. 19 (1981), 419.

Ardealite, $Ca_2(HPO_4)(SO_4) \cdot 4H_2O$, [G] Mon. Translucent light yellow. Peştera Cioclovina, Transylvania, Romania. H=?, D_m= 2.300, D_c= 2.321, VIC 01. Schadler, 1931. Bull. Min. 105 (1982), 621. Am. Min. 63 (1978), 520. See also Brushite, Gypsum, Pharmacolite.

Ardennite, $Mg_{1+x}Mn_4Al_{5-x}(AsO_4)(SiO_4)_2(Si_3O_{10})(OH)_6$, [G] Orth. Translucent yellow, yellowish-brown. Ottrez, Ardennes, Belgium. H= 6-7, D_m= 3.7, D_c= 3.74, VIIIB 19. Lasaulx & Bettendorf, 1872. Am. Min. 70 (1985), 171. Acta Cryst. B24 (1968), 845.

Arfvedsonite, $Na_3(Fe,Mg)_4Fe^{3+}Si_8O_{22}(OH)_2$, [G] Mon. Vitreous greenish-black, black. Ilímaussaq, Greenland. H= 6, D_m= 3.44, D_c= 3.40, VIIID 05d. Brooke, 1823. DHZ (1963), v.2, 364. Can. Min. 14 (1976), 346. See also Magnesio-arfvedsonite.

Argentite, [D] Acanthite dimorph; not found in nature; stable above 179°C. Dana, 7th ed. (1944), 176.

Argentocuproaurite, [D] Unnecessary name for argentoan auricupride. Min. Mag. 43 (1980), 1055. Razin, 1975.

Argentojarosite, $AgFe_3(SO_4)_2(OH)_6$, [G] Rhom. Brilliant yellow/brown. Tintic Standard mine, Dividend, Utah Co., Utah, USA. H=?, D_m= 3.62, D_c= 3.660, VIB 03a. Schaller, 1923. Am. Min. 58 (1973), 936.

Argentopentlandite, $Ag(Fe,Ni)_8S_8$, [A] Cub. Metallic reddish-brown. Jáchymov (St. Joachimsthal), Západočeský kraj, Čechy (Bohemia), Czechoslovakia and/or Oktyabr and Talnaka deposits, Noril'sk (near), Siberia (N), USSR. H= 3-3.5, D_m=?, D_c= 4.69, IIA 07. Mintkenov et al., 1977. ZVMO 106 (1977), 688. Can. Min. 12 (1973), 169.

Argentopyrite, $AgFe_2S_3$, [G] Orth. Metallic grey/white. Jáchymov (Joachimsthal), Čechy (Bohemia), Czechoslovakia. H= 3.5-4, D_m= 4.25, D_c= 4.27, IIB 08. Dana, 1868. Am. Min. 39 (1954), 475. See also Sternbergite.

Argentotennantite, $(Ag,Cu)_{10}(Zn,Fe)_2(As,Sb)_4S_{13}$, [A] Cub. Resinous grey-blue; grey in reflected light. Kvartsitoȳe Gorki deposit, Kazakhstan (N), USSR. H= 3.4, D_m=?, D_c= 5.05, IID 01a. Spiridonov et al., 1986. Am. Min. 73(1988), 439 (Abst.).

Argutite, GeO_2, [A] Tet. Transparent colorless; light grey in reflected light. Argut Plane, Pyrenees, France. H=?, D_m=?, D_c= 6.28, IVD 02. Johan et al., 1983. Am. Min. 69(1984), 406 (Abst.).

Argyrodite, Ag_8GeS_6, [G] Orth. Metallic grey. Himmelsfürst mine, Freiberg, Sachsen (Saxony), Germany. H= 2.5, D_m= 6.2, D_c= 5.98, IIA 11. Weisbach, 1886. N.Jb.Min.Mh. (1978), 269. Struct. Repts. 43A (1977), 901. See also Canfieldite.

Arhbarite, $Cu_2AsO_4(OH) \cdot 6H_2O$, [A] Mon. Vitreous blue. Arhbar mine, Bou Azzer, Morocco. H=?, D_m=?, D_c=?, VIID 02. Schmetzer et al., 1982. N.Jb.Min.Mh. (1982), 529.

Aristarainite, $Na_2Mg[B_6O_8(OH)_4]_2 \cdot 4H_2O$, [A] Mon. Vitreous colorless. Tincalayu Borax deposit, Salar Del Hombre Muerto, Salta, Argentina. H= 3.5, D_m= 2.027, D_c= 2.005, Vc 12a. Hurlbut & Erd, 1974. Am. Min. 59 (1974), 647. Am. Min. 62 (1977), 979.

Armalcolite, $(Mg, Fe)Ti_2O_5$, [A] Orth. Opaque grey. Tranquillity Base, Moon. H=?, D_m=?, D_c= 4.94, IVC 11. Anderson et al., 1970. Am. Min. 73 (1988), 1377. See also Kennedyite, Pseudobrookite.

Armangite, $Mn_{26}As_{18}O_{50}(CO_3)(OH)_4$, [G] Rhom. Black. Långban mine, Filipstad (near), Värmland, Sweden. H= 4, D_m= 4.43, D_c= 4.406, VIID 31. Aminoff & Mauzelius, 1920. Dana, 7th ed.(1951), v.2, 1031. Am. Min. 64 (1979), 748.

Armenite, $BaCa_2Al_6Si_9O_{30} \cdot 2H_2O$, [A] Hex. Translucent colorless. Armen mine, Kongsberg, Buskerud, Norway. H= 7-8, D_m= 2.77, D_c= 2.769, VIIIC 10. Neumann, 1939. Min. Mag. 51 (1987), 317. Sov.Phys.Cryst. 19 (1974), 460.

Armstrongite, $CaZrSi_6O_{15} \cdot 2.5H_2O$, [A] Mon. Vitreous brown. Khan-Bogdinskiĭ massif, Gobi, Mongolia. H= 4.6, D_m= 2.58, D_c= 2.696, VIIID 11. Vladykin et al., 1973. Powd. Diff. 2 (1987), 2. Sov.Phys.Cryst. 23 (1978), 539.

Arrojadite, $KNa_4Ca(Fe, Mn)_{14}Al(PO_4)_{12}(OH)_2$, [G] Mon. Vitreous/greasy dark green. Nickel Plate mine, Keystone, Pennington Co., South Dakota, USA. H= 5, D_m= 3.553, D_c= 3.538, VIIA 06. Guimaraes, 1925. Min. Mag. 43 (1979), 227. Am. Min. 66 (1981), 1034. See also Dickinsonite.

Arsenbrackebuschite, $Pb_2(Fe, Zn)(AsO_4)_2 \cdot H_2O$, [A] Mon. Resinous/adamantine yellow. Tsumeb, Namibia; also Clara mine (Grube Clara), Wolfach, Schwarzwald, Baden-Württemberg, Germany. H= 4-5, D_m=?, D_c= 6.54, VIIC 17. Abraham et al., 1978. Am. Min. 63(1978), 1282 (Abst.). Min. Abstr. 81-1245.

Arsendescloizite, $PbZnAsO_4(OH)$, [A] Orth. Sub-adamantine pale yellow. Tsumeb, Namibia. H= 4, D_m=?, D_c= 6.57, VIIB 11b. Keller & Dunn, 1982. Am. Min. 68(1983), 280 (Abst.).

Arsenic, As, [G] Rhom. Metallic white. H= 3.5, D_m= 5.7, D_c= 5.786, IB 01. Dana, 7th ed.(1944), v.1, 128. J. Appl. Cryst. 2 (1969), 30. See also Arsenolamprite.

Arseniopléite, $NaCaMn(Mn, Mg)_2(AsO_4)_3$, [Q] Mon. Translucent brownish-red, cherry-red, grey. Sjögruvan, Grythyttan (near), Örebro, Sweden. H= 3-4, D_m= 4.22, D_c= 4.29, VIIA 05b. Igelstrøm, 1888. Min. Mag. 51 (1987), 281. See also Caryinite.

Arseniosiderite, $Ca_2Fe_3O_2(AsO_4)_3 \cdot 3H_2O$, [G] Mon. Sub-metallic/silky yellow, brown, black. Romanèche-Thorins, Mâcon (near), Saône et Loire, France. H= 4.5, D_m= 3.60, D_c= 3.60, VIID 15a. Dufrenoy, 1842. Am. Min. 59 (1974), 48. Inorg. Chem. 16 (1977), 1096. See also Mitridatite, Robertsite.

Arsenobismite, $Bi_2AsO_4(OH)_3$, [Q] Syst=? Translucent yellowish-brown, yellowish-green. Mammoth mine, Tintic dist., Utah, USA. H=?, D_m= 5.7, D_c=?, VIID 19. Means, 1916. Dana, 7th ed.(1951), v.2, 907.

Arsenoclasite, $Mn_5(AsO_4)_2(OH)_4$, [G] Orth. Translucent red. Långban mine, Filipstad (near), Värmland, Sweden. H= 5-6, D_m= 4.16, D_c= 4.21, VIIB 08. Aminoff, 1931. Dana, 7th ed.(1951), v.2, 801. Am. Min. 56 (1971), 1539.

Arsenocrandallite, $(Ca, Sr)Al_3(AsO_4)(AsO_3OH)(OH)_6$, [A] Rhom. Vitreous blue, bluish-green. Neubulach, Schwarzwald (N), Baden-Württemberg, Germany. H= 5.5, D_m= 3.25, D_c= 3.30, VIIB 15a. Walenta, 1981. Am. Min. 67(1982), 854 (Abst.).

Arsenoflorencite-(Ce), $(Ce, La)Al_3(AsO_4)_2(OH)_6$, [A] Rhom. Translucent colorless, light brown. Kimba, Eyre Peninsula, South Australia, Australia. H= 3.5, D_m= 4.096, D_c= 4.091, VIIB 15b. Nickel & Temperley, 1987. Min. Mag. 51 (1987), 605.

Arsenogoyazite, $(Sr, Ca, Ba)Al_3(AsO_4)(AsO_3OH)(OH)_6$, [A] Rhom. Vitreous white, yellowish, pale green, greyish-green. Clara mine (Grube Clara), Wolfach, Schwarzwald, Baden-Württemberg, Germany. H= 4, D_m= 3.35, D_c= 3.33, VIIB 15a. Walenta & Dunn, 1984. Am. Min. 71(1986), 845 (Abst.).

Arsenohauchecornite, $Ni_{18}Bi_3AsS_{16}$, [A] Tet. Metallic bronze. Vermilion mine, Sudbury (near), Ontario, Canada. H= 5.5, D_m= 6.35, D_c= 6.616, IIC 13. Gait & Harris, 1980. Min. Mag. 43 (1980), 877. Can. Min. 27 (1989), 137.

Arsenolamprite, As, [G] Orth. Metallic grey. Marienberg, Sachsen (Saxony), Germany. H= 2, D_m= 5.4, D_c= 5.577, IB 01. Hintze, 1886. Am. Min. 45(1960), 479 (Abst.). See also Arsenic.

Arsenolite, As_2O_3, [G] Cub. Vitreous/silky white. H= 1.5, D_m= 3.87, D_c= 3.88, IVC 02. Dana, 1854. Dana, 7th ed.(1944), v.1, 543. See also Claudetite, Senarmontite.

Arsenopalladinite, $Pd_8(As,Sb)_3$, [G] Tric. Metallic yellowish creamy white in reflected light. Itabira, Minas Gerais, Brazil. H=?, D_m= 10.4, D_c= 11.02, IIG 04. Bannister et al., 1955. CIM SV 23 (1981), 83.

Arsenopyrite, FeAsS, [G] Mon. Metallic white. H= 5.5-6, D_m= 6.07, D_c= 6.226, IIC 09. Glocker, 1847. Am. Min. 46 (1961), 1448. Zts. Krist. 179 (1978), 335.

Arsenosulvanite, $Cu_3(As,V)S_4$, [G] Cub. Metallic bronze-yellow. Yakutiya, Siberia, USSR. H= 3.5, D_m= 4.1, D_c= 4.41, IIB 04. Betekhtin, 1941. Am. Min. 40(1955), 368 (Abst.). See also Sulvanite.

Arsenpolybasite, $(Ag,Cu)_{16}(As,Sb)_2S_{11}$, [Q] Mon. Neuer Morgenstern mine, Freiberg, Sachsen (Saxony), Germany. H=?, D_m= 6.2, D_c= 6.08, IID 11. Frondel, 1963. Can. Min. 8 (1965), 172. See also Antimonpearceite.

Arsentsumebite, $CuPb_2(AsO_4)(SO_4)(OH)$, [G] Mon. Translucent green. Tsumeb, Namibia. H=?, D_m= 6.46, D_c= 6.392, VIIB 14. Vésignié, 1935. Am. Min. 51(1966), 258 (Abst.). See also Tsumebite.

Arsenuranospathite, $HAl(UO_2)_4(AsO_4)_4 \cdot 40H_2O$, [G] Tet. Translucent pale yellow. Sophia mine, Menzenschwand, Wittichen, Schwarzwald, Germany. H= 2, D_m=?, D_c= 2.54, VIID 20a. Walenta, 1978. Min. Mag. 42 (1978), 117.

Arsenuranylite, $Ca(UO_2)_4(AsO_4)_2(OH)_4 \cdot 6H_2O$, [G] Orth. Translucent orange. Unspecified locality, USSR. H=?, D_m=?, D_c= 4.25, VIID 21. Belova, 1958. Am. Min. 44(1959), 208 (Abst.).

Arthurite, $CuFe_2(AsO_4)_2(OH)_2 \cdot 4H_2O$, [G] Mon. Translucent pale olive-green. Hingston Downs Consols, Callington (N), Cornwall, England. H=?, D_m= 3.2, D_c= 3.376, VIID 27. Davis & Hey, 1964. Min. Mag. 33 (1964), 937. Struct. Repts. 44A (1978), 349. See also Earlshannonite, Ojuelaite, Whitmoreite.

Artinite, $Mg_2CO_3(OH)_2 \cdot 3H_2O$, [G] Mon. Vitreous/satiny white. Campo Franscia, Val Lanterna, Val Malenco, Val Tellina, Sondrio, Lombardia, Italy. H= 2.5, D_m= 2.03, D_c= 2.03, VbD 01. Brugnatelli, 1902. Dana, 7th ed.(1951), v.2, 263. Acta Cryst. B33 (1977), 3951.

Arzakite, $Hg_3S_2(Br,Cl)_2$, [Q] Mon. Vitreous/adamantine brown, reddish-brown. Arzak deposit, Tuva, USSR. H= 2-2.5, D_m=?, D_c= 7.69, IIF 01. Vasil'ev et al., 1984. Dokl.Earth Sci. 290(1986), 177. See also Lavrentievite.

Arzrunite, $Cu_4Pb_2SO_4(OH)_4Cl_6 \cdot 2H_2O$, [Q] Syst=? Translucent blue, bluish-green. Buena Esperanza mine, Challacollo, Tarapacá, Chile. H=?, D_m=?, D_c=?, IIID 04. Dannenberg, 1887. Strunz (1970), 297.

Asbecasite, $Ca_3Be_2(Ti,Sn)As_6Si_2O_{20}$, [A] Hex. Translucent yellow. Cherbadung, Binntal, Valais (Wallis), Switzerland. H= 6.5-7, D_m= 3.70, D_c= 3.71, VIID 31. Graeser, 1966. Am. Min. 52(1967), 1583 (Abst.). Am. Min. 66 (1981), 819.

Asbolane, $Mn(O,OH)_2 \cdot (Co,Ni,Ca)_x(OH)_{2x} \cdot nH_2O$, [G] Hex. Dull black. H=?, D_m=?, D_c=?, IVF 05b. Breithaupt, 1847. ZVMO 116 (1987), 210. Am. Min. 67 (1982), 417 (Abst.).

Aschamalmite, $Pb_6Bi_2S_9$, [A] Mon. Metallic grey; creamy white in reflected light. Ascham Alm, Untersulzbachtal, Salzburg, Austria. H= 3.5, D_m=?, D_c= 7.33, IID 09d. Mumme et al., 1983. Am. Min. 69(1984), 810 (Abst.).

Ashanite, $(Nb,Ta,Fe,Mn,V)_4O_8$, [A] Orth. Semi-metallic dark brown. Altai Mts., Mongolia, China (NW). H= 5.5-6, D_m= 6.61, D_c= 6.60, IVD 07. Zhang et al., 1980. Am. Min. 66(1981), 217 (Abst.). See also Ixiolite.

Ashcroftine-(Y), $K_5Na_5(Y,Ca)_{12}Si_{28}O_{70}(OH)_2(CO_3)_8 \cdot 8H_2O$, [G] Tet. Translucent pink. Narsarsuk, Greenland (S). H=?, D_m= 2.61, D_c= 2.60, VIIIC 07. Hey & Bannister, 1932. Min. Mag. 37 (1969), 515. Am. Min. 72 (1987), 1176.

Ashoverite, $Zn(OH)_2$, [A] Tet. Vitreous colorless. Milltown, Ashover, Derbyshire, England. H=?, D_m= 3.3, D_c= 3.44, IVF 03a. Clark et al., 1988. Min. Mag. 52 (1988), 699.

Asisite, $Pb_7SiO_8Cl_2$, [A] Tet. Adamantine yellow, yellow-green. Asis farm, Kombat mine, Otavi (E), Namibia. H= 3.5, D_m=?, D_c= 8.041, IIIC 06. Rouse et al., 1988. Am. Min. 73 (1988), 643. Am. Min. 73 (1988), 643.

Asselbornite, $(Pb,Ba)(UO_2)_6(BiO)_4(AsO_4)_2(OH)_{12} \cdot 3H_2O$, [A] Cub. Greasy/adamantine brown, yellow. Weisser Hirsch-Walpurgis Vein, Neustädtel, Schneeberg, Sachsen (Saxony), Germany. H=?, D_m=?, D_c= 5.7, VIID 19. Sarp et al., 1983. Am. Min. 69(1984), 565 (Abst.).

Astrophyllite, $(K,Na)_3(Fe,Mn)_7Ti_2Si_8(O,OH)_{31}$, [G] Tric. Sub-metallic/pearly bronze-yellow, golden-yellow. Loven Island, Brevik (near), Langesundfjord, Norway. H= 3, D_m= 3.3, D_c= 3.41, VIIID 25. Scheerer, 1854. Min. Mag. 45 (1982), 149. Acta Cryst. 22 (1967), 673. See also Kupletskite.

Atacamite, $Cu_2Cl(OH)_3$, [G] Orth. Adamantine/vitreous green. Atacama, Chile. H= 3-3.5, D_m= 3.77, D_c= 3.76, IIIC 01. Gallitzen, 1801. Dana, 7th ed.(1951), v.2, 69. Acta Cryst. C42 (1986), 1277. See also Paratacamite, Botallackite.

Atelestite, $Bi_2OAsO_4(OH)$, [G] Mon. Resinous/adamantine yellow, yellowish-green. Neuhilfe mine, Schneeberg, Sachsen (Saxony), Germany. H= 4.5-5, D_m= 6.82, D_c= 7.275, VIIB 17. Breithaupt, 1832. Can. Min. 7 (1963), 547. Fortsch.Min. 59(1981) Bh1, 126.

Athabascaite, Cu_5Se_4, [A] Orth. Metallic light grey/bluish-grey. Martin Lake mine, Beaverlodge Lake, Saskatchewan, Canada. H= 2.5, D_m=?, D_c= 6.59, IIA 13. Harris et al., 1970. Can. Min. 10 (1970), 207.

Atheneite, $(Pd,Hg)_3As$, [A] Hex. Metallic white in reflected light. Itabira, Minas Gerais, Brazil. H= 5, D_m= 10.2, D_c= 10.16, IIG 04. Clark et al., 1974. Min. Mag. 39 (1974), 528.

Atlasovite, $Cu_6FeBiO_4(SO_4)_5 \cdot KCl$, [A] Tet. Vitreous dark brown. Tolbachik volcano, Kamchatka, USSR. H= 2-2.5, D_m= 4.20, D_c= 4.12, VIB 12. Popova et al., 1987. Am. Min. 73(1988), 927 (Abst.).

Atokite, $(Pd,Pt)_3Sn$, [A] Cub. Metallic light cream. Atok mine, Merensky Reef, Bushveld Igneous Complex, South Africa. H= 4.5, D_m=?, D_c= 14.19, IA 10. Mihalik et al., 1975. Can. Min. 13 (1975), 146.

Attakolite, $(Ca,Mn,Fe)_3Al_6(PO_4)_5(SiO_4)_2 \cdot 3H_2O$, [G] Orth. Translucent pale red. Westanå, Kristianstad, Sweden. H=?, D_m= 3.229, D_c= 3.38, VIIB 13. Blomstrand, 1868. Am. Min. 51(1966), 534 (Abst.).

Aubertite, $CuAl(SO_4)_2Cl \cdot 14H_2O$, [A] Tric. Translucent blue. Quetena mine, Calama, Antofagasta, Chile. H=?, D_m= 1.815, D_c= 1.85, VID 09. Cesbron et al., 1978. Min. Abstr. 80-2891. Acta Cryst. B35 (1979), 2499. See also Svyazhinite.

Augelite, $Al_2PO_4(OH)_3$, [G] Mon. Vitreous colorless, white, yellowish, pale rose. Vestanå mine, Nästum, Skåne, Sweden. H= 4.5-5, D_m= 2.696, D_c= 2.702, VIIB 09. Blomstrand, 1868. Dana, 7th ed.(1951), v.2, 871. Am. Min. 53 (1968), 1096.

Augite, $(Ca,Mg,Fe)_2(Si,Al)_2O_6$, [G] Mon. Vitreous/resinous brown, green, black. H= 5.5-6, D_m= 3.3, D_c= 3.31, VIIID 01a. Werner, 1792. DHZ, 2nd ed.(1978), v.2A, 294. MSA Spec. Paper 2 (1969), 31.

Aurichalcite, $(Zn,Cu)_5(CO_3)_2(OH)_6$, G Orth. Silky/pearly pale green, greenish-blue, blue. H= 1-2, D_m= 3.96, D_c= 3.94, VbB 10. Böttger, 1839. Can. Min. 8 (1965), 385.

Auricupride, Cu_3Au, G Cub. Metallic yellow. Oktyabr deposit, Talnakh, Noril'sk (near), Siberia (N), USSR. H= 3.5, D_m= 11.5, D_c= 13.77, IA 01c. Ramdohr, 1950. ZVMO 106 (1977), 540. See also Tetra-auricupride.

Aurocuproite, D Unnecessary name for palladian auricupride. Min. Mag. 43 (1980), 1055. Razin, 1975.

Aurorite, $(Mn,Ag,Ca)Mn_3O_7 \cdot 3H_2O$, A Tric. Cream-white to medium grey in reflected light. Aurora mine, Treasure Hill, Hamilton, Nevada, USA. H= < 3, D_m=?, D_c= 3.88, IVF 05a. Radtke et al., 1967. Econ. Geol. 62 (1967), 186. Am. Min. 64 (1979), 1197.

Aurostibite, $AuSb_2$, G Cub. Metallic white. Giant Yellowknife mine, Northwest Territories and Chesterville, Ontario, Canada. H= 3, D_m= 9.98, D_c= 9.909, IIC 05. Graham & Kaiman, 1952. Am. Min. 37 (1952), 461.

Austinite, $CaZnAsO_4(OH)$, G Orth. Sub-adamantine/silky colorless, white, yellowish. Utah, USA. H= 4-4.5, D_m= 4.13, D_c= 4.323, VIIB 11b. Staples, 1935. Am. Min. 56 (1971), 1359. N.Jb.Min.Mh. (1988), 159. See also Conichalcite.

Autunite, $Ca(UO_2)_2(PO_4)_2 \cdot 10H_2O$, G Tet. Vitreous/pearly yellow, greenish-yellow, pale green. L'Ouche d'Jau, Saint-Symphorien-de-Marmagne, Autun, Saône-et-Loire, France. H= 2-2.5, D_m= 3.1, D_c= 3.14, VIID 20a. Brooke & Miller, 1852. Am. Min. 46 (1961), 812. See also Meta-autunite, Pseudo-autunite.

Avicennite, Tl_2O_3, G Cub. Metallic greyish-black. Dzhuzumli, Mt. Zirabulak, Bukhara, Tadzhikistan, USSR. H=?, D_m=?, D_c= 10.35, IVC 03. Karpova et al., 1958. Am. Min. 44(1959), 1324 (Abst.).

Avogadrite, $(K,Cs)BF_4$, G Orth. Translucent colorless, white, yellowish, reddish. Mt. Vesuvius, Napoli, Campania, Italy. H=?, D_m= 2.505, D_c= 2.507, IIIB 01. Zambonini, 1926. Dana, 7th ed.(1951), v.2, 97. Acta Cryst. B25 (1969), 2161.

Awaruite, Ni_3Fe, G Cub. Metallic white. Awarua Bay, New Zealand. H= 5, D_m= 8.0, D_c= 8.323, IA 05. Skey, 1885. Can. Min. 6 (1959), 307.

Axinite, A group name for borosilicates with the general formula $(Ca,Fe,Mg,Mn)_3Al_2BSi_4O_{15}(OH)$. See Ferro-axinite, Magnesio-axinite, Manganaxinite, Tinzenite.

Azoproite, $(Mg,Fe)_2(Fe,Ti,Mg)O_2BO_3$, A Orth. Adamantine black. Tazheranskiĭ massif, Baikal, Siberia, USSR. H= 5.5, D_m= 3.63, D_c= 3.63, Vc 01c. Konev et al., 1970. Am. Min. 56(1971), 360 (Abst.).

Azorpyrrhite, D Inadequate data. Am. Min. 62 (1977), 403. Hubbard, 1886.

Azurite, $Cu_3(CO_3)_2(OH)_2$, G Mon. Vitreous blue. H= 3.5-4, D_m= 3.773, D_c= 3.778, VbB 12. Beudant, 1824. Dana, 7th ed.(1951), v.2, 264. Zts. Krist. 135 (1972), 416.

B

Babefphite, $BaBePO_4(F,O)$, [A] Tric. Vitreous/greasy white. Siberia, USSR. H= 3.5, D_m= 4.31, D_c= 4.325, VIIB 01. Nazarova et al., 1966. Am. Min. 51(1966), 1547 (Abst.). Sov.Phys.Cryst. 25 (1980), 28.

Babingtonite, $Ca_2(Fe,Mn)FeSi_5O_{14}(OH)$, [G] Tric. Vitreous dark greenish-black. Arendal, Norway. H= 5.5-6, D_m= 3.36, D_c= 3.39, VIIID 13. Lévy, 1824. Can. Min. 19 (1981), 269. Sov.Phys.Cryst. 20(1975), 446. See also Manganbabingtonite.

Baddeleyite, ZrO_2, [G] Mon. Greasy/vitreous colorless, yellow, green, reddish, etc. Sri Lanka and Brazil. H= 6.5, D_m= 5.82, D_c= 5.836, IVD 16a. Fletcher, 1892. Am. Min. 40 (1955), 275. Acta Cryst. 18 (1965), 983.

Badenite, [D] A mixture of bismuth, safflorite and modderite. Min. Mag. 47 (1983), 411. Poni, 1900.

Bafertisite, $Ba(Fe,Mn)_2Ti(Si_2O_7)(O,OH,F)_2$, [G] Mon. Translucent red, yellowish-red, light brown. Bayan-Obo deposit, Baotou (Paotow), Inner Mongolia, China. H= 5, D_m= 4.1, D_c= 4.35, VIIIB 10. Peng, 1959. Am. Min. 57(1972), 1005 (Abst.). Struct. Repts. 28 (1963), 268.

Baghdadite, $Ca_3ZrO_2(Si_2O_7)$, [A] Mon. Vitreous coloriess. Dupezeh Mt., Hero Town (near), Qala-Dizeh, Iraq (NE). H= 6, D_m= 3.46, D_c= 3.48, VIIIB 08. Al-Hermezi et al., 1986. Min. Mag. 50 (1986), 119.

Bahianite, $Sb_3Al_5O_{14}(OH)_2$, [A] Mon. Translucent tan/cream. Serra Das Almas and Serra Da Mangaheira, Paramirim, Bahia, Brazil. H=?, D_m= 5.1, D_c= 5.26, IVD 15. Moore et al., 1978. Min. Mag. 42 (1978), 179. N.Jb.Min.Abh. 126 (1976), 113.

Baileychlore, $(Zn,Fe,Al,Mg)_6(Si,Al)_4O_{10}(OH)_8$, [A] Tric. Translucent green, yellow-green. Lynd Co., Chillagoe, Red Dome deposit, Queensland, Australia. H=?, D_m= 3.18, D_c= 3.195, VIIIE 09a. Rule & Radke, 1988. Am. Min. 73 (1988), 135.

Baiyuneboite-(Ce), $NaBaCe_2(CO_3)_4F$, [Q] Hex. Greasy/adamantine yellow. Bayan Obo, Inner Mongolia, China. H= 4.5, D_m= 4.30, D_c= 4.45, VbB 04. Fu & Su, 1987. Am. Min. 75(1990), 240 (Abst.). Am. Min. 75(1990), 240 (Abst.).

Bakerite, $Ca_4B_4(BO_4)(SiO_4)_3(OH)_3{\cdot}H_2O$, [G] Mon. Porcelanous white. Furnace Creek, Death Valley, Inyo Co., California, USA. H= 4.5, D_m= 2.88, D_c= 2.94, VIIIA 23. Giles, 1903. Am. Min. 41 (1956), 689.

Balangeroite, $(Mg,Fe)_{21}Si_8O_{27}(OH)_{20}$, [A] Mon. Vitreous/greasy brown. Balangero serpentinite, Lanzo massif, Lanzo Valley, Piemonte, Italy. H=?, D_m= 2.98, D_c= 3.098, VIIIB 24. Compagnoni et al., 1983. Am. Min. 68 (1983), 214. Am. Min. 72 (1987), 382. See also Gageite.

Balavinskite, [D] Inadequate data. Min. Mag. 38 (1971), 103. Yarzhemskii, 1966.

Balipholite, $LiBaMg_2Al_3(Si_2O_6)_2(OH)_4F_4$, [G] Orth. Silky pale yellowish-white. Hsianhual area, Linwu, Hunan, China. H=?, D_m= 3.34, D_c= 3.403, VIIID 03. Am. Min. 61(1976), 338 (Abst.). Min. Abstr. 88M/1796.

Balkanite, $Ag_5Cu_9HgS_8$, [A] Orth. Metallic grey. Sedmochislenitsi mine, Vratsa district, Balkan Peninsula, Bulgaria. H= 2.5, D_m= 6.32, D_c= 6.41, IIA 08. Atanassov & Kirov, 1973. Am. Min. 58 (1973), 11.

Balyakinite, $CuTeO_3$, [A] Orth. Translucent greyish-green, bluish-green. Pionersk deposit, Sayan (E) and Aginsk deposit, Kamchatka, USSR. H= 2.8, D_m= 5.6, D_c= 5.64, VIG 11. Spiridonov, 1980. Am. Min. 66(1981), 436 (Abst.).

Bambollaite, $Cu(Se,Te)_2$, [A] Tet. Metallic brownish-grey. Moctezuma mine, Moctezuma, Sonora, Mexico. H=?, D_m= 5.64, D_c= 4.95, IIC 07. Harris & Nuffield, 1972. Can. Min. 11 (1972), 738.

Banalsite, $Na_2BaAl_4Si_4O_{16}$, [G] Orth. Translucent white. Benallt mine, Rhiw, Caernavonshire, Lleyn Peninsula, Gwynedd, Wales. H=?, D_m= 3.065, D_c= 3.045,

VIIIF 02. Smith et al., 1944. Am. Min. 30(1945), 85 (Abst.). Min. Jour. 7 (1973), 262.

Bandylite, $CuB(OH)_4Cl$, G Tet. Vitreous blue. Quetena mine, Calama, Antofagasta, Chile. H= 2.5, D_m= 2.81, D_c= 2.28, Vc 03a. Palache & Foshag, 1938. Am. Min. 44 (1959), 875. Struct. Repts. 13 (1950), 346.

Bannermanite, $(Na,K)_xV_6O_{15}$, A Mon. Sub-metallic black. Izalco Volcano, MR and L Fumaroles, El Salvador. H=?, D_m= 3.5, D_c= 3.55, IVF 09. Hughes & Finger, 1983. Am. Min. 68 (1983), 634. Am. Min. 68 (1983), 634.

Bannisterite, $KCaMn_{21}(Si,Al)_{32}O_{76}(OH)_{16}\cdot 12H_2O$, G Mon. Dark brown. Franklin, Sussex Co., New Jersey, USA. H=?, D_m= 2.83, D_c= 2.84, VIIIE 07c. Lindberg et al., 1968. Am. Min. 66 (1981), 1063.

Baotite, $Ba_4(Ti,Nb)_8O_{16}(SiO_3)_4Cl$, G Tet. Vitreous light brown, black. Baotou (Paotow), Inner Mongolia, China. H= 6, D_m= 4.42, D_c= 4.50, VIIIC 04. Peng, 1959. Am. Min. 46(1961), 466 (Abst.). Sov.Phys.Cryst. 14(1969), 508.

Bararite, $(NH_4)_2SiF_6$, G Rhom. Vitreous white. Barari, Bengal, India. H= 2.5, D_m= 2.152, D_c= 2.144, IIIB 02a. Frondel, 1951. Dana, 7th ed.(1951), v.2, 106. See also Cryptohalite.

Baratovite, $KLi_3Ca_7(Ti,Zr)_2(SiO_3)_{12}F_2$, A Mon. Pearly white. Dara-Pioz massif, Tadzhikistan, USSR. H= 3.5, D_m= 2.92, D_c= 2.912, VIIIC 05. Dusmatov et al., 1975. ZVMO 104 (1975), 580. Am. Min. 64 (1979), 383.

Barbertonite, $Mg_6Cr_2CO_3(OH)_{16}\cdot 4H_2O$, G Hex. Translucent pink/violet. Barberton, Transvaal, South Africa. H= 1.5-2, D_m= 2.10, D_c= 2.11, IVF 03c. Frondel, 1941. Am. Min. 26 (1941), 295. See also Stichtite.

Barbosalite, $Fe_3(PO_4)_2(OH)_2$, G Mon. Translucent dark blue-green, green. Sapucaia Pegmatite, Galileia, Minas Gerais, Brazil. H= 6, D_m= 3.60, D_c= 3.71, VIIB 05. Lindberg & Pecora, 1955. Min. Mag. 43 (1979), 505. Acta Cryst. 12 (1959), 695. See also Lazulite, Scorzalite.

Barentsite, $Na_7AlH_2(CO_3)_4F_4$, A Tric. Vitreous colorless. Mt. Restinyon, Khibina massif (NE), Kola Peninsula, USSR. H= 3, D_m= 2.56, D_c= 2.55, IIID 02. Khomyakov et al., 1983. Am. Min. 69(1984), 565 (Abst.).

Bariandite, $V_5O_{12}\cdot 6H_2O$, A Mon. Translucent dark greenish-brown. Mounana mine, Franceville, Haut-Ogooué, Gabon. H=?, D_m= 2.7, D_c= 3.05, IVF 09. Cesbron & Vachey, 1971. Am. Min. 57(1972), 1555 (Abst.).

Barićite, $(Mg,Fe)_3(PO_4)_2\cdot 8H_2O$, A Mon. Vitreous/pearly colorless, pale blue. Rapid Creek, Big Fish River–Blow River Area, Yukon Territory, Canada. H= 1.5-2, D_m= 2.42, D_c= 2.448, VIIC 10. Sturman & Mandarino, 1976. Can. Min. 14 (1976), 403.

Bariomicrolite, $Ba(Ta,Nb)_2(O,OH)_7$, A Cub. Chi-Chico, São João del Rei, Minas Gerais, Brazil. H= 4.5-5, D_m= 5.7, D_c= 5.60, IVC 09a. Hogarth, 1977. Am. Min. 62 (1977), 403.

Bario-orthojoaquinite, $(Ba,Sr)_4Fe_2Ti_2O_2(SiO_3)_8\cdot H_2O$, A Orth. Vitreous yellow-brown. Benitoite Gem mine, San Benito Co., California, USA. H=?, D_m= 3.959, D_c= 3.962, VIIIC 05. Wise, 1982. Am. Min. 67 (1982), 809. See also Joaquinite-(Ce), Orthojoaquinite-(Ce).

Bariopyrochlore, $(Ba,Sr)_2(Nb,Ti)_2(O,OH)_7$, A Cub. Translucent yellowish-grey. Panda Hill, Mbeya (near), Tanzania. H= 4.5-5, D_m= 4.1, D_c= 4.01, IVC 09a. Hogarth, 1977. Am. Min. 62(1977), 403.

Barite, $BaSO_4$, G Orth. Vitreous/resinous colorless, white, yellow, brown, grey, etc. H= 3-3.5, D_m= 4.50, D_c= 4.468, VIA 07. Karsten, 1800. Am. Min. 63 (1978), 506. Can. Min. 15 (1977), 522.

Barium-alumopharmacosiderite, D Inadequate data. Min. Mag. 38 (1971), 103. Walenta, 1966.

Barium-pharmacosiderite, D Name rejected by CNMMN. Min. Rec. 16 (1985), 121. Walenta, 1966.

Barnesite, $Na_2V_6O_{16} \cdot 3H_2O$, G Mon. Translucent dark red. Cactus Rat mines, Thompson (near), Grand Co., Utah, USA. H=?, D_m= 3.15, D_c= 3.21, IVF 09. Weeks et al., 1963. Am. Min. 48 (1963), 1187.

Barrerite, $(Na,K,Ca)_5(Si,Al)_{24}O_{48} \cdot 17H_2O$, A Orth. Cagliari, Capo Pula, Efisia Tower (S), Sardegna, Italy. H=?, D_m= 2.13, D_c= 2.114, VIIIF 13. Passaglia & Pongiluppi, 1974. Min. Mag. 40 (1975), 208. Bull. Min. 98 (1975), 331. See also Stellerite.

Barringerite, $(Fe,Ni)_2P$, A Hex. Metallic white. Ollague Meteorite, Tops, New South Wales, Australia. H=?, D_m=?, D_c= 6.92, IC 05. Buseck, 1969. Science 165 (1969), 169. Struct. Repts. 45A (1979), 86.

Barringtonite, $MgCO_3 \cdot 2H_2O$(?), Q Tric. Transparent colorless. Barrington Tops, New South Wales, Australia. H=?, D_m=?, D_c= 2.825, VbC 01. Nashar, 1965. Min. Mag. 34 (1965), 370.

Barroisite, $NaCa(Mg,Fe)_3Al_2(Si_7Al)O_{22}(OH)_2$, A Mon. H=?, D_m=?, D_c=?, VIIID 05c. Murgoci, 1922. Am. Min. 63 (1978), 1023. See also Ferrobarroisite, Aluminobarroisite.

Bartelkeite, $PbFeGe_3O_8$, A Mon. Translucent colorless, pale green. Tsumeb, Namibia. H= 4, D_m=?, D_c= 4.97, VIIIH 01. Keller et al., 1981. Am. Min. 67(1982), 413 (Abst.).

Bartonite, $K_6Fe_{20}S_{26}(Cl,S)$, A Tet. Submetallic blackish-brown; yellow in reflected light. Coyote Peak, Humboldt Co., California, USA. H= 3, D_m= 3.305, D_c= 3.366, IIF 02. Czamanske et al., 1981. Am. Min. 66 (1981), 369. Am. Min. 66 (1981), 376.

Barylite, $BaBe_2Si_2O_7$, G Orth. Translucent colorless. Långban mine, Filipstad (near), Värmland, Sweden. H= 6-7, D_m= 4.0, D_c= 4.00, VIIIB 03. Blomstrand, 1876. Am. Min. 47 (1962), 758, 764. Am. Min. 62 (1977), 167.

Barysilite, $Pb_8Mn(Si_2O_7)_3$, G Rhom. Translucent white. Harstigen mine, Pajsberg, Värmland, Sweden. H= 3, D_m= 6.72, D_c= 6.84, IIIB 04. Sjögren & Lundström, 1888. Am. Min. 54 (1969), 510. Acta Cryst. 20 (1966), 357.

Barytocalcite, $BaCa(CO_3)_2$, G Mon. Vitreous/resinous colorless, white, greyish, greenish. Blagill mine, Alston Moor, Cumberland, England. H= 4, D_m= 3.689, D_c= 3.65, VbA 04. Brooke, 1824. Dana, 7th ed.(1951), v.2, 220. Struct. Repts. 24 (1960), 425. See also Alstonite, Paralstonite.

Barytolamprophyllite, $Na_6Ba_3Ti_7O_4(Si_2O_7)_4(F,OH,O)_4$, G Mon. Vitreous dark brown. Lovozero massif, Kola Peninsula, USSR. H= 2-3, D_m= 3.64, D_c= 3.70, VIIIB 10. Peng & Chang, 1965. Am. Min. 51(1966), 1549 (Abst.). See also Lamprophyllite.

Basaluminite, $Al_4SO_4(OH)_{10} \cdot 4H_2O$, G Mon. Translucent white. Irchester, Northamptonshire, England. H=?, D_m= 2.10, D_c= 2.211, VID 03. Bannister & Hollingworth, 1948. Min. Mag. 43 (1980), 931.

Basiliite, D A mixture of hausmannite and feitknechtite. Am. Min. 58 (1973), 562 (Abst.). Igelström, 1892.

Bassanite, $CaSO_4 \cdot 0.5H_2O$, G Orth. Translucent white. Mt. Vesuvius, Napoli, Campania, Italy. H=?, D_m= 2.70, D_c= 2.73, VIC 01. Zambonini, 1910. Am. Min. 38 (1953), 1266.

Bassetite, $Fe(UO_2)_2(PO_4)_2 \cdot 8H_2O$, G Mon. Transparent yellow. Basset mine, Redruth, Cornwall, England. H=?, D_m= 3.4, D_c= 3.6, VIID 20c. Hallimond, 1915. Min. Mag. 30 (1954), 343.

Bastnäsite-(Ce), $(Ce,La)CO_3F$, G Hex. Vitreous/greasy yellow, reddish-brown. Bastnäs, Riddarhyttan, Västmanland, Sweden. H= 4-4.5, D_m= 4.9, D_c= 5.02,

VbB 04. Huot, 1841. Can. Min. 16 (1978), 361. Am. Min. 38 (1953), 932. See also Hydroxylbastnasite-(Ce).

Bastnäsite-(La), $(La, Ce)CO_3(F, OH)$, [A] Hex. H=?, D_m=?, D_c=?, VbB 04. Levinson, 1966. Am. Min. 51 (1966), 152.

Bastnäsite-(Y), $(Y, Ce)CO_3F$, [G] Hex. Translucent brick-red. Kazakhstan, USSR. H=?, D_m= 4.0, D_c= 4.72, VbB 04. Mineev et al., 1970. Am. Min. 57(1972), 594 (Abst.).

Batavite, $Mg_{0.3}(Mg, Al)_3(Si_3Al)O_{10}(OH)_2 \cdot 4H_2O$, [Q] Mon. Bayern (Bavaria), Germany. H=?, D_m=?, D_c=?, VIIIE 08e. Weinschenk, 1897. Strunz (1970), 447.

Batisite, $(Na, K)_2BaTi_2(Si_2O_7)_2$, [G] Orth. Dark brown. Inagli massif, Aldan Region, Yakutiya, USSR. H= 5.9, D_m= 3.432, D_c= 3.46, VIIID 15. Kravchenko et al., 1960. N.Jb.Min.Mh. (1987), 107. Struct. Repts. 27 (1962), 705.

Baumhauerite, $Ag_{0.6}Pb_{11.6}As_{15.7}S_{36}$, [G] Tric. Metallic grey; white in reflected light. Lengenbach quarry, Binntal, Valais (Wallis), Switzerland. H= 3, D_m= 5.329, D_c= 5.340, IID 05b. Solly, 1902. Dana, 7th ed.(1944), v.1, 460. Zts. Krist. 129 (1969), 178. See also Baumhauerite II.

Baumhauerite II, [D] An artificial product. Min. Mag. 39(1974), 906 (Abst.). Rosch & Hellner, 1959.

Baumite, [D] A mixture of several serpentines and chlorites. Am. Min. 75 (1990), 705. Frondel & Ito, 1975.

Bauranoite, $BaU_2O_7 \cdot 4\text{–}5H_2O$, [A] Syst=? Translucent reddish-brown. USSR. H= 5, D_m= 5.3, D_c=?, IVF 12. Rogova et al., 1973. Am. Min. 58(1973), 1111 (Abst.).

Bavenite, $Ca_4(Al, Be)_4Si_9O_{26}(OH)_2$, [G] Orth. Translucent white. Baveno, Lago Maggiore, Piemonte, Italy. H= 5.5, D_m= 2.7, D_c= 2.75, VIIID 27. Artini, 1901. Dana/Ford (1932), 656. Acta Cryst. 20 (1966), 301.

Bayankhanite, $Cu_{3-8}HgS_{3-5}$, [Q] Syst=? Metallic pale brown/bluish-grey. Idermeg-Bayan-Khan-Ula, Mongolia. H= 3-3.5, D_m=?, D_c=?, IIA 08. Vasil'ev, 1984. Am. Min. 71(1986), 1543 (Abst.).

Bayerite, $\alpha\text{–}Al(OH)_3$, [G] Mon. Translucent white. Hartrurim formation, Israel. H=?, D_m=?, D_c= 2.50, IVF 01. Fricke, 1928. Min. Mag. 33 (1963), 723. Zts. Krist. 125 (1967), 317. See also Doyleite, Gibbsite, Nordstrandite.

Bayldonite, $(Cu, Zn)_3Pb(AsO_4)_2(OH)_2 \cdot H_2O$, [G] Mon. Resinous green, yellow-green. Penberthy Croft mine, St. Hilary, Cornwall, England. H= 4.5, D_m= 5.65, D_c= 5.707, VIIB 14. Church, 1865. Am. Min. 66 (1981), 148. Acta Cryst. B35 (1979). 819.

Bayleyite, $Mg_2(UO_2)(CO_3)_3 \cdot 18H_2O$, [G] Mon. Translucent whitish. Hillside mine, Bagdad, Yavapai Co., Arizona, USA. H=?, D_m= 2.05, D_c= 2.076, VbD 04. Axelrod et al., 1951. Min. Abstr. 87M/2144. Min. Abstr. 87M/0308.

Baylissite, $K_2Mg(CO_3)_2 \cdot 4H_2O$, [A] Mon. Colorless. Gerstenegg (Kabelstollen Gerstenegg-Grimsel I), Grimsel, Bern, Switzerland. H=?, D_m= 2.06, D_c= 2.04, VbC 04. Walenta, 1976. Min. Abstr. 77-2183. Austral.J.Chem. 30(1977), 1379.

Bazhenovite, $Ca_8S_5(S_2O_3)(OH)_2 \cdot 20H_2O$, [A] Mon. Vitreous orange. Chelyabinsk coal basin, Ural Mts. (S), USSR. H= 2, D_m= 1.83, D_c= 1.845, IIE 04. Chesnokov et al., 1987. Am. Min. 74 (1989),500.

Bazirite, $BaZrSi_3O_9$, [A] Hex. Transparent colorless. Rockall Island, Inverness-shire, Scotland. H=?, D_m=?, D_c= 3.880, VIIIC 01. Young et al., 1978. Min. Mag. 42 (1978), 35. N.Jb.Min.Mh. (1987), 16. See also Benitoite, Pabstite.

Bazzite, $Be_3(Sc, Fe)_2Si_6O_{18}$, [G] Hex. Translucent blue. Baveno, Lago Maggiore, Piemonte, Italy. H= 6.5, D_m= 2.819, D_c= 2.82, VIIIC 06. Artini, 1915. Am. Min. 52(1967), 563 (Abst.). Acta Cryst. 9 (1956), 181. See also Beryl.

Bearsite, $Be_2AsO_4(OH) \cdot 4H_2O$, [G] Mon. Translucent white. Kazakhstan, USSR. H=?, D_m=?, D_c= 2.199, VIID 01. Kopchenova & Sidorenko, 1962. Am. Min. 48(1963), 210 (Abst.). See also Moraesite.

Beaverite, $CuPbFe_2(SO_4)_2(OH)_6$, [G] Rhom. Earthy yellow. Horn Silver mine, Frisco, Beaver Co., Utah, USA. H=?, D_m= 4.36, D_c= 4.310, VIB 03a. Butler & Schaller, 1911. Can. Min. 23 (1985), 47.

Beckelite, $(Ce, Ca)_5(SiO_4)_3(OH, F)$, [Q] Syst=? Waxy yellow. Mariupol, Sea of Azov (north shore), USSR. H= 5, D_m= 4.15, D_c=?, VIIIA 29. Morozewicz, 1904. Min. Zhurn. 9 (1) (1978), 78.

Becquerelite, $Ca(UO_2)_6O_4(OH)_6 \cdot 8H_2O$, [G] Orth. Adamantine amber, brownish-yellow. Shinkolobwe, Shaba, Zaïre. H= 2-3, D_m= 5.2, D_c= 5.119, IVF 12. Schoep, 1922. Am. Min. 69(1984), 214 (Abst.). Am. Min. 72 (1987), 1230. See also Billietite, Compreignacite.

Behierite, $(Ta, Nb)BO_4$, [G] Tet. Adamantine white. Manjaka, Madagascar. H= 7-7.5, D_m= 7.86, D_c= 7.91, Vc 03b. Mrose & Rose, 1961. Am. Min. 47(1962), 414 (Abst.).

Behoite, β-$Be(OH)_2$, [A] Orth. Vitreous colorless. Rode Ranch pegmatite, Llano Co., Texas, USA. H= 4, D_m= 1.92, D_c= 1.92, IVF 03a. Ehlmann & Mitchell, 1970. Am. Min. 55 (1970), 1.

Beidellite, $(Na, Ca)_{0.3}Al_2(Si, Al)_4O_{10}(OH)_2 \cdot nH_2O$, [G] Orth. Waxy white, reddish, brownish-grey. Beidell, Saguache Co., Colorado, USA. H=?, D_m=?, D_c=?, VIIIE 08a. Larsen & Wherry, 1925. Am. Min. 47 (1962), 137.

Bellidoite, Cu_2Se, [A] Tet. Metallic creamy-white. Habří, Moravia (W), Czechoslovakia. H= 1.5-2, D_m=?, D_c= 7.026, IIA 01b. De Montreuil, 1975. Econ. Geol. 70 (1975), 384. See also Berzelianite.

Bellingerite, $Cu_3(IO_3)_6 \cdot 2H_2O$, [G] Tric. Translucent light green. Chuquicamata, Antofagasta, Chile. H= 4, D_m= 4.89, D_c= 4.932, IVG 01. Berman & Wolfe, 1940. Dana, 7th ed.(1951), v.2, 313. Acta Cryst. B30 (1974), 965.

Bellite, $(Pb, Ag)_5(CrO_4, AsO_4, SiO_4)_3Cl$, [G] Hex. Translucent red, yellow, orange. Magnet, Tasmania, Australia. H= 2.5, D_m= 5.5, D_c=?, VIE 02. Petterd, 1905. Bull. Mineral. 103 (1980), 469.

Belovite-(Ce), $NaSr_3Ce(PO_4)_3(OH)$, [G] Rhom. Vitreous/greasy yellow. Mt. Karnasurt, Lovozero massif, Kola Peninsula, USSR. H= 5, D_m= 4.19, D_c= 4.131, VIIB 16. Borodin & Kazakova, 1954. Am. Min. 40(1955), 367 (Abst.). Min. Zhurn. 9 [2] (1987), 45.

Belyankinite, $Ca_{1-2}(Ti, Zr, Nb)_5O_{12} \cdot 9H_2O$(?), [Q] Amor. Vitreous/oily/pearly light yellow, brownish-yellow. Kola Peninsula, USSR. H= 2-3, D_m= 2.4, D_c=?, IVF 17. Gerasimovsky & Kazakova, 1950. Am. Min. 37(1952), 882 (Abst.). See also Manganbelyankinite.

Bementite, $Mn_5Si_4O_{10}(OH)_6$, [G] Orth. Pearly greyish-yellow. Franklin Furnace, Sussex Co., New Jersey, USA. H=?, D_m= 2.98, D_c= 3.06, VIIIE 10b. Koenig, 1887. Am. Min. 56 (1971), 416.

Benavidesite, $Pb_4(Mn, Fe)Sb_6S_{14}$, [A] Mon. Metallic grey. Uchucchacua, Oyon, Cajatambo Prov., Peru. H= 2.7, D_m=?, D_c= 5.60, IID 03. Oudin et al., 1982. Am. Min. 68(1983), 280 (Abst.). See also Jamesonite.

Benitoite, $BaTiSi_3O_9$, [G] Hex. Translucent blue, colorless. Benito Gem mine, San Benito Co., California, USA. H= 6.3, D_m= 3.6, D_c= 3.683, VIIIC 01. Louderback, 1907. N.Jb.Min.Mh. (1987), 16. Zts. Krist. 129 (1969), 222. See also Bazirite, Pabstite.

Benjaminite, $Ag_{2.3}Cu_{0.5}Pb_{0.4}Bi_{6.8}S_{12}$, [A] Mon. Metallic grey. Outlaw mine, Nye Co., Nevada, USA. H= 3.5, D_m= 6.34, D_c= 6.68, IID 09c. Shannon, 1925. Can. Min. 13 (1976), 394. Can. Min. 17 (1979), 607.

Benleonardite, $Ag_8(Sb,As)Te_2S_3$, A Tet. Metallic pale blue in reflected light. Bambolla mine, Moctezuma, Sonora, Mexico. H= 3, D_m=?, D_c= 7.79, IID 11. Stanley et al., 1986. Min. Mag. 50 (1986), 681.

Benstonite, $(Ba,Sr)_6(Ca,Mn)_6Mg(CO_3)_{13}$, G Rhom. Translucent white, ivory. National Lead Company, Baroid Division Pit, Hot Springs Co., Arkansas, USA. H= 3-4, D_m= 3.596, D_c= 3.695, VbA 03b. Lippmann, 1962. Am. Min. 47 (1962), 585. Min. Abstr. 80-1322.

Bentorite, $Ca_6(Cr,Al)_2(SO_4)_3(OH)_{12}\cdot 26H_2O$, A Hex. Vitreous bright violet. Hatrurim formation, Israel (S). H= 2, D_m= 2.025, D_c= 2.021, VID 07. Gross, 1980. Am. Min. 66(1981),637 (Abst.).

Beraunite, $Fe^{2+}Fe^{3+}_5(PO_4)_4(OH)_5\cdot 6H_2O$, G Mon. Vitreous reddish-brown, red, dark greenish-brown, green. Hrbek mine, Sveti Dobrotivá, Čechy (Bohemia), Czechoslovakia. H= 3.5-4, D_m= 3.0, D_c= 2.97, VIID 05. Breithaupt, 1841. Can. Min. 27 (1989), 441. Acta Cryst. 22 (1967), 173.

Berborite, $Be_2BO_3(OH,F)\cdot H_2O$, A Rhom. Vitreous colorless. Unspecified locality, USSR (NW). H= 3, D_m= 2.200, D_c= 2.050, Vc 01d. Nefedov, 1967. Am. Min. 53(1978), 348 (Abst.).

Berdesinskiite, V_2TiO_5, A Mon. Black; reddish-brown in reflected light. Lasamba Hill, Kwale Dist., Voi Coast, Kenya. H=?, D_m=?, D_c= 4.538, IVD 05. Bernhardt et al., 1981. Am. Min. 67 (1982), 1074.

Bergenite, $(Ba,Ca)_2(UO_2)_3(PO_4)_2(OH)_4\cdot 5.5H_2O$, G Mon. Translucent yellow. Bergen, Vogtland, Sachsen (Saxony), Germany. H=?, D_m=?, D_c= 4.09, VIID 21. Bultemann & Moh, 1959. Am. Min. 66(1981), 1102 (Abst.).

Bergslagite, $CaBeAsO_4(OH)$, A Mon. Vitreous colorless, whitish, greyish. Långban mine, Filipstad (near), Värmland, Sweden. H= 5, D_m= 3.40, D_c= 3.40, VIIB 01. Hansen et al., 1984. Am. Min. 70(1985), 436 (Abst.). Zts. Krist. 166 (1984), 73. See also Herderite, Hydroxylherderite.

Berlinite, $AlPO_4$, G Rhom. Vitreous colorless, greyish, pale rose. Vestanå mine, Nästum, Skåne, Sweden. H= 6.5, D_m= 2.64, D_c= 2.62, VIIA 01. Blomstrand, 1868. Sov. Phys. Cryst. 31(1986),712. Struct. Repts. 31A (1966), 183. See also Quartz.

Bermanite, $Mn_3(PO_4)_2(OH)_2\cdot 4H_2O$, G Mon. Vitreous/resinous reddish-brown. 7-U-7 Ranch, Bagdad Copper mine (near), Hillside, Yavapai Co., Arizona, USA. H= 3.5, D_m= 2.84, D_c= 2.867, VIID 05. Hurlbut, 1936. Am. Min. 53 (1968), 416. Am. Min. 61 (1976), 1241.

Bernardite, $TlAs_5S_8$, A Mon. Metallic black. Alšar (Allchar), Roždeh (near), Makedonija (Macedonia), Yugoslavia. H= 2, D_m= 4.5, D_c= 4.11, IID 14. Pasava et al., 1989. Min. Mag. 53 (1989), 531. Min. Mag. 53 (1989), 531.

Berndtite-2T, SnS_2, A Hex. Translucent yellow-brown. Cerro de Potosí, Bolivia. H=?, D_m=?, D_c= 4.47, IIC 11. Moh & Berndt, 1964. Min. Mag. 54 (1990), 137. Am. Min. 63 (1978), 289. See also Berndtite-4H.

Berndtite-4H, SnS_2, P Hex. Translucent yellow-brown. Panasqueira mine, Portugal. H=?, D_m=?, D_c= 4.46, IIC 11. Moh, 1966. Min. Mag. 54 (1990), 137. Bull. Min. 109 (1986), 143. See also Berndtite-2T.

Berryite, $(Ag,Cu)_5Pb_3Bi_7S_{16}$, A Mon. Metallic bluish-grey; white/grey-white in reflected light. Missouri mine, Hall's Valley, Park Co., Colorado, USA. H= 3.3, D_m= 6.7, D_c= 7.11, IID 16. Nuffield & Harris, 1966. Can. Min. 8 (1966), 407.

Berthierine-1M, $(Fe,Al)_3(Si,Al)_2O_5(OH)_4$, G Mon. Green. Chazelle, Pontgibaud, Puy-de-Dôme, France. H=?, D_m=?, D_c= 3.03, VIIIE 10b. Beudant, 1832. Rev. Min. 19 (1988), 169.

Berthierine-1H, $(Fe,Al)_3(Si,Al)_2O_5(OH)_4$, G Hex. Green. H=?, D_m=?, D_c= 3.04, VIIIE 10b. Beudant, 1832. Rev. Min. 19 (1988), 169. See also Berthierine-1M.

Berthierite, $FeSb_2S_4$, G Mon. Metallic grey. Chazelle, Pontgibaud, Puy-de-Dôme, France. H= 2-3, D_m= 4.64, D_c= 4.658, IID 06b. Haidinger, 1827. Am. Min. 40 (1955), 226. Zts. Krist. 186 (1989), 31.

Bertossaite, $(Li,Na)_2CaAl_4(PO_4)_4(OH,F)_4$, A Orth. Vitreous pale pink. Buranga pegmatite, Rwanda. H= 6, D_m= 3.10, D_c= 3.10, VIIB 13. v. Knorring & Mrose, 1966. Can. Min. 8(1966), 668 (Abst.). See also Palermoite.

Bertrandite, $Be_4Si_2O_7(OH)_2$, G Orth. Vitreous/pearly colorless, pale yellow. Petit-Port and Barbin, Nantes, Loire-Atlantique, France. H= 6-7, D_m= 2.60, D_c= 2.603, VIIIB 07. Damour, 1883. Dana, 6th ed.(1892), 545. Am. Min. 72 (1987), 979.

Beryl, $Be_3Al_2Si_6O_{18}$, G Hex. Vitreous green, blue, yellow, white, rose, etc. H= 7.5-8, D_m= 2.8, D_c= 2.66, VIIIC 06. Am. Min. 73 (1988), 826. Acta Cryst. B28 (1972), 1899. See also Bazzite.

Beryllite, $Be_3SiO_4(OH)_2 \cdot H_2O$, Q Orth. Silky white. Kola Peninsula (?), USSR. H=?, D_m= 2.196, D_c=?, VIIIA 13. Kuzmenko. 1954. Am. Min. 40(1955), 787 (Abst.).

Beryllonite, $NaBePO_4$, G Mon. Vitreous white, pale yellowish. McKean Mt., Stoneham (near), Oxford Co., Maine, USA. H= 5.5-6, D_m= 2.81, D_c= 2.803, VIIA 01. Dana, 1888. Dana, 7th ed.(1951), v.2, 677. Struct. Repts. 39A (1973), 283.

Berzelianite, $Cu_{2-x}Se$, G Cub. Metallic white. Skrikerum, Sweden. H= 2, D_m= 6.71, D_c= 7.23, IIA 01b. Beudant, 1832. Am. Min. 35 (1950), 337. Struct. Repts. 42A (1976), 75. See also Bellidoite.

Berzeliite, $NaCa_2(Mg,Mn)_2(AsO_4)_3$, G Cub. Resinous yellow, orange. Långban mine, Filipstad (near), Värmland, Sweden. H= 4.5-5, D_m= 4.08, D_c= 4.068, VIIA 07. Kühn, 1840. Dana, 7th ed.(1951), v.2, 681. Acta Cryst. B32 (1976), 1581. See also Manganberzeliite, Garnet.

Beta-alumohydrocalcite, D Inadequate data. Min. Mag. 36 (1967), 133. Morawiecki, 1961.

Beta-brocenite, D = Beta-fergusonite-(Ce). Min. Mag. 43 (1980), 1055.

Beta-fergusonite-(Ce), $(Ce,La,Nd)NbO_4$, G Mon. Vitreous/greasy red, reddish-brown. China (N). H=?, D_m= 5.34, D_c=?, IVD 13. Kuo et al., 1973. Am. Min. 62(1977), 397 (Abst.). See also Fergusonite-(Ce).

Beta-fergusonite-(Nd), $(Nd,Ce)NbO_4$, G Mon. Vitreous/greasy brownish-red, red, yellowish-brown. Bayan Obo, Inner Mongolia, China. H=?, D_m=?, D_c= 6.45, IVD 13. Sun et al., 1984. Am. Min. 69(1984), 406 (Abst.). See also Fergusonite-(Nd).

Beta-fergusonite-(Y), $YNbO_4$, G Mon. USSR (E). H=?, D_m= 5.65, D_c= 5.57, IVD 13. Gorshevskaya et al., 1961. Am. Min. 46(1961), 1516 (Abst.). Sov.Phys.Cryst. 26 (1981), 35. See also Fergusonite-(Y).

Betafite, $(Ca,U)_2(Ti,Nb,Ta)_2O_6(OH)$, G Cub. Waxy/vitreous/semi-metallic greenish-brown, yellow, black. Betafo, Madagascar. H= 4-5.5, D_m= 4.5, D_c=?, IVC 09a. Lacroix, 1912. Can. Min. 6 (1961), 610.

Beta-iridisite, $Ir_{0.75}S_2$, Q Cub. Metallic greyish-white. Unspecified locality, China. H= 6.5, D_m=?, D_c= 7.88, IIC 05. Yu, 1973. Am. Min. 74(1989), 1215 (Abst.).

Beta-lomonosovite, D Inadequate data. Min. Mag. 36 (1967), 133. Gerasimovsky & Kazakova, 1962.

Beta-moissanite, SiC, P Cub. Metallic green/black; grey in reflected light. Bridger salt deposit, Wyoming, USA. H=?, D_m=?, D_c= 3.219, IC 04. Regis & Sand, 1958. Acta Cryst. B25 (1969), 477. Struct. Repts. 13 (1950), 166. See also Moissanite-6H, Moissanite-5H.

Beta-roselite, $Ca_2(Co,Mg)(AsO_4)_2\cdot 2H_2O$, [G] Tric. Vitreous dark rose-red. Schneeberg, Sachsen (Saxony), Germany. H= 3.5-4, D_m= 3.71, D_c= 4.21, VIIC 12. Frondel, 1955. Bull. Min. 83 (1960), 118. See also Roselite.

Beta-uranophane, $Ca(UO_2)_2(SiO_3OH)_2\cdot 5H_2O$, [G] Mon. Vitreous/waxy yellowish-green, yellow, brownish-yellow. Jáchymov (St. Joachimsthal), Západočeský kraj, Čechy (Bohemia), Czechoslovakia. H=?, D_m=?, D_c= 4.117, VIIIA 25. Novaček, 1935. Am. Min. 66 (1981), 610. Am. Min. 71 (1986), 1489. See also Uranophane.

Betekhtinite, $Cu_{10}(Pb,Fe)S_6$, [G] Orth. Metallic cream. Mansfeld, Germany. H= > 3, D_m= 6.14, D_c= 5.73, IIA 01a. Schüller & Wohlmann, 1955. Am. Min. 41(1956), 371 (Abst.). Acta Cryst. 12 (1959), 646.

Betpakdalite, $(H,K)_6Ca_4Fe_6As_4Mo_{16}O_{74}\cdot 28H_2O$, [G] Mon. Waxy/vitreous yellow. Bet-Pak-Dal desert, Kazakhstan, USSR. H= 3, D_m= 3.0, D_c= 2.70, VIID 15b. Ermilova & Senderova, 1961. Am. Min. 47(1962), 172 (Abst.). Am. Min. 70(1985), 1333 (Abst.).

Beudantite, $PbFe_3(AsO_4,SO_4)_2(OH)_6$, [G] Rhom. Vitreous/resinous black, dark green, brown. Grube Louise, Horhausen, Rheinland-Pfalz, Germany. H= 3.5-4.5, D_m= 4.48, D_c= 4.49, VIB 03b. Lévy, 1826. N.Jb.Min.Mh. (1989), 27. Can. Min. 26 (1988), 923.

Beusite, $(Mn,Fe,Ca,Mg)_3(PO_4)_2$, [A] Mon. Vitreous reddish-brown. Los Aleros, San Luis, Argentina. H= 5, D_m= 3.702, D_c= 3.715, VIIA 03. Hurlbut & Aristarain, 1968. Am. Min. 53 (1968), 1799. Am. Min. 67 (1982), 826. See also Graftonite.

Beyerite, $(Ca,Pb)Bi_2O_2(CO_3)_2$, [G] Tet. Vitreous, earthy yellow, white. Stewart mine, Pala, San Diego Co., California, USA, and, Schneeberg, Sachsen (Saxony), Germany. H= 3, D_m= 6.56, D_c= 6.58, VbB 05. Frondel, 1943. Am. Min. 54 (1969), 1720. Struct. Repts. 11 (1947), 319.

Bezsmertnovite, $(Au,Ag)_4Cu(Te,Pb)$, [A] Orth. Metallic yellow. Kazakhstan, USSR. H= 4.5, D_m=?, D_c= 16.3, IA 01d. Spiridonov & Chvileva, 1979. Am. Min. 66(1981), 878 (Abst.).

Bianchite, $(Zn,Fe)SO_4\cdot 6H_2O$, [G] Mon. Vitreous white, yellowish. Predil mine, Tarvisio, Paibl, Trentino-Alto Adige, Italy. H= 2.5, D_m= 2.031, D_c= 2.00, VIC 03b. Andreatta, 1930. Per. Mineral. 54 (1985), 9.

Bicchulite, $Ca_2Al_2SiO_6(OH)_2$, [A] Cub. Translucent colorless. Bicchu, Okayama, Japan. H=?, D_m= 2.75, D_c= 2.824, VIIIF 07. Henmi et al., 1973. Am. Min. 59(1974), 1330 (Abst.). Zts. Krist. 146 (1977), 35. See also Kamaishilite.

Bideauxite, $AgPb_2(F,OH)_2Cl_3$, [A] Cub. Adamantine colorless. Mammoth-St. Anthony mine, Tiger, Pinal Co., Arizona, USA. H= 3, D_m= 6.274, D_c= 6.256, IIIB 10. Williams, 1970. Min. Mag. 37 (1970), 637.

Bieberite, $CoSO_4\cdot 7H_2O$, [G] Mon. Vitreous red. Bieber, Hanau (near), Hessen, Germany. H= 2, D_m= 1.96, D_c= 1.90, VIC 03c. Andreatta, 1930. Per. Mineral. 54 (1985), 1.

Bijvoetite-(Y), $(Y,Dy)_2(UO_2)_4(CO_3)_4(OH)_6\cdot 11H_2O$, [A] Orth. Vitreous yellow. Shinkolobwe, Shaba, Zaïre. H= 2, D_m= 3.95, D_c= 3.907, VbD 04. Deliens & Piret, 1982. Can. Min. 20 (1982), 231.

Bikitaite (triclinic), $LiAlSi_2O_6\cdot H_2O$, [P] Tric. Bikita, Fort Victoria, Zimbabwe. H=?, D_m=?, D_c= 2.298, VIIIE 29. Hurlbut, 1957. N.Jb.Min.Mh. (1986), 241. See also Bikitaite (monoclinic).

Bikitaite (monoclinic), $LiAlSi_2O_6\cdot H_2O$, [G] Mon. Transparent colorless, white. Bikita, Fort Victoria, Zimbabwe. H= 6, D_m= 2.28, D_c= 2.28, VIIIE 29. Hurlbut, 1957. Am. Min. 43 (1958), 768. Am. Min. 59 (1974), 71. See also Bikitaite (triclinic).

Bilibinskite, $Au_3Cu_2Pb\cdot nTeO_2$, [A] Cub. Semi-metallic brown. Unspecified locality, Kazakhstan or far eastern USSR. H= 4.5, D_m=?, D_c=?, IA 01d. Spiridonov et al., 1978. Dokl. Earth Sci. 267 (1982), 145.

Bilinite, $Fe_3(SO_4)_4 \cdot 22H_2O$, [G] Mon. Translucent white, yellowish. Schwaz, Bilina, Čechy (Bohemia), Czechoslovakia. H= 2, D_m= 1.875, D_c= 1.993, VIC 06. Sebor, 1913. JCPDS 25-1153.

Billietite, $Ba(UO_2)_6O_4(OH)_6 \cdot 4H_2O$, [G] Orth. Translucent deep golden yellow. Shinkolobwe, Shaba, Zaïre. H=?, D_m= 5.28, D_c= 5.093, IVF 12. Vaes, 1947. Am. Min. 45 (1960), 1026. Am. Min. 72 (1987), 1230. See also Becquerelite, Compreignacite.

Billingsleyite, $Ag_7(Sb,As)S_6$, [A] Orth. Metallic grey. North Lily mine, East Tintic Dist., Utah, USA. H= 2.5, D_m= 5.9, D_c= 5.90, IID 04. Frondel & Honea, 1968. Am. Min. 53 (1968), 1791.

Bindheimite, $Pb_2Sb_2O_6(O,OH)$, [G] Cub. Resinous, earthy yellow, brown, grey white, etc. Nerchinsk, Siberia, USSR. H= 4-4.5, D_m= 7.32, D_c= 7.50, IVC 08. Dana, 1868. JCPDS 18-687.

Biotite, $K(Mg,Fe)_3(Si_3Al)O_{10}(OH,F)_2$, [G] Mon. Black, dark brown, reddish-brown. H= 2.5-3, D_m= 3.0, D_c= 2.81, VIIIE 05b. Hausmann, 1847. DHZ (1962), v.3, 55. Am. Min. 60 (1975), 1030. See also Phlogopite.

Biphosphammite, $H_2(NH_4,K)PO_4$, [G] Tet. Earthy/vitreous white, brown. Guañape Islands, Peru. H=?, D_m= 2.04, D_c= 1.98, VIIA 14. Shepard, 1870. Min. Mag. 38 (1972), 965. See also Archerite.

Biringuccite, $Na_2B_5O_8(OH) \cdot H_2O$, [G] Mon. Translucent colorless. Lardarello, Val di Cecina, Toscana, Italy. H=?, D_m= 2.32, D_c= 2.297, Vc 10. Cipriani, 1961. Am. Min. 48(1963), 709 (Abst.). Am. Min. 59 (1974), 1005.

Birnessite, $(Na,Ca,K)(Mg,Mn)Mn_6O_{14} \cdot 5H2O$, [G] Hex. Black. Birness, Aberdeenshire, Scotland. H= 1.5, D_m= 3.0, D_c=?, IVF 05b. Jones & Milne, 1956. Clays Cl. Mins. 28(1980), 346. Am. Min. 75 (1990), 477.

Bisbeeite, [D] = chrysocolla. Min. Mag. 43 (1980), 1054. Schaller, 1915.

Bischofite, $MgCl_2 \cdot 6H_2O$, [G] Mon. Vitreous colorless, white. Leopoldshall, Prussia, Germany. H= 1-2, D_m= 1.604, D_c= 1.59, IIIC 10a. Ochsenius, 1877. NBS Monogr. 11 (1974), 37. See also Nickelbischofite.

Bismite, α–Bi_2O_3, [G] Mon. Sub-adamantine, earthy yellow, greyish-green. Colavi, Potosí, Bolivia. H= 4.5, D_m= 8.64, D_c= 9.37, IVC 02. Dana, 1868. NBS Monogr. 3 (1964), 17.

Bismoclite, BiOCl, [G] Tet. Greasy/earthy creamy-white, greyish, yellowish-brown. Jackal's Water, Namaqualand, South Africa. H= 2-2.5, D_m= 7.717, D_c= 7.70, IIIC 16. Mountain, 1935. Dana, 7th ed.(1951), v.2, 60. See also Daubréeite, Zavaritskite.

Bismuth, Bi, [G] Rhom. Metallic reddish-white; creamy-white in reflected light. H= 2-2.5, D_m= 9.8, D_c= 9.806, IB 01. Agricola, 439. ZVMO 95 (1966), 489.

Bismuthinite, Bi_2S_3, [G] Orth. Metallic grey/white. H= 2, D_m= 6.78, D_c= 6.81, IIC 02. Beudant, 1832. Dana, 7th ed.(1944), v.1, 275. Struct. Repts. 43A (1977), 33. See also Guanajuatite, Stibnite.

Bismutite, $Bi_2O_2(CO_3)$, [G] Tet. Vitreous yellow, yellowish-white, green, grey, brown. Ullersreuth, Vogtland, Sachsen (Saxony), Germany. H= 2.5-3.5, D_m= 8.15, D_c= 8.28, VbB 05. Breithaupt, 1841. Dana, 7th ed.(1951), v.2, 259. Struct. Repts. 11 (1947), 308.

Bismutoferrite, $Fe_2Bi(SiO_4)_2(OH)$, [G] Mon. Translucent yellow, green. Schneeberg, Sachsen (Saxony), Germany. H= 6, D_m= 3.0, D_c= 5.092, VIIIE 10c. Frenzel, 1871. Am. Min. 43 (1958), 656. Sov.Phys.Crust. 22 (1977), 419. See also Chapmanite.

Bismutohauchecornite, $Ni_9Bi_2S_8$, [A] Tet. Metallic brownish-yellow. Oktyabr deposit, Talnakh, Noril'sk (near), Siberia (N), USSR. H=?, D_m=?, D_c=?, IIC 13. Just, 1980. Min. Mag. 43 (1980), 873. See also Hauchecornite.

Bismutomicrolite, $(Bi,Ca)(Ta,Nb)_2O_6(OH)$, [A] Cub. Resinous yellow, pink, brown. Wampewo Hill, Busiro County, Buganda, Uganda. H= 5, D_m= 6.5, D_c= 6.83, IVC 09a. Hogarth, 1977. Am. Min. 62 (1977), 403.

Bismutostibiconite, $Bi(Sb,Fe)_2O_7$, [A] Cub. Earthy yellow, yellowish-brown, greenish. Clara mine (Grube Clara), Wolfach, Schwarzwald, Baden-Württemberg, Germany. H= 4-5, D_m=?, D_c= 7.38, IVC 08. Walenta, 1983. Am. Min. 69(1984), 1190 (Abst.).

Bismutotantalite, $Bi(Ta,Nb)O_4$, [G] Orth. Sub-metallic black. Busiro Co., Gamba Hill, Uganda. H= 5, D_m= 8.26, D_c= 8.76, IVD 14a. Wayland & Spencer, 1929. Am. Min. 48 (1963), 1348. See also Stibiocolumbite, Stibiotantalite.

Bityite, $LiCaAl_2(Si_2BeAl)O_{10}(OH)_2$, [G] Mon. Transparent colorless. Mt. Bity, Madagascar. H= 5.5, D_m= 3.0, D_c= 3.05, VIIIE 06. Lacroix, 1908. Dana/Ford (1932), 657. Am. Min. 68 (1983), 130.

Bixbyite, Mn_2O_3, [G] Cub. Metallic/sub-metallic black. Simpson (SW), Thomas Range, Jaub Co., Utah, USA. H= 6-6.5, D_m= 4.945, D_c= 5.026, IVC 03. Penfield & Foote, 1897. Am. Min. 74 (1989), 1325. Acta Cryst. B27 (1971), 821.

Bjarebyite, $(Ba,Sr)(Mn,Fe,Mg)_2Al_2(PO_4)_3(OH)_3$, [A] Mon. Sub-adamantine green. Palermo pegmatite, North Groton, New Hampshire, USA. H= 4, D_m= 3.95, D_c= 4.02, VIIB 19. Moore et al., 1973. Am. Min. 59(1974), 873 (Abst.). Am. Min. 59 (1974), 567.

Blakeite, $Fe,TeO_3(?)$, [Q] Syst=? Dull reddish-brown, deep brown. Mohawk mine, Goldfield, Nevada, USA. H=?, D_m=?, D_c=?, VIG 02. Frondel & Pough, 1944. Am. Min. 29 (1944), 211.

Blatterite, $(Mn,Mg)_2(Mn,Sb)O_2(BO_3)$, [A] Orth. Metallic/sub-metallic black. Kittelgruvan mine, Värmland, Sweden. H= 6, D_m= 4.7, D_c= 4.352, Vc 01c. Raade et al., 1988. N.Jb.Min.Mh. (1988), 121.

Blixite, $Pb_2Cl(O,OH)_2$, [G] Orth. Vitreous pale yellow. Långban mine, Filipstad (near), Värmland, Sweden. H= 3, D_m= 3.75, D_c= 3.78, IIIC 07. Gabrielson et al.,1958. Am. Min. 45(1960), 908 (Abst.).

Blödite, $Na_2Mg(SO_4)_2 \cdot 4H_2O$, [A] Mon. Vitreous colorless, bluish-green, reddish. Chuquicamata, Antofagasta, Chile. H= 2.5-3, D_m= 2.25, D_c= 2.226, VIC 10. John, 1811. Dana, 7th ed.(1951), v.2, 447. Can. Min. 23 (1985), 669. See also Nickelblödite.

Blossite, α-$Cu_2V_2O_7$, [A] Orth. Metallic black; white in reflected light. Izalco Volcano, El Salvador. H=?, D_m= 3.96, D_c= 4.051, VIIA 12. Hughes et al., 1987. Am. Min. 72 (1987), 397. Am. Min. 72 (1987), 397.

Bobfergusonite, $Na_2Mn_5FeAl(PO_4)_6$, [A] Mon. Resinous green-brown, red-brown. Cross Lake, Manitoba, Canada. H= 4, D_m= 3.54, D_c= 3.57, VIIA 05a. Ercit et al., 1986. Can. Min. 24 (1986), 599. Can. Min. 24 (1986), 605.

Bobierrite, $Mg_3(PO_4)_2 \cdot 8H_2O$, [G] Mon. Vitreous colorless, white. Mexillones, Chile. H= 2-2.5, D_m= 2.195, D_c= 2.133, VIIC 10. Dana, 1868. Am. Min. 48 (1963), 635. Am. Min. 71 (1986), 1229. See also Manganese-hörnesite, Vivianite.

Bogdanovite, $Au_5(Cu,Fe)_3(Te,Pb)_2$, [A] Cub. Semi-metallic brown. Unspecified locality, Kazakhstan or far Eastern USSR. H= 4.5, D_m=?, D_c=?, IA 01d. Spiridonov et al., 1979. Am. Min.64(1979), 1329 (Abst.).

Bøggildite, $Na_2Sr_2Al_2(PO_4)F_9$, [G] Mon. Translucent flesh red. Ivigtut, Greenland (SW). H= 4-5, D_m= 3.66, D_c= 3.692, IIID 01. Pauly, 1956. Am. Min. 41(1956), 959 (Abst.). Can. Min. 20 (1982), 263.

Boggsite, $Na_3Ca_8(Si,Al)_{96}O_{192} \cdot 70H_2O$, [A] Orth. Vitreous/dull white, colorless. Goble, Oregon, USA. H=?, D_m=?, D_c=?, VIIIF 12. Howard et al, 1990. Am. Min. 75 (1990), 501.

Bohdanowiczite, $AgBiSe_2$, [A] Rhom. Metallic creamy-yellow. Kletna, Poland. H= 2.5, D_m=?, D_c= 7.72, IID 07. Banas & Ottemann, 1967. Min. Mag. 43 (1979), 131. See also Matildite, Volynskite.

Böhmite, $AlO(OH)$, G Orth. H=?, D_m= 3.03, D_c= 3.081, IVF 04. Lapparent, 1927. Dana, 7th ed.(1944), v.1, 645. Clays Cl. Mins. 29 (1981), 435. See also Diaspore.

Bokite, $KAl_3Fe_6V_{20}O_{76} \cdot 30H_2O$(?), Q Syst=? Semi-metallic/dull black. Balasauskandyk, Kara Tau, Kazakhstan, USSR. H= 3, D_m= 3.0, D_c=?, VIIC 24. Ankinovich, 1963. Am. Min. 48(1963), 1180 (Abst.).

Boldyrevite, $NaCaMgAl_3F_{14} \cdot 4H_2O$, Q Amor. Translucent light yellow. Bilyukai crater, Klyuchevskaya volcano, Kamchatka, USSR. H=?, D_m=?, D_c=?, IIIC 12. Gagarin & Cuomo, 1949. Am. Min. 36(1951), 641 (Abst.).

Boléite, $Ag_9Cu_{24}Pb_{26}Cl_{62}(OH)_{48}$, G Cub. Vitreous blue. Boléo, Santa Rosalia (near), Baja California, Mexico. H= 3-3.5, D_m= 5.054, D_c= 5.062, IIIC 04. Mallard & Cumenge, 1891. Min. Rec. 5 (1973), 280. Struct. Repts. 39A (1973), 195.

Boleslavite, D Superfluous name for fine-grained galena. Embrey & Fuller (1980), 42. Haranczyk, 1961.

Bolivarite, $Al_2PO_4(OH)_3 \cdot 4\text{-}5H_2O$, G Syst=? Vitreous pale greenish-yellow. Pontevedra, Spain. H= 2.5, D_m= 2.05, D_c=?, VIID 10. Navarro & Barea, 1921. Min. Mag. 38 (1971), 418.

Boltwoodite, $(H_3O)K(UO_2)SiO_4$, G Mon. Pearly/vitreous/silky pale yellow. Pick's Delta mine, San Rafael Swell, Emery Co., Utah, USA. H= 3.5-4, D_m= 3.6, D_c= 4.372, VIIIA 25. Frondel & Ito, 1956. Min. Abstr. 84M/3122. Am. Min. 66 (1981), 610.

Bonaccordite, $Ni_2FeO_2(BO_3)$, A Orth. Reddish-brown; light grey (brownish tinge) in refl. light. Scotia Talc mine, Bon Accord, Barberton Dist., Transvaal, South Africa. H= 7, D_m=?, D_c= 5.17, Vc 01c. DeWaal et al., 1974. Am. Min. 61(1976), 502 (Abst.).

Bonattite, $CuSO_4 \cdot 3H_2O$, G Mon. Translucent pale blue. Elba, Toscana, Italy. H=?, D_m= 2.663, D_c= 2.68, VIC 03a. Garavelli, 1957. Can. Min. 7 (1962), 245. Acta Cryst. B24 (1968), 508.

Bonchevite, $(Pb,Cu)_3Bi_{11}S_{18}$, Q Orth. Metallic grey. Rhodope Mts., Bulgaria. H= 2.5, D_m= 6.92, D_c= 7.23, IID 08. Kostov, 1958. Min. Mag. 49 (1985), 135.

Bonshtedtite, $Na_3(Fe,Mg,Mn)(PO_4)(CO_3)$, A Mon. Vitreous colorless (rose, yellowish, greenish tints). Kovdor massif and Khibina massif, Vuonnemi river, Kola Peninsula, USSR. H= 4, D_m= 3.0, D_c= 2.92, VbB 06. Khomyakov et al., 1982. Am. Min. 68(1983), 1038 (Abst.). See also Bradleyite, Sidorenkite.

Boothite, $CuSO_4 \cdot 7H_2O$, G Mon. Silky/pearly blue. Alma mine, Leona Heights, Alameda Co., California, USA. H= 2-2.5, D_m= 2.1, D_c=?, VIC 03c. Schaller, 1903. Dana, 7th ed.(1951), v.2, 504.

Boracite (high), $\beta\text{-}Mg_3B_7O_{13}Cl$, P Cub. Vitreous colorless, white, grey, green yellow, etc. H= 7-7.5, D_m= 2.9, D_c= 2.940, Vc 14. Werner, 1789. Dana, 7th ed.(1951), v.2, 378. Am. Min. 58 (1973), 691. See also Boracite (low).

Boracite (low), $\alpha\text{-}Mg_3B_7O_{13}Cl$, G Orth. Vitreous colorless, white, grey, yellow, green, etc. Luneberg, Germany. H= 7-7.5, D_m= 2.9, D_c= 2.945, Vc 14. Werner, 1789. Am. Min. 58 (1973), 691. Zts. Krist. 138 (1973), 64. See also Ericaite, Chambersite.

Borax, $Na_2B_4O_5(OH)_4 \cdot 8H_2O$, G Mon. Vitreous/resinous/earthy colorless, white, greyish, etc. Ladakh, Kashmir and Rudok, Tibet. H= 2-2.5, D_m= 1.715, D_c= 1.710, Vc 07a. Wallerius, 1748. Dana, 7th ed.(1951), v.2, 339. Acta Cryst. B34 (1978), 3502.

Borcarite, $Ca_4MgB_4O_6(CO_3)_2(OH)_6$, G Mon. Vitreous greenish-blue, bluish-green. Siberia, USSR. H= 4, D_m= 2.77, D_c= 2.788, Vc 07a. Pertzev et al., 1965. Am. Min. 50(1965), 2097 (Abst.). Min. Abstr. 77-1501.

Bořickyite, D Probably = delvauxite. Am. Min. 65(1980), 813 (Abst.). Dana, 1868.

Borishanskiite, $Pd(As,Pb)_2$, A Orth. Metallic greyish-white. Oktyabr deposit, Talnakh, Noril'sk (near), Siberia (N), USSR. H= 3.5-4, D_m=?, D_c= 10.2, IIC 07. Razin et al., 1975. Am. Min. 61(1976), 502 (Abst.). See also Polarite.

Bornemanite, $Na_7BaTi_2NbSi_4O_{17}(PO_4)(F,OH)$, A Orth. Pearly pale yellow. Yubileĭnaya (Jubilejnaja) pegmatite, Mt. Karnasurt, Lovozero, Kola Peninsula, USSR. H= 3.5-4, D_m=?, D_c= 3.49, VIIIB 11. Men'shikov et al., 1975. Am. Min. 61(1976), 338 (Abst.).

Bornhardtite, Co_3Se_4, G Cub. Metallic rose. Trogtal quarry, Lautenthal (near), Harz, Niedersachsen, Germany. H= 4, D_m=?, D_c= 6.17, IIC 01. Ramdohr & Schmitt, 1955. Am. Min. 41(1956), 164 (Abst.).

Bornite, Cu_5FeS_4, G Tet. Metallic copper-red/brown; pinkish-brown in reflected light. H= 3, D_m= 5.07, D_c= 5.076, IIA 02. Haidinger, 1845. Am. Min. 46 (1961), 1270. Acta Cryst. B31 (1975), 2268.

Borovskite, Pd_3SbTe_4, A Cub. Metallic dark grey; white in reflected light. Khautovaar deposit, Karelia, USSR. H= 2.5, D_m=?, D_c= 8.12, IID 07. Yalovoi et al., 1973. Am. Min. 59,(1974), 873 (Abst.).

Bostwickite, $CaMn_6Si_3O_{16}\cdot 7H_2O$, A Syst=? Vitreous/sub-metallic dark red. Franklin, Sussex Co., New Jersey, USA. H= 1, D_m= 2.93, D_c=?, VIIID 13. Dunn & Leavens, 1983. Min. Mag. 47 (1983), 387.

Botallackite, $Cu_2Cl(OH)_3$, G Mon. Translucent pale bluish green to green. Botallack mine, Cornwall, England. H=?, D_m= 3.6, D_c= 3.598, IIIC 01. Church, 1865. Min. Mag. 29 (1950), 34. Min. Mag. 49 (1985), 87. See also Atacamite, Paratacamite.

Botryogen, $MgFe(SO_4)_2(OH)\cdot 7H_2O$, G Mon. Vitreous light/dark orange-red. Falun, Kopparberg, Sweden. H= 2-2.5, D_m= 2.19, D_c= 2.23, VID 04a. Haidinger, 1828. Dana, 7th ed.(1951), v.2, 617. Acta Cryst. B24 (1968), 760. See also Zincobotryogen.

Boulangerite, $Pb_5Sb_4S_{11}$, G Orth. Metallic bluish-grey; grey-white in reflected light. H= 2.5-3, D_m= 6.23, D_c= 6.181, IID 05b. Thaulow, 1837. Dana, 7th ed.(1944), v.1, 420. Min. Abstr. 80-0161.

Bournonite, $CuPbSbS_3$, G Orth. Metallic steel-grey; white in reflected light. Wheal Boys, Endellion, Cornwall, England. H= 2.5-3, D_m= 5.8, D_c= 5.84, IID 01e. Jameson, 1805. Min. Abstr. 81-2424. Zts. Krist. 131 (1970), 397. See also Seligmannite, Součekite.

Boussingaultite, $(NH_4)_2Mg(SO_4)_2\cdot 6H_2O$, G Mon. Transparent colorless, yellowish-pink. Travale, Montieri, Grosseto, Toscana, Italy. H= 2, D_m= 1.722, D_c= 1.718, VIC 11. Bechi, 1864. Dana, 7th ed.(1951), v.2, 455. Acta Cryst. 17 (1964), 1478. See also Nickel-boussingaultite.

Bowieite, $(Rh,Ir,Pt)_2S_3$, A Orth. Metallic grey/grey-brown. Salmon River, Goodnews Bay, Alaska, USA. H= 6-7, D_m=?, D_c= 6.9, IIC 15. Desborough & Criddle, 1984. Can. Min. 22 (1984), 543. Acta Cryst. 23 (1967), 832.

Boyleite, $(Zn,Mg)SO_4\cdot 4H_2O$, A Mon. Earthy white. Kropbach quarry, Münstertal, Schwarzwald (S), Germany. H= 2, D_m=?, D_c= 2.34, VIC 02. Walenta, 1978. Am. Min. 64(1979), 241 (Abst.).

Brabantite, $CaTh(PO_4)_2$, A Mon. Translucent grey-brown, reddish-brown. Brabant pegmatite, Karibib Dist., Namibia. H= 5.5, D_m= 4.9, D_c= 5.27, VIIA 11. Rose, 1980. Am. Min. 66(1981), 878 (Abst.).

Bracewellite, $CrO(OH)$, A Orth. Deep red, black. Merume River, Mazaruni dist., Guyana. H=?, D_m= 4.47, D_c= 4.286, IVF 04. Milton et al., 1967. Am. Min. 62(1977), 593 (Abst.). See also Grimaldiite, Guyanaite, Diaspore, Goethite.

Brackebuschite, $Pb_2(Mn,Fe)(VO_4)_2\cdot H_2O$, G Mon. Sub-metallic dark brown, black. Venus mine, Sierra de Córdoba, Córdoba, Argentina. H=?, D_m= 6.05, D_c= 6.11, VIIC 17. Döring, 1880. Min. Mag. 39 (1973), 69. Am. Min. 40 (1955), 597.

Bradleyite, $Na_3Mg(PO_4)(CO_3)$, G Mon. Translucent colorless, white, grey. John Hay Jr. Well #1, Sweetwater Co., Wyoming, USA. H=?, D_m= 2.72, D_c= 2.72, VbB 06. Fahey, 1941. JCPDS 22-478. See also Bonshtedtite, Sidorenkite.

Braggite, (Pt, Pd, Ni)S, G Tet. Metallic grey. Potgietersrust dist., Transvaal, South Africa. H= 5, D_m=?, D_c= 9.38, IIB 16. Bannister, 1932. Am. Min. 63 (1978), 832. Acta Cryst. B29 (1973), 1446. See also Vysotskite, Cooperite.

Braitschite-(Ce), $(Ca, Na_2)_7(Ce, La)_2B_{22}O_{43} \cdot 7H_2O$, A Hex. Vitreous colorless, white. Cave Creek mines, Grand Co., Utah, USA. H=?, D_m= 2.903, D_c= 2.837, Vc 16. Raup et al., 1968. Am. Min. 53 (1968), 1081.

Brammallite, $(Na, H_3O)(Al, Mg, Fe)_2(Si, Al)_4O_{10}(OH)_2$, G Mon. White. Llandebie, Dyfed, Wales. H=?, D_m=?, D_c= 2.83, VIIIE 07a. Bannister, 1943. Dokl.Earth Sci. 208(1973), 157.

Brandtite, $Ca_2(Mn, Mg)(AsO_4)_2 \cdot 2H_2O$, G Mon. Vitreous colorless, white. Harstig, Sweden and/or Rappold mine and Daniel mine, Schneeberg, Sachsen (Saxony), Germany. H= 3.5, D_m= 3.67, D_c= 3.70, VIIC 12. Nordenskiöld, 1888. Can. Min. 15 (1977), 36. Struct. Repts. 16 (1952), 289. See also Roselite.

Brannerite, $(U, Ca, Y, Ce)(Ti, Fe)_2O_6$, G Mon. Black. Kelly Gulch, Custer Co., Idaho, USA. H= 4.5, D_m= 5.0, D_c= 6.37, IVD 12. Hess & Wells, 1920. Can. Min. 6 (1960), 483. Can. Min. 20 (1982), 271. See also Thorutite, Orthobrannerite.

Brannockite, $KLi_3Sn_2Si_{12}O_{30}$, A Hex. Transparent colorless. Foote Mineral Company spodumene mine, Kings Mt., Cleveland Co., North Carolina, USA. H=?, D_m= 2.98, D_c= 2.994, VIIIC 10. White et al., 1973. Am. Min. 58(1973), 1111 (Abst.). Am. Min. 73 (1988), 595.

Brass, β-(Cu, Zn), Q Orth. Metallic yellow. Unspecified locality, Siberia, USSR. H=?, D_m=?, D_c=?, IA 02. Okrugin et al., 1981. ZVMO 110 (1981), 186.

Brassite, $Mg(AsO_3OH) \cdot 4H2O$, A Orth. Translucent white. Jáchymov (St. Joachimsthal), Západočeský kraj, Čechy (Bohemia), Czechoslovakia. H=?, D_m= 2.28, D_c= 2.326, VIIC 11. Fontan et al., 1973. Am. Min. 60(1975), 945 (Abst.). Acta Cryst. B32 (1976), 1460.

Braunite, Mn_7SiO_{12}, G Tet. Sub-metallic black, grey. H= 6-6.5, D_m= 4.8, D_c= 4.860, IVC 03. Haidinger, 1831. Am. Min. 61 (1976), 1226. Am. Min. 61 (1976), 1226. See also Neltnerite, Braunite II.

Braunite II, $Ca(Mn, Fe)_{14}SiO_{24}$, P Tet. Submetallic black; yellow-brown in reflected light. Black Rock mine, Kalahari, Cape Province, South Africa. H= 6, D_m= 4.75, D_c= 4.85, IVC 03. de Villiers & Herbstein, 1967. Am. Min. 71(1986), 1543 (Abst.). Am. Min. 65 (1980), 756. See also Braunite.

Bravoite, D A varietal name for nickeloan pyrite. Am. Min. 74 (1989), 1168. Hillebrand, 1907.

Brazilianite, $NaAl_3(PO_4)_2(OH)_4$, G Mon. Vitreous yellow. Divino River, Rio Doce and Rio São Mathews, Minas Gerais, Brazil. H= 5.5, D_m= 2.98, D_c= 2.998, VIIB 12. Pough & Henderson, 1945. Am. Min. 33 (1948), 135. Acta Cryst. B30 (1974), 1311.

Bredigite, $(Ca, Ba)Ca_{13}Mg_2(SiO_4)_8$, G Orth. Scawt Hill, Antrim Co., Northern Ireland. H=?, D_m= 3.42, D_c= 3.412, VIIIA 05. Tilley & Vincent, 1948. Min. Mag. 28 (1948), 255. Am. Min. 61 (1976), 74.

Breithauptite, NiSb, G Hex. Metallic copper-red/violet. St. Andreasberg, Harz, Germany. H= 5.5, D_m= 8.23, D_c= 8.629, IIB 09a. Frobel, 1840. Econ. Geol. 43 (1948), 408.

Brenkite, $Ca_2CO_3F_2$, A Orth. Translucent colorless. Schellkopf, Brenk, Eifel, Rheinland-Pfalz, Germany. H= 5, D_m= 3.10, D_c= 3.126, IIID 02. Hentschel et al., 1978. Am. Min. 64(1979), 241 (Abst.). Struct. Repts. 46A (1980), 305.

Brewsterite, $(Sr, Ba, Ca)(Al_2Si_6)O_{16} \cdot 5H_2O$, G Mon. Vitreous/pearly white, yellow, grey. Whitesmith mine, Strontian, Argyllshire, Scotland. H= 5, D_m= 2.45, D_c= 2.40, VIIIF 17. Brooke, 1822. Am. Min. 72 (1987), 645. Acta Cryst. C41 (1985), 492.

Brezinaite, Cr_3S_4, [A] Mon. Metallic brownish-grey. Tucson Meteorite. H=?, D_m=?, D_c= 4.12, IIB 09c. Bunch & Fuchs, 1969. Am. Min. 54 (1969), 1509.

Brianite, $Na_2CaMg(PO_4)_2$, [A] Mon. Transparent colorless. Dayton meteorite, Montgomery Co., Ohio, USA. H= 4-5, D_m=?, D_c= 3.127, VIIA 04. Fuchs et al., 1967. Am. Min. 60 (1975), 717.

Briartite, $Cu_2(Fe,Zn)GeS_4$, [A] Tet. Metallic grey. Prince Leopold mine, Kipushi, Shaba, Zaïre. H= 3.5-4, D_m=?, D_c= 4.35, IIB 03a. Francotte et al., 1965. Am. Min. 51(1966), 1816 (Abst.). Struct. Repts. 45A (1979), 59.

Brindleyite, $(Ni,Al)_3(Si,Al)_2O_5(OH)_4$, [A] Mon. Earthy green. Marmara mine, Megare, Greece. H= 2.5-3, D_m= 3.17, D_c= 3.16, VIIIE 10b. Maksimovic & Bish, 1978. Am. Min. 63 (1978), 484.

Britholite-(Ce), $(Ce,Ca)_5(SiO_4,PO_4)_3(OH,F)$, [G] Hex. Resinous brown. Naujakasik, Ilímaussaq, Greenland (S). H= 5.5, D_m= 3.86, D_c= 3.95, VIIIA 29. Winther, 1900. Am. Min. 49 (1964), 937. See also Apatite.

Britholite-(Y), $Ca_2Y_3(SiO_4)_3(OH)$, [A] Hex. Dark reddish-brown. Abukuma range, Fukushima, Japan. H= 6, D_m= 4.35, D_c= 4.08, VIIIA 29. Levinson, 1966. Am. Min. 53 (1968), 890. See also Apatite.

Brochantite, $Cu_4SO_4(OH)_6$, [G] Mon. Vitreous green. Bank mines, Sverdlovsk (Ekaterinburg), Ural Mts., USSR. H= 3.5-4, D_m= 3.97, D_c= 3.98, VIB 01. Lévy, 1824. Dana, 7th ed.(1951), v.2, 541. Struct. Repts. 23 (1959), 451.

Brockite, $(Ca,Th,Ce)PO_4 \cdot H_2O$, [G] Hex. Greasy/vitreous reddish-brown. Bassick mine, Wet Mountains, Custer Co., Colorado, USA. H=?, D_m= 3.9, D_c= 4.54, VIIC 19. Fisher & Meyrowitz, 1962. Am. Min. 47 (1962), 1346.

Brokenhillite, $(Mn,Fe)_8Si_6O_{15}(OH,Cl)_{10}$, [Q] Hex. Broken Hill, New South Wales, Australia. H=?, D_m=?, D_c= 3.02, VIIIE 12. Czank, 1988. Am. Min. 74(1989), 1399 (Abst.). Acta Cryst.A43 Sup.(1987) C155.

Bromargyrite, AgBr, [G] Cub. Resinous/adamantine colorless, grey, yellow, greenish-brown. Mexico; Chile. H= 2.5, D_m= 6.474, D_c= 6.50, IIIA 02. Legmerie, 1859. Dana, 7th ed.(1951), v.2, 11.

Bromellite, BeO, [G] Hex. Transparent white. Långban mine, Filipstad (near), Värmland, Sweden. H= 9, D_m= 3.017, D_c= 3.087, IVA 03. Aminoff, 1925. Dana, 7th ed.(1944), v.1, 506. Struct. Repts. 53A(1986), 123. See also Zincite.

Brongniardite, $Ag_2PbSb_5S_5$, [Q] Mon. Metallic white in reflected light. Potosí, Bolivia. H=?, D_m=?, D_c= 5.263, IID 07. Damour, 1849. ZVMO 118 (1989), 47.

Bronze-n', Cu_6Sn_5, [Q] Hex. Metallic white. Panasqueira, Beira Baixa, Portugal. H= 2, D_m=?, D_c=?, IA 02. Clark, 1972. Min. Abstr. 73-811..

Brookite, TiO_2, [G] Orth. Metallic adamantine brown, black. Snowden, Wales. H= 5.5-6, D_m= 4.14, D_c= 4.133, IVD 02. Lévy, 1825. Dana, 7th ed.(1944), v.1, 589. Can. Min. 17 (1979), 77. See also Anatase, Rutile.

Brostenite, [D] A mixture of birnessite and todorokite. Min. Abstr. 74-3408. Poni, 1900.

Brownmillerite, $Ca_2(Al,Fe)_2O_5$, [A] Orth. Translucent reddish-brown. Ettringer Bellerberg, Mayen, Laacher See Area, Eifel, Rheinland-Pfalz, Germany. H=?, D_m= 3.76, D_c= 3.68, IVA 08. Hentschel, 1964. Am. Min. 50(1965), 2106 (Abst.). Acta Cryst. B27 (1971), 2311.

Brucite, $Mg(OH)_2$, [G] Rhom. Waxy/vitreous white, pale grey, grey, blue, etc. Hoboken, Hudson Co., New Jersey, USA. H= 2.5, D_m= 2.39, D_c= 2.38, IVF 03a. Beudant, 1824. Am. Min. 50 (1965), 1893. Struct. Repts. 32A (1967), 242.

Brüggenite, $Ca(IO_3)_2 \cdot H_2O$, [A] Mon. Vitreous colorless, bright yellow. Pampa del Pique, Lautaro, Antofagasta, Chile. H= 3.5, D_m= 4.24, D_c= 4.267, IVG 01. Mrose et al., 1971. Am. Min. 57(1972), 1911 (Abst.).

Brugnatellite, $Mg_6FeCO_3(OH)_{13} \cdot 4H_2O$, G Hex. Pearly flesh-pink, yellowish, brownish-white. Val Malenco, Valtellina, Sondrio, Lombardia, Italy. H= 2, D_m= 2.14, D_c= 2.21, IVF 03c. Artini, 1909. Dana, 7th ed.(1944), v.1, 660.

Brunogeierite, $Fe_2(Ge,Fe)O_4$, A Cub. Tsumeb, Namibia. H=?, D_m=?, D_c=?, IVB 01. Ottemann & Nuber, 1972. Am. Min. 58(1973), 348 (Abst.).

Brushite, $Ca(PO_3OH) \cdot 2H_2O$, G Mon. Vitreous/pearly colorless, pale yellow. Aves Island, Caribbean Sea. H= 2.5, D_m= 2.328, D_c= 2.318, VIIC 15. Moore, 1864. Dana, 7th ed.(1951), v.2, 704. Struct. Repts. 37A (1971), 293. See also Ardealite, Pharmacolite, Gypsum.

Buchwaldite, $NaCaPO_4$, A Orth. Translucent white. Agpalilik fragment, Cape York Meteorite, Greenland. H=?, D_m=?, D_c= 3.21, VIIA 02. Olsen et al., 1977. Am. Min. 62 (1977), 362.

Buddingtonite, $(NH_4)(Si_3Al)O_8 \cdot 0.5H_2O$, A Mon. Vitreous, colorless. Sulphur Bank Quicksilver mine, Lake Co., California, USA. H= 5.5, D_m= 2.32, D_c= 2.38, VIIIF 03. Erd et al., 1964. Am. Min. 49 (1964), 831.

Buergerite, $NaFe_3Al_6(BO_3)_3Si_6O_{18}(O,F)_4$, A Rhom. Dark brown. Mexquitic, San Luis Potosi, Mexico. H=?, D_m= 3.31, D_c= 3.29, VIIIC 08. Donnay et al., 1966. Am. Min. 51 (1966), 198. Am. Min. 56 (1971), 101.

Bukovite, $Cu_3FeTl_2Se_4$, A Tet. Metallic greyish-brown. Bukov deposit, Moravia, Czechoslovakia. H= 2, D_m=?, D_c= 7.40, IIA 10. Johan & Kvacek, 1971. Bull. Min. 94 (1971), 529. See also Thalcusite, Murunskite.

Bukovskýite, $Fe_2(AsO_4)(SO_4)(OH) \cdot 7H_2O$, A Tric. Earthy pale yellowish-green, greyish-green. Kaňk, Kutná Hora (Near), Středočeský kraj, Čechy (Bohemia), Czechoslovakia. H=?, D_m= 2.334, D_c= 2.336, VIID 04. Novak et al., 1967. Min. Abstr. 87M/2138.

Bulachite, $Al_2AsO_4(OH)_3 \cdot 3H_2O$, A Orth. Silky white. Neubulach, Schwarzwald (N), Baden-Württemberg, Germany. H= 2, D_m= 2.60, D_c= 2.55, VIID 10. Walenta, 1985. Am. Min. 70(1985), 214 (Abst.).

Bultfonteinite, $Ca_2SiO_3(OH)F \cdot H_2O$, G Tric. Vitreous colorless, pale pink. Bultfontein mine, Kimberley, South Africa. H= 4.5, D_m= 2.73, D_c= 2.74, VIIIA 10. Parry et al., 1932. Min. Mag. 23 (1932), 145. Acta Cryst. 16 (1963), 551.

Bunsenite, NiO, G Cub. Vitreous green. Johanngeorgenstadt, Sachsen (Saxony), Germany. H= 5.5, D_m= 6.90, D_c= 6.79, IVA 04. Dana, 1868. Dana, 7th ed.(1944), v.1, 501.

Burangaite, $(Na,Ca)_2Fe_2Al_{10}(PO_4)_8(O,OH)_{12} \cdot 4H_2O$, A Mon. Translucent bluish, bluish-green. Buranga pegmatite, Rwanda. H= 5, D_m= 3.05, D_c= 3.31, VIID 28. von Knorring et al., 1977. Am. Min. 63(1978),793 (Abst.). See also Dufrenite, Natrodufrenite.

Burbankite, $(Na,Ca)_3(Sr,Ba,Ce)_3(CO_3)_5$, G Hex. Translucent pale yellow. Vermiculite Prospects, Big Sandy Creek, Hill Co., Montana, USA. H= 3.5, D_m= 3.50, D_c= 3.51, VbA 05. Pecora & Kerr, 1953. Am. Min. 62 (1977), 158. N.Jb.Min.Mh. (1985), H4, 161. See also Carbocernaite, Khanneshite.

Burckhardtite, $Pb_2(Fe,Mn)Te(Si_3Al)O_{12}(OH)_2 \cdot H_2O$, A Mon. Adamantine/pearly violet-red, pale pink. Moctezuma mine, Moctezuma, Sonora, Mexico. H= 2, D_m=?, D_c= 4.96, VIIIE 25. Gaines et al., 1979. Am. Min. 64 (1979), 355.

Burkeite, $Na_4SO_4(CO_3,SO_4)$, G Orth. Vitreous/greasy white, pale buff, greyish. Searles Lake, San Bernardino Co., California, USA. H= 3.5, D_m= 2.57, D_c= 2.61, VIB 05. Teeple, 1921. Min. Mag. 43 (1979), 341. N.Jb.Min.Mh. (1988), 203.

Bursaite, $Pb_5Bi_4S_{11}$, G Orth. Metallic grey. Ulu-Dağ, Bursa, Turkey. H= 2.5-3.5, D_m=?, D_c= 6.56, IID 08. Tolun, 1954-55. N.Jb.Min. Abh. 158 (1988), 293.

Burtite, $CaSn(OH)_6$, [A] Rhom. Transparent colorless. El Hamman, Morocco. H= 3, D_m= 3.28, D_c= 3.22, IVF 06. Sonnet, 1981. Can. Min. 19 (1981), 397.

Buryktalskite, [D] Mixture of Mn oxides. Min. Mag. 33 (1962), 261.

Bustamite, $CaMnSi_2O_6$, [G] Tric. Translucent pale pink, brownish-red. Mexico. H= 5.5-6.5, D_m= 3.3, D_c= 3.326, VIIID 08. Brongniart, 1826. Am. Min. 63 (1978), 274. Zts. Krist. 117 (1962), 701. See also Ferrobustamite.

Butlerite, $FeSO_4(OH)\cdot 2H_2O$, [G] Mon. Vitreous orange. United Verde mine, Jerome, Yavapai Co., Arizona, USA. H= 2.5, D_m= 2.55, D_c= 2.53, VID 02. Lausen, 1928. Dana, 7th ed.(1951), v.2, 608. Am. Min. 56 (1971), 751. See also Parabutlerite.

Bütschliite, $K_2Ca(CO_3)_2$, [G] Rhom. H=?, D_m=?, D_c= 2.607, VbA 05. Milton & Axelrod, 1947. Am. Min. 59 (1974), 353. Acta Cryst. C40 (1984), 1299. See also Fairchildite, Eitelite.

Buttgenbachite, $Cu_{18}(NO_3)_2(OH)_{32}Cl_3\cdot H_2O$, [G] Hex. Vitreous blue. Likasi, Shaba, Zaïre. H= 3, D_m= 3.42, D_c= 3.46, IIIC 02. Schoep, 1925. Min. Mag. 29 (1950), 280. Min. Mag. 39 (1973), 264. See also Connellite.

Byelorussite-(Ce), $NaBa_2Ce_2MnTi_2Si_8O_{26}(F,OH)\cdot H_2O$, [A] Orth. Vitreous yellow, pale yellow-brown. Gomel region, Byelorussia, USSR. H= 5.5-6, D_m= 3.92, D_c= 4.09, VIIIC 05. Shpanov et al., 1989. ZVMO 118 (1989), 100.

Byströmite, $MgSb_2O_6$, [G] Tet. Translucent blue-grey. La Fortuna mine, El Antimonio, Sonora, Mexico. H= 7, D_m= 5.7, D_c= 6.08, IVD 04. Mason & Vitaliano, 1952. Am. Min. 37 (1952), 53. See also Ordoñezite.

Bytownite, $(Ca,Na)(Si,Al)_4O_8$, [G] Tric. Translucent colorless, white, greenish. Ottawa, Canada. H= 6-6.5, D_m= 2.72, D_c= 2.73, VIIIF 03. Thomson, 1835. DHZ (1963), v.4, 94. Acta Cryst. 21 (1966), 782. See also Plagioclase.

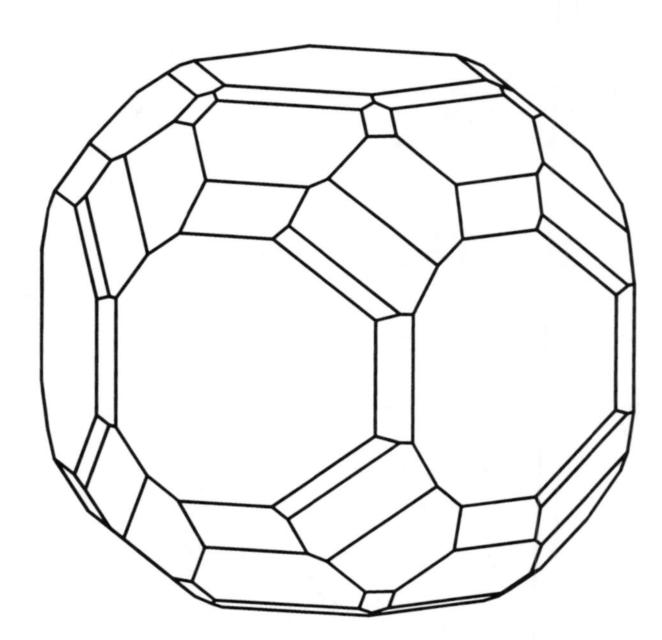

Fluorapatite from St. Gothard showing the forms $\{10\bar{1}1\}$, $\{0001\}$, $\{21\bar{3}0\}$, $\{21\bar{3}1\}$, $\{31\bar{4}1\}$, $\{11\bar{2}1\}$, $\{11\bar{2}2\}$, $\{20\bar{2}1\}$ and $\{10\bar{1}2\}$. (After Haidinger)

C

Cabriite, Pd_2CuSn, [A] Orth. Metallic white. Oktyabr deposit, Talnakh, Noril'sk (near), Siberia (N), USSR. H= 4, D_m=?, D_c= 10.7, IA 10. Evstigneeva & Genkin, 1983. Can. Min. 21 (1983), 481.

Cacoclasite, [D] A mixture of grossular, calcite and other minerals. Can. Min. 8 (1966), 527. Lewis, 1884.

Cacoxenite, $Fe_{24}AlO_6(PO_4)_{17}(OH)_{12} \cdot 75H_2O$, [G] Hex. Silky yellow, brownish/reddish-yellow, greenish. Hrbek mine, Sv. Dobrotivá, Čechy (Bohemia), Czechoslovakia. H= 3-4, D_m= 2.3, D_c= 2.217, VIID 05. Steinmann, 1825. Dana, 7th ed.(1951), v.2, 997. Nature 306 (1983), 356.

Cadmium, Cd, [A] Hex. Metallic white. Ust'Khannin intrusive, Viluĭ River basin, USSR. H=?, D_m=?, D_c= 8.642, IA 03. Oleĭnikov et al., 1979. Am. Min.65(1980), 1065 (Abst.).

Cadmoselite, β-CdSe, [G] Hex. Resinous/adamantine black. Unspecified locality, USSR. H=?, D_m=?, D_c= 5.66, IIB 06. Bur'yanova et al., 1957. Am. Min. 43(1958), 623 (Abst.). Acta Cryst. A33 (1977), 355. See also Greenockite, Wurtzite.

Cadwaladerite, $AlCl(OH)_2 \cdot 4H_2O$, [G] Amor. Vitreous yellow. Cerro Pintados, Tarapacá, Chile. H=?, D_m= 1.66, D_c=?, IIIC 12. Gordon, 1941. Dana, 7th ed.(1951), v.2, 77.

Cafarsite, $(Ca,Mn)_8(Ti,Fe)_{6.5}(AsO_3)_{12} \cdot 2H_2O$, [A] Cub. Translucent dark brown. Monte Leone, Cherbadung, Binntal, Valais (Wallis), Switzerland. H= 5.5-6, D_m= 3.90, D_c= 3.49, VIID 31. Graeser, 1966. Am. Min. 52(1967), 1584 (Abst.). Min. Abstr. 78-1499.

Cafetite, $(Ca,Mg)(Fe,Al)_2Ti_4O_{12} \cdot 4H_2O$, [G] Orth. Adamantine pale yellow, colorless. Afrikanda, Kola Peninsula, USSR. H= 4-5, D_m= 3.28, D_c= 3.19, IVF 20. Kukharenko et al., 1959. Am. Min. 71 (1986), 1045.

Cahnite, $Ca_2[B(OH)_4](AsO_4)$, [G] Tet. Vitreous colorless, white. Franklin Furnace, Sussex Co., New Jersey, USA. H= 3, D_m= 3.156, D_c= 3.175, Vc 03c. Palache, 1921. Min. Mag. 32 (1960), 666. Am. Min. 46 (1961), 1077.

Calaverite, $AuTe_2$, [G] Mon. Metallic yellow/white; creamy-white in reflected light. Stanislaus mine, Calaveras Co., California, USA. H= 2.5-3, D_m= 9.24, D_c= 9.32, IIC 04. Genth, 1861. Dana, 7th ed.(1944), v.1, 335. Acta Cryst. B44 (1988), 486.

Calciborite, CaB_2O_4, [G] Orth. Translucent white. Ural Mts., USSR. H= 3.5, D_m= 2.88, D_c= 2.880, Vc 01g. Petrova, 1956. Am. Min. 41(1956), 815 (Abst.). Min. Abstr. 81-2438.

Calcioancylite-(Ce), $(Ca,Sr)Ce(CO_3)_2(OH) \cdot H_2O$, [Q] Orth. Vitreous/greasy yellow, yellowish-brown, brown, grey. Kola Peninsula, USSR. H= 4-4.5, D_m=?, D_c=?, VbD 03. Fersman, 1922. Min. Mag. 20(1925), 448 (Abst.). See also Ancylite-(Ce), Gysinite-(Nd).

Calciobetafite, $(Ca,Na)_2(Nb,Ti)_2(O,OH)_7$, [A] Cub. Translucent reddish-brown. Campi Flegrei, Monte di Procida, Campania, Italy. H=?, D_m=?, D_c= 4.81, IVC 09a. Mazzi & Munno, 1983. Am. Min. 68 (1983),262. See also Zirkelite, Polymignite.

Calciocopiapite, $CaFe_4(SO_4)_6(OH)_2 \cdot 20H_2O$, [G] Tric. Dashkesan, Middle Caucasus, USSR. H=?, D_m=?, D_c= 2.159, VID 04b. Kashkaĭ & Aliev, 1960. Can. Min. 23 (1985), 53.

Calcioferrite, $Ca_4Mg(Fe,Al)_4(PO_4)_6(OH)_4 \cdot 13H_2O$, [G] Syst=? Pearly yellow, greenish-yellow, green, yellowish-white. Battenberg, Bayern (Bavaria), Germany. H= 2.5, D_m= 2.53, D_c=?, VIID 35. Blum, 1858. Min. Rec. 16 (1985), 477.

Calciohilairite, $CaZrSi_3O_9 \cdot 3H_2O$, [A] Rhom. Vitreous white. Liberty Bell and Washington Pass, Okanogan Co., Washington, USA. H= 4, D_m= 2.68, D_c= 2.74, VIIIC 12. Boggs, 1988. Am. Min. 73 (1988), 1191.

Calciosamarskite, [D] A uranian yttropyrochlore. Am. Min. 62 (1977), 403. Ellsworth, 1928.

Calciotantalite, [D] A mixture of pyrochlore-microlite, columbite-tantalite and possibly wodginite. Min. Mag. 38 (1972), 765. Simpson, 1907.

Calciotantite, $CaTa_4O_{11}$, [A] Hex. Adamantine colorless. Unspecified locality, Kola Peninsula, USSR. H= 6.5-7, D_m=?, D_c= 7.46, IVD 17. Voloshin et al., 1982. Am. Min. 68(1983), 471 (Abst.).

Calciouranoite, $(Ca, Ba, Pb, K, Na,)U_2O_7{\cdot}5H_2O$, [A] Syst=? Translucent brown, orange. U-Mo ore deposit, USSR. H=?, D_m= 4.62, D_c=?, IVF 12. Rogova et al., 1974. Am. Min. 60(1975), 161 (Abst.).

Calciovolborthite, $CaCuVO_4(OH)$, [G] Orth. Vitreous/pearly greenish-yellow, dark green, olive-green. Friedrichroda, Thüringen, Germany. H= 3-4, D_m= 3.75, D_c= 3.82, VIIB 11b. Credner, 1848. Bull. Min. 79 (1956), 219. N.Jb.Min.Mh. (1989), 300. See also Conichalcite.

Calcite, $CaCO_3$, [G] Rhom. Vitreous colorless, white, yellow, pink, etc. H= 3, D_m= 2.710, D_c= 2.711, VbA 02. Freiesleben, 1836. Rev. Min. 11 (1983). Zts. Krist. 156 (1981), 233. See also Aragonite, Vaterite, Rhodochrosite.

Calcium catapleiite, $CaZrSi_3O_9{\cdot}H_2O$, [G] Hex. Vitreous/dull pale yellow, cream. Burpala, Baikal (N), USSR. H= 4.5-5, D_m= 2.77, D_c=?, VIIIC 03. Portnov et al., 1964. Am. Min. 49(1964), 1153 (Abst.). See also Catapleiite.

Calciumhilgardite-2M(Cc), [D] Unnecessary name for hilgardite-4M. Min. Mag. 33 (1962), 261. Braitsch, 1959.

Calciumhilgardite-3Tc, [D] Unnecessary name for hilgardite-3Tc. Min. Mag. 33 (1962), 261. Braitsch, 1959.

Calcium-pharmacosiderite, $CaFe_4(AsO_4)_3(OH)_5{\cdot}5H_2O$, [Q] Cub. Menzenschwand, Krunkelbachtal, Schwarzwald (S), Germany. H=?, D_m=?, D_c=?, VIID 14b. Walenta, 1966. Min. Mag. 39(1974), 909 (Abst.).

Calcjarlite, $Na(Ca, Sr)_3Al_3(F, OH)_{16}$, [G] Mon. Translucent white. Yeniseĭ region, USSR. H= 4, D_m= 3.51, D_c= 3.32, IIIC 08. Povarennykh, 1973. ZVMO 99 (1970), 458. See also Jarlite.

Calclacite, $Ca(CH_3COO)Cl{\cdot}5H_2O$, [G] Mon. Translucent white. Museums, Ca-rich objects in oak boxes. H=?, D_m= 1.5, D_c= 1.55, IXA 03. Van Tassel, 1945. Dana, 7th ed.(1951), v.2, 1107. Acta Cryst. 11 (1958), 745.

Calcurmolite, $Ca(UO_2)_3(MoO_4)_3(OH)_2{\cdot}11H_2O$, [G] Syst=? Translucent yellow. Unspecified locality, USSR. H=?, D_m=?, D_c=?, VIF 02. Povarennykh, 1962. Am. Min. 49(1964), 1152 (Abst.).

Calcybeborosilite-(Y), $(Y, Ca)_2(B, Be)_2Si_2O_8(OH)_2$, [G] Mon. Greasy greenish-grey, colorless. Alai Mts., Tadzhikistan, USSR. H= 2.5, D_m= 3.78, D_c=?, VIIIA 23. Semenov et al., 1963. Am. Min. 49(1964), 443 (Abst.).

Calderite, $(Mn, Ca)_3(Fe, Al)_2(SiO_4)_3$, [G] Cub. Translucent dark yellowish/reddish. Katkamsandi, Hazaribagh (NW), India (W). H=?, D_m= 4.05, D_c= 4.17, VIIIA 06a. Piddington, 1851. Can. Min. 17 (1979), 569.

Caledonite, $Cu_2Pb_5(SO_4)_3(CO_3)(OH)_6$, [G] Orth. Resinous green, bluish-green. Leadhills, Lanarkshire, Scotland. H= 2.5-3, D_m= 5.76, D_c= 5.689, VIB 11. Beudant, 1832. Min. Mag. 40 (1976), 536. Acta Cryst. B29 (1973), 1986.

Calkinsite-(Ce), $(Ce, La)_2(CO_3)_3{\cdot}4H_2O$, [G] Orth. Translucent pale yellow. Rocky Boy's Reservation, Big Sandy Creek, Bearpaw Mtns., Hill Co.(SE), Montana, USA. H= 2.5, D_m= 3.28, D_c= 3.27, VbC 03b. Pecora & Kerr, 1953. Am. Min. 38 (1953), 1169.

Callaghanite, $Cu_2Mg_2CO_3(OH)_6{\cdot}2H_2O$, [G] Mon. Transparent blue. Gabbs, Nye Co., Nevada, USA. H= 3-3.5, D_m= 2.71, D_c= 2.65, VbD 01. Beck & Bruns, 1954. Am. Min. 39 (1954), 630. Acta Cryst. 11 (1958), 169.

Calomel, α–HgCl, [G] Tet. Adamantine colorless, white, grey, yellow, brown. Landsberg (Moschellandsberg), Obermoschel (near), Rheinland-Pfalz, Germany. H= 1.5, D_m= 7.15, D_c= 7.162, IIIA 04. Dana, 7th ed.(1951), v.2, 25. Zts. Krist. 187 (1989), 305.

Calumetite, $Cu(OH,Cl)_2 \cdot 2H_2O$, [G] Syst=? Pearly azure blue, powder blue. Centennial mine, Calumet, Houghton Co., Michigan, USA. H= 2, D_m=?, D_c=?, IVF 19. Williams, 1963. Am. Min. 48 (1963), 614.

Calzirtite, $Ca_2Zr_5Ti_2O_{16}$, [G] Tet. Semi-metallic/adamantine dark brown. Meimecha-Kotui, Siberia (E), USSR. H= 6-7, D_m= 5.01, D_c= 5.24, IVD 16a. Zdorik et al., 1961. Am. Min. 46(1961), 1515 (Abst.). Am. Min. 71 (1986), 815.

Cameronite, $AgCu_7Te_{10}$, [A] Tet. Metallic grey. Good Hope mine, Vulcan, Gunnison Co., Colorado, USA. H= 3.5-4, D_m=?, D_c= 7.14, IIB 14. Roberts et al., 1986. Can. Min. 24 (1986), 379.

Caminite, $MgSO_4 \cdot xMg(OH)_2 \cdot yH_2O$, [A] Tet. Translucent colorless, white. East Pacific Rise, 21 deg. N. H= 2.5, D_m= 2.5, D_c= 2.145, VID 01. Haymon & Kastner, 1983. Am. Min. 71 (1986), 819. Am. Min. 71 (1986), 819.

Campigliaite, $Cu_4Mn(SO_4)_2(OH)_6 \cdot 4H_2O$, [A] Mon. Vitreous light blue. Temperino mine, Campiglia Marittima, Toscana, Italy. H=?, D_m= 3.00, D_c= 3.06, VID 14. Menchetti & Sabelli, 1982. Am. Min. 67 (1982), 385. Am. Min. 67 (1982), 388.

Canaphite, $Na_2CaP_2O_7 \cdot 4H_2O$, [A] Mon. Vitreous colorless. Haledon, Passaic Co., New Jersey, USA. H= 2, D_m= 2.24, D_c= 2.27, VIIC 11. Peacor et al., 1985. Am. Min. 73 (1988), 168. Am. Min. 73 (1988), 168.

Canasite, $K_3Na_3Ca_5Si_{12}O_{30}(O,OH,F)_4$, [G] Mon. Vitreous greenish-yellow. Yukspor Mts., Khibina tundra, Kola Peninsula, USSR. H=?, D_m= 2.71, D_c= 2.63, VIIID 18. Dorfman et al., 1959. Acta Cryst. A43 Supp. C-159. Min. Zhurn. 10 (1988), 31.

Canavesite, $Mg_2(HBO_3)(CO_3) \cdot 5H_2O$, [A] Mon. Vitreous/silky. Brosso Canavese, Piemonte, Italy. H=?, D_m=?, D_c= 1.790, Vc 15. Ferraris et al., 1978. Can. Min. 16 (1978), 69.

Cancrinite, $(Na,Ca)_8(Si_6Al_6)O_{24}(CO_3)_2$, [G] Hex. Sub-vitreous/pearly white, grey, yellow, green, blue, etc. H= 5-6, D_m= 2.4, D_c= 2.37, VIIIF 05. Rose, 1839. DHZ (1963), v.4, 310. Can. Min. 20 (1982), 239.

Canfieldite, Ag_8SnS_6, [G] Orth. Metallic grey/black; greyish-white in reflected light. Avilargus, La Paz (near), Bolivia. H= 2.5, D_m= 6.2, D_c= 6.276, IIA 11. Penfield, 1893. N.Jb.Min.Mh. (1978), 269. See also Argyrodite.

Cannizzarite, $Pb_{46}Bi_{54}S_{127}$, [G] Mon. Metallic creamy white in reflected light. Vulcano Island, Eolie (Lipari) Islands, Sicilia (Sicily), Italy. H= 3, D_m= 6.7, D_c= 6.95, IID 09e. Zambonini et al.,1925. N.Jb.Min.Abh. 158 (1988), 293. Acta Cryst. B35 (1979), 133.

Cappelenite-(Y), $Ba(Y,Ce)_6B_6Si_3O_{24}F_2$, [G] Rhom. Vitreous/greasy greenish-brown. Lille Arö, Langesundfjord, Telemark, Norway. H= 6-6.5, D_m= 4.407, D_c= 4.43, VIIIE 18. Brögger, 1885. Dana, 6th ed. (1892), 413. Am. Min. 69 (1984), 190.

Caracolite, $Na_3Pb_2(SO_4)_3Cl$, [G] Hex. Vitreous colorless/greyish. Beatriz mine, Caracoles, Chile. H= 4.5, D_m= 5.1, D_c= 4.50, VIB 08. Websky, 1886. Per. Mineral. 54 (1985), 34. Struct. Repts. 34A (1969), 311.

Carboborite, $Ca_2Mg[B(OH)_4]_2(CO_3)_2 \cdot 4H_2O$, [G] Mon. Vitreous colorless. Unspecified locality, China. H=?, D_m= 2.12, D_c= 2.15, Vc 03e. Hsieh et al., 1964. Am. Min. 50(1965), 262 (Abst.). Bull. Min. 104 (1981), 578.

Carbocernaite, $(Sr,Ce,Ba)(Ca,Na)(CO_3)_2$, [G] Orth. Vitreous/greasy colorless, white, yellowish, rose, brown. Vuorijärvi, Kola Peninsula, USSR. H= 3, D_m= 3.53,

D_c= 3.60, VbA 05. Bulakh et al., 1961. Am. Min. 46(1961), 1202 (Abst.). Am. Min. 68(1983), 1251 (Abst.). See also Burbankite.

Carboirite, $FeAl_2GeO_5(OH)_2$, [A] Tric. Vitreous green. Carboire, Ariège, Pyrenees, France. H= 6, D_m=?, D_c= 3.95, VIIIH 01. Johan et al., 1983. Am. Min. 69(1984), 406 (Abst.). See also Chloritoid.

Carbonate-cyanotrichite, $Cu_4Al_2CO_3(OH)_{12} \cdot 2H_2O$, [G] Syst=? Lustrous silky pale blue/azure. Kara Tau, Dzhambul (near), Kazakhstan, USSR. H= 2, D_m= 2.66, D_c=?, IVF 03e. Ankinovich et al., 1963. Am. Min. 49(1964), 441 (Abst.). See also Cyanotrichite.

Carbonate-fluorapatite, $Ca_5(PO_4,CO_3)_3F$, [G] Hex. Translucent colorless. H= 5, D_m= 3.12, D_c= 3.10, VIIB 16. McConnell, 1973. Min. Mag. 41 (1977), M4. See also Fluorapatite.

Carbonate-hydroxylapatite, $Ca_{10}(PO_4)_3(CO_3)_3(OH)_2$, [G] Hex. Ödegaard, Bamble, Norway. H= 5, D_m= 3.053, D_c= 2.87, VIIB 16. Fleischer, ?. Science, 155 (1967), 1409. See also Hydroxylapatite.

Carbonate-vishnevite, $Na_8(Si,Al)_{12}O_{24}(CO_3) \cdot H_2O$, [Q] Hex. Translucent pale violet, light blue. Lovozero massif, Kola Peninsula, USSR. H=?, D_m= 2.4, D_c= 2.32, VIIIF 05. Semyonov et al., 1984. Am. Min. 73(1988), 927 (Abst.). See also Hydroxyl-vishnevite.

Carletonite, $KNa_4Ca_4Si_8O_{18}(CO_3)_4(F,OH) \cdot H_2O$, [A] Tet. Mont Saint-Hilaire, Rouville Co., Québec, Canada. H=?, D_m= 2.45, D_c= 2.426, VIIIE 02. Chao, 1971. Am. Min. 56 (1971), 1855. Am. Min. 57 (1972), 765.

Carlfriesite, $CaTe_3O_8$, [A] Mon. Translucent bright primrose yellow. Bambollita mine, Moctezuma, Sonora, Mexico. H= 3.5, D_m= 6.3, D_c= 5.699, VIG 06. Williams & Gaines, 1975. Min. Mag. 40 (1975), 127. Am. Min. 63 (1978), 847.

Carlhintzeite, $Ca_2AlF_7 \cdot H_2O$, [A] Tric. Vitreous colorless. Hagendorf pegmatite, Waidhaus, Oberpfalz, Bayern (Bavaria), Germany. H=?, D_m= 2.86, D_c= 2.89, IIIB 04. Dunn et al., 1979. Can. Min. 17 (1979), 103.

Carlinite, Tl_2S, [A] Rhom. Metallic grey. Carlin mine, Elko (NW), Eureka Co., Nevada, USA. H= 1.5-2, D_m= 8.1, D_c= 8.55, IIA 17. Radtke & Dickson, 1975. Am. Min. 60 (1975), 559.

Carlosturanite, $(Mg,Fe,Ti)_{21}(Si,Al)_{12}O_{28}(OH)_{34} \cdot H_2O$, [A] Mon. Vitreous/pearly light brown. Varaita Valley, Sempeyre (near), Piemonte, Italy. H=?, D_m= 2.63, D_c= 2.606, VIIIE 10b. Compagnoni et al., 1985. Am. Min. 70 (1985), 767.

Carlsbergite, CrN, [A] Cub. Metallic light grey. Agpalilik fragment, Cape York meteorite, Greenland. H= > 7, D_m= 5.9, D_c= 6.1, IC 01. Buchwald & Scott, 1971. Nat.Phys.Sci. 233 (1971), 113.

Carminite, $PbFe_2(AsO_4)_2(OH)_2$, [G] Orth. Vitreous/pearly red, reddish-brown. Luise mine, Horhausen, Rheinland-Pfalz, Germany. H= 3.5, D_m= 5.22, D_c= 5.46, VIIB 13. Sandberger, 1850. Dana, 7th ed.(1951), v.2, 912. Am. Min. 48 (1963), 1.

Carnallite, $KMgCl_3 \cdot 6H_2O$, [G] Orth. Greasy colorless, white, reddish. Stassfurt deposit, Sachsen (Saxony), Germany. H= 2.5, D_m= 1.602, D_c= 1.587, IIIB 08. Rose, 1856. Min. Mag. 29 (1951), 667. Am. Min. 70 (1985), 1309.

Carnevallite, [D] Inadequate data. Min. Mag. 43 (1980), 1055. Geier & Ottemann, 1970.

Carnotite, $K_2(UO_2)_2(VO_4)_2 \cdot 3H_2O$, [G] Mon. Dull/earthy/pearly/silky yellow. Montrose Co., Colorado, USA. H=?, D_m= 4.95, D_c= 4.99, VIIID 23. Friedel & Cumenge, 1899. Am. Min. 43 (1958), 799. Am. Min. 50 (1965), 825. See also Margaritasite, Tyuyamunite.

Carobbiite, KF, [G] Cub. Transparent colorless. Mt. Vesuvius, Napoli, Campania, Italy. H=?, D_m= 2.505, D_c= 2.528, IIIA 02. Strunz, 1956. Am. Min. 42(1957), 117 (Abst.).

Carpholite, $MnAl_2Si_2O_6(OH)_4$, [G] Orth. Silky yellow. Slavkov, Čechy (Bohemia), Czechoslovakia. H= 5-5.5, D_m= 2.935, D_c= 3.071, VIIID 03. Werner, 1817. Am. Min. 74 (1989), 1084. Struct. Repts. 45A (1979), 369. See also Ferrocarpholite, Magnesiocarpholite.

Carrboydite, $(Ni, Al)_8(SO_4)_{1.6}(OH)_{16} \cdot 8.5H_2O$, [A] Hex. Translucent yellowish-green, blue-green. Carr Boyd mine, Kalgoorlie, Western Australia, Australia. H=?, D_m= 2.50, D_c= 2.692, IVF 03e. Nickel & Clarke, 1976. Min. Mag. 44 (1981),333.

Carrollite, $Cu(Co, Ni)_2S_4$, [G] Cub. Metallic grey; white in reflected light. Patapsco mine, Finksburg, Carroll Co., Maryland, USA. H= 4.5-5.5, D_m= 4.83, D_c= 4.85, IIC 01. Faber, 1852. Min. Mag. 43 (1980), 733.

Caryinite, $Na(Ca, Pb)(Ca, Mn)(Mn, Mg)_2(AsO_4)_3$, [G] Mon. Greasy brown, yellowish-brown. Långban mine, Filipstad (near), Värmland, Sweden. H= 4, D_m= 4.29, D_c= 4.406, VIIA 05b. Lundström, 1874. Min. Mag. 51 (1987), 281.

Caryopilite, $(Mn, Mg)_3Si_2O_5(OH)_4$, [G] Mon. Brown. Harstigen mine, Pajsberg, Värmland, Sweden. H= 3-3.5, D_m= 2.9, D_c= 2.81, VIIIE 10b. Hamberg, 1889. Can. Min. 20 (1982), 1. See also Friedelite.

Cascandite, $Ca(Sc, Fe)Si_3O_8(OH)$, [A] Tric. Vitreous pale pink. Cava Diverio, Baveno, Italy. H=?, D_m=?, D_c= 3.10, VIIID 08. Mellini et al., 1982. Am. Min. 67 (1982), 599. Am. Min. 67 (1982), 604.

Cassedanneite, $Pb_5(VO_4)_2(CrO_4)_2 \cdot H_2O$, [A] Mon. Resinous orange-red. Berezov, Sverdlovsk, Ural Mts., USSR. H= 3.5, D_m=?, D_c= 6.52, VIID 30. Cesbron et al., 1988. Am. Min. 73(1988), 1493 (Abst.).

Cassidyite, $Ca_2(Ni, Mg)(PO_4)_2 \cdot 2H_2O$, [A] Tric. Translucent pale/bright green. Wolf Creek Meteorite Crater, Western Australia, Australia. H=?, D_m=?, D_c= 3.1, VIIC 12. White et al, 1958. Am. Min. 52 (1967), 1190.

Cassiterite, SnO_2, [G] Tet. Adamantine/metallic brown, yellow, grey, white, etc. H= 6-7, D_m= 6.99, D_c= 6.99, IVD 02. Beudant, 1832. NBS Circ. 539, v.1 (1953), 54.

Castaingite, [D] Inadequate data; may be molybdenite. Min. Mag. 36 (1967), 133. Schüller & Ottemann, 1963.

Caswellsilverite, $NaCrS_2$, [A] Rhom. Metallic yellowish-grey/light grey. Norton Co. enstatite (meteorite). H= 2, D_m= 3.21, D_c= 3.24, IIB 17. Okada & Keil, 1982. Am. Min. 67 (1982), 132.

Catapleiite, $Na_2ZrSi_3O_9 \cdot 2H_2O$, [G] Mon. Translucent colorless, grey, blue. Låven, Langesundfjord, Norway. H= 5-6, D_m= 2.8, D_c= 2.768, VIIIC 03. Weibye & Sjögren, 1850. DHZ, 2nd ed.(1986), v.1B, 364. Min. Abstr. 83M/0153. See also Calcium catapleiite, Gaidonnayite.

Cattierite, CoS_2, [G] Cub. Metallic white. Shinkolobwe, Shaba, Zaïre. H= 4, D_m= 4.80, D_c= 4.819, IIC 05. Kerr, 1945. Am. Min. 30 (1945), 488. Zts. Krist. 150 (1979), 165. See also Pyrite, Vaesite.

Cavansite, $Ca(VO)Si_4O_{10} \cdot 4H_2O$, [A] Orth. Vitreous greenish-blue. Charles W. Chapman Quarry, Columbia Co., Oregon, USA; Lake Owyhee State Park, Malheur Co., Oregon, USA. H= 3-4, D_m= 2.3, D_c= 2.33, VIIIE 26. Staples et al., 1973. Am. Min. 58 (1973), 405. Am. Min. 58 (1973), 412. See also Pentagonite.

Caysichite-(Y), $(Ca, Yb, Er)_4Y_4Si_8O_{20}(CO_3)_6(OH) \cdot 7H_2O$, [A] Orth. Translucent colorless, white, pale yellow, greenish. Evans-Lou mine, St-Pierre-de-Wakefield (near), Portland Twp., Papineau Co., Québec, Canada. H= 4.5, D_m= 3.03, D_c= 3.029, VIIID 18. Hogarth et al., 1974. Can. Min. 12 (1974), 293. Can. Min. 16 (1978), 81.

Cebaite-(Ce), $Ba_3Ce_2(CO_3)_5F_2$, [G] Mon. Vitreous/waxy orange-yellow, yellow, dark yellow. Bayan Obo, Inner Mongolia, China. H= 4.5-5, D_m= 4.5, D_c= 4.81, VbB 04. Zhang & Tao, 1983. Am. Min. 70(1985), 214 (Abst.). Acta Cryst. B40 (1984), 454.

Cebaite-(Nd), $Ba_3(Nd,Ce)_2(CO_3)_5F_2$, Q Syst=? Vitreous beige. Bayan Obo, Inner Mongolia, China (N). H=?, D_m=?, D_c=?, VbB 04. Zhang & Tao, 1986. Am. Min. 73(1988), 1493 (Abst.).

Cebollite, $Ca_5Al_2(SiO_4)_3(OH)_4$, G Syst=? White, greenish-grey. Iron Hill Area, Powder Horn (near), Beaver Creek, Gunnison Co., Colorado, USA. H= 5, D_m= 2.96, D_c=?, VIIIA 26. Larsen & Schaller, 1914. Min. Mag. 43 (1980), 583.

Čechite, $Pb(Fe,Mn)VO_4(OH)$, A Orth. Sub-metallic/resinous black. Vranciče, Příbram (near), Stredočský kraj, Čechy (Bohemia), Czechoslovakia. H= 4.5-5, D_m= 5.88, D_c= 5.99, VIIB 11b. Mrázek & Táborský, 1981. Am. Min. 67(1982), 1074 (Abst.). N. Jb. Min. Mh. (1989), 34.

Celadonite, $K(Mg,Fe)(Fe,Al)Si_4O_{10}(OH)_2$, G Mon. Translucent green. Baldo Mt., Verona (near), Italy. H=?, D_m=?, D_c= 3.03, VIIIE 05a. Glocker, 1847. Min. Mag. 42 (1978), 373. Min. Zhurn. 8 (3) (1986), 32.

Celestine, $SrSO_4$, G Orth. Vitreous pale blue, white, reddish, greenish, brownish. H= 3-3.5, D_m= 3.97, D_c= 3.982, VIA 07. Werner, 1798. Am. Min. 63 (1978), 506. Can. Min. 13 (1975), 181.

Celsian, $BaAl_2Si_2O_8$, G Mon. Translucent colorless, white, yellow. Jakobsberg mine, Värmland, Sweden. H= 6-6.5, D_m= 3.37, D_c= 3.389, VIIIF 03. Sjögren, 1895. DHZ (1963), v.4, 166. Am. Min. 61 (1976), 414. See also Paracelsian, Hyalophane, Orthoclase.

Cerianite-(Ce), CeO_2, G Cub. Translucent greenish-amber. Firetown area, Nemegos (NE), Lackner Township, Sudbury, Ontario, Canada. H=?, D_m=?, D_c= 7.22, IVD 16b. Graham, 1955. NBS Monogr. 20 (1983), 38. See also Thorianite, Uraninite.

Ceriopyrochlore-(Ce), $(Ce,Ca,Y)_2(Nb,Ta)_2O_6(OH,F)$, A Cub. Resinous brown. Wausau, Marathon Co., Wisconsin, USA. H= 5.5.5, D_m= 4.13, D_c=?, IVC 09a. Hogarth, 1977. Am. Min. 62 (1977), 403.

Cerite-(Ce), $(Ce,La,Ca)_9(Fe,Mg)(SiO_4)_6(SiO_3OH)(OH)_3$, G Rhom. Resinous brown. Bastnäs, Riddarhyttan, Västmanland, Sweden. H= 5, D_m= 4.75, D_c= 5.65, VIIIA 21. Goddard & Glass, 1940. Am. Min. 43 (1958), 460. Am. Min. 68 (1983), 996.

Černýite, Cu_2CdSnS_4, A Tet. Metallic light grey. Tanco mine, Bernic Lake, Manitoba, Canada and Hugo mine, Keystone, Pennington Co., South Dakota, USA. H= 4, D_m=?, D_c= 4.779, IIB 03a. Kissin et al., 1978. Can. Min. 16 (1978), 139. Can. Min. 16 (1978), 147.

Cerolite, D A mixture of serpentine and stevensite. Am. Min. 50 (1965), 2111. Breithaupt, 1823.

Cerotungstite-(Ce), $CeW_2O_6(OH)_3$, A Mon. Translucent orange-yellow. Kirwa mine, Kigezi Dist., Uganda. H= 1, D_m=?, D_c= 6.244, IVF 15. Sahama et al., 1970. Min. Rec. 12 (1981), 81. See also Yttrotungstite-(Y).

Cerphosphorhuttonite, D Unnecessary name for cerian phosphatian huttonite. Min. Mag. 36 (1968), 1144. Pavlenko et al., 1965.

Ceruléite, $Cu_2Al_7(AsO_4)_4(OH)_{13} \cdot 12H_2O$, G Tric. Clayey turquoise-blue. Huanaco, Chile. H= 5-6, D_m= 2.70, D_c= 2.734, VIID 08. Dufet, 1900. Am. Min. 62(1977), 598 (Abst.).

Cerussite, $PbCO_3$, G Orth. Adamantine/vitreous colorless, white, grey. H= 3-3.5, D_m= 6.55, D_c= 6.577, VbA 04. Haidinger, 1845. Min. Mag. 36 (1968), 632. Zts. Krist. 139 (1974), 215.

Cervandonite-(Ce), $(Ce,Nd,La)(Fe,Ti,Al)_3(Si,As)_3O_{13}$, A Mon. Adamantine black. Pizzo Cervandone, Val Dévero, Piemonte, Italy and Cherbadung, Binntal, Valais (Wallis), Switzerland. H= 5, D_m=?, D_c= 4.9, VIIIG 01. Armbruster et al., 1988. Min. Abstr. 89M/4847.

Cervantite, Sb_2O_4, G Orth. Greasy/earthy yellow, reddish-white. Cervantes, Galicia, Spain. H= 4-5, D_m= 6.5, D_c= 6.64, IVD 14a. Dana, 1850. Am. Min. 47(1962), 1221 (Abst.). Acta Cryst. B33 (1977), 1271.

Cervelleite, Ag_4TeS, A Cub. Metallic white/grey in reflected light. Bambolla mine, Moctezuma, Sonora, Mexico. H= 1.7, D_m=?, D_c= 8.53, IIA 03. Criddle et al., 1989. Europ. Jour. Min. 1 (1989), 371.

Cesanite, $Na_3Ca_2(SO_4)_3(OH)$, A Hex. Translucent colorless/white. Cesano #1 Well, Cesano Geothermal Field, Lazio (Latium), Italy. H= 2-3, D_m= 2.786, D_c= 2.831, VIB 08. Cavarretta et al., 1981. Min. Mag. 47 (1983), 59. Min. Mag. 47 (1983), 59. See also Apatite.

Cesarolite, $PbMn_3O_6(OH)_2$, G Syst=? Dull/sub-metallic grey. Sidi Amor ben Salaam, Tunisia. H= 4.5, D_m= 5.29, D_c=?, IVF 05b. Buttgenbach & Gillet, 1920. Dana, 7th ed.(1944), v.1, 744.

Cesbronite, $Cu_5(TeO_3)_2(OH)_6 \cdot 2H_2O$, A Orth. Transparent green. La Oriental mine, Moctezuma, Sonora, Mexico. H= 3, D_m= 4.45, D_c= 4.455, VIG 04. Williams, 1974. Min. Mag. 39 (1974), 744.

Cesium Kupletskite, $Cs_3(Mn,Fe)_7(Ti,Nb)_2Si_8O_{24}(O,OH,F)_7$, A Tric. Dull gold-brown. Alai alkalic province, Alai Mts., Tadzhikistan, USSR. H= 4, D_m= 3.68, D_c= 3.62, VIIID 25. Efimov et al., 1971. Am. Min. 57(1972), 328 (Abst.). See also Kupletskite.

Cesplumtantite, $(Cs,Na)_2(Pb,Sb)_3Ta_8O_{24}$, A Tet. Adamantine colorless. Manono, Zaïre. H= 6.7, D_m=?, D_c= 6.87, IVC 09a. Voloshin et al., 1986. Am. Min. 74(1989), 501 (Abst.).

Cesstibtantite, $(Cs,Na)SbTa_4O_{12}$, A Cub. Adamantine colorless/grey. Kola Peninsula, USSR. H= 3-7, D_m= 6.5, D_c= 6.35, IVC 09a. Voloshin et al., 1982. Am. Min. 67(1982), 413 (Abst.). See also Natrobistantite.

Cetineite, $K_{3.5}(Sb_2O_3)_3(SbS_3)(OH)_{0.5} \cdot 2H_2O$, A Hex. Translucent orange-red. Cetine mine, Rosia, Siena, Toscana, Italy. H= 3.3, D_m=?, D_c= 4.223, IIE 02. Sabelli & Vezzalini, 1987. N.Jb.Min.Mh. (1987), 419. Am. Min. 73 (1988), 398.

Chabazite, $Ca(Al_2Si_4)O_{12} \cdot 6H_2O$, G Tric. Vitreous white, pink, brick-red. H= 4-5, D_m= 2.1, D_c= 2.03, VIIIF 15. Bosc d'Antic, 1788. Natural Zeolites (1985), 175. Acta Cryst. B38 (1982), 602. See also Herschelite.

Chabournéite, $Tl_{21-x}Pb_{2x}(Sb,As)_{91-x}S_{147}$, A Tric. Submetallic/greasy black; white in reflected light. Jas Roux, Valgaudemar, Hautes-Alpes, France and Abuta, Hokkaido, Japan. H= 3, D_m= 5.10, D_c= 5.121, IID 14. Johann et al., 1981. Am. Min. 67(1982), 621 (Abst.).

Chaidamuite, $ZnFe(SO_4)_2(OH) \cdot 4H_2O$, A Mon. Vitreous brown, yellow-brown. Xitieshan, Chaidamu, Qinhai, China. H= 2.5-3, D_m= 2.722, D_c= 2.72, VID 04a. Li et al., 1986. Am. Min. 73(1988), 1493 (Abst.).

Chalcanthite, $CuSO_4 \cdot 5H_2O$, G Tric. Vitreous blue. H= 2.5, D_m= 2.286, D_c= 2.284, VIC 03a. Kobell, 1853. Dana, 7th ed.(1951), v.2, 488. Sov.Phys.Cryst. 28(1983), 383.

Chalcoalumite, $CuAl_4SO_4(OH)_{12} \cdot 3H_2O$, G Mon. Dull/vitreous turquoise-green, pale blue, bluish-grey. Bisbee, Cochise Co., Arizona, USA. H= 2.5, D_m= 2.29, D_c= 2.25, IVF 03e. Larsen & Vassar, 1925. Min. Rec. 2 (1971), 126. See also Mbobomkulite, Nickelalumite.

Chalcocite, Cu_2S, G Hex. Metallic blackish-grey; grey in reflected light. H= 2.5-3, D_m= 5.7, D_c= 5.789, IIA 01a. Dana, 1868. Dana, 7th ed.(1944), v.1, 187. Zts. Krist. 150 (1979), 299. See also Chalcocite, tetragonal.

Chalcocite (tetragonal), $Cu_{1.96}S$, Q Tet. Mina Maria, Quebrada Puquios, Atacama Province, Chile. H=?, D_m=?, D_c= 5.770, IIA 01a. Clark & Sillitoe, 1971. Min. Abstr. 72-1369. Acta Cryst. 17 (1964), 311. See also Chalcocite.

Chalcocyanite, $CuSO_4$, G Orth. Translucent colorless, green, brown, yellow, blue. Mt. Vesuvius, Napoli, Campania, Italy. H= 3.5, D_m= 3.65, D_c= 3.89, VIA 01. Frondel, 1951. Dana, 7th ed.(1951), v.2, 429. Min. Petrol. 39 (1988), 201.

Chalcolamprite, D Impure pyrochlore. Am. Min. 62 (1977), 403. Flink, 1898.

Chalcomenite, $CuSeO_3 \cdot 2H_2O$, G Orth. Vitreous blue. Cerro de Cacheuta, Mendoza, Argentina. H= 2-2.5, D_m= 3.35, D_c= 3.346, VIG 01. Des Cloizeaux & Damour, 1881. Am. Min. 49 (1964), 1481 . N.Jb.Min.Mh. (1989), 551. See also Clinochalcomenite, Teineite.

Chalconatronite, $Na_2Cu(CO_3)_2 \cdot 3H_2O$, G Mon. Greenish-blue. On artifacts, Egypt. H=?, D_m= 2.29, D_c= 2.31, VbC 04. Frondel & Gettens, 1955. Am. Min. 40(1955), 943 (Abst.). Zts. Krist. 148 (1978), 165.

Chalcophanite, $(Zn, Fe, Mn)Mn_3O_7 \cdot 3H_2O$, G Rhom. Metallic black. Passaic mine, Sterling Hill, Ogdensburg, Sussex Co., New Jersey, USA. H= 2.5, D_m= 4.00, D_c= 3.86, IVF 05a. Moore, 1875. Dana, 7th ed.(1944), v.1, 739. Am. Min. 73 (1988), 1401.

Chalcophyllite, $Cu_9Al(AsO_4)_2(SO_4)_{1.5}(OH)_{12} \cdot 18H_2O$, G Rhom. Vitreous/sub-adamantine green, bluish-green. H= 2, D_m= 2.69, D_c= 2.684, VIID 09. Breithaupt, 1841. Min. Abstr. 80-4170. Zts. Krist. 151 (1980), 129.

Chalcopyrite, $CuFeS_2$, G Tet. Metallic yellow. H= 3.5-4, D_m= 4.2, D_c= 4.148, IIB 02. Henckel, 1725. Can. Min. 18 (1980), 157. Can. Min. 13 (1975), 168. See also Eskebornite.

Chalcosiderite, $CuFe_6(PO_4)_4(OH)_8 \cdot 4H_2O$, G Tric. Vitreous light green. Wheal Phoenix mine, Cornwall, England and/or Sayn (Siegen), Westphalia, Germany. H= 4.5, D_m= 3.22, D_c= 3.603, VIID 08. Ullmann, 1814. Am. Min. 50 (1965), 227. N.Jb.Min.Mh. (1989), 227. See also Turquoise.

Chalcostibite, $CuSbS_2$, G Orth. Metallic grey. Wolfsberg, Harz, Germany. H= 3-4, D_m= 4.95, D_c= 5.01, IID 06b. Glocker, 1847. Can. Min. 17 (1979), 601. See also Emplectite.

Chalcothallite, $(Cu, Fe, Ag)_{6.3}(Tl, K)_2SbS_4$, G Tet. Metallic grey/black; light grey in reflected light. Nakalaq, Ilímaussaq Intrusion, Greenland (S). H= 2-2.5, D_m= 6.6, D_c= 6.69, IID 06a. Kovalenker et al., 1978. Am. Min. 53(1968), 1775 (Abst.). N.Jb.Min.Abh. 138 (1980), 122.

Chambersite, $Mn_3B_7O_{13}Cl$, G Orth. Translucent colorless/deep purple. Chambers Co., Mont Belvieu, Barkers Hill Salt Dome (NW), Texas, USA. H= 7, D_m= 3.49, D_c= 3.479, Vc 14. Honea & Beck, 1962. Am. Min. 47 (1962), 665. See also Boracite, Ericaite.

Chaméanite, $(Cu, Fe)_4As(Se, S)_4$, A Cub. Metallic grey. Chaméane, Vernet-la-Varenne, Livardois Mtns., Puy-de-Dôme, France. H= 4, D_m=?, D_c= 6.17, IIB 03b. Johan et al., 1982. Am. Min. 67(1982), 1074 (Abst.).

Chamosite, $(Fe, Mg, Al)_6(Si, Al)_4O_{10}(OH, O)_8$, G Mon. Greenish-grey, black. Chamoson, St. Maurice, Valais (Wallis), Switzerland. H= 3, D_m= 3.19, D_c= 3.28, VIIIE 09a. Berthier, 1820. DHZ (1962), v.3, 164. Struct. Repts. 12 (1949), 280. See also Orthochamosite, Clinochlore.

Changbaiite, $PbNb_2O_6$, G Rhom. Adamantine/pearly colorless, cream-white, pale brown, etc. Changbai Mt., Kirin (E), China. H= 5.3, D_m= 6.48, D_c= 6.51, IVD 11. Am. Min. 64(1979), 242 (Abst.).

Chantalite, $CaAl_2SiO_4(OH)_4$, A Tet. Vitreous colorless, white. Covur Yokusutepe Hill, Doğanbaba, Burdur, Taurus Mts., Turkey (SW). H=?, D_m= 2.9, D_c= 2.97, VIIIA 26. Sarp et al., 1977. Am. Min. 63(1978), 1282 (Abst.). Zts. Krist. 150 (1979), 53.

Chaoite, C, G Hex. Lamellae in graphite. Mottingen, Ries Crater, Germany. H=?, D_m=?, D_c= 3.43, IB 02a. El Goresy, 1970. Am. Min. 55(1970), 1067 (Abst.). See also Diamond, Lonsdaleite, Graphite-2H, Graphite-3R.

Chapmanite, $Fe_2Sb(SiO_4)_2(OH)$, [G] Mon. Translucent olive-green. Keeley mine, South Lorrain Twp., Timiskaming Dist., Ontario, Canada. H=?, D_m= 3.58, D_c= 4.29, VIIIE 10c. Walker, 1924. Am. Min. 43 (1958), 656. Sov.Phys.Cryst. 22 (1977), 419. See also Bismutoferrite.

Charlesite, $Ca_6Al_2(SO_4)_2B(OH)_4(OH,O)_{12} \cdot 26H_2O$, [A] Rhom. Vitreous colorless. Franklin mine, Franklin, Sussex Co., New Jersey, USA. H= 2.5, D_m= 1.77, D_c= 1.79, VID 07. Dunn et al., 1983. Am. Min. 68 (1983), 1033.

Charoite, $(K,Na)_5(Ca,Ba,Sr)_8Si_{18}O_{46}(OH,F) \cdot nH_2O$, [A] Mon. Translucent violet. Charo River, Murun Massif, Aldan (NE), Yakutiya, USSR. H= 5, D_m= 2.54, D_c= 2.77, VIIID 18. Rogova et al., 1978. Am. Min. 73(1988), 198 (Abst.).

Chatkalite, $Cu_6FeSn_2S_8$, [A] Tet. Metallic pale rose. Chatkal-Kuramin Mts., Uzbekistan (E), USSR. H= 4, D_m=?, D_c= 5.00, IIB 03b. Kovalenker et al., 1981. Am. Min. 67(1982), 621 (Abst.). See also Mawsonite.

Chavesite, Ca,Mn,PO_4, [Q] Tric. Translucent colorless. Boqueirão pegmatite, Borborema, Paraíba, Brazil. H= 3, D_m=?, D_c=?, VIIC 12. Murdoch, 1958. Am. Min. 43 (1958), 1148.

Chayesite, $K(Mg,Fe)_4FeSi_{12}O_{30}$, [A] Hex. Vitreous blue. Moon Canyon, Utah, USA. H=?, D_m=?, D_c= 2.68, VIIIC 10. Velde et al., 1989. Am. Min. 74 (1989), 1368.

Chekhovicite, $Bi_2Te_4O_{11}$, [A] Mon. Adamantine grey. Zod, Armenia, USSR. H= 4, D_m= 6.88, D_c= 7.005, VIG 05. Spiridonov et al., 1987. Am. Min. 74(1989), 1400 (Abst.).

Chelkarite, $CaMgB_2O_4Cl_2 \cdot 7H_2O$(?), [G] Orth. Transparent colorless. Chelkar, Kazakhstan, USSR. H=?, D_m=?, D_c= 1.96, Vc 02. Avrova et al., 1968. Am. Min. 56(1971), 1122 (Abst.).

Chenevixite, $Cu_2Fe_2(AsO_4)_2(OH)_4 \cdot H_2O$, [G] Mon. Earthy greenish-yellow, olive-green, dark green. Wheal Garland mine, Cornwall, England. H= 3.5-4.5, D_m= 4.59, D_c= 4.594, VIID 33. Adam, 1866. Min. Mag. 41 (1977), 27. See also Luetheite.

Chenite, $CuPb_4(SO_4)_2(OH)_6$, [A] Tric. Vitreous blue. Strathclyde, Leadhills, Lanarkshire, Scotland. H= 2.5, D_m= 5.98, D_c= 6.044, VIB 06. Paar et al., 1986. Min. Mag. 50 (1986), 129. N.Jb.Min.Mh. (1988), 259.

Cheralite, $(Th,Ca,Ce)(P,Si)O_4$, [G] Mon. Translucent green. Travancore, India (SW). H= 5, D_m= 5.28, D_c= 5.39, VIIA 11. Bowie & Horne, 1953. Min. Mag. 43 (1980), 885.

Cherepanovite, RhAs, [A] Orth. Metallic orange. Koryk-Kamchatsk fold zone, USSR. H= 6, D_m=?, D_c= 9.72, IIB 09d. Rudashevskiĭ et al., 1985. Am. Min. 71(1986), 1544 (Abst.).

Chernikovite, $(H_3O)(UO_2)PO_4 \cdot 3H_2O$, [A] Tet. Vitreous pale yellow. Unspecified locality, USSR. H=?, D_m= 3.399, D_c= 3.26, VIID 20b. Atencio, 1988. Min. Rec. 19 (1988), 249.

Chernovite-(Y), $YAsO_4$, [A] Tet. Vitreous colorless, pale yellow. Nyarta-syu-yu River, Telpos-iz (E), Ural Mts., USSR. H= 4.5, D_m=?, D_c= 4.866, VIIA 10. Goldin et al., 1967. Am. Min. 53(1968), 1777 (Abst.). See also Xenotime-(Y), Wakefieldite-(Y).

Chernykhite, $(Ba,Na)(V,Al)_2(Si,Al)_4O_{10}(OH)_2$, [A] Mon. Pearly olive/dark green. Kara Tau, Kazakhstan, USSR. H= 2-4.5, D_m= 3.15, D_c= 3.07, VIIIE 05a. Ankinovich et al., 1973. Am. Min. 58(1973), 966 (Abst.). Sov.Phys.Cryst. 19 (1974), 70.

Chervetite, $Pb_2V_2O_7$, [G] Mon. Adamantine colorless, grey, brown. Mounana mine, Franceville, Haut-Ogooué, Gabon. H= < 3, D_m= 6.31, D_c= 6.38, VIIA 12. Bariand et al., 1963. Am. Min. 48(1963), 1416 (Abst.). Bull. Min. 90 (1967), 279.

Chessexite, $Na_4Ca_2Mg_3Al_8(SiO_4)_2(SO_4)_{10}(OH)_{10} \cdot 40H_2O$, [A] Orth. Silky white. Maine mine, Autun, Saône-et-Loire, France. H=?, D_m=?, D_c= 2.21, VID 10. Sarp & Deferne, 1982. Am. Min. 69(1984), 406 (Abst.).

Chesterite, $(Mg,Fe)_{17}Si_{20}O_{54}(OH)_6$, [A] Orth. Transparent colorless, light pinkish-brown. Carleton mine, Chester, Windham Co., Vermont, USA. H=?, D_m=?, D_c= 3.09, VIIID 12. Veblen & Burnham, 1978. Am. Min. 63 (1978), 1000. Am. Min. 63 (1978), 1053.

Chestermanite, $Mg_2(Fe,Mg,Al,Sb)O_2BO_3$, [A] Orth. Vitreous/silky greyish-green, black. Kaiser Peak quadrangle, Twin Lakes region, Fresno Co., California, USA. H= 6, D_m= 3.72, D_c= 3.650, Vc 01c. Erd & Foord, 1988. Can. Min. 26 (1988), 911.

Chevkinite-(Ce), $(Ce,La)_4(Ti,Fe)_5O_8(Si_2O_7)_2$, [G] Mon. Vitreous black. Ilmen Mts., Ural Mts., USSR. H= 5-5.5, D_m= 4.53, D_c= 4.99, VIIIB 16. Rose, 1839. Am. Min. 63 (1978), 499. Am. Min. 59 (1974), 1277. See also Perrierite, Strontio-chevkinite.

Chiavennite, $CaBe_2MnSi_5O_{13}(OH)_2 \cdot 2H_2O$, [A] Orth. Vitreous pale orange-yellow. Chiavenna, Sondrio, Lombardia, Italy; also Langangen, Langesundfjord, Telemark, Norway. H= 3, D_m= 2.64, D_c= 2.657, VIIIA 02. Bondi et al., 1983. Am. Min. 68 (1983), 623.

Childrenite, $(Fe,Mn)AlPO_4(OH)_2 \cdot H_2O$, [G] Orth. Vitreous/resinous brown, yellowish-brown. Tavistock (near), Devon, England. H= 5, D_m= 3.25, D_c= 3.167, VIID 07. Brooke, 1823. Am. Min. 35 (1950), 793. Min. Abstr. 85M/0189. See also Eosphorite.

Chiolite, $Na_5Al_3F_{14}$, [G] Tet. Vitreous white, colorless. Miask, Ilmen Mts., Ural Mts., USSR. H= 3.5-4, D_m= 2.998, D_c= 2.989, IIIB 05. Hermann & Auerbach, 1846. Dana, 7th ed.(1951), v.2, 123.

Chkalovite, $Na_2BeSi_2O_6$, [G] Orth. Vitreous white. Lovozero massif, Kola Peninsula, USSR. H= 6, D_m= 2.662, D_c= 2.680, VIIIB 02b. Gerasimovsky, 1938. Am. Min. 25(1940), 380 (Abst.). Min. Abstr. 77-1471.

Chloraluminite, $AlCl_3 \cdot 6H_2O$, [G] Rhom. Translucent white, yellowish. Mt. Vesuvius, Napoli, Campania, Italy. H=?, D_m=?, D_c= 1.67, IIIC 10a. Scacchi, 1873. NBS Circ. 539, v.7 (1957), 3.

Chlorapatite, $Ca_5(PO_4)_3Cl$, [G] Hex. Translucent green, blue, pink, etc. H= 5, D_m= 3.181, D_c= 3.199, VIIB 16. Rammelsberg, 1860. Can. Min. 10 (1970), 252. Am. Min. 74 (1989), 870.

Chlorargyrite, AgCl, [G] Cub. Resinous/adamantine colorless, grey, yellowish. H= 2.5, D_m= 5.556, D_c= 5.55, IIIA 02. Weisbach, 1875. Dana, 7th ed.(1951), v.2, 11.

Chlorellestadite, $Ca_5(SiO_4,SO_4,PO_4)_3Cl$, [A] Hex. Crestmore, Riverside Co., California, USA. H=?, D_m=?, D_c= 3.21, VIIIA 29. Rouse & Dunn, 1982. Am. Min. 67 (1982), 90. See also Hydroxylellestadite, Fluorellestadite.

Chlorhastingsite, [D] Unncessary name for chlorian hastingsite. Min. Mag. 38 (1971), 103. Krutov & Vinogradova, 1966.

Chlorite, A group name for sheet silicates with the general formula $(Mg,Al,Fe,Li,Mn,Ni)_{4-6}(Si,Al,B,Fe)_4O_{10}(OH,O)_8$.

Chloritoid (monoclinic), $(Fe,Mg,Mn)Al_2SiO_5(OH)_2$, [P] Mon. Dark green, grey, greenish-black. Kosoĭbrok, Sverdlovsk (Ekaterinburg), Ural Mts., USSR. H= 6.5, D_m= 3.56, D_c= 3.57, VIIIA 22. Rose, 1837. DHZ, 2nd ed.(1982), v.1A, 867. Acta Cryst. B32 (1975), 780. See also Chloritoid (triclinic), Carboirite, Ottrelite, Magnesiochloritoid.

Chloritoid (triclinic), $(Fe,Mg,Mn)Al_2SiO_5(OH)_2$, [G] Tric. Translucent dark green, grey, greenish-black. Kosoĭbrok, Sverdlovsk (Ekaterinburg), Ural Mts., USSR. H= 6.5, D_m= 3.5, D_c= 3.56, VIIIA 22. Rose, 1837. DHZ, 2nd ed.(1982), v.1A, 867. Am. Min. 65 (1980), 534. See also Chloritoid (monoclinic).

Chlormagaluminite, $(Mg,Fe)_4Al_2(OH)_{12}Cl_2 \cdot 2H_2O$, [A] Hex. Translucent colorless, yellow-brown. Kapaev pipe, Angara River, Siberian platform (S), USSR. H=?, D_m= 2.0, D_c= 2.06, IVF 03c. Kashaev et al., 1982. Am. Min. 68(1983), 849 (Abst.).

Chlormanasseite, [D] Name changed to chlormagaluminite. Am. Min. 64(1979), 1329 (Abst.). Feokistov et al., 1978.

Chlormanganokalite, K_4MnCl_6, [G] Rhom. Vitreous yellow. Mt. Vesuvius, Napoli, Campania, Italy. H= 2.5, D_m= 2.31, D_c= 2.310, IIIB 06. Johnston-Lavis, 1906. Strunz (1970), 163.

Chlorocalcite, $KCaCl_3$, [G] Orth. Transparent white. Mt. Vesuvius, Napoli, Campania, Italy. H= 2.5-3, D_m=?, D_c= 2.155, IIIB 05. Scacchi, 1872. NBS Monogr. 7 (1969), 36.

Chloromagnesite, $MgCl_2$, [G] Rhom. Translucent colorless, white. Mt. Vesuvius, Napoli, Campania, Italy. H=?, D_m= 2.325, D_c= 2.44, IIIA 06. Scacchi, 1873. Dana, 7th ed.(1951), v.2, 41.

Chlorophoenicite, $(Mn,Mg,Zn)_3Zn_2AsO_4(OH,O)_6$, [G] Mon. Vitreous light greyish-green, pink, light purplish-red. Franklin, Sussex Co., New Jersey, USA. H= 3.5, D_m= 3.46, D_c= 3.47, VIIB 10a. Foshag & Gage, 1924. Can. Min. 19 (1981), 333. Am. Min. 53 (1968), 1110. See also Magnesium-chlorophoenicite, Jarosewichite.

Chlorothionite, $K_2CuSO_4Cl_2$, [G] Orth. Translucent blue. Mt. Vesuvius, Napoli, Campania, Italy. H= 2.5, D_m= 2.67, D_c= 2.678, VIB 02. Scacchi, 1872. Dana, 7th ed.(1951), v.2, 547. Zts. Krist. 144 (1976), 226.

Chloroxiphite, $CuPb_3O_2Cl_2(OH)_2$, [G] Mon. Resinous/adamantine olive green, yellowish-green. Morehead Quarry, Mendip Hills, Somerset, England. H= 2.5, D_m= 6.93, D_c= 6.84, IIIC 04. Spencer & Mountain, 1923. Dana, 7th ed.(1951), v.2, 84. Min. Mag. 41 (1977), 357.

Choloalite, $CuPb(TeO_3)_2 \cdot H_2O$, [A] Cub. Translucent green. La Oriental mine, Moctezuma, Sonora, Mexico. H= 3, D_m= 6.4, D_c= 6.498, VIG 11. Williams, 1981. Min. Mag. 44 (1981), 55.

Chondrodite, $(Mg,Fe)_5(SiO_4)_2(F,OH)_2$, [G] Mon. Translucent yellow, brown, red, greyish-green. Monte Somma, Mt. Vesuvius, Napoli, Campania, Italy. H= 6.5, D_m= 3.177, D_c= 3.18, VIIIA 16. d'Ohsson, 1817. DHZ, 2nd ed.(1982), v.1A, 379. Am. Min. 55 (1970), 1182.

Christite, $TlHgAsS_3$, [A] Mon. Adamantine red. Carlin mine, Elko (NW), Eureka Co., Nevada, USA. H= 1-2, D_m= 6.2, D_c= 6.37, IID 06a. Radtke et al., 1977. Am. Min. 62 (1977), 421. See also Routhierite.

Chromatite, $CaCrO_4$, [G] Tet. Translucent yellow. Jerusalem-Jericho Highway, Jordan. H=?, D_m=?, D_c= 3.14, VIE 01. Eckhardt & Heimbach, 1963. NBS Circ. 539, v.7 (1957), 13.

Chromdisthene, [D] Unnecessary name for chromian kyanite. Min. Mag. 38 (1971), 103. Sobolev & Sobolev, 1967.

Chromdravite, $NaMg_3(Cr,Fe)_6(BO_3)_3Si_6O_{18}(OH)_4$, [A] Rhom. Translucent dark green. Onega depression, Karelia, USSR. H=?, D_m= 3.40, D_c= 3.40, VIIIC 08. Rumantseva, 1983. Am. Min. 69(1984), 210 (Abst.).

Chromferide, $Fe_{1.5}Cr_{0.2}$, [A] Cub. Metallic light grey. Unspecified locality, Ural Mts. (S), USSR. H= 4, D_m=?, D_c= 6.69, IA 05. Novgorodova et al., 1986. ZVMO 115 (1986), 355.

Chromite, $FeCr_2O_4$, [G] Cub. Metallic black; greyish-white in reflected light. H= 5.5, D_m= 4.7, D_c= 5.09, IVB 01. Haidinger, 1845. Dana, 7th ed.(1944), v.1, 709. See also Magnesiochromite, Hercynite, Donathite.

Chromium, Cr, [A] Cub. Metallic white. Unspecified locality, Sichuan, China. H= 9, D_m= 7.21, D_c= 7.190, IA 05. Yue et al., 1981. Am. Min 67(1982), 854 (Abst.).

Chromsteigerite, [D] Unnecessary name for chromian steigerite. Min. Mag. 36 (1967), 133. Ankinovich, 1963.

Chrysoberyl, $BeAl_2O_4$, [G] Orth. Vitreous green, yellow, greenish-brown. H= 8.5, D_m= 3.75, D_c= 3.695, IVB 04. Werner, 1790. Dana, 7th ed.(1944), v.1, 718. Sov.Phys.Cryst. 30(1985), 277.

Chrysocolla, $(Cu,Al)_2H_2Si_2O_5(OH)_4 \cdot nH_2O$, [G] Orth. Vitreous/earthy green, bluish-green, blue. H= 2-4, D_m= 2, D_c=?, VIIID 04. Theophrastus, 315 B.C. Am. Min. 54(1969), 993 (Abst.).

Chrysotile, A subgroup name for sheet silicates of the serpentine group with the formula $Mg_3Si_2O_5(OH)_4$. See Clinochrysotile, Orthochrysotile.

Chudobaite, $(Mg,Zn)_5(AsO_4)_2(AsO_3OH)_2 \cdot 10H_2O$, [G] Tric. Translucent pink. Tsumeb, Namibia. H= 2.5-3, D_m= 2.94, D_c= 2.90, VIIC 28. Strunz, 1960. Am. Min. 45(1960), 1130 (Abst.). Am. Min. 62(1977), 599 (Abst.). See also Geigerite.

Chukhrovite-(Ce), $Ca_3(Ce,Y)Al_2(SO_4)F_{13} \cdot 10H_2O$, [Q] Cub. Translucent white, pale brown. Clara mine (Grube Clara), Wolfach, Schwarzwald, Baden-Württemberg, Germany. H=?, D_m=?, D_c=?, IIIC 11. Walenta, 1978. Am. Min. 65 (1980),1065 (Abst.).

Chukhrovite-(Y), $Ca_3(Y,Ce)Al_2(SO_4)F_{13} \cdot 10H_2O$, [G] Cub. Vitreous colorless. Kara-Oba deposit, Central Kazakhstan, USSR. H= 3, D_m= 2.35, D_c=?, IIIC 11. Ermilova et al., 1960. Am. Min. 45(1960), 1132 (Abst.). Am. Min. 66 (1981), 392.

Churchite-(Nd), $NdPO_4 \cdot 2H_2O$, [G] Mon. Dull/vitreous yellowish, pinkish-white. Kazakhstan (NW), USSR. H= 3.2, D_m= 3.13, D_c= 3.84, VIIC 05. Podpovina et al., 1983. Dokl. Earth Sc. 268(1983), 139. See also Churchite-(Y).

Churchite-(Y), $(Y,Er)PO_4 \cdot 2H_2O$, [G] Mon. Vitreous/pearly grey, reddish. Cornwall, England. H= 3, D_m= 3.265, D_c= 3.25, VIIC 05. Williams, 1865. Min. Mag. 30 (1953), 211. See also Churchite-(Nd).

Chursinite, Hg_2AsO_4, [A] Mon. Adamantine yellow, brown, orange-brown. Khaidarkan deposit, Kirgizia, USSR. H= 3, D_m=?, D_c= 9.06, VIIA 14. Vasil'ev et al., 1984. Am. Min. 70 (1985), 871 (Abst.).

Chvaleticeite, $(Mn,Mg)SO_4 \cdot 6H_2O$, [A] Mon. Vitreous white, pinkish, yellowish-green. Chvaletice, Čechy (Bohemia), Czechoslovakia. H= 1.5, D_m= 1.84, D_c= 1.84, VIC 03b. Pašava et al., 1986. Am. Min. 72(1987), 1023 (Abst.).

Chvilevaite, $Na(Cu,Fe,Zn)_2S_2$, [A] Hex. Metallic bronze. Akatuya deposit, Transbaikal, USSR. H= 3, D_m=?, D_c= 3.94, IIB 17. Kacholovskaya et al., 1988. Am. Min. 74(1989), 946 (Abst.).

Cinnabar, HgS, [G] Rhom. Translucent adamantine red. H= 2-2.5, D_m= 8.09, D_c= 8.129, IIB 01b. Theophrastus, 315 B.C. Dana, 7th ed.(1944), v.1, 251. Bull. Min. 96 (1973), 218. See also Metacinnabar, Hypercinnabar.

Cirrolite, $Ca_3Al_2(PO_4)_3(OH)_3$, [Q] Syst=? Translucent pale yellow. Vestanå mine, Nästum, Skåne, Sweden. H= 5-6, D_m= 3.08, D_c=?, VIIB 12. Blomstrand, 1868. Dana, 7th ed.(1951), v.2, 845.

Clairite, $(NH_4)_2(Fe,Mn)_3(SO_4)_4(OH)_3 \cdot 3H_2O$, [A] Tric. Translucent yellow. Lone Creek Fall Cave, Sabie (near), Transvaal (E), South Africa. H=?, D_m= 2.31, D_c= 2.32, VID 13. Martini, 1983. Am. Min. 71(1986), 229 (Abst.).

Claraite, $(Cu,Zn)_3CO_3(OH)_4 \cdot 4H_2O$, [A] Hex. Translucent blue. Clara mine (Grube Clara), Wolfach, Schwarzwald, Baden-Württemberg, Germany. H= 2, D_m= 3.35, D_c= 3.34, VbD 01. Walenta & Dunn, 1982. Am. Min. 68(1983), 471 (Abst.).

Claringbullite, $Cu_4Cl(OH)_7 \cdot nH_2O$, [A] Hex. Pearly blue. Nchanga Open Pit, Zambia. H=?, D_m= 3.9, D_c= 3.99, IIIC 02. Fejer et al, 1977. Min. Mag. 41 (1977), 433.

Clarkeite, $(Na, Ca, Pb)_2U_2(O, OH)_7$, G Syst=? Waxy dark reddish-brown, brown. Deer Park mine, Spruce Pine, Mitchell Co., North Carolina, USA. H= 4-4.5, D_m= 6.29, D_c=?, IVD 10b. Ross et al., 1931. Am. Min. 41 (1956), 127.

Claudetite, As_2O_3, G Mon. Vitreous colorless, white. San Domingo mines, Portugal. H= 2.5, D_m= 4.15, D_c= 3.96, IVC 02. Dana, 1868. Dana, 7th ed.(1944), v.1, 545. Struct. Repts. 41A (1975), 213. See also Arsenolite.

Clausthalite, PbSe, G Cub. Metallic grey. Charlotte mine, Clausthal, Harz, Niedersachsen, Germany. H= 2.5-3, D_m= 7.8, D_c= 8.264, IIB 11. Beudant, 1832. Dana, 7th ed.(1944), v.1, 204. See also Galena, Altaite.

Cliffordite, UTe_3O_9, A Cub. Adamantine bright yellow. San Miguel mine, Moctezuma, Sonora, Mexico. H= 4, D_m= 6.57, D_c= 6.76, VIG 06. Gaines, 1969. Am. Min. 54 (1969), 697.

Clinobehoite, $Be(OH)_2$, A Mon. Vitreous/pearly colorless, white. Murzinsk region, Ural Mts., USSR. H= 2-3, D_m= 1.93, D_c=?, IVF 03a. Voloshin et al., 1989. Min. Zhurn. 11(5) (1989), 88. Zts. Krist. 185 (1988), 612.

Clinobisvanite, $BiVO_4$, A Mon. Earthy/sub-vitreous yellow, orange. Yinnietharra Station, Pyramid Hill (S), Western Australia, Australia. H=?, D_m=?, D_c= 6.95, VIIA 13. Bridge & Pryce, 1974. Min. Mag. 39 (1974), 847. See also Dreyerite, Pucherite.

Clinochalcomenite, $CuSeO_3 \cdot 2H_2O$, G Mon. Vitreous bluish-green. China. H= 2, D_m= 3.28, D_c= 3.42, VIG 01. Luo et al., 1980. Am. Min. 66(1981), 217 (Abst.). See also Chalcomenite.

Clinochlore, $(Mg, Al)_6(Si, Al)_4O_{10}(OH)_8$, G Tric. Pearly green, olive-green, yellowish, white. H= 2-2.5, D_m= 2.7, D_c= 2.88, VIIIE 09a. Blake, 1851. Clays Cl. Mins. 35 (1987), 129. Clays Cl. Mins. 38 (1990), 216. See also Chamosite.

Clinochrysotile, $Mg_3Si_2O_5(OH)_4$, G Mon. Yellow, white, grey, green. H= 2.5, D_m= 2.55, D_c= 2.56, VIIIE 10b. Kobell, 1834. Rev. Min. 19 (1988), 91. Acta Cryst. 9 (1956), 855. See also Antigorite, Lizardite, Orthochrysotile, Parachrysotile, Pecoraite.

Clinoclase, $Cu_3AsO_4(OH)_3$, G Mon. Vitreous greenish-blue, blue. Wheal Gorland, Cornwall, England. H= 2.5-3, D_m= 4.38, D_c= 4.42, VIIB 09. Breithaupt, 1830. Dana, 7th ed.(1951), v.2, 787. Acta Cryst. 18 (1965), 777.

Clinoenstatite, $MgSiO_3$, G Mon. Vitreous/resinous colorless, yellow. H= 5-6, D_m= 3.21, D_c= 3.189, VIIID 01a. Wahl, 1906. DHZ, 2nd ed. (1978), v.2A, 30. Zts. Krist. 114 (1960), 120. See also Clinoferrosilite, Enstatite, Kanoite.

Clinoferrosilite, $(Fe, Mg)SiO_3$, G Mon. Vitreous/resinous green, brown. Obsidian Cliff, Yellowstone National Park, Wyoming, USA. H= 5-6, D_m= 3.96, D_c= 4.00, VIIID 01a. Bowen, 1935. DHZ, 2nd ed.(1978), v.2A, 30. IMA (1966), 334 (Abst.). See also Clinoenstatite, Orthoferrosilite, Kanoite.

Clinohedrite, $CaZnSiO_4 \cdot H_2O$, G Mon. Translucent colorless, white, amethystine. Trotter mine, Franklin Furnace, Sussex Co., New Jersey, USA. H= 5.5, D_m= 3.33, D_c= 3.414, VIIIA 13. Penfield & Foote, 1898. Dana/Ford (1932), 633. Zts. Krist. 144 (1976), 377.

Clinoholmquistite, $Li_2(Al, Mg, Fe)_5Si_8O_{22}(OH, F)_2$, A Mon. Siberia, USSR. H=?, D_m=?, D_c=?, VIIID 05a. Ginzburg, 1965. Am. Min. 63 (1978), 1023. See also Holmquistite, Magnesioclinoholmquistite, Ferroclinoholmquistite.

Clinohumite, $(Mg, Fe)_9(SiO_4)_4(F, OH)_2$, G Mon. Translucent yellow, brown, red, reddish-brown. Monte Somma, Mt. Vesuvius, Napoli, Campania, Italy. H= 6, D_m= 3.260, D_c= 3.26, VIIIA 16. Des Cloizeaux, 1876. DHZ, 2nd ed.(1982), v.1A, 380. Am. Min. 58 (1973), 43.

Clinojimthompsonite, $(Mg, Fe)_5Si_6O_{16}(OH)_2$, A Mon. Transparent colorless, pale pinkish-brown. Carlton mine, Chester, Windham Co., Vermont, USA. H=?,

D_m=?, D_c= 3.03, VIIID 12. Veblen & Burnham, 1978. Am. Min. 63 (1978), 1000. Am. Min. 63 (1978), 1053. See also Jimthompsonite.

Clinokurchatovite, $Ca(Mg,Fe,Mn)B_2O_5$, A Mon. USSR. H= 4.5, D_m= 3.07, D_c= 3.27, Vc 02. Malinko & Pertsev, 1983. Am. Min. 69(1984), 810 (Abst.). See also Kurchatovite.

Clinophosinaite, $Na_3Ca(SiO_3)(PO_4)$, A Mon. Vitreous pale lilac. Yukspor Mts., Lovozero massif; also Mt. Koashva, Khibina massif, Kola Peninsula, USSR. H= 4, D_m= 2.86, D_c= 2.839, VIIIC 13. Khomyakov et al. 1981. Am. Min. 67(1982), 414 (Abst.). Sov.Phys.Cryst. 25(1980), 138. See also Phosinaite.

Clinoptilolite, $(Na,K)_6(Al_6Si_{30})O_{72}\cdot 20H_2O$, G Mon. Translucent colorless, reddish, brick-red. Hoodoo Mts., Wyoming, USA. H= 3.5, D_m=?, D_c= 2.13, VIIIF 13. Schaller, 1932. Min. Abstr. 87M/4741. Zts. Krist. 145 (1977), 216.

Clinosafflorite, $(Co,Fe,Ni)As_2$, A Mon. Metallic white. Cobalt, Ontario, Canada. H=?, D_m=?, D_c= 7.46, IIC 08. Radcliffe & Berry, 1971. Can. Min. 10 (1971), 877. See also Safflorite.

Clinotyrolite, $Ca_2Cu_9(AsO_4,SO_4)_4(OH,O)_{10}\cdot 10H_2O$, A Mon. Silky green. Dongchuan mine, Yunnan, China. H=?, D_m= 3.22, D_c= 3.214, VIID 09. Ma et al., 1980. Min. Abstr. 80-4909. See also Tyrolite.

Clinoungemachite, $K_3Na_9Fe(SO_4)_6(OH)_3\cdot 9H_2O$, G Mon. Chuquicamata, Antofagasta, Chile. H=?, D_m=?, D_c=?, VID 13. Peacock & Bandy, 1938. Dana 7th ed.(1951), v.2, 597.

Clinozoisite, $Ca_2Al_3(Si_2O_7)(SiO_4)(O,OH)_2$, G Mon. Translucent colorless, pale yellow-grey, green. Goslerwand, Prägratten, Tyrol, Austria. H= 6.5, D_m= 3.3, D_c= 3.313, VIIIB 15. Weinschenk, 1896. DHZ, 2nd ed.(1986), v.1B, 44. Am. Min. 53 (1968), 1882. See also Epidote, Zoisite.

Clintonite, $Ca(Mg,Al)_3(Al,Si)_4O_{10}(OH,F)_2$, G Mon. Translucent colorless, yellow, green, reddish-brown. Amity, New York, USA. H= 3.5-6, D_m= 3.1, D_c= 3.12, VIIIE 06. Mather, 1843. Am. Min. 71 (1986), 1194. Am. Min. 73 (1988), 365.

Coalingite, $Mg_{10}Fe_2CO_3(OH)_{24}\cdot 2H_2O$, A Rhom. Translucent brown. New Idria serpentine, Coalinga (near), Fresno Co., California, USA. H=?, D_m= 2.32, D_c= 2.26, IVF 03e. Mumpton et al., 1965. Am. Min. 50 (1965), 1893. Min. Mag. 38 (1971), 286.

Cobaltaustinite, $CaCoAsO_4(OH)$, A Orth. Dull green. Dome Rock, South Australia, Australia. H= 4.5, D_m=?, D_c= 4.24, VIIB 11b. Nickel & Birch, 1988. Austral. Min. 3 (1988), 53.

Cobaltite, $CoAsS$, G Orth. Metallic white. H= 5.5, D_m= 6.33, D_c= 6.302, IIC 06a. Beudant, 1832. Am. Min. 50 (1965), 1002. Am. Min. 67 (1982), 1048.

Cobaltkoritnigite, $(Co,Zn)(AsO_3OH)\cdot H_2O$, A Tric. Vitreous purple. Erzgebirge, Sachsen (Saxony), Germany. H=?, D_m=?, D_c= 3.44, VIID 32. Schmetzer et al., 1981. Am. Min. 67(1982), 414 (Abst.). See also Koritnigite.

Cobaltomelane, D Mixture of Mn oxides. Min. Mag. 33 (1962), 261. Ginzberg & Rukavishnikova 1951.

Cobaltomenite, $CoSeO_3\cdot 2H_2O$, G Mon. Translucent red, pink. Cerro de Cacheuta, Mendoza, Argentina. H=?, D_m=?, D_c= 3.42, VIG 01. Bertrand, 1882. Can. Min. 12 (1974), 304. See also Ahlfeldite.

Cobalt pentlandite, $(Co,Ni,Fe)_9S_8$, G Cub. Metallic yellow. Varislahti deposit, Karelia (N), Finland. H= 4-4.5, D_m=?, D_c= 4.69, IIA 07. Kouvo et al., 1959. Am. Min. 44 (1959), 897. Can. Min. 13 (1975), 75. See also Pentlandite.

Cobalt-zippeite, $Co_2(UO_2)_6(SO_4)_3(OH)_{10}\cdot 16H_2O$, A Syst=? Translucent tan, brownish-yellow, orange-yellow. Happy Jack mine, Emery Co., Utah, USA. H= 2, D_m=?, D_c=?, VID 08. Frondel et al., 1976. Can. Min. 14 (1976), 429. See also Zippeite, Sodium-zippeite, Nickel-zippeite, Zinc-zippeite, Magnesium-zippeite.

Coccinite, HgI_2, [Q] Tet. Greasy orange-red. Casas Viejas, Mexico. H=?, D_m=?, D_c= 6.354, IIIA 05. Haidinger, 1845. DAN USSR, Ser.B (1979), 702.

Cochromite, $(Co, Ni, Fe)(Cr, Al, Fe)_2O_4$, [A] Cub. Metallic black; grey in reflected light. Bon Accord, Barberton Dist., Transvaal, South Africa. H=?, D_m=?, D_c= 5.01, IVB 01. DeWaal, 1978. Am. Min. 65(1980), 811 (Abst.).

Cocinerite, [D] A mixture of chalcocite, silver and other minerals. Am. Min. 52 (1967), 1214. Hough, 1919.

Coconinoite, $Fe_2^{3+}Al_2(UO_2)_2(PO_4)_4(SO_4)(OH)_2 \cdot 20H_2O$, [A] Orth. Translucent light creamy-yellow. Sun Valley mine, Coconino Co., Arizona, USA. H=?, D_m= 2.90, D_c= 2.92, VIID 22. Young et al., 1966. Am. Min. 51(1966), 651 (Abst.).

Coeruleolactite, $(Ca, Cu)Al_6(PO_4)_4(OH)_8 \cdot 4\text{–}5H_2O$, [G] Tric. Translucent white, light blue. Rindsberg mine, Katzenellnbogen (near), Nassau, Germany. H= 5, D_m= 2.57, D_c= 2.99, VIID 08. Petersen, 1871. Am. Min. 43(1958), 1224 (Abst.).

Coesite, SiO_2, [G] Mon. Vitreous colorless. Meteor Crater, Arizona, USA. H= 8, D_m= 3.01, D_c= 2.92, IVD 01. Sosman, 1954. Am. Min. 47 (1962), 1292. Zts. Krist. 145 (1977), 108. See also Cristobalite, Quartz, Tridymite, Stishovite.

Coffinite, $U[SiO_4, (OH)_4]$, [G] Tet. Adamantine black. La Sal No. 2 mine and others, Mesa Co., Colorado, USA. H= 5-6, D_m= 4.64, D_c= 6.9, VIIIA 07. Stieff et al., 1956. Am. Min. 43 (1958), 243. Min. Abstr. 82M/2411. See also Thorogummite.

Cohenite, Fe_3C, [G] Orth. Metallic white. Uivfaq, Disko, Greenland (W). H=?, D_m= 7.4, D_c= 7.68, IC 02. Weinschenk, 1889. Dana, 7th ed.(1944), v.1, 122. Struct. Repts. 29 (1964), 34.

Colemanite, $CaB_3O_4(OH)_3 \cdot H_2O$, [G] Mon. Vitreous/adamantine colorless, white, yellowish, grey. Furnace Creek, Death Valley, Inyo Co., California, USA. H= 4.5, D_m= 2.423, D_c= 2.419, Vc 06. Evans, 1884. Am. Min. 38 (1953), 411. Acta Cryst. 11 (1958), 761.

Collinsite, $Ca_2(Mg, Fe)(PO_4)_2 \cdot 2H_2O$, [G] Tric. Silky/sub-vitreous light brown, white. François Lake, Coast Dist., British Columbia, Canada. H= 4-5, D_m= 2.99, D_c= 2.955, VIIC 12. Poitevin, 1927. Min. Mag. 39 (1974), 577. Min. Abstr. 75-1944.

Coloradoite, HgTe, [G] Cub. Metallic grey. Keystone area, Boulder Co., Colorado, USA. H= 2.5, D_m= 8.04, D_c= 8.42, IIB 01a. Genth, 1877. Am. Min. 34 (1949), 342.

Colquiriite, $LiCaAlF_6$, [A] Rhom. Translucent white. Colquiri mine, Colquiri, Bolivia. H=?, D_m= 2.94, D_c= 2.95, IIIB 04. Walenta et al., 1980. Am. Min. 66(1981), 879 (Abst.).

Columbite, A group name for oxides with the general formula $(Fe, Mn, Mg)Nb_2O_6$. See Ferrocolumbite, Manganocolumbite, Magnocolumbite.

Colusite, $Cu_{26}V_2(As, Sn)_6S_{32}$, [G] Cub. Metallic cream. Colusa claim, Butte, Silver Bow Co., Montana, USA. H= 4.5, D_m= 4.43, D_c= 4.59, IIB 03c. Sales, pre-1932. Can. Min. 19 (1981), 423. See also Germanite, Nekrasovite.

Comancheite, $Hg_{13}O_9(Cl, Br)_8$, [A] Orth. Vitreous/resinous red, orange-red, yellow. Mariposa mine, Terlingua Dist., Brewster Co., Texas, USA. H= 2, D_m= 7.7, D_c= 8.0, IIIC 03. Roberts et al., 1981. Can. Min. 19 (1981), 393.

Combeite, $Na_2Ca_2Si_3O_9$, [G] Hex. Translucent colorless. Mt. Shahern, Volcanos Area, Rutshuru territory, Kivu, Zaïre. H=?, D_m= 2.844, D_c= 2.826, VIIIC 07. Sahama & Hytönen, 1957. Am. Min. 69(1984), 214 (Abst.). Acta Cryst. C43 (1987), 1852.

Comblainite, $Ni_6Co_2CO_3(OH)_{16} \cdot 4H_2O$, [A] Rhom. Translucent turquoise-blue. Shinkolobwe, Shaba, Zaïre. H=?, D_m= 3.05, D_c= 3.2, IVF 03b. Piret & Deliens, 1980. Am. Min. 65(1980), 1065 (Abst.).

Compreignacite, $K_2(UO_2)_6(OH)_{14} \cdot 4H_2O$, [A] Orth. Translucent yellow. Compreignac, Margnac deposit, Haute-Vienne, France. H=?, D_m= 5.03, D_c= 5.13,

IVF 12. Protas, 1964. Am. Min. 50(1965), 807 (Abst.). See also Becquerelite, Billietite.

Congolite, $(Fe,Mg,Mn)_3B_7O_{13}Cl$, G Rhom. Transparent pale red. Brazzaville, Congo. H=?, D_m=?, D_c= 3.520, Vc 14. Wendling et al., 1972. Am. Min. 57(1972), 1315 (Abst.). See also Ericaite.

Conichalcite, $CaCuAsO_4(OH)$, G Orth. Vitreous green, yellowish-green. Hinojosa de Cordoba, Andalusia, Spain. H= 4.5, D_m= 4.33, D_c= 4.338, VIIB 11b. Breithaupt & Fritzsche, 1849. Am. Min. 56 (1971), 1359. Can. Min. 7 (1963), 561. See also Austinite, Calciovolborthite.

Connellite, $Cu_{19}Cl_4SO_4(OH)_{32} \cdot 3H_2O$, G Hex. Vitreous blue. Wheal Providence, Cornwall, England. H=?, D_m= 3.36, D_c= 3.46, IVF 03d. Dana, 1850. Min. Mag. 29 (1950), 280. Am. Min. 57 (1972), 426. See also Buttgenbachite.

Cookeite, $(Al,Li)_3Al_2(Si,Al)_4O_{10}(OH)_8$, G Tric. Pearly white, yellowish-green. Hebron, Maine, USA. H= 2.5, D_m= 2.70, D_c= 2.72, VIIIE 09b. Brush, 1866. Clays Cl. Mins. 37 (1989), 193. Am. Min. 60 (1975), 1041.

Cooperite, $(Pt,Pd,Ni)S$, G Tet. Metallic grey. Bushveld, Transvaal, South Africa. H= 4-4.5, D_m=?, D_c= 10.12, IIB 16. Wartenweiler, 1928. Am. Min. 63 (1978), 832. See also Braggite.

Copiapite, $Fe_5(SO_4)_6(OH)_2 \cdot 20H_2O$, G Tric. Pearly yellow, orange, greenish-yellow. Copiapó, Atacama, Chile. H= 2.5-3, D_m= 2.1, D_c= 2.118, VID 04b. Rose, 1833. Can. Min. 23 (1985), 53.

Copper, Cu, G Cub. Metallic light rose, copper-red. H= 2.5-3, D_m= 8.95, D_c= 8.929, IA 01a. Dana, 7th ed.(1944), v.1, 99.

Coquimbite, $Fe_2(SO_4)_3 \cdot 9H_2O$, G Rhom. Vitreous pale violet, amethystine, yellowish, greenish. Coquimbo, Chile. H= 2.5, D_m= 2.11, D_c= 2.115, VIC 17. Breithaupt, 1841. Dana, 7th ed.(1951), v.2, 532. Am. Min. 55 (1970), 1534. See also Paracoquimbite.

Corderoite, $\alpha\text{-}Hg_3S_2Cl_2$, A Cub. Translucent colorless, pale yellowish-white. McDermitt mine (Old Cordero Mine), Humboldt Co., Nevada, USA. H= 2, D_m=?, D_c= 6.85, IIF 01. Foord et al., 1974. Am. Min. 59 (1974), 652. See also Lavrentievite.

Cordierite, $Mg_2Al_4Si_5O_{18}$, G Orth. Translucent greyish/lilac/dark blue. Bodenmais, Bayern (Baveria), Germany. H= 7, D_m= 2.57, D_c= 2.56, VIIIC 06. Lukas, 1813. Am. Min. 71 (1986), 746. Am. Min. 62 (1977), 67. See also Sekaninaite, Indialite.

Cordylite-(Ce), $Ba(Ce,La)_2(CO_3)_3F_2$, G Hex. Greasy/adamantine colorless, yellow. Narsarsuk, Greenland (S). H= 4.5, D_m= 4.31, D_c= 3.97, VbB 04. Flink, 1900. Can. Min. 13 (1975), 93.

Corkite, $PbFe_3(SO_4)(PO_4)(OH)_6$, G Rhom. Vitreous/resinous dark green, yellowish-green pale yellow. Cork Co., Ireland. H= 3.5-4.5, D_m= 4.295, D_c= 4.308, VIB 03b. Adam, 1869. Am. Min. 72 (1987), 178. N.Jb.Min.Mh. (1987), 71.

Cornetite, $Cu_3PO_4(OH)_3$, G Orth. Vitreous blue, greenish-blue. L'Étoile du Congo mine, Elizabethville, Shaba, Zaïre. H= 4.5, D_m= 4.10, D_c= 4.14, VIIB 09. Buttgenbach, 1916. Am. Min. 35 (1950), 365. Min. Petrol. 40 (1989), 127.

Cornubite, $Cu_5(AsO_4)_2(OH)_4$, G Tric. Vitreous/porcelanous light/dark green. Cornwall, England. H=?, D_m= 4.64, D_c= 4.85, VIIB 08. Claringbull et al., 1959. Am. Min. 70(1985), 1333 (Abst.).

Cornwallite, $Cu_5(AsO_4)_2(OH)_4$, G Mon. Translucent green, blackish-green. Wheal Garland, St. Day, Cornwall, England. H= 4.5, D_m= 4.52, D_c= 4.645, VIIB 08. Zippe, 1846. Min. Mag. 32 (1959), 1.

Coronadite, $PbMn_8O_{16}$, G Mon. Dull/submetallic dark grey; white in reflected light. J. Horse-Shoe Shaft (W), Coronado Lode, Clifton-Morenci Dist., Arizona, USA.

H= 4.5-5, D_m= 5.44, D_c= 5.35, IVD 03b. Lindgren & Hillebrand, 1904. Am. Min. 27 (1942), 48. Am. Min. 74 (1989), 913.

Corrensite, $(Mg, Fe, Al)_9(Si, Al)_8O_{20}(OH)_{10} \cdot nH_2O$, G Orth. Maulbronn (near), Zaiserweiher (near) and Göttingen (near), Huntstollen, Niedersachsen, Germany. H=?, D_m=?, D_c=?, VIIIE 08c. Lippmann, 1954. JCPDS 31-794. See also Chlorite, Vermiculite, Smectite.

Corundum, Al_2O_3, G Rhom. Adamantine/vitreous colorless, blue, red, etc. H= 9, D_m= 4.0, D_c= 3.995, IVC 04a. Dana, 7th ed.(1944), v.1, 520. Acta Cryst. B36 (1980), 228.

Corvusite, $V_7O_{17} \cdot nH_2O$(?), Q Syst=? Opaque blue-black, brown. Jack Claim, La Sal Mountains (E), Grand Co., Utah, USA; Ponto #3 Claim, Gypsum Valley (N), San Miguel Co., Colorado, USA. H= 2.5-3, D_m= 2.82, D_c=?, IVF 09. Henderson & Hess, 1933. Am. Min. 44 (1959), 322.

Cosalite, $Pb_2Bi_2S_5$, G Orth. Metallic grey; white in reflected light. Cosala mine, Sinaloa, Mexico. H= 2.5-3, D_m= 6.76, D_c= 7.17, IID 08. Genth, 1868. Dana, 7th ed.(1944), v.1, 445. Zts. Krist. 140 (1974), 114. See also Veenite.

Costibite, CoSbS, A Orth. Metallic grey. A.B.H. Consols mine, Broken Hill, New South Wales, Australia. H= 6, D_m= 6.9, D_c= 6.87, IIC 08. Cabri et al., 1970. Am. Min. 55 (1970), 10. Can. Min. 13 (1975), 188. See also Paracostibite.

Cotunnite, $PbCl_2$, G Orth. Adamantine white, yellowish, greenish. Mt. Vesuvius, Napoli, Campania, Italy. H= 2.5, D_m= 5.80, D_c= 5.956, IIIC 05. Monticelli & Covelli, 1825. Dana, 7th ed.(1951), v.2, 42. Sov.Phys.Cryst. 21(1976), 38.

Coulsonite, FeV_2O_4, G Cub. Metallic bluish-grey; grey in reflected light. Bihar, India. H= 4.5-5, D_m= 5.2, D_c= 5.15, IVB 01. Radtke, 1962. Am. Min. 47 (1962), 1284.

Cousinite, $Mg(UO_2)_2(MoO_4)_2(OH)_2 \cdot 5H_2O$(?), Q Syst=? Vitreous black. Shinkolobwe, Shaba (Katanga), Zaïre. H=?, D_m=?, D_c=?, VIF 02. Vaes, 1958. Am. Min. 44(1959), 910 (Abst.).

Covellite, CuS, G Hex. Submetallic blue. Mt. Vesuvius, Napoli, Campania, Italy. H= 1.5-2, D_m= 4.7, D_c= 4.602, IIB 15. Beudant, 1832. Zts. Krist. 184 (1988), 111. Am. Min. 61 (1976), 996. See also Klockmannite.

Cowlesite, $Ca(Al_2Si_3)O_{10} \cdot 5\text{–}6H_2O$, A Orth. Translucent colorless, white. Goble, Columbia Co., Oregon, USA. H= 5-5.5, D_m= 2.13, D_c= 2.10, VIIIF 10. Wise & Tschernich, 1975. Min. Mag. 48 (1984), 565.

Coyoteite, $NaFe_3S_5 \cdot 2H_2O$, A Tric. Metallic black; pale brownish-grey in reflected light. Coyote Peak, Humboldt Co., California, USA. H= 1.5, D_m=?, D_c= 2.879, IIE 03. Erd & Czamanske, 1983. Am. Min. 68 (1983), 245.

Craigite, D Not proven to occur naturally. Min. Mag. 43 (1980), 1055.

Crandallite, $CaAl_3(PO_4)(PO_3OH)(OH)_6$, G Rhom. Vitreous/dull/chalky yellow, white, grey. Brooklyn mine dump, Tintic Dist., Silver City (NW), Utah, USA. H= 5, D_m= 2.8, D_c= 2.997, VIIB 15a. Loughlin & Schaller. Am. Min. 65 (1980), 953. Am. Min. 59 (1974), 41.

Creaseyite, $Cu_2Pb_2(Fe, Al)_2Si_5O_{17} \cdot 6H_2O$, A Orth. Translucent green. Mammoth mine, Tiger, Pinal Co., Arizona, USA. H= 2.5, D_m= 4.1, D_c= 4.01, VIIIB 13. Williams & Bideaux, 1975. Min. Mag. 40 (1975), 227.

Crednerite, $CuMnO_2$, G Mon. Metallic black; creamy-white in reflected light. Friedrichroda, Thüringen, Germany. H= 4, D_m= 5.38, D_c= 5.49, IVA 06. Rammelsberg, 1847. Bull. Min. 89 (1966), 80.

Creedite, $Ca_3Al_2SO_4(OH)_2F_8 \cdot 2H_2O$, G Mon. Vitreous colorless, white, purple. Wagon Wheel Gap, Creede Quad., Mineral Co., Colorado, USA. H= 4, D_m= 2.72, D_c= 2.738, IIID 05. Larsen & Wells, 1916. Am. Min. 37 (1952), 785. N. Jb. Min. Mh. (1983), 69.

Crichtonite, $(Sr, La, Ce)(Ti, Fe, Mn)_{21}O_{38}$, G Rhom. Opaque black. St. Christophe, Bourg d'Oisans, Isère, France. H= 4.5-5, D_m= 4.46, D_c= 4.54, IVC 06. Bournon, 1813. Min. Mag. 37 (1969), 349. Am. Min. 61 (1976), 1203.

Criddleite, $Ag_2Au_3TlSb_{10}S_{10}$, A Mon. Metallic grey. Hemlo deposit, Hemlo, Ontario, Canada. H= 3-3.5, D_m= 6.86, D_c= 6.57, IID 17. Harris et al., 1988. Min. Mag. 52 (1988), 691.

Cristobalite, SiO_2, G Tet. Vitreous white. Cerro San Cristóbal, Pachuca, Mexico. H=?, D_m= 2.33, D_c= 2.331, IVD 01. vom Rath, 1887. Dana, 7th ed.(1962), v.3, 276. Struct. Repts. 52A (1985), 320. See also Coesite, Quartz, Stishovite, Tridymite.

Crocoite, $PbCrO_4$, G Mon. Adamantine/vitreous red, orange, yellow. Berezov, Sverdlovsk, Ural Mts., USSR. H= 2.5-3, D_m= 6.0, D_c= 6.10, VIE 01. Beudant, 1832. Dana, 7th ed.(1951), v.2, 646. Zts. Krist. 176 (1986), 75.

Cronstedtite-1H, $Fe_3(Si, Fe)_2O_5(OH)_4$, G Rhom. Dark brown, black. Cornucopia mine, Nye Co., Nevada, USA. H=?, D_m=?, D_c= 3.60, VIIIE 10b. Steinmann, 1821. Am. Min. 47 (1962), 781. Acta Cryst. 16 (1963), 1.

Cronstedtite-1M, $Fe^{2+}_2Fe^{3+}(Si, Fe)_2O_5(OH)_4$, P Mon. Black. Wheal Jane mine, Kea, Cornwall, England. H=?, D_m= 3.34, D_c= 3.59, VIIIE 10b. Steinmann, 1821. JCPDS 17-470. Acta Cryst. 17 (1964), 404.

Cronstedtite-2H, $Fe^{2+}_2Fe^{3+}(Si, Fe)_2O_5(OH)_4$, P Hex. Dark brown, black. Wheal Maudlin mine, Truro, Cornwall, England. H=?, D_m=?, D_c= 3.60, VIIIE 10b. Steinmann, 1821. Am. Min. 47 (1962), 781. Acta Cryst. 16 (1963), 1.

Cronstedtite-2M, $Fe^{2+}_2Fe^{3+}(Si, Fe)_2O_5(OH)_4$, P Mon. Dark brown, black. Wheal Maudlin mine, Truro, Cornwall, England. H=?, D_m=?, D_c= 3.60, VIIIE 10b. Steinmann, 1821. DHZ (1962), v.3, 167. Acta Cryst. 17 (1964), 404.

Cronstedtite-3R, $Fe^{2+}_2Fe^{3+}(Si, Fe)_2O_5(OH)_4$, P Rhom. Dark brown, black. Kutná Hora (Kuttenberg), Čechy (Bohemia), Czechoslovakia. H=?, D_m=?, D_c= 3.60, VIIIE 10b. Steinmann, 1821. Am. Min. 47 (1962), 781. Acta Cryst. 16 (1963), 1.

Cronstedtite-6R, $Fe^{2+}_2Fe^{3+}(Si, Fe)_2O_5(OH)_4$, P Rhom. Dark brown, black. Wheal Maudlin mine, Truro, Cornwall, England. H=?, D_m=?, D_c= 3.60, VIIIE 10b. Steinmann, 1821. DHZ (1962), v.3, 167. Acta Cryst. 16 (1963), 1.

Cronstedtite-9R, $Fe^{2+}_2Fe^{3+}(Si, Fe)_2O_5(OH)_4$, P Rhom. Dark brown, black. Cornucopia mine, Nye Co., Nevada, USA. H=?, D_m=?, D_c= 2.53, VIIIE 10b. Steinmann, 1821. Am. Min. 47 (1962), 781.

Crookesite, Cu_7TlSe_4, G Tet. Metallic grey. Skrikerum, Sweden. H= 2.5-3, D_m= 6.90, D_c= 7.443, IIA 09. Nordenskiöld, 1866. Am. Min. 35 (1950), 337. Zts. Krist. 181 (1987), 241.

Crossite, $(Na, Ca)_2(Fe, Mg, Al)_5(Si, Al)_8O_{22}(OH)_2$, G Mon. Translucent blue. Berkeley (N), Alameda Co., California, USA. H=?, D_m= 3.21, D_c= 3.22, VIIID 05d. Palache, 1894. Contrib.Min.Pet. 15(1967), 67.

Cryolite, α–Na_3AlF_6, G Mon. Vitreous/greasy white, brownish, grayish, black. Ivigtut, Greenland (SW). H= 2.5, D_m= 2.97, D_c= 2.963, IIIB 03. Abildgaard, 1799. Dana, 7th ed.(1951), v.2, 110. Can. Min. 13 (1975), 377.

Cryolithionite, $Na_3Li_3Al_2F_{12}$, G Cub. Vitreous colorless, white. Ivigtut, Greenland (SW). H= 2.5-3, D_m= 2.771, D_c= 2.771, IIIB 04. Ussing, 1904. Dana, 7th ed.(1951), v.2, 99. Am. Min. 56 (1971), 18.

Cryptohalite, $(NH_4)_2SiF_6$, G Cub. Vitreous white, grey. Mt. Vesuvius, Napoli, Campania, Italy. H= 2.5, D_m= 2.004, D_c= 2.029, IIIB 02b. Scacchi, 1873. Dana, 7th ed.(1951), v.2, 104. See also Bararite.

Cryptomelane (monoclinic), KMn_8O_{16}, P Mon. Grey, black. Chindwara, India. H=?, D_m=?, D_c= 4.398, IVD 03b. Richmond & Fleischer, 1942. Am. Min. 27 (1942), 607. Acta Cryst. B38 (1982), 1056. See also Cryptomelane (tetragonal).

Cryptomelane (tetragonal), KMn_8O_{16}, G Tet. Earthy brown; greyish-white in reflected light. Tombstone, Cochise Co., Arizona, USA. H= 5-6.5, D_m= 4.3, D_c= 4.442, IVD 03b. Richmond & Fleischer, 1942. Contr.Min.Pet. 55 (1976), 191 . Acta Cryst. B42 (1986), 162. See also Cryptomelane (monoclinic).

Cryptonickelmelane, D Mixture. Min. Mag. 33 (1962), 261. Ginzberg & Rukavishnikova 1951.

Csiklovaite, $Bi_2Te(S,Se)_2$, Q Syst=? Metallic; grey in reflected light. Csiklova, Romania. H=?, D_m=?, D_c=?, IIC 03a. Koch & Grasselly, 1948. Am. Min. 35(1950), 333 (Abst.).

Cualstibite, $Cu_6Al_3(SbO_4)_3(OH)_{12} \cdot 10H_2O$, A Hex. Vitreous blue-green. Clara mine (Grube Clara), Wolfach, Schwarzwald, Baden-Württemberg, Germany. H= 2, D_m= 3.18, D_c= 3.25, VIID 11. Walenta, 1984. Am. Min. 70(1985), 1329 (Abst.).

Cubanite, $CuFe_2S_3$, G Orth. Metallic brass/bronze-yellow. Barracanao, Cuba. H= 3.5, D_m= 4.1, D_c= 4.028, IIB 08. Breithaupt, 1843. Dana, 7th ed.(1944), v.1, 243. Zts. Krist. 140 (1974), 218.

Cumengite, $Cu_{20}Pb_{21}Cl_{42}(OH)_{40}$, G Tet. Vitreous blue. Amelia mine, Boléo, Santa Rosalia (near), Baja California, Mexico. H= 2.5, D_m= 4.67, D_c= 4.66, IIIC 04. Mallard, 1893. Min. Rec. 5 (1974), 280. Min. Mag. 50 (1986), 157.

Cummingtonite, $(Mg,Fe)_7Si_8O_{22}(OH)_2$, A Mon. Translucent dark green, brown. Cummington, Hampshire Co., Massachusetts, USA. H= 5-6, D_m= 3.2, D_c= 3.195, VIIID 05a. Dewey, 1824. Am. Min. 63 (1978), 1023. Am. Min. 74 (1989), 1091. See also Magnesiocummingtonite, Grünerite.

Cupalite, $(Cu,Zn)Al$, A Orth. Metallic grey. Listvenitovȳí, Koriakskhiye Mts., USSR (E). H= 4-4.5, D_m=?, D_c= 5.12, IA 02. Razin et al., 1985. ZVMO 114 (1985), 90.

Cuprite, Cu_2O, G Cub. Adamantine/sub-metallic red. H= 3.5-4, D_m= 6.14, D_c= 6.103, IVA 02. Haidinger, 1845. Dana, 7th ed.(1944), v.1, 491. Zts. Krist. 182 (1988), 158.

Cuproadamite, $(Cu,Zn)_2AsO_4(OH)$, Q Syst=? Translucent pale green. Cap Garonne, Var, France. H=?, D_m=?, D_c=?, VIIB 04. Lacroix, 1910. Min. Mag. 52 (1988), 552. See also Adamite.

Cuproartinite, D Probably nakauriite. Am. Min. 67 (1982), 156. Oswald and Crook, 1979.

Cuprobismutite, $Cu_4Bi_7S_{12}$, G Mon. Dark bluish-grey. Missouri mine, Hall's Valley, Park Co., Colorado, USA. H=?, D_m= 6.31, D_c= 6.813, IID 06a. Hillebrand, 1884. Am. Min. 37 (1952), 447. Acta Cryst. B31 (1975), 703. See also Paderaite, Hodrushite.

Cuprocopiapite, $CuFe_4(SO_4)_6(OH)_2 \cdot 20H_2O$, G Tric. Pearly yellow, orange, greenish-yellow. Chuquicamata, Chile. H= 2.5-3, D_m= 2.13, D_c= 2.26, VID 04b. Bandy, 1938. Can. Min. 23 (1985), 53.

Cuprohydromagnesite, D Probably nakauriite. Am. Min. 67 (1982), 156. Oswald & Crook, 1979.

Cuproiridsite, $CuIr_2S_4$, A Cub. Metallic grey. Konder massif, Aldan shield, Siberia, USSR (E). H= 5.5, D_m=?, D_c= 7.24, IIC 01. Rudashevskii et al., 1985. Am. Min. 71(1986), 1277 (Abst.). See also Cuprorhodsite, Malanite.

Cupropavonite, $AgCu_{1.8}Pb_{1.2}Bi_5S_{10}$, A Mon. Metallic grey; white in reflected light. Alaska mine, San Juan Co., Colorado, USA. H=?, D_m=?, D_c= 6.83, IID 09c. Karup-Møller & Makovicky, 1979. Am. Min. 65(1980), 206 (Abst.).

Cuprorhodsite, $CuRh_2S_4$, A Cub. Metallic grey. Konder massif, Aldan shield, Siberia, USSR (E). H= 5, D_m=?, D_c= 6.74, IIC 01. Rudashevskiĭ et al., 1985. Am. Min. 71(1986), 1277 (Abst.). See also Cuproiridsite, Malanite.

Cuprorivaite, $CaCuSi_4O_{10}$, [G] Tet. Vitreous blue. Mt. Vesuvius, Napoli, Campania, Italy. H= 5, D_m= 3.05, D_c= 3.09, VIIIE 01. Minguzzi, 1938. Am. Min. 47 (1962), 409. Acta Cryst. 12 (1959), 733.

Cuprosklodowskite, $Cu(UO_2)_2(SiO_3OH)_2 \cdot 6H_2O$, [G] Tric. Translucent green, greenish-yellow. Kalongwe, Shaba, Zaïre. H=?, D_m= 3.83, D_c= 3.815, VIIIA 25. Buttgenbach, 1933. Am. Min. 66 (1981), 610. Am. Min. 60 (1975), 448. See also Sklodowskite.

Cuprospinel, $(Cu, Mg)Fe_2O_4$, [A] Cub. Metallic black; grey in reflected light. Consolidated Rambler mine, Baie Verte (near), Newfoundland, Canada. H= 6.5, D_m=?, D_c= 5.42, IVB 01. Nickel, 1973. Can. Min. 11 (1973), 1003.

Cuprostibite, $Cu_2(Sb, Tl)$, [G] Tet. Metallic grey; violet-rose in reflected light. Ilímaussaq Intrusion, Greenland (S). H= 4, D_m=?, D_c= 8.42, IIG 01. Sørensen et al., 1969. Am. Min. 55(1970), 1810 (Abst.).

Cuprotungstite, $Cu_3(WO_4)_2 \cdot 2H_2O$, [G] Tet. Vitreous/waxy/earthy green. La Paz, Baja California, Mexico. H=?, D_m=?, D_c= 7.06, VIF 03. Adam, 1869. Min. Mag. 43 (1979), 448.

Curetonite, $Ba_4TiAl_3(PO_4)_4(O, OH)_6$, [A] Mon. Translucent yellow-green, green. Barite mine, Golconda (near), Humboldt Co., Nevada, USA. H= 3.5, D_m= 4.42, D_c= 4.31, VIIB 19. Williams, 1979. Am. Min. 65(1980), 206 (Abst.).

Curiénite, $Pb(UO_2)_2(VO_4)_2 \cdot 5H_2O$, [A] Orth. Translucent yellow. Mounana mine, Franceville, Haut-Ogooué, Gabon. H=?, D_m= 4.88, D_c= 4.94, VIID 23. Cesbron & Morin, 1968. Bull. Min. 91 (1968), 453. Bull. Min. 94 (1971), 8. See also Francevillite.

Curite, $Pb_{6.5}(UO_2)_{16}O_{16}(OH)_{12}(H_2O, OH)_4$, [G] Orth. Adamantine orange-red. Shinkolobwe, Shaba, Zaïre. H= 4-5, D_m= 7.37, D_c= 7.435, IVF 14. Schoep, 1921. Dana, 7th ed.(1944), v.1, 629. Struct. Repts. 48A (1981), 246.

Cuspidine, $Ca_4Si_2O_7(F, OH)_2$, [G] Mon. Vitreous pale rose-red. Monte Somma, Mt. Vesuvius, Napoli, Campania, Italy. H= 5-6, D_m= 2.96, D_c= 2.978, VIIIB 06. Scacchi, 1876. Dana, 6th ed.(1892), 533. Min. Jour. 8 (1977), 286.

Cuzticite, $Fe_2TeO_6 \cdot 3H_2O$, [A] Hex. Translucent yellow. Bambolla mine, Moctezuma, Sonora, Mexico. H= 3, D_m= 3.9, D_c= 4.01, VIG 10. Williams, 1982. Min. Mag. 46 (1982), 257.

Cyanochroite, $K_2Cu(SO_4)_2 \cdot 6H_2O$, [G] Mon. Vitreous greenish-blue. Mt. Vesuvius, Napoli, Campania, Italy. H=?, D_m= 2.224, D_c= 2.232, VIC 11. Dana 1868. Per. Mineral. 54 (1985), 1.

Cyanophyllite, $Cu_5Al_2(SbO_4)_3(OH)_2 \cdot 12H_2O$, [A] Orth. Pearly/silky greenish-blue. Clara mine (Grube Clara), Wolfach, Schwarzwald, Baden-Württemberg, Germany. H= 2, D_m= 3.10, D_c= 3.12, VIID 11. Walenta, 1981. Am. Min. 66(1981), 1274 (Abst.).

Cyanotrichite, $Cu_4Al_2SO_4(OH)_{12} \cdot 2H_2O$, [G] Orth. Silky blue. Moldava, Banat, Romania. H=?, D_m= 2.85, D_c= 2.88, IVF 03e. Glocker, 1839. JCPDS 11-131. See also Carbonate-cyanotrichite.

Cyclowollastonite, [D] An artificial product; not known naturally. Min. Mag. 43 (1980), 1055. Strunz, 1970.

Cylindrite, $FePb_3Sn_4Sb_2S_{14}$, [G] Tet. Metallic blackish-grey; grey-white in reflected light. Mina Santa Cruz, Poopó, Bolivia. H= 2.5, D_m= 5.46, D_c=?, IID 13. Frenzel, 1893. Min. Abstr. 76-3298. See also Incaite, Potosiite, Cylindrite.

Cymrite, $Ba(Si, Al)_4O_8 \cdot H_2O$, [G] Mon. Vitreous/satiny colorless, green, brown. Benallt mine, Rhiw, Caernavonshire, Lleyn Peninsula, Gwynedd, Wales. H= 2-3, D_m= 3.413, D_c= 3.49, VIIIE 24. Campbell Smith et al., 1949. Am. Min. 49 (1964), 158. Sov.Phys.Cryst. 20 (1975), 171.

Cyrilovite, $NaFe_3(PO_4)_2(OH)_4 \cdot 2H_2O$, [G] Tet. Translucent orange, brownish-yellow. Cyrilov (near), Velké Meziříčí, Moravia (W), Czechoslovakia. H=?, D_m= 3.088, D_c= 3.115, VIID 13. Novotný & Staněk, 1953. Am. Min. 42 (1957), 586. Min. Abstr. 88M/1837. See also Wardite.

D

Dachiardite, $(Na,K,Ca_{0.5})_4(Al_4Si_{20})O_{48}\cdot18H_2O$, [G] Mon. Transparent colorless. San Piero di Campo, Elba, Toscana, Italy. H= 4-4.5, D_m= 2.165, D_c= 2.20, VIIIF 12. D'Achiardi, 1906. Natural Zeolites (1985), 233. Zts. Krist. 166 (1984), 63.

Dadsonite, $Pb_{21}Sb_{23}S_{55}Cl$, [A] Mon. Metallic grey; greenish-white in reflected light. Yellowknife, Madoc, Ontario, Canada; Wolfsberg, Harz, Germany; Pershing Co., Nevada, USA. H= 2.7, D_m= 5.68, D_c= 5.51, IIF 03. Jambor, 1969. Can. Min 17 (1979), 601.

Dalyite, $K_2ZrSi_6O_{15}$, [G] Tric. Vitreous colorless. Green Mountain, Ascension Island. H= 7.5, D_m= 2.84, D_c= 2.798, VIIID 11. Van Tassel, 1952. Min. Mag. 29 (1952), 850. Zts. Krist. 121 (1965), 349. See also Davanite.

Danalite, $Be_3Fe_4(SiO_4)_3S$, [G] Cub. Vitreous/resinous pink, grey. Rockport, Cape Ann, Essex Co., Massachusetts, USA. H= 5.5-6, D_m= 3.43, D_c= 3.40, VIIIF 08. Cooke, 1866. Dana, 6th ed. (1892), 435. Am. Min. 70 (1985), 186. See also Genthelvite, Helvite.

Danbaite, $CuZn_2$, [A] Cub. Metallic white. Danba, Sichuan, China. H= 4, D_m=?, D_c= 7.36, IA 02. Yue et al., 1983. Am. Min. 69(1984), 566 (Abst.).

Danburite, $CaB_2Si_2O_8$, [G] Orth. Vitreous/greasy pale yellow, colorless, yellowish-brown. Danbury, Connecticut, USA. H= 7-7.2, D_m= 3.0, D_c= 3.003, VIIIB 18. Shepard, 1839. Dana, 6th ed. (1892), 490. Am. Min. 59 (1974), 79.

Danielsite, $(Cu,Ag)_{14}HgS_8$, [A] Orth. Metallic grey. Coppin Pool, Western Australia, Australia. H= 1.5-2, D_m=?, D_c= 6.54, IIA 08. Nickel, 1987. Am. Min. 73 (1988), 187.

Dannemorite, $Mn_2(Fe,Mg)_5Si_8O_{22}(OH)_2$, [G] Mon. Translucent yellowish-brown, greenish-grey. Dannemor, Sweden. H=?, D_m= 3.34, D_c= 3.64, VIIID 05a. Kenngott, 1855. JCPDS 23-302. See also Tirodite.

D'Ansite, $Na_{21}Mg(SO_4)_{10}Cl_3$, [G] Cub. Hall, Innsbruck (near), Tyrol, Austria. H=?, D_m= 2.655, D_c= 2.66, VIB 02. Autenrieth & Braune, 1958. Am. Min. 43(1958), 1221 (Abst.). Struct. Repts. 26 (1961), 453.

Daomanite, $CuPtAsS_2$, [Q] Orth. Metallic pale yellowish-green. Dao and Ma Districts, China. H= 3.5, D_m=?, D_c= 7.303, IIC 09. Yu et al., 1974. Am. Min. 65(1980), 408 (Abst.).

Daqingshanite-(Ce), $(Sr,Ca,Ba)_2(Ce,La)PO_4(CO_3,OH,F)_3$, [A] Rhom. Greasy/vitreous pale yellow. Bayan Obo, Inner Mongolia, China. H= 4.5, D_m= 3.81, D_c= 3.71, VbA 03b. Ren et al., 1983. Am. Min. 69(1984), 811 (Abst.). Min. Abstr. 86M/1443. See also Huntite.

Darapiosite, $(K,Na)_3Li(Mn,Zn)_2ZrSi_{12}O_{30}$, [G] Hex. Translucent colorless, white, brownish, bluish. Dara-Pioz, Tadzhikistan (N), USSR. H= 5, D_m= 2.92, D_c= 2.78, VIIIC 10. Semenov et al., 1975. Am. Min. 61(1976), 1053 (Abst.).

Darapskite, $Na_3(SO_4)(NO_3)\cdot H_2O$, [G] Mon. Transparent colorless. Atacama desert, Chile. H= 2.5, D_m= 2.20, D_c= 2.20, Va 03. Dietze, 1891. Am. Min. 55 (1970), 1500. Struct. Repts. 32A (1967), 332.

Datolite, $CaBSiO_4(OH)$, [G] Mon. Vitreous white, greyish, pale green, red, yellow, etc. H= 5-5.5, D_m= 3.0, D_c= 2.999, VIIIA 23. Esmark, 1806. Dana, 6th ed.(1892), 502. Am. Min. 58 (1973), 909.

Daubréeite, $BiO(OH,Cl)$, [G] Tet. Greasy/earthy creamy-white, greyish, yellowish-brown. Constancia mine, Tazna, Bolivia. H= 2-2.5, D_m= 6.5, D_c= 7.56, IIIC 16. Domeyko, 1876. Dana, 7th ed.(1951), v.2, 60. See also Bismoclite, Zavaritskite.

Daubréelite, $FeCr_2S_4$, [G] Cub. Metallic black; greenish-grey in reflected light. Coahuila Meteorite (Bolson de Mapimi), Coahuila, Mexico. H=?, D_m= 3.81,

D_c= 3.87, IIC 01. Smith, 1876. Dana, 7th ed.(1944), v.1, 265. Struct. Repts. 11 (1947), 289.

Davanite, $K_2TiSi_6O_{15}$, A Tric. Vitreous colorless. Murun massif, Davan Spring (near), Yakutiya (W), USSR. H= 5, D_m= 2.76, D_c= 2.725, VIIID 11. Lazebnik et al., 1984. Am. Min. 72 (1987), 1014. See also Dalyite.

Davidite-(Ce), $(Ce, La)(Y, U)(Ti, Fe)_{20}O_{38}$, A Rhom. Vitreous black. Tuftane, Iveland, Norway. H=?, D_m=?, D_c=?, IVC 06. Levinson, 1966. Am. Min. 51 (1966), 152. Am. Min. 64 (1979), 1010. See also Davidite-(La), Davidite-(Y).

Davidite-(La), $(La, Ce, Ca)(Y, U)(Ti, Fe)_{20}O_{38}$, G Rhom. Vitreous black. Radium Hill mine, Olary Province, South Australia, Australia. H=?, D_m= 4.42, D_c= 4.72, IVC 06. Mawson, 1906. Am. Min. 46 (1961), 700. Am. Min. 64 (1979), 1010. See also Davidite-(Ce), Davidite-(Y).

Davidite-(Y), $Y(Ti, Fe)_{21}O_{38}$, A Rhom. Vitreous black. Vishnevye Gory, Ural Mts., USSR. H=?, D_m=?, D_c=?, IVC 06. Levinson, 1966. Am. Min. 51 (1966), 152. See also Davidite-(La), Davidite-(Ce).

Davisonite, D Discredited as mixture of apatite and crandallite. Am. Min. 71 (1986), 1515. Larsen & Shannon, 1930.

Davreuxite, $MnAl_6Si_4O_{17}(OH)_2$, G Mon. Translucent creamy-white, pale rose. Ottrez, Ardennes, Belgium. H=?, D_m= 3.3, D_c= 3.364, VIIIB 21. De Koninck, 1878. Am. Min. 69 (1984), 777. Am. Min. 69 (1984), 783.

Davyne, $(Na, Ca, K)_8(Si_6Al_6)O_{24}(Cl, SO_4, CO_3)_{2-3}$, G Hex. Vitreous/pearly colorless, white. Monte Somma, Mt. Vesuvius, Napoli, Campania, Italy. H= 5.5, D_m= 2.4, D_c= 2.43, VIIIF 05. Monticelli & Covelli, 1825. Bull. Min. 91 (1968), 34.

Dawsonite, $NaAlCO_3(OH)_2$, G Orth. Vitreous/silky colorless, white. Siena, Toscana, Italy and/or Trenton limestone, McGill University, Montreal, Québec, Canada. H= 3, D_m= 2.44, D_c= 2.430, VbB 02. Harrington, 1874. Can. Min. 8 (1965), 377. N.Jb.Min.Mh. (1977), 381.

Dayingite, D Rejected by CNMMN in 1974. Am. Min. 67(1982), 1081 (Abst.).

Deerite, $(Fe, Mn)_6(Fe, Al)_3(Si_6O_{17})O_3(OH)_5$, A Mon. Black. Laytonville quarry, Laytonville (S), Mendocino Co., California, USA. H=?, D_m= 3.837, D_c= 3.82, VIIID 20. Agrell et al., 1964. Min. Mag. 43 (1979), 251. Am. Min. 62 (1977), 990.

Defernite, $Ca_3CO_3(OH, Cl)_4 \cdot H_2O$, A Orth. Vitreous colorless, red, rose-brown. Güneyce-Ikizdere, Trabzon, Pontides (E), Turkey. H=?, D_m= 2.34, D_c= 2.31, VbD 01. Sarp et al., 1980. Am. Min. 65(1980), 1066 (Abst.). Am. Min. 73 (1988), 888.

Delafossite, $CuFeO_2$, G Rhom. Metallic black; brown-white in reflected light. Sverdlovsk (Ekaterinburg), Ural Mts., USSR. H= 5.5, D_m= 5.41, D_c= 5.507, IVA 06. Friedel, 1873. Min. Mag. 35 (1966), 731. Struct. Repts. 37A (1971), 262. See also Mcconnellite.

Delhayelite, $(Na, K)_{10}Ca_5Al_6Si_{32}O_{80}Cl_6 \cdot 18H_2O$, G Orth. Colorless. Mt. Shaheru, Kivu, Zaïre. H=?, D_m= 2.60, D_c= 2.71, VIIIE 15. Sahama & Hytönen, 1959. Min. Mag. 32 (1959), 6.

Delindeite, $(Na, K)_{2.7}(Ba, Ca)_4Ti_6Si_8O_{26}(OH)_{14}$, A Mon. Resinous/pearly light pinkish-grey. Diamond Jo quarry, Magnet Cove, Hot Springs (near), Arkansas, USA. H=?, D_m= 3.3, D_c= 3.70, VIIIB 10. Appleman et al., 1987. Min. Mag. 51 (1987), 417.

Dellaite, $Ca_6Si_3O_{11}(OH)_2$, A Tric. Ardnamurchan, Kilchoan, Scotland. H=?, D_m=?, D_c= 2.97, VIIIA 05. Agrell, 1965. JCPDS 29-376.

Delrioite, $SrCaV_2O_6(OH)_2 \cdot 3H_2O$, G Mon. Vitreous/pearly pale yellow-green. Jo Dandy mine, Paradox Valley, Montrose Co., Colorado, USA. H= 2, D_m= 3.1, D_c= 3.16, VIID 32. Thompson & Sherwood, 1959. Am. Min. 55 (1970), 185.

Deltaite, D A mixture of crandallite and hydroxylapatite. Am. Min. 46 (1961), 467 (Abst.). Larsen & Shannon, 1930.

Delvauxite, $(Ca,Mg)(Fe,Al)_4(PO_4)_2(OH)_8 \cdot 4\text{–}5H_2O$, Q Amor. Translucent yellowish-brown, brownish-black, reddish. Berneau, Visé, Liège, Belgium. H= 2.5, D_m= 1.85, D_c=?, VIID 13. Dumont, 1838. Am. Min. 65(1980), 813 (Abst.).

Demesmaekerite, $Cu_5Pb_2(UO_2)_2(SeO_3)_6(OH)_6 \cdot 2H_2O$, A Tric. Translucent green. Musonoi, Shaba, Zaïre. H= 3-4, D_m= 5.28, D_c= 5.42, VIG 03. Cesbron et al., 1965. Am. Min. 51(1966), 1815 (Abst.). Acta Cryst. C39 (1983), 824.

Denisovite, $(K,Na)Ca_2Si_3O_8(F,OH)$, A Mon. Pearly greenish-grey. Eveslogchorr Mts. and Yukspor Mts., Khibina massif, Kola Peninsula, USSR. H= 4-5, D_m= 2.76, D_c= 2.81, VIIIE 16. Men'shikov, 1984. Am. Min. 70(1985), 1329.

Denningite, $(Ca,Mn)(Mn,Zn)Te_4O_{10}$, G Tet. Adamantine colorless/pale green. Moctezuma, Sonora, Mexico. H= 4, D_m= 5.05, D_c= 5.07, VIG 05. Mandarino et al., 1961. TMPM 10 (1965), 241.

Derbylite, $Fe_4Ti_3SbO_{13}(OH)$, G Mon. Resinous black. Tripuhy, Ouro Preto, Minas Gerais, Brazil. H= 5, D_m= 4.53, D_c= 4.798, IVC 10. Hussak & Prior, 1895. Can. Min. 21 (1983), 513. Am. Min. 62(1977), 396 (Abst.). See also Tomichite.

Derriksite, $Cu_4(UO_2)(SeO_3)_2(OH)_6$, A Orth. Translucent green. Musonoi, Shaba, Zaïre. H=?, D_m=?, D_c= 4.61, VIG 03. Cesbron et al., 1971. Am. Min. 57(1972), 1912 (Abst.). Acta Cryst. C39 (1983), 1605.

Dervillite, Ag_2AsS_2, G Mon. Metallic white. Ste.-Marie-aux-Mines, Vosges Mtns., Haut-Rhin, Alsace, France. H= 1.3, D_m=?, D_c= 5.62, IID 07. Weil, 1941. Am. Min. 68(1983), 1041 (Abst.).

Desautelsite, $Mg_6Mn_2CO_3(OH)_{16} \cdot 4H_2O$, A Rhom. Bright orange. Cedar Hill Quarry, Lancaster Co., Pennsylvania, USA. H= 2, D_m= 2.13, D_c= 2.10, IVF 03b. Dunn et al., 1979. Am. Min. 64 (1979), 127.

Descloizite, $PbZnVO_4(OH)$, G Orth. Greasy brownish-red, orange, blackish-brown. H= 3-3.5, D_m= 6.2, D_c= 6.202, VIIB 11b. Damour, 1854. Dana, 7th ed.(1951), v.2, 811. Acta Cryst. B35 (1979), 717. See also Mottramite.

Despujolsite, $Ca_3Mn(SO_4)_2(OH)_6 \cdot 3H_2O$, A Hex. Vitreous yellow. Tachgagalt, Morocco. H= 2.5, D_m= 2.46, D_c= 2.54, VID 06. Gaudefroy et al., 1968. Am. Min. 54(1969), 326 (Abst.). See also Fleischerite, Schaurteite.

Devilline, $CaCu_4(SO_4)_2(OH)_6 \cdot 3H_2O$, G Mon. Vitreous/pearly green, bluish-green. Cornwall, England. H= 2.5, D_m= 3.13, D_c= 3.06, VID 14. Pisani, 1864. Am. Min. 54(1969), 329 (Abst.). Acta Cryst. B28 (1972), 1182.

Deweylite, D A mixture of serpentine and other silicate minerals.. Min. Mag. 42 (1978), 75. Emmons, 1826.

Dewindtite, $Pb_2(UO_2)_4(PO_4)_3(OH)_3 \cdot 7H_2O$, G Orth. Translucent yellow. Kasolo, Shaba, Zaïre. H=?, D_m= 5.03, D_c= 5.06, VIID 21. Schoep, 1928. Am. Min. 39 (1954), 444.

Dhanrasite, D Unnecessary name for Sn-bearing garnet. Min. Mag. 38 (1971), 103. Murthy, 1967.

Diaboleite, $CuPb_2Cl_2(OH)_4$, G Tet. Transparent blue. Higher Pitts mine, Mendip Hills, Priddy, Somerset, England. H= 2.5, D_m= 5.42, D_c= 5.410, IIIC 04. Spencer, 1923. Min. Mag. 36 (1968), 933. Zeits. Krist. 134 (1971), 69.

Diadochite, $Fe_2(PO_4)(SO_4)(OH) \cdot 5H_2O$, G Tric. Earthy/sub-vitreous yellow, yellowish-brown, brown, etc. Ansbach, Gräfenthal, Saalfeld, Thüringen, Germany. H= 3-4, D_m= 2.2, D_c= 2.36, VIID 04. Breithaupt, 1837. JCPDS 24-528. See also Sarmientite.

Diamond, C, G Cub. Adamantine colorless, pale tints. H= 10, D_m= 3.51, D_c= 3.514, IB 02b. Pliny, 77. Dana, 7th ed.(1944), v.1, 146. Struct. Repts. 11 (1947), 187. See also Chaoite, Lonsdaleite, Graphite-2H, Graphite-3R.

Diaoyudaoite, $NaAl_{11}O_{17}$, A Hex. Vitreous colorless/light green. Diaoyudao Island (near), Taiwan. H= 7.6, D_m= 3.30, D_c= 3.21, IVC 05a. Shen et al., 1986. Acta Min. Sin. 6(3)(1986), 224.

Diaphorite, $Ag_3Pb_2Sb_3S_8$, G Mon. Metallic grey; white/greyish-white in reflected light. Vranciče, Příbram (near), Stŗedočský kraj, Čechy (Bohemia), Czechoslovakia and Braunsdorf, Freiberg (near), Sachsen (Saxony), Germany. H= 2.5-3, D_m= 6.04, D_c= 6.018, IID 07. Zepharovich, 1871. ZVMO 118 (1989), 47. Zts. Krist. 110 (1958), 169.

Diaspore, α-AlO(OH), G Orth. Vitreous white, colorless, greyish, brown, yellow, etc. Mramorskoi, Kossoïbrod (S), Ural Mts., USSR. H= 6.5-7, D_m= 3.4, D_c= 3.376, IVF 04. Haüy, 1801. Dana, 7th ed.(1944), v.1, 675. Min. Abstr. 80-2872. See also Böhmite, Bracewillite, Goethite.

Dickinsonite, $KNa_4Ca(Mn,Fe)_{14}Al(PO_4)_{12}(OH)_2$, G Mon. Vitreous olive/grass green, yellowish-green, brownish-green. Branchville, Fairfield Co., Connecticut, USA. H= 3.5-4, D_m= 3.41, D_c= 3.426, VIIA 06. Brush & Dana, 1878. Am. Min. 50 (1965), 1647. Am. Min. 66 (1981), 1034. See also Arrojadite.

Dickite, $Al_2Si_2O_5(OH)_4$, G Mon. Pearly/earthy white, grey, yellowish, brownish, bluish, etc. American Belle mine, San Juan Co., Colorado, USA. H= 2-2.5, D_m= 2.6, D_c= 2.600, VIIIE 10a. Ross & Kerr, 1930. Rev. Min. 19 (1988), 29. Acta Cryst. B27 (1971), 248. See also Halloysite, Kaolinite, Nacrite.

Dienerite, Ni_3As, Q Cub. Metallic greyish-white. Radstadt, Salzburg, Austria. H=?, D_m=?, D_c=?, IIG 03. Doelter, 1926. Dana, 7th ed.(1944), v.1, 175.

Dietrichite, $(Zn,Fe,Mn)Al_2(SO_4)_4 \cdot 22H_2O$, G Mon. Silky dirty white, brownish-yellow. Bořeň, Bilina (near), Čechy (Bohemia), Czechoslovakia. H= 2, D_m=?, D_c= 1.86, VIC 06. Schröckinger, 1878. JCPDS 25-1173.

Dietzeite, $Ca_2CrO_4(IO_3)_2$, G Mon. Transparent golden yellow. Atacama Desert, Antofagasta, Chile. H= 3.5, D_m= 3.62, D_c= 3.61, IVG 01. Osann, 1894. Dana, 7th ed.(1951), v.2, 318.

Digenite, Cu_9S_5, G Rhom. Blue/black; blue in reflected light. H= 2.5-3, D_m= 5.6, D_c= 5.715, IIA 01a. Breithaupt, 1844. Dana, 7th ed.(1944), v.1, 180. Am. Min. 48 (1963), 110.

Dimorphite I, α-As_4S_3, G Orth. Translucent adamantine orange-yellow. Solfatara, Phlegraean fields, Italy. H= 1.5, D_m= 2.58, D_c= 3.51, IIB 19. Scacchi, 1850. Dana, 7th ed.(1944), v.1, 197. Zts. Krist. 138 (1973), 161. See also Dimorphite II.

Dimorphite II, β-As_4S_3, P Orth. Translucent adamantine orange-yellow. Solfatara, Phlegraean fields, Italy. H=?, D_m=?, D_c= 3.60, IIB 19. Frankel & Zoltai, 1973. Dana, 7th ed.(1944), v.1, 197. Zts. Krist. 138 (1973), 161. See also Dimorphite I.

Diomignite, $Li_2B_4O_7$, A Tet. Tanco Pegmatite, Bernic Lake, Manitoba, Canada. H=?, D_m=?, D_c= 2.437, Vc 07b. London et al., 1987. Can. Min. 25 (1987), 173. Acta Cryst. 15 (1962), 190.

Diopside, $CaMgSi_2O_6$, G Mon. Vitreous/resinous white, yellowish, greyish, pale green. H= 5.5-6.5, D_m= 3.3, D_c= 3.278, VIIID 01a. d'Andrada, 1800. DHZ, 2nd ed.(1978), v.2A, 198. MSA Spec. Paper, 2 (1969), 31. See also Hedenbergite, Johannsenite.

Dioptase, $CuSiO_3 \cdot H_2O$, G Rhom. Vitreous green. Altyn-Tübe, Kirghese Steppes, Khirghesia, USSR. H= 5, D_m= 3.3, D_c= 3.296, VIIIC 09. Haüy, 1797. Zts. Krist. 187 (1989), 15. Am. Min. 62 (1977), 807.

Dittmarite, $(NH_4)MgPO_4 \cdot H_2O$, G Orth. Skipton Caves, Ballarat (near), Victoria, Australia. H=?, D_m=?, D_c= 2.19, VIIC 14. MacIvor, 1887. Am. Min. 57(1972), 1316 (Abst.).

Dixenite, $CuFeMn_{14}(AsO_4)(AsO_3)_5(SiO_4)_2(OH)_6$, G Rhom. Resinous/metallic black. Långban mine, Filipstad (near), Värmland, Sweden. H= 3-4, D_m= 4.20, D_c= 4.363, VIIB 10c. Flink, 1920. Zts. Krist. 127 (1968), 309. Am. Min. 66 (1981), 1263.

Dixeyite, [D] Inadequate data. Min. Mag. 33 (1962), 261. Marmo, 1959.

Djerfisherite, $K_6Na(Fe, Cu)_{24}S_{26}Cl$, [A] Cub. Submetallic khaki-olive. Kota-Kota meteorite, Marimba district, Malawi, and St. Mark's meteorite, St. Mark's Mission Station, Transkei, Cape Province, South Africa. H=?, D_m=?, D_c= 3.57, IIF 02. Fuchs, 1966. Am. Min. 64 (1979), 776. Min. Abstr. 81-1228. See also Thalfenisite.

Djurleite, $Cu_{31}S_{16}$, [G] Mon. Metallic grey. Barranca de Cobre, Chihuahua, Mexico. H= 2.5-3, D_m= 5.7, D_c= 5.740, IIA 01a. Roseboom, 1962. Am. Min. 47 (1962), 1181. Zts. Krist. 150 (1979), 299.

Dolerophanite, Cu_2OSO_4, [G] Mon. Brown/black. Mt. Vesuvius, Napoli, Campania, Italy. H= 3, D_m= 4.17, D_c= 4.156, VIB 07. Scacchi, 1873. Am. Min. 46 (1961), 146. Struct. Repts. 52A (1985), 307.

Dollaseite-(Ce), $Ca(Ce, La, Nd)Mg_2AlSi_3O_{11}(OH)F$, [A] Mon. Translucent brown. Östanmössa mine, Norberg, Sweden. H=?, D_m= 3.9, D_c= 4.01, VIIIB 15. Peacor & Dunn, 1988. Am. Min. 73 (1988), 838. Am. Min. 73 (1988), 838.

Dolomite, $CaMg(CO_3)_2$, [G] Rhom. Vitreous colorless, white, grey, greenish. H= 3.5-4, D_m= 2.85, D_c= 2.876, VbA 03a. Saussure, 1792. Rev. Min. 11 (1983). Am. Min. 62 (1977), 772. See also Ankerite, Kutnohorite.

Doloresite, $V_3O_4(OH)_4$, [G] Mon. Sub-metallic black; grey in reflected light. Dolores River, Colorado Plateau, Colorado, USA. H=?, D_m= 3.3, D_c= 3.436, IVF 07. Stern et al., 1957. Am. Min. 45 (1960), 1144. Acta Cryst. 11 (1958), 56.

Domeykite, α–Cu_3As, [G] Cub. Metallic white/grey. Algodones mines, Coquimbo (Calabozo), Chile. H= 3-3.5, D_m= 7.5, D_c= 7.92, IIG 01. Haidinger, 1845. Dana, 7th ed.(1944), v.1, 172. Zts. Krist. 145 (1977), 334.

Donathite, $(Fe, Mg)(Cr, Fe)_2O_4$, [A] Tet. Metallic black. Hestmandö Island, Norway. H= 6.5-7, D_m= 5, D_c= 5.08, IVB 02b. Seeliger & Mücke, 1969. Am. Min. 54(1969), 1218 (Abst.). See also Chromite.

Donbassite, $Al_2(Si_3Al)O_{10}(OH)_2 \cdot Al_{2.33}(OH)_6$, [G] Mon. Pearly white. Donetz basin, Ukraine, USSR. H= 2.5, D_m= 2.628, D_c= 2.658, VIIIE 09b. Lazarenko, 1940. Clays Cl. Mins. 37 (1989), 193.

Donharrisite, $Ni_9Hg_3S_9$, [A] Mon. Metallic brown; creamy white in reflected light. Erasmus mine, Leogang, Salzburg, Austria. H= 2, D_m=?, D_c= 5.18, IIA 08. Paar et al., 1989. Can. Min. 27 (1989), 257.

Donnayite-(Y), $NaSr_3CaY(CO_3)_6 \cdot 3H_2O$, [A] Tric. Vitreous yellow, colorless, white, grey, brown, etc. Mont Saint-Hilaire, Rouville Co., Québec, Canada. H= 3, D_m= 3.30, D_c= 3.266, VbC 03a. Chao et al., 1978. Can. Min. 16 (1978), 335. See also McKelveyite-(Y), Weloganite.

Donpeacorite, $(Mn, Mg)Mg(SiO_3)_2$, [A] Orth. Translucent pale buff. Balmat, St. Lawrence Co., New York, USA. H=?, D_m=?, D_c= 3.403, VIIID 02. Petersen et al., 1984. Am. Min. 69 (1984), 472. Am. Min. 69 (1984), 472.

Dorfmanite, $Na_2(PO_3OH) \cdot 2H_2O$, [A] Orth. Powdery white. Khibina massif and Lovozero massif, Kola Peninsula, USSR. H= 1-1.5, D_m= 1.99, D_c= 2.06, VIIC 15. Kapustin et al., 1980. Am. Min. 66(1981), 217 (Abst.).

Dorrite, $CaMgFe_2Al_2SiO_{10}$, [A] Tric. Sub-metallic dark brown. Durham Ranch, Gillette (S), Reno Junction (NE), Wyoming, USA. H= 5, D_m=?, D_c= 3.96, VIIID 07. Cosca et al., 1988. Am. Min. 73 (1988), 1440.

Dosulite, [D] Inadequate data. Min. Mag. 43 (1980), 1055. Yoshimura, 1967.

Douglasite, $K_2FeCl_4 \cdot 2H_2O$, [Q] Mon. Vitreous light green. Douglashall, Westerefeln (near), Stassfurt (NW), Sachsen (Saxony), Germany. H=?, D_m=?, D_c=?, IIIB 07. Precht, 1880. Dana, 7th ed.(1951), v.2, 100.

Downeyite, SeO_2, [A] Tet. Adamantine colorless. Forestville, Pennsylvania, USA. H=?, D_m=?, D_c= 4.146, IVD 06. Finkelman & Mrose, 1977. Am. Min. 62 (1977), 316.

Doyleite, $Al(OH)_3$, [A] Tric. Vitreous/pearly white, creamy-white, bluish-white. Mont Saint-Hilaire, Rouville Co., Québec, Canada. H= 2.5-3, D_m= 2.48, D_c= 2.482, IVF 01. Chao et al., 1985. Can. Min. 23 (1985), 21. See also Bayerite, Gibbsite, Nordstrandite.

Dravite, $NaMg_3Al_6(BO_3)_3Si_6O_{18}(OH)_4$, [G] Rhom. Vitreous black, brown, greenish-black. Drave, Carinthia, Austria. H= 7, D_m= 3.1, D_c= 3.122, VIIIC 08. Tschermak, 1883. DHZ, 2nd ed.(1986), v.1B, 559. See also Schorl, Elbaite.

Dresserite, $Ba_2Al_4(CO_3)_4(OH)_8 \cdot 3H_2O$, [A] Orth. Vitreous/silky white. Francon quarry, St.-Michel Dist., Montreal Island, Québec, Canada. H= 2.5-3, D_m= 2.96, D_c= 3.06, VbD 02. Jambor et al., 1969. Can. Min. 10 (1969), 84. See also Strontiodresserite.

Dreyerite, $BiVO_4$, [A] Tet. Adamantine orange-yellow, brownish-yellow. Hirschhorn, Rheinland-Pfalz, Germany. H= 2-3, D_m=?, D_c= 6.25, VIIA 13. Dreyer & Tillmanns, 1981. Am. Min. 67(1982), 622 (Abst.). N.Jb.Min.Mh. (1981), 151. See also Clinobisvanite, Pucherite.

Drugmanite, $Pb_2(Fe,Al)(PO_4)(PO_3OH)(OH)_2$, [A] Mon. Adamantine pale yellow. Richelle, Liège (N), Belgium. H= < 6, D_m=?, D_c= 5.55, VIIC 17. van Tassel et al., 1979. Min. Mag. 43 (1979), 463. Bull. Min. 111 (1988), 431.

Drysdallite, $Mo(Se,S)_2$, [A] Hex. Metallic white/grey. Solwezi (SE), Kapijimpanga, Northwestern Province, Zambia. H= 2, D_m=?, D_c= 6.25, IIC 10. Čech et al., 1973. Am. Min. 59(1974), 1139 (Abst.). See also Molybdenite, Tungstenite.

Dufrenite, $Ca_{0.5}Fe_6(PO_4)_4(OH)_6 \cdot 2H_2O$, [G] Mon. Vitreous/silky dark green, olive-green, olive-brown. Anglar (near), Dept. of Haute Vienne, France; Hirschberg, Westphalia, Germany. H= 3.5-4.5, D_m= 3.2, D_c= 3.407, VIID 18. Brongniart, 1833. Dana, 7th ed.(1951), v.2, 873. Am. Min. 55 (1970), 135. See also Burangaite, Natrodufrenite.

Dufrénoysite, $Pb_2As_2S_5$, [G] Mon. Metallic grey; white in reflected light. Lengenbach quarry, Binntal, Valais (Wallis), Switzerland. H= 3, D_m= 5.53, D_c= 5.65, IID 02. Damour, 1845. Dana, 7th ed.(1944), v.1, 442. Zts. Krist. 130 (1969), 15.

Duftite, $PbCuAsO_4(OH)$, [G] Orth. Vitreous olive-green, grey-green. Tsumeb, Namibia. H= 3, D_m= 6.40, D_c= 6.57, VIIB 11b. Pufahl, 1920. Bull. Min. 79 (1956), 7. Zts. Krist. 185 (1988), 610.

Dugganite, $Pb_3(Zn,Cu)_3(TeO_6)(AsO_4)(OH)_3$, [A] Hex. Adamantine white, green. Emerald mine, Tombstone, Cochise Co., Arizona, USA. H= 3, D_m= 6.33, D_c= 6.33, VIG 10. Williams, 1978. Am. Min. 63 (1978), 1016.

Duhamelite, $Cu_4Pb_2Bi(VO_4)_4(OH)_3 \cdot 8H_2O$, [A] Orth. Translucent greenish-yellow, olive-green. Lousy Gulch, Payson (near), Gila Co., Arizona, USA. H= 3, D_m= 5.80, D_c= 5.99, VIID 30. Williams, 1981. Min. Mag. 44 (1981), 151.

Dumontite, $Pb_2(UO_2)_3(PO_4)_2O_2 \cdot 5H_2O$, [G] Mon. Translucent yellow. Shinkolobwe, Shaba, Zaïre. H=?, D_m= 5.65, D_c= 5.66, VIID 22. Schoep, 1924. Bull. Min. 81 (1958), 63. Bull. Min. 111 (1988), 439. See also Hügelite.

Dumortierite, $Al_{27}B_4Si_{12}O_{69}(OH)_3$, [G] Orth. Vitreous blue, greenish-blue. Beaunan (near), France. H= 7, D_m= 3.265, D_c= 3.340, VIIIA 24. Gonnard, 1881. Am. Min. 71 (1986), 786. Struct. Repts. 44A(1978), 305. See also Holtite.

Dundasite, $PbAl_2(CO_3)_2(OH)_4 \cdot H_2O$, [G] Orth. Vitreous/silky white. Dundas, Tasmania, Australia. H= 2, D_m= 3.25, D_c= 3.573, VbD 02. Petterd, 1893. Bull. Min. 83 (1960), 121. Min. Mag. 38 (1972), 564.

Durangite, $NaAlAsO_4F$, [G] Mon. Vitreous orange-red, light/dark green. Durango, Mexico. H= 5, D_m= 4.0, D_c= 3.616, VIIB 11a. Brush, 1869. Dana, 7th ed.(1951), v.2, 829. Zts. Krist. 99 (1938), 38. See also Lacroixite.

Duranusite, As_4S, [A] Orth. Metallic greyish-white. Duranus, Alpes-Maritimes, France. H= 2, D_m=?, D_c= 4.50, IB 01. Johan et al., 1973. Bull. Min. 96 (1973), 131.

Dussertite, $BaFe_3(AsO_4)(AsO_3OH)(OH)_6$, [G] Rhom. Translucent green. Djebel Debar, Qacentina (Constantine), Algeria. H= 3.5, D_m= 3.75, D_c= 4.09, VIIB 15a. Barthoux, 1925. JCPDS 35-621.

Duttonite, $VO(OH)_2$, [G] Mon. Transparent pale brown. Peanut mine, Montrose Co., Colorado, USA. H= 2.5, D_m=?, D_c= 3.24, IVF 07. Thompson et al., 1957. Am. Min. 42 (1957), 455. Acta Cryst. 11A (1958), 56.

Dwornikite, $(Ni,Fe)SO_4 \cdot H_2O$, [A] Mon. Translucent white. Minasragra, Junin, Peru. H=?, D_m=?, D_c= 3.34, VIC 01. Milton et al., 1982. Min. Mag. 46(1982), 351.

Dypingite, $Mg_5(CO_3)_4(OH)_2 \cdot 5H_2O$, [A] Syst=? Pearly white. Dypingdal, Snarum, Norway (S). H=?, D_m=?, D_c= 2.15, VbD 01. Raade, 1970. Am. Min. 55 (1970), 1457.

Dyscrasite, Ag_3Sb, [G] Orth. Metallic white. Wenzelgang, Wolfach (near), Baden, Germany. H= 3.5-4, D_m= 9.712, D_c= 9.720, IIG 02. Beudant, 1832. Dana, 7th ed.(1944), v.1, 173. Can. Min. 14 (1976), 139.

Dzhalindite, $In(OH)_3$, [G] Cub. Translucent yellow-brown. Dzhalind ore deposit, Little Khingan Ridge, Siberia, USSR. H=?, D_m=?, D_c= 4.34, IVF 06. Genkin & Murav'eva, 1963. Am. Min. 49(1964), 439 (Abst.). See also Sohngeite.

Dzhezkazganite, [D] Inadequate data. Min. Mag. 36 (1967), 133. Poplavko et al., 1962.

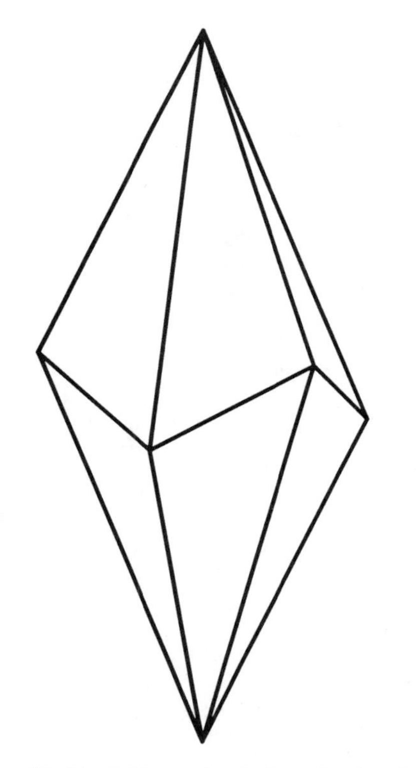

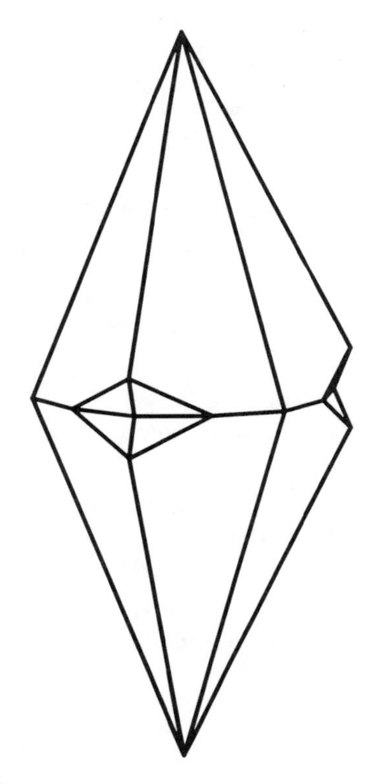

(Left) Calcite scalenohedron showing the form $\{21\bar{3}1\}$.

(Right) Calcite contact twin having a composition plane $\{0001\}$ and the form $\{21\bar{3}1\}$.

E

Eakerite, $Ca_2SnAl_2Si_6O_{18}(OH)_2 \cdot 2H_2O$, [A] Mon. Translucent colorless, white. Foote Mineral Company spodumene mine, Kings Mt. (near), Cleveland Co., North Carolina, USA. H= 5.5, D_m= 2.93, D_c= 2.65, VIIIE 27. Leavens et al., 1970. Am. Min. 56(1971), 637 (Abst.). Am. Min. 61 (1976), 956.

Earlandite, $Ca_3(C_6H_5O_7)_2 \cdot 4H_2O$, [G] Mon. Translucent white, pale yellow. Weddell Sea, Antarctica. H=?, D_m= 1.95, D_c= 1.96, IXA 02. Bannister, 1936. JCPDS 28-2003.

Earlshannonite, $(Mn,Fe)Fe_2(PO_4)_2(OH)_2 \cdot 4H_2O$, [A] Mon. Vitreous reddish-brown. Foote Mineral Company spodumene mine, Kings Mt. (near), Cleveland Co., North Carolina, USA; Hagendorf pegmatite, Waidhaus, Oberpfalz, Bayern (Bavaria), Germany. H= 3-4, D_m= 2.90, D_c= 2.92, VIID 27. Peacor et al., 1984. Can. Min. 22 (1984), 471. See also Arthurite, Ojuelaite, Whitmoreite.

Ecandrewsite, $(Zn,Fe,Mn)TiO_3$, [A] Rhom. Sub-metallic dark brown/black; greyish-white in refl. light. Melbourne Rockwell mine, Little Broken Hill, Broken Hill (near), New South Wales, Australia. H= 5.5, D_m=?, D_c= 4.99, IVC 04b. Birch et al., 1988. Min. Mag. 52 (1988), 237.

Ecdemite, $Pb_6As_2O_7Cl_4$, [G] Tet. Vitreous greenish-yellow, yellow. Långban mine, Filipstad (near), Värmland, Sweden. H= 2.5-3, D_m= 7.14, D_c= 7.32, IIID 03. Nordenskiöld, 1877. JCPDS 23-343.

Eckermannite, $Na_3(Mg,Fe)_4AlSi_8O_{22}(OH)_2$, [G] Mon. Vitreous dark bluish-green. Norra Kärr, Gränna, Sweden. H= 5-6, D_m= 3.00, D_c= 2.97, VIIID 05d. Adamson, 1942. JCPDS 20-386. See also Ferro-eckermannite.

Eclarite, $(Cu,Fe)Pb_9Bi_{12}S_{28}$, [A] Orth. Metallic whitish-grey. Hollersbach Tal, Barenbad, Salzburg, Austria. H= 2.5-3.5, D_m= 6.85, D_c= 6.88, IID 08. Paar et al., 1983. Am. Min. 70(1985), 214 (Abst.). Min. Abstr. 84M/3840.

Edenite, $NaCa_2(Mg,Fe)_5(Si_7Al)O_{22}(OH)_2$, [G] Mon. Translucent green, black. Edenville, New York, USA. H= 5-6, D_m= 3.06, D_c= 3.06, VIIID 05b. Glocker, 1839. JCPDS 23-1405. See also Ferro-edenite.

Edgarbaileyite, $Hg_6Si_2O_7$, [A] Mon. Vitreous/resinous yellow, olive-green, dark green-brown. Socrates mine, San Benito Co., California, USA. H= 4, D_m= 9.4, D_c= 9.11, VIIIA 30. Roberts et al., 1990. Min. Rec. 8 (1990), 215.

Edingtonite (orthorhombic), $Ba(Al_2Si_3)O_{10} \cdot 4H_2O$, [P] Orth. Translucent colorless, white, pink, brown. Böhlet, Västergötland, Sweden. H=?, D_m= 2.78, D_c= 2.80, VIIIF 10. Haidinger, 1825. DHZ (1963), v.4, 359. Acta Cryst. B32 (1976), 1623.

Edingtonite (tetragonal), $Ba(Al_2Si_3)O_{10} \cdot 3.5H_2O$, [G] Tet. Translucent colorless, white, pink, brown. Old Kilpatrick, Dumbartonshire, Scotland. H=?, D_m= 2.7, D_c= 2,764, VIIIF 10. Haidinger, 1825. N.Jb.Min.Mh. (1984), 1333. Am. Min. 70(1985), 1333 (Abst.). See also Edingtonite (orthorhombic).

Efremovite, $(NH_4)_2Mg_2(SO_4)_3$, [A] Cub. Dull grey/white. Chelyabinsk coal basin, Ural Mts. (S), USSR. H= 2, D_m=?, D_c= 2.52, VIA 02. Shcherbakova & Bazhenova, 1989. ZVMO 118(3) (1989), 84.

Eggletonite, $Na_2Mn_8(Si,Al)_{12}O_{29}(OH)_7 \cdot 11H_2O$, [A] Mon. Vitreous dark/golden brown. Big Rock Quarry, Little Rock, Pulaski Co., Arkansas, USA. H= 3-4, D_m= 2.76, D_c= 2.76, VIIIE 07b. Peacor et al., 1984. Min. Mag. 48 (1984), 93. See also Ganophyllite.

Eglestonite, $Hg_6OCl_3(OH)$, [G] Cub. Adamantine/resinous yellow, orange-yellow, brown. Terlingua, Brewster Co., Texas. H= 2.5, D_m= 8.4, D_c= 8.65, IIIC 03. Moses, 1903. Am. Min. 62(1977), 396 (Abst.). Am. Min. 62(1977), 396 (Abst.).

Ehrleite, $Ca_2ZnBe(PO_4)_2(PO_3OH) \cdot 4H_2O$, [A] Tric. Vitreous white, colorless. Tip Top Pegmatite, Custer (near), Custer Co., South Dakota, USA. H= 3.5, D_m= 2.64,

D_c= 2.996, VIIC 09. Robinson et al., 1985. Can. Min. 23 (1985), 507. Can. Min. 25 (1987), 767.

Eifelite, $KNa_2Mg_{4.5}Si_{12}O_{30}$, A Hex. Vitreous colorless, light green/yellow. Ettringer Bellerberg, Mayen, Eifel, Rheinland-Pfalz, Germany. H=?, D_m=?, D_c= 2.67, VIIIC 10. Abraham et al., 1980. Am. Min. 66(1981), 218 (Abst.). Contr.Min.Pet. 82(1983), 252.

Eitelite, $Na_2Mg(CO_3)_2$, G Rhom. Transparent colorless. Green River Formation, Duchesne Co., Utah, USA. H= 3.5, D_m= 2.737, D_c= 2.732, VbA 05. Milton et al., 1954. Am. Min. 40 (1955), 326. Am. Min. 58 (1973), 211. See also Roedderite, Bütschliite.

Ekanite, $Ca_2ThSi_8O_{20}$, G Tet. Vitreous yellow, dark red, green. Ratnapura Dist., Eheliyagoda, Sri Lanka. H= 4.5, D_m= 3.08, D_c= 3.36, VIIIE 17. Anderson et al., 1961. Am. Min. 46(1961), 1516 (Abst.). Can. Min. 20 (1982), 65. See also Iraqite, Steacyite.

Ekaterinite, $Ca_2B_4O_7(Cl,OH)_2 \cdot 2H_2O$, A Hex. Pearly white. Korshunovskoe deposit, Irkutsk, Siberia, USSR. H= 1, D_m= 2.44, D_c= 2.23, Vc 07b. Malinko et al., 1980. Am. Min. 66(1981), 437 (Abst.).

Ekmanite, $(Fe,Mg)_3(Si,Al)_4O_{10}(OH)_2 \cdot 2H_2O$, Q Orth. Translucent green, greyish-white, black. Brunsjögruvan, Grythyttan, Örebro, Sweden. H=?, D_m=?, D_c=?, VIIIE 07b. Igelström, 1865. Strunz (1970), 443.

Elbaite, $Na(Li,Al)_3Al_6(BO_3)_3Si_6O_{18}(OH)_4$, G Rhom. Vitreous blue, green, yellow, red, colorless. Elba, Toscana, Italy. H= 7, D_m= 3.05, D_c= 3.13, VIIIC 08. Vernadsky, 1913. DHZ, 2nd ed.(1986), v.1B, 559. Struct. Repts. 39A (1973), 340. See also Dravite.

Electrum, (Au,Ag), G Cub. Metallic yellow to white depending on Au:Ag ratio. H= 2.5-3, D_m=?, D_c=?, IA 01a. Pliny, 77. Min. Abstr. 88M/4285. See also Gold, Silver.

Ellenbergerite, $Mg_6TiAl_6Si_8O_{28}(OH)_{10}$, A Hex. Vitreous purple, lilac. Parigi, Martiniana Po, Doria Maria Massif, Piemonte, Italy. H= 6.5, D_m= 3.15, D_c= 3.13, VIIIA 24. Chopin et al., 1986. Contr.Min.Petr. 92 (1986), 316.

Ellisite, Tl_3AsS_3, A Rhom. Metallic grey; light purplish-grey in reflected light. Carlin deposit, Elko (NW), Eureka Co., Nevada, USA. H= 2, D_m= 7.18, D_c= 7.043, IID 14. Dickson et al., 1979. Am. Min. 64 (1979), 701. Zts. Krist. 151 (1980), 249.

Ellweilerite, D = sodium uranospinite. Min. Mag. 33 (1962), 261. Bultemann, 1960.

Elpasolite, K_2NaAlF_6, G Cub. Vitreous/greasy colorless. Cincinnati mine, St. Peter's Dome, Pikes Peak Region, El Paso Co., Colorado, USA. H= 2.5, D_m= 2.995, D_c= 3.008, IIIB 03. Cross & Hillebrand, 1885. Dana, 7th ed.(1951), v.2, 114. N.Jb.Min.Mh. (1987), 481.

Elpidite, $Na_2ZrSi_6O_{15} \cdot 3H_2O$, G Orth. Translucent white, brick-red. Narsarsuk, Greenland (S). H= 7, D_m= 2.54, D_c= 2.59, VIIID 11. Lindstrom, 1894. Dana/Ford (1932), 581. Am. Min. 58 (1973), 106.

Elroquite, D A mixture of quartz and ferrian variscite. Can. Min. 7 (1963), 676. Shepard, 1877.

Elyite, $CuPb_4SO_4(OH)_8$, A Mon. Silky violet. Silver King mine, Ward, White Pine Co., Nevada, USA. H= 2, D_m= 6, D_c= 6.321, VIB 06. Williams, 1972. Am. Min. 57 (1972), 364.

Embreyite, $Pb_5(CrO_4)_2(PO_4)_2 \cdot H_2O$, A Mon. Dull/resinous orange. Berezov, Sverdlovsk, Ural Mts., USSR. H= 3.5, D_m= 6.45, D_c= 6.41, VIID 30. Williams, 1972. Min. Mag. 38 (1972), 790.

Emeleusite, $Na_2LiFeSi_6O_{15}$, A Orth. Vitreous colorless, creamy-pinkish. Igdlutalik, Julianehåb, Greenland (S). H= 5-6, D_m= 2.76, D_c= 2.81, VIIID 14. Upton et al., 1978. Min. Mag. 42 (1978), 31. Zts. Krist. 147 (1978), 297.

Emmonsite, $Fe_2(TeO_3)_3 \cdot 2H_2O$, [G] Tric. Vitreous yellowish-green. Tombstone, Cochise Co., Arizona, USA. H= 5, D_m= 4.549, D_c= 4.719, VIG 02. Hillebrand, 1885. Min. Rec. 3 (1972), 82. Struct. Repts. 39A (1973), 323.

Emplectite, $CuBiS_2$, [G] Orth. Metallic white. Johanngeorgenstadt, Sachsen (Saxony), Germany. H= 2, D_m= 6.38, D_c= 6.393, IID 06b. Kengott, 1855. Dana, 7th ed.(1944), v.1, 435. Zts. Krist. 141 (1975), 387. See also Chalcostibite.

Empressite, AgTe, [A] Orth. Metallic pale bronze. Empress Josephine mine, Bonanza, Kerber Creek Dist., Saguache Co., Colorado, USA. H= 3-3.5, D_m= 7.51, D_c= 7.61, IIB 05. George, 1914. Am. Min. 49 (1964), 325.

Enargite, Cu_3AsS_4, [G] Orth. Metallic greyish-black; grey/rose-brown in reflected light. H= 3, D_m= 4.45, D_c= 4.40, IIB 03a. Breithaupt, 1850. Dana, 7th ed.(1944), v.1, 389. Acta Cryst. B26 (1970), 1878. See also Luzonite.

Endellite, $Al_2Si_2O_5(OH)_4 \cdot 2H_2O$, [G] Mon. Porcelanous white. Djebal Deber, Libya. H=?, D_m=?, D_c= 2.12, VIIIE 11. Alexander et al., 1943. Am. Min. 40 (1955), 1110. See also Halloysite.

Englishite, $K_3Na_2Ca_{10}Al_{15}(PO_4)_{21}(OH)_7 \cdot 26H_2O$, [G] Mon. Vitreous/pearly colorless. Clay Canyon, Fairfield (near), Utah Co., Utah, USA. H= 3, D_m= 2.68, D_c= 2.69, VIID 15c. Larsen & Shannon, 1930. Can. Min. 22 (1984), 469. Min. Mag. 40 (1976), 469.

Enstatite, $MgSiO_3$, [G] Orth. Pearly/vitreous colorless, grey, green, yellow, brown. H= 5-6, D_m= 3.209, D_c= 3.204, VIIID 02. Kenngott, 1855. DHZ, 2nd ed. (1978), v.2A, 20. Zts. Krist. 176 (1986), 159. See also Orthoferrosilite, Hypersthene, Clinoenstatite.

Eosphorite, $(Mn,Fe)AlPO_4(OH)_2 \cdot H_2O$, [G] Orth. Vitreous/resinous pink, rose-red. Branchville, Fairfield Co., Connecticut, USA. H= 5, D_m= 3.06, D_c= 3.04, VIID 07. Brush & Dana, 1878. Am. Min. 35 (1950), 793. Acta Cryst. 13 (1960), 384. See also Childrenite.

Ephesite, $NaLiAl_2(Si_2Al_2)O_{10}(OH)_2$, [G] Tric. Translucent pink. Gumuch-dagh, Ephesus, Izmir (near), Turkey. H=?, D_m=?, D_c= 2.95, VIIIE 06. Smith, 1851. Am. Min. 52 (1967), 1689. N.Jb.Min.Mh. (1987), 275.

Epididymite, $NaBeSi_3O_7(OH)$, [G] Orth. Transparent colorless. Narsarsuk, Greenland (S). H= 5.5, D_m= 2.55, D_c= 2.559, VIIID 26. Flink, 1893. Dana/Ford (1932), 535. Am. Min. 55 (1970), 1541. See also Eudidymite.

Epidote, $Ca_2FeAl_2(Si_2O_7)(SiO_4)(O,OH)_2$, [G] Mon. Vitreous yellowish-green, green, brownish-green, black. H= 6, D_m= 3.4, D_c= 3.469, VIIIB 15. Haüy, 1801. DHZ, 2nd ed.(1986), v.1B, 44. Am. Min. 56 (1971), 447. See also Clinozoisite.

Epigenite, [D] A mixture of tennantite, chalcopyrite and pyrite. Min. Mag. 47 (1983), 411. Sandberger, 1869.

Epistilbite, $NaCa_3(Al_6Si_{18})O_{48} \cdot 16H_2O$, [G] Mon. Vitreous colorless, white, yellowish. Isle of Skye, Scotland and/or Berufjord, Iceland. H= 4-4.5, D_m= 2.25, D_c= 2.282, VIIIF 12. Rose, 1826. Am. Min. 59 (1974), 1055. Zts. Krist. 173 (1985), 257. See also Goosecreekite.

Epistolite, $Na_5TiNb_2(Si_2O_7)_2(O,F)_4 \cdot 5H_2O$, [G] Tric. Translucent white, greyish, brownish. Narsarsuk, Greenland (S). H= 1-1.5, D_m= 2.9, D_c= 3.00, VIIIB 11. Bøggild, 1900. N.Jb.Min. Abh. (1986), 289.

Epsomite, $MgSO_4 \cdot 7H_2O$, [G] Orth. Vitreous/silky/earthy colorless, pink, greenish. Epsom, Surrey, England. H= 2-2.5, D_m= 1.677, D_c= 1.67, VIC 03d. Delamétherie, 1806. Dana, 7th ed.(1951), v.2, 509. Acta Cryst. B40 (1984), 218. See also Goslarite, Morenosite.

Erdite, $NaFeS_2 \cdot 2H_2O$, [A] Mon. Metallic copper-red. Coyote Peak, Humboldt Co., California, USA. H= 1.8, D_m=?, D_c= 2.22, IIB 17. Czamanske et al., 1980. Am. Min. 65 (1980), 509. Am. Min. 65 (1980), 516.

Ericaite, $(Fe,Mg,Mn)_3B_7O_{13}Cl$, [G] Orth. Translucent light green, red, black. Wathlingen-Hänigsen, Hannover (NE), Germany. H=?, D_m= 3.2, D_c=?, Vc 14. Werner, 1950. Am. Min. 41(1956), 372 (Abst.). See also Boracite, Congolite, Chambersite.

Ericssonite, $BaFeMn_2O(Si_2O_7)(OH)$, [A] Mon. Translucent reddish-black. Långban mine, Filipstad (near), Värmland, Sweden. H= 4.5, D_m= 4.21, D_c= 4.38, VIIIB 10. Moore, 1971. Am. Min. 56(1971), 2157 (Abst.). See also Orthoericssonite.

Eriochalcite, $CuCl_2 \cdot 2H_2O$, [G] Orth. Vitreous blue-green. Mt. Vesuvius, Napoli, Campania, Italy. H= 2.5, D_m= 2.47, D_c= 2.52, IIIC 10b. Scacchi, 1884. NBS Monogr. 18 (1981). Zts. Krist. 189 (1989), 13.

Erionite, $K_2NaCa_{1.5}Mg(Al_8Si_{28})O_{72} \cdot 28H_2O$, [G] Hex. Translucent white. Durkee, Baker Co., Oregon, USA. H=?, D_m= 2.070, D_c= 2.095, VIIIF 15. Eakle, 1898. Natural Zeolites (1985), 200. Bull. Min. 92 (1969), 250.

Erlianite, $Fe_6Si_6O_{15}(OH,O)_8$, [A] Orth. Silky black. Harhada deposit, Jining-Erlian railway (near), Inner Mongolia, China. H= 3.7, D_m= 3.11, D_c= 3.11, VIIIG 01. Feng & Yang, 1986. Min. Mag. 50 (1986), 285.

Erlichmanite, OsS_2, [A] Cub. Metallic white. MacIntosh mine, Trinity River, Willow Creek, Humboldt Co., California, USA. H=?, D_m=?, D_c= 9.59, IIC 05. Snetsinger, 1971. Am. Min. 56 (1971), 1501. See also Laurite.

Ernstite, $(Mn,Fe)AlPO_4(OH,O)_2$, [A] Mon. Translucent yellow-brown. Karibib (near), Davib-East, Namibia. H= 3-3.5, D_m= 3.07, D_c= 3.086, VIID 07. Seeliger & Mücke, 1970. Am. Min. 56(1971), 637 (Abst.).

Ertixiite, $Na_2Si_4O_9$, [A] Cub. Vitreous colorless, white. Altai mine, Ürümqi City, Fuyun Co., Xinjiang, China. H= 5.8-6.5, D_m= 2.35, D_c= 2.34, VIIIA 28. Zhang et al., 1985. Min. Abstr. 86M/2251.

Erythrite, $Co_3(AsO_4)_2 \cdot 8H_2O$, [G] Mon. Adamantine/pearly/earthy red, pink. Schneeberg, Sachsen (Saxony), Germany. H= 1.5-2.5, D_m= 3.09, D_c= 3.14, VIIC 10. Beudant, 1832. JCPDS 33-413. See also Annabergite, Hörnesite.

Erythrosiderite, $K_2FeCl_5 \cdot H_2O$, [G] Orth. Vitreous red. Mt. Vesuvius, Napoli, Campania, Italy. H=?, D_m= 2.372, D_c= 2.300, IIIB 07. Scacchi, 1872. Dana, 7th ed.(1951), v.2, 101. See also Kremersite.

Eskebornite, $CuFeSe_2$, [G] Tet. Metallic brown. Eskeborn adit, Harz, Germany. H= 2.5, D_m=?, D_c= 5.474, IIB 02. Ramdohr, 1949. N.Jb.Min.Mh. (1988), 337. See also Chalcopyrite.

Eskimoite, $Ag_7Pb_{10}Bi_{15}S_{36}$, [A] Mon. Ivigtut, Greenland (SW). H= 3.6, D_m=?, D_c= 7.103, IID 09b. Makovicky & Karup-Møller, 1977. Am. Min. 64(1979), 243 (Abst.).

Eskolaite, Cr_2O_3, [G] Rhom. Black; grey in reflected light. Outokumpu, Finland. H= 9, D_m= 5.18, D_c= 5.218, IVC 04a. Kouvo & Vuorelainen. Am. Min. 43 (1958), 1098.

Esperite, $Ca_3PbZn_4(SiO_4)_4$, [A] Mon. Greasy white. Franklin, Sussex Co., New Jersey, USA. H= 5+, D_m= 4.28, D_c= 4.25, VIIIA 02. Moore & Ribbe, 1965. Am. Min. 50 (1965), 1170. See also Larsenite.

Esseneite, $CaFe(AlSi)O_6$, [A] Mon. Vitreous reddish-brown. Durham Ranch, Gillette (S), Reno Junction (NE), Wyoming, USA. H= 6, D_m=?, D_c= 3.54, VIIID 01a. Cosca & Peacor, 1987. Am. Min. 72 (1987), 148. Am. Min. 72 (1987), 148.

Ettringite, $Ca_6Al_2(SO_4)_3(OH)_{12} \cdot 26H_2O$, [G] Rhom. Transparent colorless. Ettringer Bellerberg, Ettringen, Mayen, Eifel, Rheinland-Pfalz, Germany. H= 2-2.5, D_m= 1.77, D_c= 1.77, VID 07. Lehman, 1874. Am. Min. 45 (1960), 1137. Acta Cryst. B26 (1970), 386.

Eucairite, $CuAgSe$, [G] Orth. Metallic white/grey. Skutterum copper mine, Smaland, Sweden. H= 2.5, D_m= 7.7, D_c= 7.91, IIA 04. Berzelius, 1818. Dana, 7th ed.(1944), v.1, 183. Zts. Krist. 108 (1957), 389.

Euchlorine, $KNaCu_3O(SO_4)_3$, G Mon. Translucent green. Mt. Vesuvius, Napoli, Campania, Italy. H=?, D_m= 3.27, D_c= 3.28, VIB 07. Scacchi, 1869. N.Jb.Min.Mh. (1989), 541.

Euchroite, $Cu_2AsO_4(OH)\cdot 3H_2O$, G Orth. Vitreous green. Lubietová (Libethen), Banská Bystrica (Neusohl), Středoslovenský kraj, Slovensko (Slovakia), Czechoslovakia. H= 3.5-4, D_m= 3.389, D_c= 3.47, VIID 02. Breithaupt, 1823. Am. Min. 36 (1951), 484. Acta Cryst. C45 (1989), 1479.

Euclase, $BeAlSiO_4(OH)$, G Mon. Vitreous pale green, blue. H= 7.5, D_m= 3.095, D_c= 3.115, VIIIA 13. Haüy, 1792. Dana/Ford (1932), 619. Struct. Repts. 27 (1962), 579.

Eucryptite, $LiAlSiO_4$, G Rhom. Translucent colorless, white. Branchville, Fairfield Co., Connecticut, USA. H=?, D_m= 2.667, D_c= 2.66, VIIIA 01. Brush & Dana, 1880. Dana, 6th ed. (1892), 426. Zts. Krist. 172 (1985), 147.

Eudialyte, $Na_{16}Ca_6Fe_3Zr_3(Si_3O_9)_2(Si_9O_{27})_2(OH,Cl)_4$, G Rhom. Vitreous rose/bluish/brownish-red, brown. Kangerdluarsuk, Ilímaussaq, Greenland (S). H= 5-5.5, D_m= 2.92, D_c= 2.77, VIIIC 03. Stromeyer, 1819. DHZ, 2nd ed.(1986), v.1B, 348. Min. Zhurn. 10(1) (1988), 48.

Eudidymite, $Na_2Be_2Si_6O_{15}\cdot H_2O$, G Mon. Vitreous white. Ovre-Arö island, Langesundfjord, Norway. H= 6, D_m= 2.553, D_c= 2.566, VIIID 26. Brögger, 1887. Can. Min. 7 (1963), 643. Am. Min. 57 (1972), 1345. See also Epididymite.

Eugsterite, $Na_4Ca(SO_4)_3\cdot 2H_2O$, A Mon. Transparent colorless. Konya Basin, Turkey; also Kenya. H=?, D_m=?, D_c=?, VIC 12. Vergouwen, 1981. Am. Min. 66 (1981),632.

Eulytite, $Bi_4(SiO_4)_3$, G Cub. Translucent dark brown, grey, yellow, colorless. Rotor mine, Schneeberg, Sachsen (Saxony), Germany. H= 4.5, D_m= 6.1, D_c= 6.76, VIIIA 09. Breithaupt, 1827. Dana/Ford (1932), 591. Zts. Krist. 123 (1966), 73.

Euxenite-(Y), $(Y,Ca,Ce,U,Th)(Nb,Ta,Ti)_2O_6$, G Orth. Submetallic/greasy black. Jölster, Söndfjord, Norway (W). H= 5.5-6.5, D_m= 5.0, D_c=?, IVD 10b. Scheerer, 1870. Can. Min. 15 (1977), 92. Zts. Krist. 152 (1980), 69. See also Tanteuxenite-(Y), Yttrocrasite-(Y).

Evansite, $Al_3PO_4(OH)_6\cdot 6H_2O$(?), Q Amor. Vitreous/resinous/waxy colorless, white, various tints. Železník, Sirk (near), Slovakia, Czechoslovakia. H= 3-4, D_m= 1.9, D_c=?, VIID 26. Forbes, 1864. Dana, 7th ed.(1951), v.2, 923.

Eveite, $Mn_2AsO_4(OH)$, A Orth. Translucent apple-green. Långban mine, Filipstad (near), Värmland, Sweden. H= 4, D_m= 3.76, D_c= 3.75, VIIB 04. Moore, 1968. Am. Min. 55(1970), 319 (Abst.). Am. Min. 53 (1968), 1841. See also Adamite.

Evenkite, $C_{24}H_{50}$, G Mon. Waxy colorless, yellowish. Evenki region, Lower Tunguska River, Siberia, USSR. H= 1, D_m= 0.920, D_c= 0.926, IXB 01. Skropyshev, 1953. Am. Min. 50(1965), 2109 (Abst.).

Ewaldite, $Ba(Ca,Y,Na,K)(CO_3)_2$, A Hex. Translucent bluish-green. Green River Formation, Uintah Co., Wyoming, USA. H=?, D_m= 3.25, D_c= 3.37, VbA 05. Donnay & Donnay, 1971. Am. Min. 56(1971), 2156 (Abst.).

Eylettersite, $(Th,Pb)_{1-x}Al_3(PO_4,SiO_4)_2(OH)_6$(?), A Rhom. Translucent white, creamy. Kobokobo pegmatite, Kinshasa, Kivu, Zaïre. H=?, D_m= 3.4, D_c= 3.5, VIIB 15b. Van Wambeke, 1972. Am. Min. 59(1974), 208 (Abst.).

Ezcurrite, $Na_2B_5O_7(OH)_3\cdot 2H_2O$, G Tric. Vitreous colorless. Tincalayu Borax mine, Salta, Argentina. H= 3-3.5, D_m= 2.053, D_c= 2.049, Vc 09. Muessig & Allen, 1957. Am. Min. 52 (1967), 1048. Am. Min. 58 (1973), 110.

Eztlite, $Pb_2Fe_6Te_4O_{15}(OH)_{10}\cdot 8H_2O$, A Mon. Brilliant red. Bambolla mine, Moctezuma, Sonora, Mexico. H= 3, D_m= 4.5, D_c= 4.61, VIG 08. Williams, 1982. Min. Mag. 46 (1982), 257.

F

Fabianite, $CaB_3O_5(OH)$, G Mon. Translucent colorless. Diepholz (near), Rehden, Germany. H= 6, D_m= 2.796, D_c= 2.788, Vc 06. Gaertner et al., 1962. Am. Min. 48(1963), 212 (Abst.). Zts. Krist. 132 (1970), 241.

Faheyite, $Be_2(Mn,Mg,Na)Fe_2^{3+}(PO_4)_4 \cdot 6H_2O$, G Hex. White, bluish-white, brownish-white. Sapucaia Pegmatite, Galileia (near), Minas Gerais, Brazil. H=?, D_m= 2.660, D_c= 2.670, VIIC 09. Lindberg & Murata, 1953. Am. Min. 49 (1964), 395.

Fahleite, $CaZn_5Fe_2(AsO_4)_6 \cdot 14H_2O$, A Orth. Silky/pearly yellow, grey, bright green. Tsumeb, Namibia. H=?, D_m=?, D_c= 3.16, VIIC 26. Medenbach et al., 1988. N.Jb.Min.Mh. (1988), 167.

Fairbankite, $PbTeO_3$, A Tric. Resinous/adamantine colorless. Grand Central mine, Tombstone, Cochise Co., Arizona, USA. H= 2, D_m=?, D_c= 7.45, VIG 11. Williams, 1979. Min. Mag. 43 (1979), 453. See also Plumbotellurite.

Fairbanksite, D Inadequate data. Min. Mag. 36 (1968), 1144. Morgan, 1965.

Fairchildite, $K_2Ca(CO_3)_2$, G Hex. Coolin, Kaniksu National Forest, Bonner Co., Idaho, USA; Grand Canyon National Park, Coconino Co., Arizona, USA. H=?, D_m=?, D_c= 2.441, VbA 05. Milton & Axelrod, 1947. Am. Min. 32 (1947), 607. Zts. Krist. 157 (1981), 199. See also Butschliite.

Fairfieldite, $Ca_2(Mn,Fe)(PO_4)_2 \cdot 2H_2O$, G Tric. Pearly/sub-adamantine white, yellow, greenish-white. Branchville, Fairfield Co., Connecticut, USA. H= 3.5, D_m= 3.08, D_c= 3.095, VIIC 12. Brush & Dana, 1879. Dana, 7th ed.(1951), v.2, 720. Acta Cryst. B26 (1970), 333.

Falcondoite, $(Ni,Mg)_4Si_6O_{15}(OH)_2 \cdot 6H_2O$, A Orth. Translucent whitish-green. Bonao, Dominican Republic. H= 2-3, D_m= 1.9, D_c= 2.54, VIIIE 13. Springer, 1976. Can. Min. 14 (1976), 407. See also Sepiolite.

Falkmanite, $Pb_{5.4}Sb_{3.6}S_{11}$, Q Mon. Metallic white. Bayerland mine, Bayern (Bavaria), Germany. H= 3-3.5, D_m=?, D_c= 6.310, IID 05b. Ramdohr & Ödman, 1940. Can. Min. 25 (1987), 15. See also Boulangerite.

Famatinite, Cu_3SbS_4, G Tet. Metallic grey; pink in reflected light. Sierra de Famatina, La Rioja, Argentina. H= 3.5, D_m= 4.52, D_c= 4.660, IIB 03a. Stelzner, 1873. Dana, 7th ed.(1944), v.1, 387. Am. Min. 42 (1957), 766.

Farringtonite, $Mg_3(PO_4)_2$, G Mon. Waxy white, yellow. Springwater pallasite meteorite, Springwater, Saskatchewan, Canada. H=?, D_m= 2.74, D_c= 2.76, VIIA 03. Dufresne & Roy, 1961. Am. Min. 58 (1973), 949.

Fassaite, D Discredited as Fe-Al diopside or augite. Min. Mag. 52 (1988), 535.

Faujasite, $Na_{20}Ca_{12}Mg_8(Al_{60}Si_{132})O_{384} \cdot 235H_2O$, G Cub. Vitreous/adamantine colorless, white. Sasbach, Kaiserstuhl, Germany. H= 5, D_m= 1.93, D_c= 1.91, VIIIF 15. Damour, 1842. Natural Zeolites (1985), 214. Am. Min. 49 (1964), 697.

Faustite, $(Zn,Cu)Al_6(PO_4)_4(OH)_8 \cdot 4H_2O$, G Tric. Waxy/dull apple-green. Copper King mine, Maggie Creek Dist., Eureka Co., Nevada, USA. H= 5.5, D_m= 2.92, D_c= 2.99, VIID 08. Erd et al., 1953. Am. Min. 38 (1953), 964.

Fayalite, Fe_2SiO_4, G Orth. Translucent pale yellow, greenish-yellow, yellow-amber. Fayal Island, Azores, Portugal. H= 6.5, D_m= 4.392, D_c= 4.399, VIIIA 03. Gmelin, 1840. DHZ, 2nd ed.(1982), v.1A, 2. Am. Min. 60 (1975), 1092. See also Forsterite, Tephroite.

Fedorite, $(K,Na)_{2.5}(Ca,Na)_7Si_{16}O_{38}(OH,F)_2 \cdot H_2O$, G Tric. Vitreous colorless, pink. Kola Peninsula, USSR. H=?, D_m= 2.43, D_c= 2.46, VIIIE 14. Kukharenko et al., 1965. Am. Min. 52(1967), 561 (Abst.). Sov.Phys.Cryst. 28 (1983), 95.

Fedorovskite, $Ca_2(Mg,Mn)_2B_4O_7(OH)_6$, A Orth. Translucent brown. Solongo deposit, Buryat, Ural Mts., USSR. H= 4.5, D_m= 2.7, D_c= 2.81, Vc 07b. Malinko et al., 1977. Am. Min. 62(1977), 173 (Abst.). See also Roweite.

Fedotovite, $K_2Cu_3O(SO_4)_3$, A Mon. Transparent green. Tolbachik volcano, Kamchatka, USSR. H= 2.5, D_m= 3.205, D_c= 3.17, VIB 07. Vergasova et al., 1988. Am. Min. 75(1990), 241 (Abst.).

Feitknechtite, β-MnO(OH), G Tet. Black, brownish-black. Franklin, Sussex Co, New Jersey, USA. H=?, D_m= 3.80, D_c= 3.80, IVF 04. Bricker, 1965. Am. Min. 50 (1965), 1296. See also Groutite, Manganite.

Feldspar, A group name for framework silicates with the general formula $(K,Na,Ca,Ba,NH_4)(Si,Al)_4O_8$.

Felsöbányaite, $Al_4SO_4(OH)_{10}\cdot5H_2O$, G Hex. Pearly yellow, white. Baia Sprie (Felsöbánya), Romania. H= 1.5, D_m= 2.33, D_c= 2.205, IVF 02. Kenngott, 1853. Am. Min. 50(1965), 812 (Abst.).

Fenaksite, $KNaFeSi_4O_{10}$, G Tric. Pearly light rose. Khibina, Kola Peninsula, USSR. H= 5-5.5, D_m= 2.74, D_c= 2.74, VIIID 18. Dorfman et al., 1959. Am. Min. 45(1960), 252 (Abst.). Struct. Repts. 35A (1970), 480.

Fenghuanglite, D =thorian britholite-(Ce); name also given as Feng-huang-shih. Min. Mag. 33 (1962), 261. Peng, 1959.

Ferberite, $FeWO_4$, G Mon. Sub-metallic/adamantine black. Sierra Almagrera, Spain (S). H= 4-4.5, D_m= 7.40, D_c= 7.55, IVD 08. Breithaupt, 1863. Dana, 7th ed.(1951), v.2, 1064. Zts. Krist. 127 (1968), 61. See also Hübnerite, Sanmartinite.

Ferchromide, $Cr_{1.5}Fe_{0.2}$, A Cub. Metallic grey. Urali Meridionali, Ural Mts. (S), USSR. H= 6-6.5, D_m=?, D_c= 6.18, IA 05. Novgorodova et al., 1986. ZVMO 115 (1986), 355.

Ferdisilicite, $FeSi_2$, Q Tet. Metallic steel-grey. Zachativsk station (near), Donets region, USSR. H= 6.25, D_m=?, D_c= 5.05, IC 03. Gevork'yan et al., 1969. Am. Min. 54(1969), 1737 (Abst.) . Struct. Repts. 24 (1960), 77.

Fergusonite-(Ce), $(Ce,Nd,La)NbO_4\cdot0.3H_2O$, Q Syst=? Dark red, black. Novopoltavsk massif, USSR. H=?, D_m= 5.48, D_c=?, IVD 13. Kapustin, 1986. Am. Min. 74(1989), 946 (Abst.). See also beta-Fergusonite-(Ce).

Fergusonite-(Nd), $(Nd,Ce)(Nb,Ti)O_4$, Q Syst=? Adamantine dark brown. Bayan Obo, Inner Mongolia, China. H=?, D_m=?, D_c=?, IVD 13. Zhang & Tao, 1987. Am. Min. 74(1989), 946 (Abst.). See also beta-Fergusonite-(Nd).

Fergusonite-(Y), $YNbO_4$, G Tet. Vitreous/sub-metallic black, brownish-black. Qeqertaussaq, Julianehåb dist., Greenland (S). H= 5.5-6.5, D_m= 5.7, D_c= 6.53, IVD 13. Haidinger, 1827. Strunz (1970), 209. See also Formanite-(Y), beta-Fergusonite-(Y).

Fermorite, $(Ca,Sr)_5(AsO_4,PO_4)_3(OH)$, G Hex. Greasy pale pinkish-white. Chindwara Dist., Sitapar, India (Central). H= 5, D_m= 3.518, D_c= 3.46, VIIB 16. Prior & Smith, 1910. Dana, 7th ed.(1951), v.2, 904.

Fernandinite, D Discredited as a mixture of bariandite, roscoelite, gypsum. Min. Mag. 53 (1989), 511. Schaller, 1915.

Feroxyhyte, δ-FeO(OH), A Hex. Pacific Ocean (iron-manganese concretions). H=?, D_m=?, D_c= 4.20, IVF 04. Chukhrov et al., 1976. Am. Min. 62(1977), 1057 (Abst.). See also Akaganéite, Goethite, Lepidocrocite.

Ferrarisite, $Ca_5(AsO_3OH)_2(AsO_4)_2\cdot9H_2O$, A Tric. Transparent colorless. Gabe Gottes vein, Rautenthal, Ste.-Marie-aux-Mines, Vosges Mtns., Haut-Rhin, Alsace, France. H=?, D_m= 2.63, D_c= 2.57, VIIC 02. Bari et al., 1980. Am. Min. 66(1981), 637 (Abst.). Bull. Min. 103 (1980), 541. See also Guerinite.

Ferrazite, $(Pb,Ba)_3(PO_4)_2\cdot8H_2O(?)$, Q Syst=? Translucent dark yellowish-white. Brazil. H=?, D_m= 3.1, D_c=?, VIIC 16. Lee & de Moraes, 1919. Dana, 7th ed.(1951), v.2, 832.

Ferri-annite, $K(Fe,Mg)_3(Si,Fe)_4O_{10}(OH)_2$, G Mon. Dales Gorge member, Wittenoom, Western Australia, Australia. H=?, D_m=?, D_c= 3.54, VIIIE 05b. Wones, 1963. Am. Min. 67 (1982), 1179. Acta Cryst. 17 (1964), 1369. See also Annite.

Ferri-barroisite, $NaCaMg_3Fe_2^{3+}(Si_7Al)O_{22}(OH)_2$, [A] Mon. H=?, D_m=?, D_c=?, VIIID 05c. Leake et al., 1978. Am. Min. 63 (1978), 1023. See also Ferrobarroisite.

Ferricopiapite, $(Fe,Al,Mg)Fe_5(SO_4)_6(OH)_2 \cdot 20H_2O$, [G] Tric. Pearly yellow, orange. Atacama, Chile. H= 2.5-3, D_m= 2.1, D_c= 2.123, VID 04b. Berry, 1938. Can. Min. 23 (1985), 53. Am. Min. 58 (1973), 314.

Ferridravite, $(Na,K)Mg_3Fe_6(BO_3)_3Si_6O_{18}(OH)_4$, [A] Rhom. Resinous black. San Francisco mine, Aranibar, Alto Chapare, Villa Tunari (near), Cochabamba, Bolivia. H= 7, D_m= 3.26, D_c= 3.33, VIIIC 08. Walenta & Dunn, 1979. Am. Min. 64 (1979), 945.

Ferrierite (monoclinic), $KNa_3Mg(Al_5Si_{31})O_{72} \cdot 18H_2O$, [P] Mon. Altoona, Washington, USA. H=?, D_m=?, D_c= 2.16, VIIIF 12. Gramlich-Meier et al., 1985. Am. Min. 70 (1985), 619. See also Ferrierite (orthorhombic).

Ferrierite (orthorhombic), $(Mg,K,Ca)_{4.4}(Si,Al)_{36}O_{72} \cdot 18H_2O$, [G] Orth. White. Kamloops Lake, British Columbia, Canada. H= 3, D_m= 2.15, D_c= 2.13, VIIIF 12. Graham, 1918. Min. Mag. 50 (1986), 63. Zts. Krist. 178 (1987), 249. See also Ferrierite (monoclinic).

Ferrihydrite, $Fe_{4-5}(OH,O)_{12}$, [A] Hex. Yellow-brown, dark brown. USSR. H=?, D_m= 3.96, D_c=?, IVF 04. Chukhrov et al., 1973. Am. Min. 60(1975), 485 (Abst.). Clays Cl. Mins. 36 (1988), 111.

Ferri-katophorite, $Na_2CaFe^{2+}4Fe^{3+}(Si_7Al)O_{22}(OH)_2$, [A] Mon. H=?, D_m=?, D_c=?, VIIID 05c. Leake et al., 1978. Am. Min. 63 (1978), 1023. See also Aluminokatophorite.

Ferrimolybdite, $Fe_2(MoO_4)_3 \cdot 7H_2O$, [G] Orth. Adamantine/silky/earthy yellow. Siberia, USSR. H= 1-2, D_m= 4.46, D_c=?, VIF 03. Pilipenko, 1914. Am. Min. 48 (1963), 14.

Ferrinatrite, $Na_3Fe(SO_4)_3 \cdot 3H_2O$, [G] Rhom. Vitreous greyish-white, greenish, bluish-green, amethystine. Mina la Compania, Sierra Gorda (south of), Atacama Desert, Chile. H= 2.5, D_m= 2.562, D_c= 2.55, VIC 09. Mackintosh, 1889. N. Jb. Min. Mh. (1987), 171. Min. Mag. 41 (1977), 375.

Ferriphlogopite, $KMg_3(Si_3Fe)O_{10}(OH)_2$, [Q] Mon. Yesseĭ complex, Maĭmecha-Kotuĭ, USSR. H=?, D_m= 2.953, D_c= 2.90, VIIIE 05b. Skosyreva et al., 1985. Dokl.Earth Sci. 285(1985), 129. Am. Min. 47 (1962), 886.

Ferripyrophyllite, $FeSi_2O_5(OH)$, [Q] Mon. Strassenschacht, Germany; Mt. Tologay, Kazakhstan, USSR. H=?, D_m=?, D_c= 3.05, VIIIE 04. Chukhrov et al., 1979. Min. Abstr. 80-3525. See also Pyrophyllite.

Ferrisicklerite, $Li_{1-x}(Fe,Mn)PO_4$, [G] Orth. Sub-translucent yellowish/dark brown. Morocco. H= 4, D_m= 3.41, D_c= 3.53, VIIA 02. Quensel, 1937. Bull. Min. 99 (1976), 274. Acta Cryst. B32 (1976), 2761. See also Sicklerite.

Ferristrunzite, $Fe_3(PO_4)_2(OH)_3 \cdot 5H_2O$, [A] Tric. Vitreous light brownish-yellow. Canal Willebroeck, Charleroi, Blaton, Hainaut, Belgium. H=?, D_m= 2.5, D_c= 2.55, VIID 05. Peacor et al., 1987. Am. Min. 74(1989), 502 (Abst.).

Ferrisymplesite, $Fe_3(AsO_4)_2(OH)_3 \cdot 5H_2O$, [Q] Syst=? Resinous amber-brown. Hudson Bay mine, Cobalt, Ontario, Canada. H=?, D_m=?, D_c=?, VIID 10. Walker & Parsons, 1924. Dana, 7th ed.(1951), v.2, 753.

Ferri-taramite, $Na_2CaFe_3^{2+}Fe_2^{3+}(Si_6Al_2)O_{22}(OH)_2$, [A] Mon. H=?, D_m=?, D_c= 3.47, VIIID 05c. Leake et al., 1978. Am. Min. 63 (1978), 1023. Can. Min. 16 (1978), 53.

Ferrithorite, [D] Mixture of thorite and Fe hydroxide. Am. Min. 73(1988), 198 (Abst.). Starik et al., 1941.

Ferri-tschermakite, $Ca_2Mg_3Fe_2^{3+}(Si_6Al_2)O_{22}(OH)_2$, [A] Mon. H=?, D_m=?, D_c=?, VIIID 05b. Winchell, 1949. Am. Min. 63 (1978), 1023. See also Ferro-alumino-tschermakite.

Ferritungstite, $(K,Ca)_{0.2}(W,Fe)_2(O,OH)_6 \cdot H_2O$, [G] Cub. Ocherous yellow. Germania Tungsten mine, Deer Trail Mining Dist., Stevens Co., Washington, USA. H=?, D_m= 5.02, D_c=?, VIF 03. Schaller, 1911. Min. Rec. 12 (1981), 81. GAC-MAC 13 (1988), A37 (Abst.).

Ferri-winchite, $NaCaMg_4Fe^{3+}Si_8O_{22}(OH)_2$, [A] Mon. H=?, D_m=?, D_c=?, VIIID 05c. Leake et al., 1978. Am. Min. 63 (1978), 1023.

Ferro-actinolite, $Ca_2(Fe,Mg)_5Si_8O_{22}(OH)_2$, [G] Mon. Translucent dark green, black. H= 5-6, D_m= 3.27, D_c= 3.49, VIIID 05b. Sundius, 1946. Am. Min. 63 (1978), 1023. Zts. Krist. 133 (1971), 273. See also Tremolite, Actinolite.

Ferroalluaudite, $(Na,Ca)Fe(Fe,Mn,Mg)_2(PO_4)_3$, [A] Mon. Vitreous greenish-black. Pleasant Valley Pegmatite, Custer (near), Custer Co., South Dakota, USA. H= 5, D_m= 3.5, D_c= 3.61, VIIA 05b. Moore & Ito, 1979. Am. Min. 50 (1965), 713. See also Alluaudite, Hagendorfite, Maghagendorfite.

Ferro-alumino-barroisite, $NaCaFe_3Al_2(Si_7Al)O_{22}(OH)_2$, [A] Mon. H=?, D_m=?, D_c=?, VIIID 05c. Leake et al., 1978. Am. Min. 63 (1978), 1023. See also Ferro-alumino-tschermakite.

Ferro-alumino-tschermakite, $Ca_2Fe_3Al_2(Si_6Al_2)O_{22}(OH)_2$, [A] Mon. Translucent green. H= 5-6, D_m=?, D_c= 3.302, VIIID 05b. Leake et al., 1978. Am. Min. 63 (1978), 1023. Min. Mag. 39 (1973), 36. See also Ferro-tschermakite.

Ferro-alumino-winchite, $NaCaFe_4AlSi_8O_{22}(OH)_2$, [A] Mon. H=?, D_m=?, D_c=?, VIIID 05c. Leake et al., 1978. Am. Min. 63 (1978), 1023. See also Winchite, Ferrowinchite.

Ferroalunite, [D] Unnecessary name for ferrian alunite. Min. Mag. 36 (1968), 1144. Gvakhariya & Nazarov, 1963.

Ferro-anthophyllite, $(Fe,Mg)_7Si_8O_{22}(OH)_2$, [A] Orth. Translucent greyish-green. Coeur d'Alene, Idaho, USA. H=?, D_m=?, D_c=?, VIIID 06. Shannon, 1921. Am. Min. 63 (1978), 1023. See also Magnesio-anthophyllite, Anthophyllite.

Ferro-axinite, $Ca_2FeAl_2(BO_3OH)(SiO_3)_4$, [A] Tric. Translucent lilac-brown. H= 6.5-7, D_m= 3.3, D_c= 3.31, VIIIC 04. Schaller, 1909. DHZ, 2nd ed.(1986), v.1B, 603. Acta Cryst. 5 (1952), 202. See also Manganaxinite.

Ferrobarroisite, $NaCa(Fe,Mg)_5(Si_7Al)O_{22}(OH)_2$, [A] Mon. Translucent green. H= 5-6, D_m=?, D_c=?, VIIID 05c. Leake, 1978. Am. Min. 63 (1978), 1023. See also Barroisite.

Ferrobustamite, $Ca(Fe,Ca,Mn)Si_2O_6$, [G] Tric. Scawt Hill, Antrim Co., Northern Ireland. H=?, D_m=?, D_c= 3.07, VIIID 08. Rapoport & Burnham, 1973. Zts. Krist. 138 (1973), 419. Am. Min. 62 (1977), 1216. See also Bustamite.

Ferrocarpholite, $(Fe,Mg)Al_2Si_2O_6(OH)_4$, [G] Orth. Translucent dark green. Tomate (W), Celebes (E), Indonesia. H= 5.5, D_m= 3.04, D_c= 3.05, VIIID 03. de Roever, 1951. Am. Min. 36 (1951), 736. Acta Cryst. 9 (1956), 773. See also Carpholite, Magnesiocarpholite.

Ferroclinoholmquistite, $Li_2(Fe,Mg)_3Al_2Si_8O_{22}(OH)_2$, [A] Mon. H=?, D_m=?, D_c=?, VIIID 05a. Leake et al., 1978. Am. Min. 63 (1978), 1023. See also Ferroholmquistite, Magnesioclinoholmquistite, Clinoholmquistite.

Ferrocolumbite, $FeNb_2O_6$, [G] Orth. Black, brownish-black; greyish-white in reflected light. Western Australia, Australia. H= 6, D_m= 5.20, D_c= 5.430, IVD 10a. Simpson, 1928. N.Jb.Min.Mh. (1985), 372. See also Ferrotantalite, Manganocolumbite, Magnocolumbite.

Ferro-eckermannite, $Na_3(Fe,Mg)_4AlSi_8O_{22}(OH)_2$, [A] Mon. H=?, D_m=?, D_c=?, VIIID 05d. Phillips & Layton, 1964. Am. Min. 63 (1978), 1023. See also Eckermannite.

Ferro-edenite, $NaCa_2(Fe,Mg)_5(Si_7Al)O_{22}(OH)_2$, [G] Mon. Translucent green. H= 5-6, D_m= 3.42, D_c= 3.53, VIIID 05b. Sundius, 1946. Strunz (1970), 418. See also Edenite.

Ferro-ferri-barroisite, $NaCaFe_3^{2+}Fe_2^{3+}(Si_7Al)O_{22}(OH)_2$, [A] Mon. H= 5-6, D_m=?, D_c= 3.50, VIIID 05c. Leake et al., 1978. Am. Min. 63 (1978), 1023.

Ferro-ferri-tschermakite, $Ca_2Fe_3^{2+}Fe_2^{3+}(Si_6Al_2)O_{22}(OH)_2$, [A] Mon. H=?, D_m= 3.45, D_c=?, VIIID 05b. Leake et al., 1978. Am. Min. 63 (1978), 1023.

Ferro-ferri-winchite, $NaCaFe_4^{2+}Fe^{3+}Si_8O_{22}(OH)_2$, [A] Mon. H=?, D_m=?, D_c=?, VIIID 05c. Leake et al., 1978. Am. Min. 63 (1978), 1023.

Ferro-gedrite, $(Fe,Mg)_5Al_2(Si_6Al_2)O_{22}(OH)_2$, [G] Orth. H=?, D_m= 3.56, D_c= 3.56, VIIID 06. Tilley, 1939. Am. Min. 42 (1957), 506. See also Magnesiogedrite, Gedrite.

Ferro-glaucophane, $Na_2(Fe,Mg)_3Al_2Si_8O_{22}(OH)_2$, [A] Mon. Translucent bluish-grey. H= 6, D_m=?, D_c= 3.224, VIIID 05d. Miyachiro, 1957. Am. Min. 63 (1978), 1023. Can. Min. 17 (1979), 1. See also Glaucophane.

Ferrohagendorfite, $(Na,Ca)_2Fe_3(PO_4)_3$, [A] Mon. North Groton, New Hampshire, USA. H= 5-5.5, D_m= 3.4, D_c=?, VIIA 05b. Moore & Ito, 1979. Min. Mag. 43 (1979), 227. See also Alluaudite, Hagendorfite.

Ferrohalotrichite, [D] Unnecessary name for halotrichite. Min. Mag. 43 (1980), 1055. Vieira de Mello, 1969.

Ferrohexahydrite, $FeSO_4 \cdot 6H_2O$, [G] Mon. Translucent colorless. Tateria (NE), USSR. H=?, D_m=?, D_c= 1.93, VIC 03b. Vlasov & Kuznetsov, 1962. ZVMO 91 (1962), 490.

Ferro-holmquistite, $Li_2(Fe,Mg)_3Al_2Si_8O_{22}(OH)_2$, [A] Orth. H=?, D_m=?, D_c=?, VIIID 06. Leake et al., 1978. Am. Min. 63 (1978), 1023. See also Ferroclinoholmquistite, Magnesioholmquistite, Holmquistite.

Ferro-kaersutite, $NaCa_2(Fe,Mg)_4Ti(Si_6Al_2)(O,OH)_{24}$, [A] Mon. Dark brown, black. H= 5-6, D_m= 3.2, D_c= 3.49, VIIID 05b. Leake et al., 1978. Am. Min. 63 (1978), 1023. See also Kaersutite.

Ferrokesterite, $Cu_2(Fe,Zn)SnS_4$, [A] Tet. Metallic steel grey; grey in reflected light. Cligga mine, Perranzabuloe, Cornwall, England. H= 4, D_m=?, D_c= 4.490, IIB 03a. Kissin & Owens, 1989. Can. Min. 27 (1989), 673.

Ferrolizardite, [D] Unnecessary name for ferroan lizardite. Min. Mag. 36 (1968), 1144. Chia & Cheng, 1964.

Ferronickelplatinum, Pt_2FeNi, [A] Tet. Metallic rosy-cream. Unspecified locality, Koryak-Kamchatka region, USSR. H= 5, D_m=?, D_c= 15.38, IA 12. Rudashevski et al., 1983. ZVMO 112 (1983), 487. See also Tulameenite.

Ferro-pargasite, $NaCa_2(Fe,Mg,Al)_5(Si_6Al_2)O_{22}(OH)_2$, [G] Mon. Translucent brown. H= 5-6, D_m=?, D_c= 3.54, VIIID 05b. Wilkinson, 1961. Am. Min. 65 (1980), 996. See also Pargasite.

Ferropseudobrookite, [D] Unnecessary name for armalcolite. MSA (1989), A236 (Abst.). Anderson et al., 1970.

Ferropyrosmalite, $(Fe,Mn)_8Si_6O_{15}(OH,Cl)_{10}$, [A] Rhom. Pearly brown, grey, green. Pegmont Pb–Zn deposit, Queensland, Australia. H= 4-4.5, D_m= 3.1, D_c= 3.10, VIIIE 12. Vaughan, 1987. Min. Mag. 51 (1987), 174.

Ferro-richterite, $Na_2Ca(Fe,Mg)_5Si_8O_{22}(OH)_2$, [A] Mon. H=?, D_m=?, D_c= 3.46, VIIID 05c. Leake et al., 1978. Am. Min. 59 (1974), 518. See also Richterite.

Ferroselite, $FeSe_2$, [G] Orth. Metallic rosy-cream. Tuva, USSR. H= 6-6.2, D_m= 7.20, D_c= 7.28, IIC 07. Burjanova & Komkov, 1955. N.Jb.Min.Mh (1972), 276. Min. Abstr. 77-1487.

Ferrosilite, $(Fe,Mg)_2(SiO_3)_2$, [A] Orth. Translucent green, dark brown. USSR. H= 5-6, D_m= 3.96, D_c= 4.00, VIIID 02. Bowen, 1935. Min. Mag. 52 (1988), 535. Am. Min. 61 (1976), 38. See also Enstatite, Hypersthene, Clinoferrosilite.

Ferrostrunzite, $Fe_3(PO_4)_2(OH)_2 \cdot 6H_2O$, [A] Tric. Vitreous light brown. Mullica Hill, New Jersey, USA. H= 4, D_m= 2.50, D_c= 2.57, VIID 05. Peacor et al., 1983. Am. Min. 69(1984), 810 (Abst.). See also Strunzite.

Ferrotantalite, $FeTa_2O_6$, Q Orth. Black, brownish-black. H= 6-6.5, D_m= 7.95, D_c=?, IVD 10a. Thomson, 1836. Can. Min. 8 (1966), 461. See also Manganotantalite, Ferrocolumbite, Tapiolite.

Ferrotapiolite, $Fe(Ta,Nb)_2O_6$, A Tet. Dark brown; grey in reflected light. Viitaniemi, Eräjärvi, Orivesi, Finland. H= 5.8, D_m=?, D_c= 7.72, IVD 04. Lahti et al., 1983. Am. Min. 70(1985), 217 (Abst.).

Ferrotellurite, $FeTeO_4$(?), Q Syst=? Translucent yellow. Keystone mine, Magnolia Dist., Boulder Co., Colorado, USA. H=?, D_m=?, D_c=?, VIG 07. Genth, 1877. Dana, 6th ed. (1892), 980.

Ferrotschermakite, $Ca_2(Fe,Mg)_5(Si_6Al_2)O_{22}(OH)_2$, A Mon. Translucent green. H= 5-6, D_m=?, D_c= 3.41, VIIID 05b. Winchell, 1945. Am. Min. 63 (1978), 1023. See also Tschermakite.

Ferrotychite, $Na_6(Fe,Mg,Mn)_2(CO_3)_4(SO_4)$, A Cub. Vitreous colorless, light yellow. Khibina massif, Kola Peninsula, USSR. H= 4, D_m= 2.79, D_c= 2.78, VbB 03. Khomyakov et al., 1981. Am. Min. 67(1982), 622 (Abst.). Min. Abstr. 81-2430. See also Tychite.

Ferrowinchite, $NaCa(Fe,Mg)_5(Si,Al)_8O_{22}(OH)_2$, A Mon. Translucent green. H= 5-6, D_m=?, D_c=?, VIIID 05c. Leake, 1978. Am. Min. 63 (1978), 1023. See also Winchite.

Ferrowyllieite, $(Na,Ca,Mn)_2(Fe,Mn,Mg)_2Al(PO_4)_3$, A Mon. Vitreous/sub-metallic bluish-green, green, greyish-green. Victory mine, Custer (near), Custer Co., South Dakota, USA. H= 4, D_m= 3.601, D_c= 3.60, VIIA 05a. Moore & Ito, 1979. Min. Mag. 43 (1979), 227. Am. Min. 59 (1974), 280. See also Wyllieite, Rosemaryite, Qingheiite.

Ferruccite, $NaBF_4$, G Orth. Translucent colorless, white. Mt. Vesuvius, Napoli, Campania, Italy. H= 3, D_m= 2.496, D_c= 2.507, IIIB 01. Carobbi, 1933. Dana, 7th ed.(1951), v.2, 98. Acta Cryst. B24 (1968), 1703.

Fersilicite, FeSi, Q Cub. Metallic white. Zachativsk station (near), Donets region, USSR. H= 6.5, D_m= 6.18, D_c= 6.170, IC 01. Gevorkyan, 1969. Am. Min. 54(1969), 1737 (Abst.). Struct. Repts. 28 (1963), 28.

Fersmanite, $(Ca,Na)_8(Ti,Nb)_4(Si_2O_7)O_8F_3$, G Mon. Vitreous dark brown, golden yellow. Khibina massif, Kola Peninsula, USSR. H= 5-5.5, D_m= 3.44, D_c= 3.18, VIIIB 09. Labuntzov, 1929. Can. Min. 15 (1977), 87. Sov.Phys.Cryst. 29 (1984), 31.

Fersmite, $(Ca,Ce,Na)(Nb,Ta,Ti)_2(O,OH,F)_6$, G Orth. Resinous black. Ural Mts., USSR. H= 4.5, D_m= 4.69, D_c=?, IVD 10b. Bohnstedt-Kupletskaya & Burova, 1946. Am. Min. 32 (1947), 373. Struct. Repts. 24 (1960), 367.

Feruvite, $(Ca,Na)(Fe,Mg,Ti)_3Al_6(BO_3)_3Si_6O_{18}(OH)_4$, A Rhom. Vitreous/dull dark brown-black. Cuvier Island, New Zealand. H= 7, D_m= 3.207, D_c= 3.21, VIIIC 08. Grice & Robinson, 1989. Can. Min. 27 (1989), 199. Can. Min. 27 (1989), 199.

Fervanite, $Fe_4(VO_4)_4 \cdot 5H_2O$, G Mon. Brilliant golden-brown. Polar Mesa, La Sal Mts. (N), Grand Co., Utah, USA; Gypsum Valley (N), San Miguel Co., Colorado, USA. H=?, D_m=?, D_c=?, VIIC 05. Hess & Henderson, 1931. Am. Min. 44 (1959), 322.

Fe-shafranovskite, D Inadequate data. Am. Min. 75(1990), 432 (Abst.). Korovyshkin et al., 1987.

Feuermineral, D Superfluous name for germanian mawsonite. Am. Min. 63(1978), 427 (Abst.). Geier & Ottemann, 1970.

Fibroferrite, $FeSO_4(OH) \cdot 5H_2O$, G Rhom. Silky/pearly pale yellow, greenish-grey, yellowish green. Tierra Amarilla, Copiapó (near), Chile. H= 2.5, D_m= 1.95, D_c= 1.996, VID 02. Rose, 1833. N. Jb. Min. Mh. (1987), 171. Min. Abstr. 83M/1237.

Fichtelite, $C_{19}H_{34}$, G Mon. Greasy white. Redwitz (near), Fichtel Mts., Bayern (Bavaria), Germany. H= 1, D_m= 1.03, D_c= 1.032, IXB 01. Bromeis, 1841. Strunz (1970), 496.

Fiedlerite, $Pb_3Cl_4(OH)_2$, G Mon. Adamantine colorless, white. Lávrion (Laurium), Attikí, Greece. H= 3.5, D_m= 5.88, D_c= 5.64, IIIC 05. vom Rath, 1887. Dana, 7th ed.(1951), v.2, 67.

Filipstadite, $(Mn,Mg)_2(Sb,Fe)O_4$, A Cub. Metallic black; grey in reflected light. Långban mine, Filipstad (near), Värmland, Sweden. H= 6, D_m=?, D_c= 4.9, IVB 01. Dunn et al., 1988. Am. Min. 73 (1988), 413.

Fillowite, $Na_2Ca(Mn,Fe)_7(PO_4)_6$, G Rhom. Sub-resinous/greasy yellow, yellowish/reddish brown. Branchville, Fairfield Co., Connecticut, USA. H= 4.5, D_m= 3.43, D_c= 3.54, VIIA 06. Brush & Dana, 1879. Am. Min. 50 (1965), 1647. Am. Min. 66 (1981), 827. See also Johnsomervilleite.

Fingerite, $Cu_{11}O_2(VO_4)_6$, A Tric. Metallic black; grey in reflected light. Izalco Volcano, El Salvador. H=?, D_m=?, D_c= 4.776, VIIA 12. Hughes & Hadidiacos, 1985. Am. Min. 70 (1985), 197.

Finnemanite, $Pb_5(AsO_3)_3Cl$, G Hex. Sub-adamantine grey, black. Långban mine, Filipstad (near), Värmland, Sweden. H= 2.5, D_m= 7.265, D_c= 7.347, VIIB 16. Aminoff, 1923. Dana, 7th ed.(1951), v.2, 1038. Min. Abstr. 81-1244.

Fischesserite, Ag_3AuSe_2, A Cub. Metallic pink. Předbořice, Středočeský kraj, Čechy (Bohemia), Czechoslovakia. H= 2, D_m=?, D_c= 9.05, IIA 16. Johan et al., 1971. Am. Min. 57(1972), 1554 (Abst). See also Petzite.

Fizélyite, $Ag_5Pb_{14}Sb_{21}S_{48}$, G Mon. Metallic grey. Herja (Kisbánya), Judeţul Maramureş, Romania. H= 2, D_m=?, D_c= 5.663, IID 09b. Krantz, 1914. Am. Min. 70 (1985), 219 (Abst.).

Flagstaffite, $C_{10}H_{22}O_3$, G Orth. Transparent colorless. Flagstaff (N), Coconino Co., Arizona, USA. H=?, D_m= 1.09, D_c= 1.088, IXB 01. Guild, 1920. Am. Min. 50(1965), 2109 (Abst.).

Fleischerite, $Pb_3Ge(SO_4)_2(OH)_6 \cdot 3H_2O$, G Hex. Silky white, pale rose. Tsumeb, Namibia. H=?, D_m= 4.55, D_c= 4.674, VID 06. Frondel & Strunz, 1960. Am. Min. 45(1960), 1313 (Abst.). Struct. Repts. 41A (1975), 345. See also Despujolsite, Schaurteite.

Fletcherite, $Cu(Ni,Co)_2S_4$, A Cub. Metallic grey. Fletcher mine, Viburnum Trend, Reynolds Co., Missouri, USA. H= 4.5-5, D_m=?, D_c= 4.76, IIC 01. Craig & Carpenter, 1977. Am. Min. 62(1977), 1057 (Abst.).

Flinkite, $Mn_3AsO_4(OH)_4$, G Orth. Vitreous/greasy greenish-brown, green. Harstigen mine, Pajsberg, Värmland, Sweden. H= 4.5, D_m= 3.87, D_c= 3.76, VIIB 10b. Hamberg, 1889. Can. Min. 7 (1963), 547. Am. Min. 52 (1967), 1603.

Florencite-(Ce), $CeAl_3(PO_4)_2(OH)_6$, G Rhom. Greasy/resinous pale yellow, colorless, pink, grey, etc. Diamantina, Minas Gerais, Brazil. H= 5-6, D_m= 3.6, D_c= 3.72, VIIB 15b. Husak & Prior, 1899. Can. Min. 18 (1980), 301.

Florencite-(La), $(La,Ce)Al_3(PO_4)_2(OH)_6$, G Rhom. Translucent colorless, pale yellow. Shituru deposit, Likasi (near), Shaba, Zaïre. H= 5, D_m= 3.52, D_c= 3.71, VIIB 15b. Lefebvre & Gasparrini, 1980. Can. Min. 18 (1980), 301.

Florencite-(Nd), $(Nd,Ce)Al_3(PO_4)_2(OH)_6$, G Rhom. Sausalito, Marin Co., California, USA. H=?, D_m=?, D_c= 3.70, VIIB 15b. Milton & Bastron, 1971. Powd. Diff. 1 (1986), 330.

Florensovite, $Cu(Cr,Sb)_2S_4$, A Cub. Metallic black; pale cream in reflected light. Slyudyanka complex, Lake Baikal region (S), USSR. H= 5, D_m=?, D_c= 4.28, IIC 01. Reznitskii et al., 1989. ZVMO 118 (1989), 57.

Fluckite, $CaMn(AsO_3OH)_2 \cdot 2H_2O$, A Tric. Translucent pale/deep rose. Ste.-Marie-aux-Mines, Vosges Mtns., Haut-Rhin, Alsace, France. H= 3.5-4, D_m= 3.05, D_c= 3.11, VIIC 15. Bari et al., 1980. Am. Min. 65(1980), 1066 (Abst.).

Fluellite, $Al_2(PO_4)F_2(OH)\cdot 7H_2O$, [G] Orth. Vitreous, colorless, white, yellow. Stenna Gwyn, St. Austell (near), Cornwall, England. H= 3, D_m= 2.18, D_c= 2.16, IIID 01. Lévy, 1824. Dana, 7th ed.(1951), v.2, 124. Am. Min. 51 (1966), 1579.

Fluoborite, $Mg_3(BO_3)(F,OH)_3$, [G] Hex. Translucent colorless, white. Norberg, Sweden. H= 3.5, D_m= 2.98, D_c= 2.915, Vc 01c. Geijer, 1926. Min. Rec. 6 (1975), 174. TMPM 21 (1974), 94.

Fluocerite-(Ce), $(Ce,La)F_3$, [G] Hex. Translucent colorless, pale yellow. Finbo and Broddbo, Dalarne, Sweden. H= 4-5, D_m= 6.0, D_c= 6.37, IIIA 09. Haidinger, 1845. Min. Mag. 47 (1983), 41. Sov.Phys.Cryst. 33(1988), 105.

Fluocerite-(La), $(La,Ce)F_3$, [G] Rhom. Vitreous pale greenish-yellow. Kazakhstan, USSR. H= 4.7, D_m= 5.93, D_c= 5.94, IIIA 09. Chistyakova & Kazakova, 1969. JCPDS 32-483.

Fluorapatite, $Ca_5(PO_4)_3F$, [G] Hex. Translucent green, blue, pink, etc. H= 5, D_m= 3.18, D_c= 3.183, VIIB 16. Rammelsberg, 1860. Am. Min. 55 (1970), 170. Am. Min. 74 (1989), 870.

Fluorapophyllite, $KCa_4Si_8O_{20}(F,OH)\cdot 8H_2O$, [A] Tet. Pearly/vitreous white, greyish. H= 4.5-5, D_m= 2.3, D_c= 2.370, VIIIE 02. Dunn & Wilson, 1978. Min. Rec. 9 (1978), 95. Am. Min. 56 (1971), 1222. See also Hydroxyapophyllite, Natroapophyllite.

Fluorellestadite, $Ca_5(SiO_4,SO_4,PO_4)_3(F,OH,Cl)$, [A] Hex. Translucent pale rose. Kopeysk, Chelyabinsk, Ural Mts. (S), USSR. H= 5, D_m= 3.07, D_c= 2.48, VIIIA 29. Rouse & Dunn, 1982. Am. Min. 67 (1982), 90. See also Hydroxylellestadite, Apatite.

Fluorite, CaF_2, [G] Cub. Vitreous colorless, green, blue, yellow, purple, etc. H= 4, D_m= 3.180, D_c= 3.180, IIIA 07. Agricola, 1529. Am. Min. 37 (1952), 910. See also Frankdicksonite.

Foggite, $CaAlPO_4(OH)_2\cdot H_2O$, [A] Orth. Translucent colorless, white. Palermo #1 mine, North Groton, Grafton Co., New Hampshire, USA. H= 4, D_m= 2.78, D_c= 2.771, VIID 13. Moore et al., 1975. Am. Min. 60 (1975), 957. Am. Min. 60 (1975), 965.

Foordite, $Sn(Nb,Ta)_2O_6$, [A] Mon. Adamantine, brownish-yellow. Lutsiro, Sebeya River, Rwanda; also Punia, Zaïre. H= 6, D_m= 6.73, D_c= 6.66, IVD 14c. Černý et al., 1988. Can. Min. 26 (1988), 889. Can. Min. 26 (1988), 899. See also Thoreaulite.

Forbesite, [D] A mixture of cobaltoan annabergite and arsenolite. Can. Min. 14 (1976), 414. Kenngott, 1868.

Formanite-(Y), $YTaO_4$, [G] Tet. Vitreous/sub-metallic black. Cooglegong, Western Australia, Australia. H= 5.5-6.5, D_m=?, D_c= 7.03, IVD 13. Berman & Frondel, 1944. Dana. 7th ed. (1944), v.1, 757. See also Fergusonite-(Y).

Fornacite, $CuPb_2(CrO_4)(AsO_4)(OH)$, [G] Mon. Translucent olive-green. Renéville, Djoué, Congo. H=?, D_m= 6.27, D_c= 6.30, VIE 02. Lacroix, 1915. Bull. Min. 103 (1980), 469. Zts. Krist. 124 (1967), 385. See also Molybdofornacite, Vauquelinite.

Forsterite, Mg_2SiO_4, [G] Orth. Translucent green, yellow. Monte Somma, Mt. Vesuvius, Napoli, Campania, Italy. H= 7, D_m= 3.222, D_c= 3.223, VIIIA 03. Lévy, 1824. DHZ, 2nd ed.(1982), v.1A, 2. Am. Min. 61 (1976), 1280. See also Fayalite, Ringwoodite, Wadsleyite.

Foshagite, $Ca_4(SiO_3)_3(OH)_2$, [G] Tric. Silky white. Crestmore, Riverside Co., California, USA. H=?, D_m= 2.73, D_c= 2.74, VIIID 08. Eakle, 1925. Am. Min. 43 (1958), 1. Acta Cryst. 13 (1960), 785.

Foshallasite, $Ca_3Si_2O_7\cdot 3H_2O$(?), [Q] Syst=? Pearly white. Kola Peninsula, USSR. H= 2.5-3, D_m= 2.5, D_c=?, VIIIB 06. Chirvinsky, 1936. Min. Abstr. 7 (1938), 10.

Foucherite, [D] Probably = delvauxite. Am. Min. 65(1980), 813 (Abst.). Lacroix, 1910.

Fourmarierite, $PbO_3(UO_2)_4(OH)_4 \cdot 4H_2O$, G Orth. Translucent reddish-orange, brown. Shinkolobwe, Shaba, Zaïre. H= 3-4, D_m= 5.74, D_c= 5.978, IVF 13. Buttgenbach, 1924. Am. Min. 45 (1960), 1026. Bull. Min. 108 (1985), 659.

Fraipontite, $(Zn, Al)_3(Si, Al)_2O_5(OH)_4$, G Mon. Silky colorless, yellowish-white. Moresnet, Vieille Montagne, Belgium (E). H=?, D_m=?, D_c= 3.54, VIIIE 10a. Cesaro, 1927. Bull. Min. 98 (1975), 235.

Francevillite, $(Ba, Pb)(UO_2)_2(VO_4)_2 \cdot 5H_2O$, G Orth. Translucent yellow. Mounana mine, Franceville, Haut-Ogooué, Gabon. H= 3, D_m= 4.55, D_c= 4.559, VIID 23. Branche et al., 1957. Dokl.Earth Sci. 220(1975), 137. N.Jb.Min.Mh. (1986), 552. See also Curienite.

Franciscanite, $Mn_6V(SiO_4)_2(O, OH)_6$, A Hex. Vitreous red, brownish-red. Pennsylvania mine, San Antonio Valley, Santa Clara Co., California, USA. H= 4, D_m= 4.1, D_c= 3.93, VIIIA 17. Dunn et al., 1986. Am. Min. 71 (1986), 1522. N.Jb.Min.Mh. (1986), 493.

Franckeite, $Pb_5Sn_3Sb_2S_{14}$, G Tric. Metallic greyish-black; gray-white in reflected light. Las Animas dist., Chocaya (SE of), Bolivia. H= 2.5-3, D_m= 6.01, D_c= 5.88, IID 13. Stelzner, 1893. Dana, 7th ed.(1944), v.1, 448. Bull. Min. 84 (1961), 350.

Francoanellite, $H_6(K, Na)_3(Al, Fe)_5(PO_4)_8 \cdot 13H_2O$, A Rhom. Earthy yellow-white. Grotte di Castellana, Puglia, Italy. H=?, D_m= 2.26, D_c= 2.11, VIIC 01. Balenzano et al., 1976. Am. Min. 61(1976), 1054 (Abst.).

Françoisite-(Nd), $(Nd, Y, Sm, Ce, Pr)(UO_2)_3(PO_4)_2O(OH) \cdot 6H_2O$, A Mon. Vitreous yellow. Kamoto-Est deposit, Kolwezi (near), Zaïre. H= 3, D_m=?, D_c= 4.63, VIID 21. Piret et al., 1988. Bull. Min. 111 (1988), 443. Bull. Min. 111 (1988), 443.

Franconite, $Na_2Nb_4O_{11} \cdot 9H_2O$, A Mon. Vitreous/silky white. Francon quarry, St.-Michel Dist., Montreal Island, Québec, Canada. H=?, D_m= 2.72, D_c= 2.71, IVF 16. Sabina et al., 1984. Can. Min. 22 (1984), 239.

Frankdicksonite, BaF_2, A Cub. Vitreous colorless. Carlin mine, Elko (NW), Eureka Co., Nevada, USA. H= 2.8, D_m= 4.89, D_c= 4.885, IIIA 07. Radtke & Brown, 1974. Am. Min. 59 (1974), 885. See also Fluorite.

Franklinfurnaceite, $Ca_2Mn_4FeZn_2Si_2O_{10}(OH)_8$, A Mon. Vitreous dark brown. Franklin, Sussex Co., New Jersey, USA. H= 3, D_m= 3.66, D_c= 3.737, VIIIE 09a. Dunn et al., 1987. Am. Min. 72 (1987), 812. Am. Min. 73 (1988), 876.

Franklinite, $(Zn, Mn, Fe)(Fe, Mn)_2O_4$, G Cub. Metallic/sub-metallic black; white in reflected light. Franklin, Sussex Co., New Jersey, USA. H= 5.5-6.5, D_m= 5.2, D_c= 5.16, IVB 01. Berthier, 1819. Am. Min. 64 (1979), 599.

Fransoletite, $Ca_3Be_2(PO_4)_2(PO_3OH)_2.4H_2O$, A Mon. Sub-vitreous colorless, white. Tip Top Pegmatite, Custer (near), Custer Co., South Dakota, USA. H= 3, D_m= 2.56, D_c= 2.53, VIIC 02. Peacor et al., 1983. Am. Min. 70(1985), 215 (Abst.).

Franzinite, $(Na, Ca)_7(Si, Al)_{12}O_{24}(SO_4, OH)_3 \cdot H_2O$, A Rhom. Pearly white. Pitigliano, Grosseto, Toscana, Italy. H= 5, D_m= 2.49, D_c= 2.52, VIIIF 05. Merlino & Orlandi, 1977. Am. Min. 62(1977), 1259 (Abst.).

Freboldite, γ–CoSe, G Hex. Metallic copper-red. Steinbruch Trogtal, Lautenthal (near), Harz, Germany. H=?, D_m=?, D_c= 7.56, IIB 09a. Strunz, 1957. Am. Min. 44(1959), 907 (Abst.).

Fredrikssonite, $Mg_2(Mn, Fe)O_2(BO_3)$, A Orth. Vitreous reddish-brown. Långban mine, Filipstad (near), Värmland, Sweden. H= 6, D_m= 3.84, D_c= 3.80, Vc 01c. Dunn et al., 1983. Am. Min. 71(1986), 227 (Abst.). See also Orthopinakiolite, Pinakiolite, Takeuchiite.

Freedite, $CuPb_8(AsO_3)_2O_3Cl_5$, A Mon. Greasy/vitreous greenish-yellow. Långban mine, Filipstad (near), Värmland, Sweden. H= 3, D_m= 7.0, D_c= 7.43, IIID 03. Dunn & Rouse, 1985. Am. Min. 70 (1985), 845. Min. Petrol. 36 (1987), 85.

Freibergite, $Cu_6(Ag,Fe)_6Sb_4S_{13}$, G Cub. Metallic grey. Freiberg, Sachsen (Saxony), Germany. H=?, D_m=?, D_c= 5.19, IID 01a. Kenngott, 1853. Am. Min. 60(1975), 489 (Abst.). Min. Mag. 50 (1986), 717. See also Tetrahedrite.

Freieslebenite, $AgPbSbS_3$, G Mon. Metallic grey; white in reflected light. Himmelsfürst mine, Freiberg, Sachsen (Saxony), Germany. H= 2-2.5, D_m= 6.1, D_c= 6.194, IID 07. Haidinger, 1845. Dana, 7th ed.(1944), v.1, 416. Zts. Krist. 139 (1974), 85. See also Laffittite, Marrite.

Fresnoite, $Ba_2TiO(Si_2O_7)$, A Tet. Vitreous yellow. Big Creek – Rush Creek area, Fresno Co., California, USA. H= 4.7, D_m= 4.446, D_c= 4.520, VIIIB 02a. Alfors et al., 1965. Min. Abstr. 81-1016. Zts. Krist. 130 (1969), 438.

Freudenbergite, $Na_2(Ti,Fe)_8O_{16}$, G Mon. Translucent blackish. Katzenbuckel, Odenwald, Germany. H=?, D_m= 4.3, D_c= 3.97, IVC 12. Frenzel, 1961. Am. Min. 46(1961), 765 (Abst.). Acta Cryst. B34 (1978), 255.

Freyalite, D A mixture of thorian melanocerite and alteration products. Am. Min. 70 (1985), 1059. Brøgger, 1890.

Friedelite, $Mn_8Si_6O_{15}(OH,Cl)_{10}$, G Mon. Translucent rose-red. Adervielle mine, Louron valley, Hautes Pyrénées, France. H= 4-5, D_m= 3.07, D_c= 3.06, VIIIE 12. Bertrand, 1876. Am. Min. 66 (1981), 1054. Can. Min. 21 (1983), 7. See also Caryopilite, McGillite.

Friedrichite, $Cu_5Pb_5Bi_7S_{18}$, A Orth. Metallic creamy yellowish-white in reflected light. Habachtal, Sedl, Oberpinzgau, Salzburg, Austria. H= 4, D_m= 6.98, D_c= 7.06, IID 05a. Chen et al., 1978. Can. Min. 16 (1978), 127.

Frigidite, D A mixture of tetrahedrite and Ni-bearing minerals. Min. Mag. 43 (1979), 99. d'Achiardi, 1881.

Fritzscheite, $Mn(UO_2)_2(VO_4)_2 \cdot 4H_2O$, G Orth. Vitreous/pearly reddish-brown, red. Johanngeorgenstadt, Sachsen (Saxony), Germany. H= 2-3, D_m=?, D_c= 4.39, VIID 23. Breithaupt, 1865. Bull. Min. 93 (1970), 320.

Frohbergite, $FeTe_2$, G Orth. Metallic pinkish-white. Robb-Montbray mine, Abitibi Co., Québec (NW), Canada. H= 4, D_m=?, D_c= 8.057, IIC 07. Thompson, 1947. Am. Min. 34 (1949), 342. Struct. Repts. 54A (1987), 71. See also Mattagamite.

Frolovite, $Ca[B(OH)_4]_2$, G Tric. Translucent white (greyish tint). Novo-Frolovsk deposit, Turinsk region, Ural Mts., USSR. H= 3.5, D_m= 2.14, D_c= 2.169, Vc 03a. Petrova, 1957. Am. Min. 43(1958), 385 (Abst.). Min. Abstr. 78-262.

Frondelite, $(Mn,Fe)Fe_4(PO_4)_3(OH)_5$, G Orth. Vitreous brown. Sapucaia Pegmatite, Conselheiro Pena, Galileia, Minas Gerais, Brazil. H= 4.5, D_m= 3.476, D_c= 3.55, VIIB 07. Lindberg, 1949. Am. Min. 34 (1949), 541. See also Rockbridgeite.

Froodite, α–$PdBi_2$, A Mon. Metallic grey. Frood mine, Sudbury Dist., Ontario, Canada. H= 2.5, D_m= 12.5, D_c= 11.52, IIC 07. Hawley & Berry, 1958. Can. Min. 6 (1958), 200.

Fukalite, $Ca_4Si_2O_6CO_3(OH,F)_2$, A Orth. Translucent white, pale brown. Bicchu, Fuka, Okayama, Japan. H=?, D_m= 2.770, D_c= 2.77, VIIIA 10. Henmi et al., 1977. Am. Min. 63(1978), 793 (Abst.).

Fukuchilite, Cu_3FeS_8, Q Cub. Submetallic pinkish-brown. Hanawa mine, Iwate, Hanawa, Honshu, Japan. H= 6, D_m=?, D_c= 4.80, IIC 05. Kajiwara, 1969. Am. Min. 74 (1989), 1168.

Fülöppite, $Pb_3Sb_8S_{15}$, G Mon. Metallic grey; white in reflected light. Săcărâmb (Nagyág), Transylvania, Romania. H= 2.5, D_m= 5.22, D_c= 5.19, IID 09a. Finaly & Koch, 1929. Dana, 7th ed.(1944), v.1, 463. Acta Cryst. B31 (1975), 151.

Furongite, $Al_{13}(UO_2)_7(PO_4)_{13}(OH)_{14} \cdot 58H_2O$, A Tric. Vitreous yellow. Hunan, China. H=?, D_m= 2.9, D_c= 2.90, VIID 21. Am. Min. 73(1988), 198 (Abst.). Acta Cryst. A37 (1981), C186.

Furutobeite, $(Cu,Ag)_6PbS_4$, A Mon. Metallic creamy-grey. Furutobe mine, Akita, Japan. H= 3, D_m=?, D_c= 6.74, IIA 09. Srigaki et al., 1981. Bull. Min. 104 (1981), 737.

G

Gabrielsonite, $PbFeAsO_4(OH)$, [A] Orth. Adamantine greenish-brown, black. Långban mine, Filipstad (near), Värmland, Sweden. H= 3.5, D_m= 6.67, D_c= 6.69, VIIB 11b. Moore, 1967. Am. Min. 53(1968), 1063 (Abst.).

Gadolinite-(Ce), $Be_2Fe(Ce,La,Nd,Y)_2Si_2O_{10}$, [G] Mon. Vitreous black. Skien, Oslo (near), Norway. H=?, D_m= 4.20, D_c= 4.90, VIIIA 23. Segalstad & Larsen, 1978. Am. Min. 63 (1978), 188.

Gadolinite-(Y), $Be_2FeY_2Si_2O_{10}$, [G] Mon. Vitreous/greasy black, greenish-black, brown. H= 6.5-7, D_m= 4.4, D_c= 4.307, VIIIA 23. Klaproth, 1800. Am. Min. 59 (1974), 700. Am. Min. 69 (1984), 948.

Gagarinite-(Y), $NaCaY(F,Cl)_6$, [G] Hex. Vitreous creamy, yellowish, rosy. Kazakhstan, USSR. H= 4.6, D_m= 4.21, D_c= 4.02, IIIA 07. Stepanov & Severov, 1961. Am. Min. 47(1962), 805 (Abst.).

Gageite-1Tc, $Mn_{21}O_3(Si_4O_{12})_2(OH)_{20}$, [P] Tric. Franklin, Sussex Co., New Jersey, USA. H=?, D_m=?, D_c= 3.61, VIIIB 24. Phillips, 1911. Am. Min. 72 (1987), 382. Am. Min. 72 (1987), 382. See also Balangeroite, Gageite-2M.

Gageite-2M, $Mn_{21}O_3(Si_4O_{12})_2(OH)_{20}$, [P] Mon. Translucent, colorless, pink. Franklin, Sussex Co., New Jersey, USA. H=?, D_m= 3.46, D_c= 3.599, VIIIB 24. Phillips, 1911. Am. Min. 72 (1987), 382. Am. Min. 72 (1987), 382. See also Balangeroite, Gageite-1Tc.

Gahnite, $ZnAl_2O_4$, [G] Cub. Vitreous dark blue-green, yellow, brown. H= 7.5-8, D_m=?, D_c= 4.62, IVB 01. von Moll, 1807. Dana, 7th ed.(1944), v.1, 689. Zts. Krist. 120 (1964), 476. See also Spinel, Hercynite.

Gaidonnayite, $Na_2ZrSi_3O_9 \cdot 2H_2O$, [A] Orth. Vitreous colorless, white, beige. Mont Saint-Hilaire, Rouville Co., Québec, Canada. H= 5, D_m= 2.67, D_c= 2.70, VIIIC 03. Chao & Watkinson, 1974. Can. Min. 12 (1974), 316. Can. Min. 23 (1985), 11. See also Catapleiite, Georgechaoite.

Gainesite, $Na_2(Be,Li)(Zr,Zn)_2(PO_4)_4$, [A] Tet. Vitreous pale bluish-lavender. Lower Pit and Twin Tunnels, Newry, Oxford Co., Maine, USA. H= 4, D_m= 2.94, D_c= 2.84, VIIA 01. Moore et al., 1983. Am. Min. 68 (1983), 1022. Am. Min. 68 (1983), 1022.

Gaitite, $Ca_2Zn(AsO_3OH)_2(OH)_2$, [A] Tric. Vitreous white, colorless. Tsumeb, Namibia. H= 5, D_m= 3.81, D_c= 3.80, VIIC 12. Sturman & Dunn, 1980. Can. Min. 18 (1980), 197. See also Talmessite.

Gajite, [D] A mixture of calcite and brucite. Min. Mag. 33 (1962), 262. Tucan, 1911.

Galaxite, $MnAl_2O_4$, [G] Cub. Vitreous red-brown. Bald Knob mine, Galax, Alleghany Co., North Carolina, USA. H= 7.5-8, D_m=?, D_c= 4.03, IVB 01. Ross & Kerr, 1932. Am. Min. 68 (1983), 449.

Galeite, $Na_{15}(SO_4)_5ClF_4$, [G] Rhom. Translucent colorless, white. Searles Lake, San Bernardino Co., California, USA. H=?, D_m= 2.605, D_c= 2.596, VIB 04. Pabst et al., 1963. Am. Min. 48 (1963), 485. Min. Mag. 40 (1975), 357.

Galena, PbS, [G] Cub. Metallic grey; white in reflected light. H= 2.6, D_m= 7.58, D_c= 7.596, IIB 11. Pliny, 77. Dana, 7th ed.(1944), v.1, 200. See also Clausthalite, Altaite.

Galenobismutite, $PbBi_2S_4$, [G] Orth. Metallic grey/white; white in reflected light. Ko mine, Nordmark, Värmland, Sweden. H= 2.5-3.5, D_m= 7.04, D_c= 7.09, IID 09e. Sjögren, 1878. Dana, 7th ed.(1944), v.1, 471. Acta Cryst. 15 (1962), 691.

Galenobornite, [D] Probably a mixture. Min. Mag. 36 (1967), 133. Satpaeva et al., 1964.

Galkhaite, $(Cs,Tl)(Hg,Cu,Zn)_6(As,Sb)_4S_{12}$, [A] Cub. Vitreous dark orange-red; grey in reflected light. Gal-Khaya, Yakutiya, USSR. H= 3, D_m= 5.4, D_c= 5.34, IID 01a. Grudzen et al., 1972. Am. Min. 59(1974), 208 (Abst.). Can. Min. 19 (1981), 571.

Gallite, $CuGaS_2$, G Tet. Metallic grey. Tsumeb, Namibia. H= 3-3.5, D_m=?, D_c= 4.40, IIB 02. Strunz et al., 1958. Am. Min. 44(1959), 906 (Abst.).

Gamagarite, $Ba_2(Fe,Mn)(VO_4)_2(OH)$, G Mon. Adamantine dark brown. Postmasburg manganese deposits, Cape Province, South Africa. H= 4.5-5, D_m= 4.62, D_c= 4.71, VIIC 17. de Villiers, 1943. Am. Min. 69 (1984), 803. N.Jb.Min.Mh. (1987), 295.

Gananite, α–BiF_3, A Cub. Semi-metallic brown/black. Laiking dist., Jiangxi, China. H= 3-3.5, D_m=?, D_c= 8.93, IIIA 09. Cheng et al., 1984. Am. Min. 73(1988), 1494 (Abst.).

Ganomalite, $Ca_5Pb_9MnSi_9O_{33}$, G Hex. Resinous/vitreous colorless, grey. Långban mine, Filipstad (near), Värmland, Sweden. H= 3, D_m= 5.74, D_c= 5.69, VIIIA 29. Nordenskiöld, 1876. Min. Mag. 49 (1985), 579. See also Nasonite.

Ganophyllite, $(K,Na)_6(Mn,Al,Mg)_{24}(Si,Al)_{40}O_{96}(OH)_{16} \cdot 21H_2O$, G Mon. Translucent brown. Harstigen mine, Pajsberg, Värmland, Sweden. H= 4-4.5, D_m= 2.84, D_c= 2.875, VIIIE 07b. Hamberg, 1890. Min. Mag. 47 (1983), 563. Min. Mag. 50 (1986), 307. See also Eggletonite.

Garavellite, $FeSbBiS_4$, A Orth. Metallic olive-grey. Valle del Frigido, Apennines Alps, Toscana, Italy. H= 3.8, D_m=?, D_c= 5.64, IID 06b. Gregorio et al., 1979. Min. Mag. 43 (1979), 99.

Garnet, A group name for orthosilicates with the general formula $(Ca,Fe,Mg,Mn)_3(Al,Fe,Mn,Cr,Ti,V)_2(SiO_4)_3$.

Garrelsite, $NaBa_3B_7Si_2O_{16}(OH)_4$, G Mon. Translucent colorless. South Ouray No. 1 Well, Uintah Co., Utah, USA. H=?, D_m= 3.68, D_c= 3.890, VIIIA 23. Milton et al., 1955. Am. Min. 59 (1974), 632. Acta Cryst. B32 (1976), 824.

Garronite, $NaCa_{2.5}(Al_6Si_{10})O_{32} \cdot 13H_2O$, G Tet. Transparent colorless. Garron Plateau, Antrim Co., Northern Ireland. H=?, D_m= 2.16, D_c= 2.18, VIIIF 14. Walker, 1962. Natural Zeolites (1985), 122.

Gartrellite, $(Cu,Fe)_2Pb(AsO_4,SO_4)_2(CO_3,H_2O)_{0.7}$, A Tric. Chalky yellow, greenish-yellow. Anticline prospect, Ashburton Downs, Western Australia and Broken Hill, New South Wales, Australia. H=?, D_m=?, D_c= 5.38, VIIC 17. Nickel et al., 1989. Austral. Min. 4 (1989), 83.

Garyansellite, $(Mg,Fe)_3(PO_4)_2(OH,H_2O)_3$, A Orth. Vitreous brown. Yukon Territory, Canada. H= 4, D_m= 3.16, D_c= 3.154, VIIC 04. Sturman & Dunn, 1984. Am. Min. 69 (1984), 207. See also Kryzhanovskite.

Gasparite-(Ce), $(Ce,La,Nd)AsO_4$, A Mon. Translucent light brown-red. Cherbadung, Binntal Valais (Wallis), Switzerland and Pizzo Cervandone, Val Dévero, Piemonte, Italy. H= 4.5, D_m=?, D_c= 5.63, VIIA 11. Graeser & Schwander, 1987. Am. Min. 73(1988), 1494 (Abst.).

Gaspéite, $(Ni,Mg,Fe)CO_3$, A Rhom. Vitreous/dull light green. Gaspé Peninsula, Lemieux Township, Gaspé-ouest Co., Québec, Canada. H= 4.5-5, D_m= 3.71, D_c= 4.358, VbA 02. Kohls & Rodda, 1966. Am. Min. 51 (1966), 677. Acta Cryst. C42 (1986), 4. See also Magnesite.

Gatumbaite, $CaAl_2(PO_4)_2(OH)_2 \cdot H_2O$, A Mon. Pearly white. Buranga pegmatite, Rwanda. H= < 5, D_m= 2.92, D_c= 2.95, VIID 13. von Knorring & Fransolet, 1977. Am. Min. 63(1978), 793 (Abst.).

Gaudefroyite, $Ca_4Mn_{3-x}(BO_3)_3(CO_3)(O,OH)_3$, A Hex. Brilliant/dull black. Tachgagalt, Morocco. H= 6, D_m= 3.4, D_c= 3.53, Vc 15. Jouravsky & Permingeat, 1964. Am. Min. 50(1965), 806 (Abst.). Sov.Phys.Cryst. 20(1975), 87.

Gaylussite, $Na_2Ca(CO_3)_2 \cdot 5H_2O$, G Mon. Translucent colorless, yellowish-white, greyish-white, white. Lagunillas, Merida (near), Venezuela. H= 2.5-3, D_m= 1.991, D_c= 1.991, VbC 04. Boussingault, 1826. Am. Min. 52 (1967), 1570. Struct. Repts. 33A (1968), 435.

Gearksutite, $CaAl(F,OH)_5 \cdot H_2O$, [G] Syst=? Earthy white. Ivigtut, Greenland (SW). H= 2, D_m= 2.768, D_c=?, IIIC 09. Dana, 1868. Dana, 7th ed.(1951), v.2, 119.

Gebhardite, $Pb_8O(As_2O_5)_2Cl_6$, [A] Mon. Adamantine brown. Tsumeb, Namibia. H= 3, D_m=?, D_c= 6.0, IIID 03. Medenbach et al. Am. Min. 70(1985), 215 (Abst.).

Gedrite, $(Mg,Fe)_5Al_2(Si_6Al_2)O_{22}(OH)_2$, [A] Orth. Translucent white, grey, green brown. Héas, Gèdres, France. H= 5.5-6, D_m=?, D_c= 3.2, VIIID 06. Dufrenoy, 1836. Am. Min. 63 (1978), 1023. Am. Min. 55 (1970), 1945. See also Magnesiogedrite, Ferrogedrite.

Geerite, Cu_8S_5, [A] Cub. Metallic bluish-white. Dekalb (near), St. Lawrence Co., New York, USA. H=?, D_m=?, D_c= 5.61, IIA 01a. Goble & Robinson, 1980. Can. Min. 18 (1980), 519. Can. Min. 23 (1985), 61.

Geffroyite, $(Cu,Fe,Ag)_9(Se,S)_8$, [A] Cub. Metallic creamy-brown. Chaméane, Vernet-la-Varenne, Livardois Mtns., Puy-de-Dôme, France. H= 4, D_m=?, D_c= 5.39, IIA 07. Johan et al., 1982. Am. Min. 67(1982), 1074 (Abst.).

Gehlenite, $Ca_2Al(Si,Al)_2O_7$, [G] Tet. Translucent colorless, greyish-green, brown. Mt. Monzoni, Fassathal, Tyrol, Austria(?). H= 5-6, D_m= 3.07, D_c= 3.054, VIIIB 02a. Fuchs, 1815. DHZ, 2nd ed.(1986), v.1B, 285. N.Jb.Min.Abh. 144 (1982), 254. See also Åkermanite.

Geigerite, $Mn_5(AsO_4)_2(AsO_3OH)_2 \cdot 10H_2O$, [A] Tric. Vitreous/pearly pale rose-red. Falotta, Sursass (Oberhalbstein), Tinizong (Tinzen), Grischum (Graubünden), Switzerland. H= 3, D_m= 3.05, D_c= 3.00, VIIC 28. Graeser et al., 1989. Am. Min. 74 (1989), 676. Am. Min. 74 (1989), 676. See also Chudobaite.

Geikielite, $MgTiO_3$, [G] Rhom. Metallic/sub-metallic brownish-black. Rakwana, Sri Lanka. H= 5-6, D_m= 4.05, D_c= 3.89, IVC 04b. Dick, 1892. NBS Circ. 539, v.5 (1955), 43. See also Ilmenite, Pyrophanite.

Gelzircon, [D] Unnecessary name for type of zircon. Min. Mag. 36 (1967), 133. Semenov, 1960.

Genkinite, $(Pt,Pd)_4Sb_3$, [A] Tet. Metallic pale brown in reflected light. Onverwacht (#330) mine, Lydenberg Dist., Transvaal, South Africa. H= 5.5, D_m=?, D_c= 9.256, IIG 04. Cabri et al., 1977. Can. Min. 15 (1977), 389.

Genthelvite, $Be_3Zn_4(SiO_4)_3S$, [G] Cub. Translucent green, red, yellow. Cheyenne Canyon (W), El Paso Co., Colorado, USA. H=?, D_m= 3.7, D_c= 3.64, VIIIF 08. Glass et al., 1944. Am. Min. 73 (1988), 1384. Am. Min. 70 (1985), 186. See also Danalite, Helvite.

Gentnerite, [D] Inadequate data. Am. Min. 52(1967), 559 (Abst.). El Goresy & Ottemann, 1966.

Geocronite, $Pb_{14}(Sb,As)_6S_{23}$, [G] Mon. Metallic grey/greyish-blue; white in reflected light. Sala and Falun, Sweden. H= 2.5, D_m= 6.45, D_c= 6.56, IID 01f. Svanberg, 1852. Am. Min. 39 (1954), 908. Am. Min. 61 (1976), 963. See also Jordanite.

Georgechaoite, $KNaZrSi_3O_9 \cdot 2H_2O$, [A] Orth. Translucent colorless. Wind Mt., Otero Co., New Mexico, USA. H= 5, D_m= 2.70, D_c= 2.689, VIIIC 03. Boggs & Ghose, 1985. Can. Min. 23 (1985), 1. Can. Min. 23 (1985), 11. See also Gaidonnayite.

Georgeite, $Cu_5(CO_3)_3(OH)_4 \cdot 6H_2O$, [A] Amor. Vitreous/earthy blue. Carr Boyd mine, Kalgoorlie, Western Australia, Australia. H=?, D_m= 2.55, D_c=?, VbD 01. Bridge et al., 1979. Min. Mag. 43 (1979), 97.

Georgiadesite, $Pb_{16}(AsO_4)_4Cl_{14}(OH)_6$, [G] Mon. Resinous white, brownish-yellow. Lávrion (Laurium), Attikí, Greece. H= 3.5, D_m= 6.3, D_c= 6.44, IIID 03. Lacroix & de Schulten, 1907. Min. Mag. 47 (1983), 219.

Gerasimovskite, $(Mn,Ca)(Nb,Ti)_5O_{12} \cdot 9H_2O(?)$, [G] Amor. Pearly brown, grey. Lovozero massif, Kola Peninsula, USSR. H= 2, D_m= 2.55, D_c=?, IVF 17. Semenov, 1957. Am. Min. 43(1958), 1220 (Abst.). See also Manganbelyankinite.

Gerdtremmelite, $(Zn,Fe)(Al,Fe)_2AsO_4(OH)_5$, A Tric. Adamantine yellowish/dark brown. Tsumeb, Namibia. H=?, D_m=?, D_c= 3.66, VIIB 05. Schmetzer, 1985. Am. Min. 71(1986), 845 (Abst.).

Gerhardtite, $Cu_2NO_3(OH)_3$, G Orth. Transparent green. United Verde mine, Jerome, Yavapai Co., Arizona, USA. H= 2, D_m= 3.41, D_c= 3.389, Va 03. Wells & Penfield, 1885. Zts. Krist. 116 (1961), 210. Struct. Repts. 49A (1982), 251.

Germanite, $Cu_{26}Fe_4Ge_4S_{32}$, G Cub. Metallic pale greyish-pink. Tsumeb, Namibia. H= 4, D_m= 4.46, D_c= 4.47, IIB 03c. Pufahl, 1922. ZVMO 104 (1975), 28. Am. Min. 69 (1984), 943. See also Colusite, Nekrasovite.

Gersdorffite-Pa3, NiAsS, A Cub. Metallic white. H= 5.5, D_m= 5.9, D_c= 5.93, IIC 06a. Löwe, 1845. Can. Min. 24 (1986), 27. See also Gersdorffite-Pca2_1, Gersdorffite-P$2_1$3.

Gersdorffite-P$2_1$3, NiAsS, A Cub. Metallic white. H= 5.5, D_m= 5.9, D_c= 5.98, IIC 06a. Löwe, 1845. Can. Min. 24 (1986), 27. See also Gersdorffite-Pa3, Gersdorffite-Pca2_1.

Gersdorffite-Pca2_1, NiAsS, A Orth. Metallic white. H= 5.5, D_m= 5.9, D_c= 5.963, IIC 06a. Löwe, 1845. Can. Min. 24 (1986), 27. Am. Min. 67 (1982), 1058. See also Gersdorffite-Pa3, Gersdorffite-P$2_1$3.

Gerstleyite, $Na_2(Sb,As)_8S_{13}\cdot 2H_2O$, G Mon. Adamantine red. Baker mine, Kramer dist., Kern Co., California, USA. H= 2.5, D_m= 3.62, D_c= 3.53, IIE 03. Frondel & Morgan, 1956. Am. Min. 41 (1956), 839. Min. Abstr. 82M/1149.

Gerstmannite, $(Mn,Mg)MgZnSiO_4(OH)_2$, A Orth. Vitreous/sub-adamantine white, pale pink. Sterling Hill mine, Ogdensburg, Sussex Co., New Jersey, USA. H= 4.5, D_m= 3.68, D_c= 3.66, VIIIA 13. Moore & Araki, 1977. Am. Min. 62 (1977), 51. Am. Min. 62 (1977), 51.

Getchellite, $SbAsS_3$, A Mon. Resinous/vitreous dark red. Getchell mine, South Pit Extension, Humboldt Co., Nevada, USA. H= 1.5-2, D_m= 3.92, D_c= 3.98, IIB 19. Weissberg, 1965. Am. Min. 50 (1965), 1817. Acta Cryst. B29 (1973), 2536.

Geversite, $PtSb_2$, G Cub. Metallic light grey. Driekop mine, Transvaal, South Africa. H= 4.5-5, D_m=?, D_c= 10.91, IIC 05. Stumpfl, 1961. Min. Mag. 32 (1961), 833.

Gianellaite, $Hg_4SO_4N_2$, A Cub. Translucent straw-yellow. Mariposa mine, Terlingua Dist., Brewster Co., Texas, USA. H= 3, D_m= 7.19, D_c= 7.13, VIB 10. Tunell et al., 1977. Am. Min. 62(1977), 1057 (Abst.).

Gibbsite, γ-$Al(OH)_3$, G Mon. Vitreous/pearly white, greyish, greenish. Richmond, Berkshire Co., Massachusetts, USA. H= 2.5-3.5, D_m= 2.40, D_c= 2.421, IVF 01. Torrey, 1822. Dana, 7th ed. (1944), v.1, 663. Zts. Krist. 139 (1974), 129. See also Bayerite, Doyleite, Nordstrandite.

Giessenite, $Cu_2Pb_{26}(Bi,Sb)_{20}S_{57}$, A Mon. Metallic greyish-black; white in reflected light. Turtschi, Giessen, Binntal, Valais (Wallis), Switzerland. H= 2.5, D_m=?, D_c= 6.96, IID 08. Graeser, 1963. Can. Min. 24 (1986), 21.

Gilalite, $Cu_5Si_6O_{17}\cdot 7H_2O$, A Mon. Translucent green, blue-green. Christmas mine, Christmas, Gila Co., Arizona, USA. H= 2, D_m= 2.82, D_c= 2.54, VIIID 04. Cesbron & Williams, 1980. Min. Mag. 43 (1980), 639.

Gillespite, $BaFeSi_4O_{10}$, G Tet. Translucent red. Dry Delta, Alaska Range, Alaska, USA. H= 4, D_m= 3.33, D_c= 3.404, VIIIE 01. Schaller, 1922. Dana/Ford (1932), 584. Am. Min. 59 (1974), 1166.

Giniite, $Fe_5(PO_4)_4(OH)_2\cdot 2H_2O$, A Mon. Vitreous/greasy blackish-green/brown. Sandamab, Usakos (near), Namibia. H= 3-4, D_m= 3.41, D_c= 3.42, VIID 05. Keller, 1980. Min. Abstr. 81-3230.

Ginorite, $Ca_2B_{14}O_{23}\cdot 8H_2O$, G Mon. Transparent white. Sasso Pisano, Val di Cecina, Pisa, Toscana, Italy. H= 3.5, D_m= 2.09, D_c= 2.14, Vc 08. D'Achiardi, 1934. JCPDS 8-116. See also Strontioginorite.

Ginzburgite, $Ca_4Be_2Al_4Si_7O_{24}(OH)_4 \cdot 3H_2O$, Q Tet. Vitreous colorless. Murzinka region, Ural Mts., USSR. H= 3-4, D_m= 2.3, D_c= 2.37, VIIIF 20. Voloshin et al., 1986. Am. Min. 73(1988), 439 (Abst.).

Giorgiosite, $Mg_5(CO_3)_4(OH)_2 \cdot 5H_2O$, Q Syst=? Powdery white. Thíra (Santorini), Greece. H=?, D_m= 2.15, D_c=?, VbD 01. Lacroix, 1905. Dana, 7th ed.(1951), v.2, 274.

Giraudite, $(Cu,Zn,Ag)_{12}(As,Sb)_4(Se,S)_{13}$, A Cub. Metallic light grey. Chaméane, Vernet-la-Varenne, Livardois Mtns., Puy-de-Dôme, France. H=?, D_m=?, D_c= 5.75, IID 01a. Johan et al., 1982. Am. Min. 67(1982), 1074 (Abst.).

Girdite, $H_2Pb_3(TeO_3)(TeO_6)$, A Mon. Chalky white. Grand Central mine, Tombstone, Cochise Co., Arizona, USA. H= 2, D_m= 5.5, D_c= 5.49, VIG 08. Williams, 1979. Min. Mag. 43 (1979), 453.

Gismondine, $Ca_2Al_4Si_4O_{16} \cdot 9H_2O$, G Mon. Vitreous colorless, whitish, greyish. Capo di Bove, Roma, Italy. H= 4.5, D_m= 2.24, D_c= 2.223, VIIIF 14. Leonhard, 1817. Bull. Min. 107 (1984), 805. Struct. Repts. 37A (1971), 348.

Gittinsite, $CaZrSi_2O_7$, A Mon. Chalky white. Kipawa River Complex, Villedieu Twp., Témiscamingue Co., Québec, Canada. H= 3.5-4, D_m=?, D_c= 3.624, VIIIB 01. Ansell et al., 1980. Can. Min. 18 (1980), 201. Can. Min. 27 (1989), 703.

Giuseppettite, $(Na,K,Ca)_{7-8}(Si,Al)_{12}O_{24}(SO_4,Cl)_{1-2}$, A Hex. Translucent pale violet-blue. Sacrofano, Lazio (Latium), Roma, Italy. H= 6-7, D_m= 2.35, D_c= 2.365, VIIIF 05. Mazzi & Tadini, 1981. Am. Min. 67(1982), 415 (Abst.).

Gladite, $CuPbBi_5S_9$, G Orth. Metallic grey. Gladhammar, Småland, Kalmar, Sweden. H= 2-3, D_m= 6.96, D_c= 6.91, IID 05a. Johansson, 1924. Dana, 7th ed.(1944), v.1, 483. Acta Cryst. B32, (1976), 2401.

Glauberite, $Na_2Ca(SO_4)_2$, G Mon. Vitreous grey, yellowish, colorless. Villarubia, Ocana (near), Toledo, Spain. H= 2.5-3, D_m= 2.80, D_c= 2.78, VIA 06. Brongniart, 1808. Dana, 7th ed.(1951), v.2, 431. Am. Min. 52 (1967), 1272.

Glaucocerinite, $(Zn,Cu)_5Al_3(SO_4)_{1.5}(OH)_{16} \cdot 9H_2O$, G Rhom. Waxy sky-blue, turquoise blue. Lávrion (Laurium), Attikí, Greece. H= 1, D_m= 2.4, D_c= 2.33, IVF 03e. Dittler & Koechlin, 1932. Min. Mag. 49 (1985), 583.

Glaucochroite, $CaMnSiO_4$, G Orth. Translucent delicate bluish-green. Franklin Furnace, Sussex Co., New Jersey, USA. H= 6, D_m= 3.4, D_c= 3.496, VIIIA 04. Penfield & Warren, 1899. Dana/Ford (1932), 599. Am. Min. 63 (1978), 365. See also Kirschsteinite, Monticellite.

Glaucodot, $(Co,Fe)AsS$, G Orth. Metallic white. Huasco, Chile. H= 5, D_m= 6.04, D_c= 6.16, IIC 09. Breithaupt & Plattner, 1849. JCPDS 5-643. See also Alloclasite.

Glauconite, $(K,Na)(Fe,Al,Mg)_2(Si,Al)_4O_{10}(OH)_2$, G Mon. Translucent colorless, yellowish-green, green, blue-green. H= 2, D_m= 2.68, D_c= 2.90, VIIIE 05a. Keferstein, 1828. JCPDS 9-439. Am. Min. 20 (1935), 699.

Glaucophane, $Na_2(Mg,Fe)_3Al_2Si_8O_{22}(OH)_2$, G Mon. Vitreous pearly grey, blue, lavender-blue, bluish-black. Syra island, Cyclades Islands, Greece. H= 6, D_m= 3.1, D_c= 3.127, VIIID 05d. Hausmann, 1845. DHZ (1963), v.2, 333. Am. Min. 53 (1968), 1156. See also Ferroglaucophane.

Glaukosphaerite, $(Cu,Ni)_2CO_3(OH)_2$, A Mon. Sub-vitreous/silky green. Hampton East Location 48, Kambalda, Western Australia, Australia. H= 3-4, D_m= 3.78, D_c= 4.22, VbB 01. Pryce & Just, 1974. Min. Mag. 39 (1974), 737.

Glucine, $CaBe_4(PO_4)_2(OH)_4 \cdot 0.5H_2O$, Q Syst=? Ural Mts., USSR. H= 5, D_m= 2.3, D_c=?, VIID 01. Grigoriev, 1963. Am. Min. 49(1964), 1152 (Abst.).

Glushinskite, $MgC_2O_4 \cdot 2H_2O$, A Mon. Translucent creamy-white. Arctic, USSR. H= 2, D_m= 1.85, D_c= 1.865, IXA 01. Nefedov, 1960. Min. Mag. 43 (1980), 837.

Gmelinite, $Na_4(Al_4Si_8)O_{24}\cdot 11H_2O$, [G] Hex. Vitreous colorless, yellowish/greenish-white, reddish. Montecchio Maggiore, Vicenza, Italy; Glenarm, Ireland. H= 4.5, D_m= 2.01, D_c= 2.02, VIIIF 15. Brewster, 1825. Natural Zeolites (1985), 168. Min. Abstr. 83M/0165.

Gobbinsite, $(Na,K)_4Ca(Al_6Si_{10})O_{32}\cdot 12H_2O$, [A] Orth. Chalky white. Antrim Co., Magee Island, Northern Ireland. H=?, D_m= 2.194, D_c= 2.22, VIIIF 14. Nawaz & Malone, 1982. Min. Mag. 46 (1982), 365. Zts. Krist. 171 (1985), 281.

Godlevskite, $(Ni,Fe)_9S_8$, [A] Orth. Metallic yellow. Talnakh, Noril'sk (near), Siberia (N), USSR. H= 4.5-5, D_m=?, D_c= 5.273, IIA 07. Kulagov et al., 1969. Can. Min. 26 (1988), 283. Acta Cryst. C43 (1987), 2255.

Godovikovite, $(NH_4)(Al,Fe)(SO_4)_2$, [A] Hex. Dull white. Schooler mine, Kladno, Czechoslovakia and/or Kopeysk (near), Chelyabinsk coal basin, Ural Mts. (S), USSR. H= 2, D_m= 2.53, D_c= 2.52, VIA 03. Shcherbakova et al., 1988. ZVMO 117 (1988), 208.

Goedkenite, $(Sr,Ca)_2Al(PO_4)_2(OH)$, [A] Mon. Translucent pale yellow. Palermo #1 mine, North Groton, Grafton Co., New Hampshire, USA. H= 5, D_m=?, D_c= 3.83, VIIB 13. Moore et al., 1975. Am. Min. 60 (1975), 957.

Goethite, α–FeO(OH), [G] Orth. Adamantine-metallic/earthy brown, yellow. H= 5-5.5, D_m= 4.28, D_c= 4.26, IVF 04. Lenz, 1806. Dana, 7th ed.(1944), v.1, 680. Struct. Repts. 33A (1968), 266. See also Akaganéite, Feroxyhyte, Lepidocrocite, Bracewellite, Diaspore.

Gold, Au, [G] Cub. Metallic yellow. H= 2.5-3, D_m= 19.3, D_c= 19.273, IA 01a. Dana, 7th ed.(1944), v.1, 90. J. Appl. Cryst. 1 (1968), 123. See also Silver.

Goldamalgam, γ–Au_2Hg_3, [Q] Cub. Metallic yellow. Hongshila, Hebei, China. H= 3, D_m=?, D_c= 16.48, IA 06. Chen et al., 1981. Am. Min. 70(1985), 215 (Abst.).

Goldfieldite, $Cu_{12}(Te,Sb,As)_4S_{13}$, [G] Cub. Metallic lead grey/ iron-black. Mohawk mine, Goldfield, Esmeralda Co., Nevada, USA. H= 3-3.35, D_m=?, D_c= 4.95, IID 01a. Ransome, 1909. JCPDS 29-531. Zts. Krist. 185 (1988), 601.

Goldichite, $KFe(SO_4)_2\cdot 4H_2O$, [G] Mon. Translucent pale yellowish-green. Dexter 7 mine, Emery Co., San Rafael Swell, Calf Mesa, Utah, USA. H= 2.5, D_m= 2.43, D_c= 2.461, VIC 07. Rosenzweig & Gross, 1955. Am. Min. 40 (1955), 469. Am. Min. 56 (1971), 1917.

Goldmanite, $Ca_3(V,Al,Fe)_2(SiO_4)_3$, [A] Cub. Translucent green, brownish-green. Laguna, Valencia Co., New Mexico, USA. H=?, D_m= 3.74, D_c= 3.75, VIIIA 06a. Moench & Meyrowitz, 1964. Am. Min. 49 (1964), 644. Am. Min. 56 (1971), 791.

Gonnardite, $(Na,Ca)_2(Si,Al)_5O_{10}\cdot 3H_2O$, [G] Tet. Translucent white. Chaux de Bergogne, Gignat, Puy-de-Dôme, France. H= 4-4.5, D_m= 2.694, D_c= 2.33, VIIIF 10. Lacroix, 1896. Min. Mag. 52 (1988), 207. N.Jb.Min.Mh. (1986), 219.

Gonyerite, $(Mn,Mg)_5Fe(Si_3Fe)O_{10}(OH)_8$, [G] Orth. Deep brown. Långban mine, Filipstad (near), Värmland, Sweden. H= 2.5, D_m= 3.01, D_c= 3.03, VIIIE 09a. Frondel, 1955. Am. Min. 40 (1955), 1090.

Goongarrite, [D] A mixture of cosalite and galena. Am. Min. 62 (1977), 398 (Abst.). Simpson, 1924.

Goosecreekite, $Ca(Al_2Si_6)O_{16}\cdot 5H_2O$, [A] Mon. Vitreous/pearly colorless, white. Luck Goose Creek Quarry, Loudoun Co., Virginia, USA. H= 4.5, D_m= 2.21, D_c= 2.227, VIIIF 17. Dunn et al., 1980. Can. Min. 18 (1980), 323. Am. Min. 71 (1986), 1494. See also Epistilbite.

Gorceixite, $BaAl_3(PO_4)(PO_3OH)(OH)_6$, [G] Mon. Vitreous/dull brown. Brazil. H= 6, D_m= 3.1, D_c= 3.41, VIIB 15a. Hussak, 1906. Am. Min. 43 (1958), 688. N.Jb.Min.Mh. (1982), 446.

Gordonite, $MgAl_2(PO_4)_2(OH)_2\cdot 8H_2O$, [G] Tric. Vitreous/pearly smoky white, colorless. Clay Canyon, Fairfield (near), Utah Co., Utah, USA. H= 3.5, D_m= 2.23,

D_c= 2.319, VIID 03. Larsen & Shannon, 1930. Am. Min. 47 (1962), 1. N.Jb.Min.Mh. (1988), 265.

Görgeyite, $K_2Ca_5(SO_4)_6 \cdot H_2O$, G Mon. Vitreous colorless, yellowish. Ischl salt mine, Oberösterreich (Upper Austria), Austria. H= 3.5, D_m= 2.75, D_c= 2.887, VIC 12. Mayrhofer, 1953. Am. Min. 39(1954), 403 (Abst.). Min. Abstr. 82M/1152.

Gormanite, $(Fe, Mg)_3Al_4(PO_4)_4(OH)_6 \cdot 2H_2O$, A Tric. Vitreous blue-green. Rapid Creek, Big Fish River–Blow River Area, Yukon Territory, Canada. H= 4-5, D_m= 3.12, D_c= 3.13, VIID 28. Sturman et al., 1981. Can. Min. 19 (1981),381. See also Souzalite.

Gortdrumite, $(Cu, Fe)_6Hg_2S_5$, A Orth. Metallic white/grey. Gortdrum deposit, Tipperary Co., Ireland. H= 3.5-4, D_m=?, D_c= 6.80, IIA 08. Steed, 1983. Min. Mag. 47(1983), 35.

Goslarite, $ZnSO_4 \cdot 7H_2O$, G Orth. Vitreous/silky colorless, green, blue. Goslar, Harz, Germany. H= 2-2.5, D_m= 1.978, D_c= 1.94, VIC 03d. Haidinger, 1845. Dana, 7th ed.(1951), v.2, 513. See also Epsomite, Morenosite.

Götzenite, $(Ca, Na)_6Ti(Si_2O_7)_2(F, OH, O)_4$, G Tric. Translucent colorless. Volcanos Area, Rutshuru territory, Kivu, Zaïre. H=?, D_m= 3.138, D_c= 3.08, VIIIB 09. Sahama & Hytönen, 1957. Am. Min. 45 (1960), 221. Sov.Phys.Cryst. 16(1971), 1026.

Goudeyite, $Cu_6(Al, Y)(AsO_4)_3(OH)_6 \cdot 3H_2O$, A Hex. Translucent yellow-green. Majuba Hill mine, Pershing Co., Nevada, USA. H= 3-4, D_m= 3.50, D_c= 3.58, VIID 17. Wise, 1978. Am. Min. 63 (1978), 704.

Gowerite, $CaB_6O_8(OH)_4 \cdot 3H_2O$, G Mon. Vitreous colorless, white. Hard Scrabble Claim and Debley mine (NNW), Death Valley National Monument, Inyo Co., California, USA. H= 3, D_m= 2.00, D_c= 2.003, Vc 10. Erd et al., 1959. Am. Min. 44 (1959), 911. Am. Min. 57 (1972), 381.

Goyazite, $SrAl_3(PO_4)(PO_3OH)(OH)_6$, G Rhom. Greasy/resinous pink, yellow. Goyaz, Brazil. H= 4.5-5, D_m= 3.26, D_c= 3.265, VIIB 15a. Damour, 1884. Min. Mag. 47 (1983), 221. Min. Jour. 13 (1987), 390.

Graemite, $CuTeO_3 \cdot H_2O$, A Orth. Translucent blue-green. Cole Shaft, Bisbee, Cochise Co., Arizona, USA. H= 3-3.5, D_m= 4.13, D_c= 4.24, VIG 11. Williams & Matter, 1975. Am. Min. 60(1975), 486 (Abst.).

Graftonite, $(Fe, Mn, Ca)_3(PO_4)_2$, G Mon. Vitreous salmon-pink, reddish-brown, dark brown. Grafton (near), Melvin Mts., New Hampshire, USA. H= 5, D_m= 3.7, D_c= 3.95, VIIA 03. Penfield, 1900. Am. Min. 53 (1968), 742. Am. Min. 67 (1982), 826. See also Beusite.

Grandidierite, $(Mg, Fe)Al_3BSiO_9$, G Orth. Translucent bluish-green. Andrahomana, Taolagnaro (Fort-Dauphin), Madagascar. H= 7.5, D_m= 2.99, D_c= 2.96, VIIIA 24. Lacroix, 1902. Bull. Min. 99 (1976), 58. Acta Cryst. B24 (1968), 1518.

Grandreefite, $Pb_2(SO_4)F_2$, A Orth. Subadamantine colorless. Grand Reef mine, Laurel Canyon, Graham Co., Arizona, USA. H= 2.5, D_m= 7.0, D_c= 7.15, IIID 04. Kampf et al., 1989. Am. Min. 74 (1989), 927. See also Pseudograndreefite.

Grantsite, $Na_4Ca_{0.7}V_{12}O_{32} \cdot 8H_2O$, G Mon. Silky/sub-adamantine dark olive-green, greenish-black. New Mexico, Colorado, Utah, USA. H=?, D_m= 2.94, D_c= 2.95, IVF 09. Weeks et al., 1964. Am. Min. 49 (1964), 1511.

Graphite-2H, C, G Hex. Metallic black/grey; whitish/bluish-grey in reflected light. H= 1-2, D_m= 2.2, D_c= 2.271, IB 02a. Werner, 1789. Min. Abst. 77-308. See also Chaoite, Diamond, Lonsdaleite, Graphite-3R.

Graphite-3R, C, P Rhom. Metallic black/grey. H=?, D_m=?, D_c= 2.27, IB 02a. Werner, 1789. Zts. Krist. 107 (1956), 337. See also Chaoite, Diamond, Lonsdaleite, Graphite-2H.

Gratonite, $Pb_9As_4S_{15}$, G Rhom. Metallic grey. Excelsior mine, Cerro de Pasco, Peru. H= 2.5, D_m= 6.22, D_c= 6.18, IID 01f. Palache & Fisher, 1939. Dana, 7th ed.(1944), v.1, 397. Zts. Krist. 128 (1969), 321.

Grayite, $(Th, Pb, Ca)PO_4 \cdot H_2O$, Q Syst=? Zimbabwe. H=?, D_m=?, D_c=?, VIIC 19. Bowie, 1957. Am. Min. 47(1962), 419 (Abst.).

Greenalite, $(Fe)_{2-3}Si_2O_5(OH)_4$, G Mon. Translucent green. Mesabi Dist., St. Louis Co., Minnesota, USA. H=?, D_m=?, D_c= 2.92, VIIIE 10b. Leith, 1903. Min. Mag. 44 (1981), 153. Can. Min. 20 (1982), 1.

Greenockite, β-CdS, G Hex. Adamantine/resinous yellow/orange. Bishopton, Renfrewshire, Scotland. H= 3-3.5, D_m= 4.9, D_c= 4.818, IIB 06. Jameson, 1840. NBS Circ. 539 v.4 (1955), 15. See also Hawleyite, Cadmoselite, Wurtzite.

Gregoryite, α-$(Na, K, Ca,)_2CO_3$, A Hex. Oldoinyo Lengai, Tanzania. H=?, D_m=?, D_c= 2.27, VbA 06. Gittins & McKie, 1980. JCPDS 25-815.

Greigite, Fe_3S_4, A Cub. Metallic creamy-white. Kramer-Four Corners Area, San Bernardino Co., California, USA. H= 6-6.5, D_m=?, D_c= 4.08, IIC 01. Skinner et al., 1964. Am. Min. 49 (1964), 543.

Griceite, LiF, A Cub. Dull/vitreous white. Poudrette quarry, Mont Saint-Hilaire, Rouville Co., Québec, Canada. H= 4.5, D_m= 2.62, D_c= 2.67, IIIA 02. Van Velthuizen & Chao, 1989. Can. Min. 27 (1989), 125.

Grimaldiite, CrO(OH), A Rhom. Translucent deep red. Merume River, Mazaruni dist., Guayana. H=?, D_m= 5.49, D_c= 5.61, IVF 04. Milton et al., 1967. Am. Min. 62 (1977), 593. See also Bracewellite, Guyanaite, Heterogenite.

Grimselite, $K_3Na(UO_2)(CO_3)_3 \cdot H_2O$, G Hex. Translucent yellow. Gerstenegg (Kabelstollen Gerstenegg-Grimsel I), Grimsel, Bern, Switzerland. H= 2-2.5, D_m= 3.30, D_c= 3.27, VbD 04. Walenta, 1972. Am. Min. 58(1973), 139 (Abst.).

Griphite, $Ca(Mn, Na, Li)_6FeAl_2(PO_4)_6(F, OH)_2$, G Cub. Resinous/vitreous yellow, dark brown, brownish-black. Everly mine, Harney City, Pennington Co., South Dakota, USA. H= 5.5, D_m= 3.64, D_c= 3.65, VIIA 07. Headden, 1891. Bull. Min. 101 (1978), 536. Bull. Min. 101 (1978), 543.

Grischunite, $NaCa_2Mn_5Fe(AsO_4)_6 \cdot 2H_2O$, A Orth. Vitreous dark red-brown. Falotta, Sursass (Oberhalbstein), Tinizong (Tinzen), Grischun (Graubünden), Switzerland. H= 5, D_m= 3.8, D_c= 4.144, VIIA 03. Graeser et al., 1984. Am. Min. 71(1986), 227 (Abst.). Am. Min. 72 (1987), 1225.

Grossular, $Ca_3Al_2(SiO_4)_3$, G Cub. Vitreous/resinous colorless, white, yellow, brown, etc. Siberia, USSR. H= 6.5-7.5, D_m= 3.53, D_c= 3.62, VIIIA 06a. Werner, 1811. DHZ, 2nd ed.(1982), v.1A, 468. Am. Min. 56 (1971), 791. See also Andradite, Hibschite, Katoite, Uvarovite.

Groutite, α-MnO(OH), G Orth. Sub-metallic/adamantine black. Cuyuna Range, Crow Wing Co., Minnesota, USA. H=?, D_m= 4.14, D_c= 4.171, IVF 04. Gruner, 1945. Am. Min. 32 (1947), 654. Acta Cryst. B24 (1968), 1233. See also Manganite, Feitknechtite.

Grovesite, $(Mn, Mg, Al)_3(Si, Al)_2(O, OH)_9$, Q Tric. Dark brown. Benallt mine, Rhiw, Caernarvonshire, Lleyn Peninsula, Gwynedd, Wales. H=?, D_m= 3.15, D_c= 3.31, VIIIE 09a. Bannister et al., 1955. Am. Min. 59 (1974), 1153.

Grumantite, $NaSi_2O_4(OH) \cdot H_2O$, A Orth. Silky/vitreous white. Mt. Alluaiv, Lovozero massif, Kola Peninsula, USSR. H= 4-5, D_m= 2.21, D_c= 2.264, VIIIE 16. Khomyakov et al., 1987. Am. Min. 73(1988), 440 (Abst.). Zts. Krist. 185 (1988), 612.

Grünerite, $(Fe, Mg)_7Si_8O_{22}(OH)_2$, G Mon. Silky brown, beige. H=?, D_m= 3.713, D_c= 3.54, VIIID 05a. Kenngott, 1853. Am. Min. 49 (1964), 963. MSA Spec. Paper 2 (1969), 117. See also Magnesiocummingtonite, Cummingtonite.

Gruzdevite, $Cu_6Hg_3Sb_4S_{12}$, A Rhom. Metallic white. Chauvai Sb-Hg deposit, Kirgizia (S), USSR. H= 4.3, D_m=?, D_c= 5.88, IID 01c. Spiridonov, 1981. Am. Min. 67(1982), 855 (Abst.). See also Aktashite, Nowackiite.

Guanajuatite, Bi_2Se_3, [G] Orth. Metallic bluish-grey; white in reflected light. Santa Catarina mine, Sierra de Santa Rosa, Guanajuato, Mexico. H= 2.5-3.5, D_m= 6.6, D_c= 8.21, IIC 02. Fernandez, 1873. Am. Min. 35 (1950), 337. See also Paraguanajuatite, Bismuthinite, Stibnite.

Guanglinite, [D] = isomertieite. Am. Min. 65(1980), 408 (Abst.).

Guanine, $C_5H_3(NH_2)N_4O$, [A] Mon. White. North Chincha Island, Peru; Murra-el-elevyn Cave, Nullarbor Plain, Western Australia, Australia. H=?, D_m=?, D_c= 1.489, IXE 01. Unger, 1844. Min. Mag. 39 (1974), 889.

Gudmundite, FeSbS, [G] Mon. Metallic white/grey; pinkish-white in reflected light. Gudmundstorp, Sala, Sweden. H= 6, D_m= 6.72, D_c= 6.94, IIC 09. Johansson, 1928. JCPDS 8-104.

Guerinite, $Ca_5(AsO_3OH)_2(AsO_4)_2 \cdot 9H_2O$, [G] Mon. Vitreous/pearly colorless, white. Daniel mine, Schneeberg, Sachsen, (Saxony), Germany. H= 1.5, D_m= 2.76, D_c= 2.74, VIIC 02. Nefedov, 1961. Am. Min. 47(1962), 416 (Abst.). Acta Cryst. B30 (1974), 1789. See also Ferrarisite.

Guettardite, $Pb(Sb,As)_2S_4$, [A] Mon. Metallic black; white in reflected light. Taylor pit, Huntingdon Twp., Hastings Co., Ontario, Canada. H= 3.5, D_m=?, D_c= 5.39, IID 15. Jambor, 1967. Can. Min. 18 (1980), 13. See also Twinnite.

Gugiaite, $Ca_2BeSi_2O_7$, [G] Tet. Vitreous colorless. Gugia, China. H= 5, D_m= 3.034, D_c= 3.113, VIIIB 02a. Peng et al., 1962. Am. Min. 48(1963), 211 (Abst.). Min. Abstr. 82M/2413. See also Jeffreyite.

Guildite, $CuFe(SO_4)_2(OH) \cdot 4H_2O$, [G] Mon. Vitreous brown, yellow. United Verde mine, Jerome, Yavapai Co., Arizona, USA. H= 2.5, D_m= 2.695, D_c= 2.717, VID 04a. Lausen, 1928. Am. Min. 55 (1970), 502. Am. Min. 63 (1978), 478.

Guilleminite, $Ba(UO_2)_3(SeO_3)_2(OH)_4 \cdot 3H_2O$, [A] Orth. Silky yellow. Musonoi, Shaba, Zaïre. H=?, D_m= 4.88, D_c= 5.08, VIG 03. Pierrot et al., 1965. Am. Min. 50(1965), 2103 (Abst.).

Gunningite, $(Zn,Mn)SO_4 \cdot H_2O$, [G] Mon. Vitreous white. Calumet mine, Keno Hill–Galena Hill area, Yukon Territory, Canada. H= 2.5, D_m= 3.195, D_c= 3.321, VIC 01. Jambor & Boyle, 1962. Can. Min. 7 (1962), 209.

Gupeiite, Fe_3Si, [A] Cub. Metallic steel-grey. Yanshan, China. H= 5, D_m=?, D_c= 7.15, IC 03. Yu, 1984. Am. Min. 71(1986), 228 (Abst.).

Gustavite, $AgPbBi_3S_6$, [A] Mon. Metallic white/greyish-white in reflected light. Ivigtut, Greenland (SW). H= 3.7, D_m=?, D_c= 7.01, IID 09b. Karup-Møller, 1970. Can. Min. 13 (1975), 411. See also Lillianite.

Gutsevichite, [D] Inadequate data. Min. Mag. 33 (1962), 261. Ankinovich, 1959.

Guyanaite, CrO(OH), [A] Orth. Translucent reddish-brown, golden brown, green. Guyana. H=?, D_m= 4.53, D_c= 4.574, IVF 04. Milton et al., 1967. Am. Min. 62(1977), 593 (Abst.). See also Bracewellite, Grimaldiite.

Gypsum, $CaSO_4 \cdot 2H_2O$, [G] Mon. Sub-vitreous colorless, white, grey, yellowish, brownish. H= 2, D_m= 2.317, D_c= 2.309, VIC 01. Theophrastus, 315 B.C. Dana, 7th ed.(1951), v.2, 482. Acta Cryst. B38 (1982), 1074. See also Ardealite, Brushite, Pharmacolite.

Gyrolite, $NaCa_{16}AlSi_{24}O_{60}(OH)_8 \cdot 14H_2O$, [G] Tric. Translucent white. Isle of Skye, Scotland. H= 3-4, D_m= 2.4, D_c= 2.46, VIIIE 14. Anderson, 1851. Dana/Ford (1932), 641. Min. Mag. 52 (1988), 377.

Gysinite-(Nd), $(Nd,Pb)CO_3(OH,H_2O)$, [A] Orth. Vitreous/greasy pink, reddish-pink. Shinkolobwe, Shaba, Zaïre. H=?, D_m=?, D_c= 5.18, VbD 03. Sarp & Bertrand, 1985. Am. Min. 70 (1985), 1314. Zts. Krist. 171 (1985), 155. See also Ancylite-(Ce), Calcio-ancylite-(Ce).

H

Haapalaite, 2[(Fe, Ni)S]•1.61[(Mg, Fe)(OH)$_2$], [A] Hex. Metallic bronze-red; light brown in reflected light. Kokka serpentinite, Outokumpu (NNW), Finland. H= 1.3, D_m=?, D_c= 3.57, IIE 01. Huhma et al., 1973. Am. Min. 58(1973), 1111 (Abst.). See also Tochilinite, Valleriite, Yushkinite.

Hafnon, $HfSiO_4$, [A] Tet. Zambezia district, Mozambique. H=?, D_m= 6.97, D_c= 6.977, VIIIA 07. Neves et al., 1974. Am. Min. 61(1976), 175 (Abst.). Am. Min. 67 (1982), 804. See also Zircon.

Hagendorfite, NaCaMn(Fe, Mn)$_2$(PO$_4$)$_3$, [G] Mon. Hagendorf (S) pegmatite, Waidhaus, Oberpfalz, Bayern (Bavaria), Germany. H= 3.5, D_m= 3.71, D_c= 3.84, VIIA 05b. Strunz, 1954. Min. Mag. 43 (1979), 227. See also Varulite, Alluaudite, Ferro-alluaudite, Maghagendorfite.

Häggite, $V_2O_2(OH)_3$, [G] Mon. Black. Carlile, Crook Co., Wyoming, USA. H=?, D_m=?, D_c= 3.53, IVF 07. Evans & Mrose, 1960. Am. Min. 45 (1960), 1144. Am. Min. 45 (1960), 1144.

Haidingerite, $Ca(AsO_3OH)•H_2O$, [G] Orth. Jáchymov (St. Joachimsthal), Západočeský kraj, Čechy (Bohemia), Czechoslovakia. H= 2-2.5, D_m= 2.95, D_c= 2.959, VIID 32. Turner, 1827. Am. Min. 64 (1979), 1248. Bull. Min. 89 (1966), 18.

Haiweeite, $Ca(UO_2)_2Si_6O_{15}•5H_2O$, [G] Mon. Pearly pale yellow, greenish-yellow. Haiwee Reservoir, Coso Mts., Inyo Co., California, USA. H= 3.5, D_m= 3.35, D_c= 4.88, VIIIA 25. McBurney, 1959. Am. Min. 66 (1981), 610.

Hakite, (Cu, Hg, Ag)$_{12}$Sb$_4$(Se, S)$_{13}$, [A] Cub. Metallic grey-brown; creamy-white in reflected light. Předbořice, Středočeský kraj, Čechy (Bohemia), Czechoslovakia. H= 4.5, D_m=?, D_c= 6.17, IID 01a. Johan & Kvacek, 1971. Am. Min. 57(1972), 1553 (Abst.).

Halite, NaCl, [G] Cub. Vitreous colorless, white, yellow, red, blue, purple. H= 2, D_m= 2.168, D_c= 2.165, IIIA 02. Glocker, 1847. Dana, 7th ed.(1951), v.2, 4.

Hallimondite, $Pb_2(UO_2)(AsO_4)_2$, [G] Tric. Sub-adamantine yellow. Schwarzwald, Germany. H= 2.5-3, D_m= 6.39, D_c= 6.40, VIID 19. Walenta, 1965. Am. Min. 50 (1965), 1143.

Halloysite-7Å, $Al_2Si_2O_5(OH)_4$, [G] Mon. Pearly/waxy/dull white, greyish, greenish, etc. H= 1-2, D_m= 2.1, D_c= 2.57, VIIIE 11. Berthier, 1826. Acta Cryst. B31 (1975), 2851. Clays Cl. Mins. 26 (1978), 25. See also Dickite, Kaolinite, Nacrite, Endellite, Halloysite-10Å.

Halloysite-10Å, $Al_2Si_2O_5(OH)_4•2H_2O$, [G] Mon. Pearly/waxy/dull white, greyish, greenish, etc. H= 1-2, D_m=?, D_c= 2.09, VIIIE 11. Berthier, 1826. Rev. Min. 19 (1988), 29. Clays Cl. Mins. 26 (1978), 25. See also Dickite, Kaolinite, Nacrite, Endellite, Halloysite-7Å.

Halotrichite, $FeAl_2(SO_4)_4•22H_2O$, [G] Mon. Vitreous colorless, white, yellowish, greenish. H= 1.5, D_m= 1.89, D_c= 1.957, VIC 06. Glocker, 1839. Dana, 7th ed.(1951), v.2, 523. See also Pickeringite.

Halurgite, $Mg_2[B_4O_5(OH)_4]_2•H_2O$, [G] Mon. Translucent white. Kungursk, Inder basin(?), USSR. H= 2.5-3, D_m= 2.19, D_c= 2.245, Vc 07a. Lobanova, 1962. Strunz (1970), 258.

Hambergite, $Be_2BO_3(OH)$, [G] Orth. Vitreous colorless, greyish, white, yellowish. Helgeraen, Langesundfjord, Telemark, Norway. H= 7.5, D_m= 2.359, D_c= 2.366, Vc 01d. Brögger, 1890. Am. Min. 50 (1965), 85. Acta Cryst. 16 (1963), 1144.

Hammarite, $Cu_2Pb_2Bi_4S_9$, [G] Orth. Metallic grey. Gladhammar, Småland, Kalmar, Sweden. H= 3-4, D_m=?, D_c= 7.05, IID 05a. Johansson, 1924. N.Jb.Min.Mh. (1990), 35. Can. Min. 14 (1976), 536.

Hancockite, $CaPbAl_3(Si_2O_7)(SiO_4)(O,OH)_2$, [G] Mon. Translucent brownish-red. Franklin Furnace, Sussex Co., New Jersey, USA. H= 6-7, D_m= 4.0, D_c= 4.327, VIIIB 15. Penfield & Warren, 1899. Dana/Ford (1932), 624. Am. Min. 56 (1971), 447.

Hanksite, $KNa_{22}(SO_4)_9(CO_3)_2Cl$, [G] Hex. Vitreous/dull colorless, yellowish, grey. Searles Lake, San Bernardino Co., California, USA. H= 3-3.5, D_m= 2.562, D_c= 2.585, VIB 05. Hidden, 1885. Dana, 7th ed.(1951), v.2, 628. Am. Min. 58 (1973), 799.

Hannayite, $(NH_4)_2Mg_3(PO_3OH)_4 \cdot 8H_2O$, [G] Tric. Translucent yellowish. Skipton Caves, Ballarat (SW of), Victoria, Australia. H=?, D_m= 2.03, D_c= 2.030, VIIC 14. vom Rath, 1878. Am. Min. 48 (1963), 635. Acta Cryst. B32 (1976), 2842.

Hannebachite, $CaSO_3 \cdot 0.5H_2O$, [A] Orth. Vitreous colorless. Hannebacher Ley, Hannebach, Eifel, Rheinland-Pfalz, Germany. H= 3.5, D_m= 2.52, D_c= 2.54, VIG 01. Hentschel et al., 1985. Am. Min. 73(1988), 928 (Abst.).

Haradaite, $SrVSi_2O_7$, [A] Orth. Vitreous green. Yamato mine, Kagoshima, Japan. H= 4.5, D_m= 3.80, D_c= 3.83, VIIID 21. Watanabe et al., 1974. Am. Min. 60(1975), 340 (Abst.). Am. Min. 56(1971), 1123 (Abst.). See also Suzukiite.

Hardystonite, $Ca_2ZnSi_2O_7$, [G] Tet. Translucent white. North Hill mine, Franklin, Sussex Co., New Jersey, USA. H= 3-4, D_m= 3.4, D_c= 3.42, VIIIB 02a. Wolff, 1899. Dana/Ford (1932), 607. Zts. Krist. 130 (1969), 427.

Harkerite, $Ca_{12}Mg_4Al(CO_3)_4(BO_3)_4(SiO_4)_4(H_2O,Cl)_{17}$, [G] Rhom. Vitreous colorless. Isle of Skye, Scotland. H=?, D_m= 2.96, D_c= 3.00, Vc 01e. Tilley, 1948. Min. Mag. 29 (1951), 621. Am. Min. 62 (1977), 263. See also Sakhaite.

Harmotome, $Ba_2(Ca_{0.5},Na)(Si_{11}Al_5)O_{32} \cdot 12H_2O$, [G] Mon. Vitreous colorless, white, grey, yellow, red, brown. Bellesgrove mine, Agrell, Strontium, Scotland. H= 4.5, D_m= 2.35, D_c= 2.416, VIIIF 14. Haüy, 1801. Natural Zeolites (1985), 134. Acta Cryst. B30 (1974), 2426. See also Phillipsite, Wellsite.

Harstigite, $Ca_6Be_4(Mn,Mg)Si_6O_{22}(OH)_2$, [G] Orth. Vitreous colorless. Harstigen mine, Pajsberg, Värmland, Sweden. H= 5.5, D_m= 3.049, D_c= 3.19, VIIIB 24. Flink, 1886. Am. Min. 53 (1968), 1418. Zts. Krist. 177 (1986), 143.

Hartite, $C_{20}H_{34}$, [G] Tric. Translucent colorless, white. Oberhart, Gloggnitz, Austria. H=?, D_m= 1.08, D_c= 1.07, IXB 01. Haidinger, 1841. JCPDS 30-2002. Acta Cryst. B34 (1978), 1311.

Hashemite, $Ba(Cr,S)O_4$, [A] Orth. Adamantine dark to light brown. Lisdan-Siwaga Fault (along the), Hashem region, Amman (near), Jordan (East Central). H= 3.5, D_m= 4.54, D_c= 4.52, VIA 07. Hauff et al., 1983. Am. Min. 68 (1983), 1223. Am. Min. 71 (1986), 1217.

Hastingsite, $NaCa_2(Fe,Mg)_5(Si_6Al_2)O_{22}(OH)_2$, [G] Mon. Dark green, black. Dungannon Twp., Hastings County, Ontario, Canada. H= 5-6, D_m= 3.50, D_c= 3.21, VIIID 05b. Adams & Harrington, 1896. Am. Min. 63 (1978), 1023. Min. Mag. 41 (1977), 43. See also Magnesiohastingsite.

Hastite, $CoSe_2$, [G] Orth. Light brownish red/dark reddish-violet. Trogtal Quarry, Lautenthal (near), Harz, Niedersachsen, Germany. H= 6, D_m=?, D_c= 7.23, IIC 07. Ramdohr & Schmitt, 1955. Am. Min. 41(1956), 164 (Abst.). See also Trogtalite.

Hatchite, $AgPbTlAs_2S_5$, [G] Tric. Metallic grey. Lengenbach quarry, Binntal, Valais (Wallis), Switzerland. H=?, D_m=?, D_c= 5.8, IID 05c. Solly & Smith, 1912. Dana, 7th ed.(1944), v.1, 487. Zts. Krist. 125 (1967), 249. See also Wallisite.

Hatrurite, Ca_3SiO_5, [G] Rhom. Translucent colorless. Hatrurim formation, Israel. H=?, D_m=?, D_c= 3.02, VIIIA 05. Gross, 1977. JCPDS 16-406.

Hauchecornite, Ni_9BiSbS_8, [A] Tet. Metallic bronze. Freidrich mine, Wissen an der Sieg, Rheinland-Pfalz, Germany. H= 5-5.5, D_m= 6.4, D_c= 6.58, IIC 13. Scheibe, 1892. Min. Mag. 43 (1980), 873. Can. Min. 12 (1974), 269. See also Bismutohauchecornite.

Hauckite, $Fe_3(Mg,Mn)_{24}Zn_{18}(SO_4)_4(CO_3)_2(OH)_{81}$, [A] Hex. Vitreous light orange, yellow. Sterling Hill mine, Ogdensburg, Sussex Co., New Jersey, USA. H= 2-3, D_m= 3.02, D_c= 3.10, IVF 03e. Dunn et al., 1980. Am. Min. 65 (1980), 192.

Hauerite, MnS_2, [G] Cub. Metallic reddish-brown/black; greyish-white in refl. light. Kalinka, Hungary. H= 4, D_m= 3.463, D_c= 3.444, IIC 05. Haidinger, 1846. Min. Abstr. 70-1595.

Hausmannite, Mn_3O_4, [G] Tet. Sub-metallic brownish-black; greyish-white in refl. light. Ilfeld, Harz, Germany and/or Raddusa, Sicily, Italy. H= 5.5, D_m= 4.84, D_c= 4.842, IVB 02a. Haidinger, 1827. Dana, 7th ed.(1944), v.1, 712. Struct. Repts. 54A (1987), 141. See also Hetaerolite.

Haüyne, $Na_3Ca(Si_3Al_3)O_{12}(SO_4)$, [G] Cub. Vitreous/greasy blue, greenish-blue, white, grey. Monte Somma, Mt. Vesuvius, Napoli, Campania, Italy. H= 5.5-6, D_m= 2.5, D_c= 2.465, VIIIF 07. Brunn-Neergard, 1807. Can. Min. 27 (1989), 173. Struct. Repts. 33A (1968), 483.

Hawleyite, α–CdS, [G] Cub. Translucent yellow. Hector-Calumet mine, Galena Hill, Yukon Territory, Canada. H=?, D_m=?, D_c= 4.87, IIB 01a. Traill & Boyle, 1955. Am. Min. 40 (1955), 555. See also Greenockite.

Hawthorneite, $BaMgTi_3Cr_4Fe_4O_{19}$, [A] Hex. Metallic black; grey in reflected light. Bulfontein, Kimberley, South Africa. H= 5.8, D_m=?, D_c= 5.02, IVC 05a. Haggerty et al., 1989. Am. Min. 74 (1989), 668. Am. Min. 74 (1989), 668. See also Magnetoplumbite.

Haxonite, $(Fe,Ni)_{23}C_6$, [A] Cub. Metallic white. Toluca and Canyon Diablo meteorites. H=?, D_m=?, D_c= 7.70, IC 02. Scott, 1971. Am. Min. 59(1974), 209 (Abst.). Nature Phys. Sci. 229(1971), 61.

Haycockite, $Cu_4Fe_5S_8$, [A] Orth. Metallic yellow. Mooihoek Farm, Transvaal, South Africa. H= 4, D_m=?, D_c= 4.35, IIB 02. Cabri & Hall, 1972. Am. Min. 57 (1972), 689. Can. Min. 13 (1975), 168.

Heazlewoodite, Ni_3S_2, [G] Rhom. Metallic yellowish-white/bronze. Waratah Dist., Heazlewood Mining Area, Tasmania, Australia. H= 4, D_m= 5.82, D_c= 5.87, IIA 06. Petterd, 1896. Am. Min. 62 (1977), 341. Acta Cryst. B36 (1980), 1179.

Hectorfloresite, $Na_9(IO_3)(SO_4)_4$, [A] Mon. Translucent colorless, white. Alianza mine, Tarapacá Province, Chile. H=?, D_m= 2.80, D_c= 2.90, VIB 13. Ericksen et al., 1989. Am. Min. 74 (1989), 1205.

Hectorite, $Na_{0.3}(Mg,Li)_3Si_4O_{10}(F,OH)_2$, [G] Mon. Translucent white. Hector Bentonite mines, San Bernardino Co., California, USA. H=?, D_m=?, D_c=?, VIIIE 08b. Strese & Hofmann, 1941. JCPDS 9-31.

Hedenbergite, $CaFeSi_2O_6$, [G] Mon. Vitreous/resinous brownish, dark/pale yellow green, black. Sweden. H= 6, D_m= 3.5, D_c= 3.655, VIIID 01a. Berzelius, 1819. DHZ, 2nd ed.(1978), v.2A, 198. Min. Abstr. 79-147. See also Diopside, Johannsenite.

Hedleyite, $Bi_{2+x}Te_{1-x}$, [G] Rhom. Metallic white. Good Hope Claim, Hedley, Similkameen Dist., British Columbia, Canada. H= 2, D_m= 8.6, D_c= 8.85, IIA 15. Warren & Peacock, 1945. Dokl. Earth Sci. 230 (1976), 153.

Hedyphane, $Ca_2Pb_3(AsO_4)_3Cl$, [G] Hex. Greasy/resinous white, yellowish-white, buff, bluish. Långban mine, Filipstad (near), Värmland, Sweden. H= 4.5, D_m= 5.85, D_c= 5.99, VIIB 16. Breithaupt, 1830. Dana, 7th ed.(1951), v.2, 900. Am. Min. 69 (1984), 920.

Heideite, $(Fe,Cr)_{1+x}(Ti,Fe)_2S_4$, [A] Mon. Metallic cream. Bustee meteorite, Gorakhpur (near), Basti dist., Uttar Pradesh, India. H= 3.5-4.5, D_m=?, D_c= 4.1, IIB 09c. Keil & Brett, 1974. Am. Min. 59 (1974), 465.

Heidornite, $Na_2Ca_3B_5O_8(SO_4)_2(OH)_2Cl$, [G] Mon. Transparent colorless. Nordhorn, Hannover, Germany. H= 4-5, D_m= 2.753, D_c= 2.698, Vc 10. Engelhardt & Fuchtbauer, 1956. Am. Min. 42(1957), 120 (Abst.). Min. Abstr. 69-1968.

Heinrichite, $Ba(UO_2)_2(AsO_4)_2 \cdot 10H_2O$, G Tet. Vitreous/pearly yellow, green. Lakeview (near), Lakeview Co., Oregon, USA and/or Schwarzwald, Germany. H= 2.5, D_m=?, D_c= 3.50, VIID 20a. Gross et al., 1958. Strunz (1970), 352.

Heliophyllite, $Pb_6As_2O_7Cl_4$, G Orth. Vitreous/greasy yellow, yellowish-green. Harstigen and/or Långban mine, Pajsberg, Värmland, Sweden. H= 2, D_m= 6.89, D_c= 7.33, IIID 03. Flink, 1888. Dana, 7th ed.(1951), v.2, 1037.

Hellandite-(Y), $(Ca,Y)_4Y_2(Al,Fe)B_4Si_4O_{20}(OH)_4$, G Mon. Vitreous/dull reddish-brown, black, grey, green yellow, etc. Lindvikskollen dike, Kragerø (near), Norway (S). H= 4.5-6.5, D_m= 3.63, D_c= 3.645, VIIID 23. Brögger, 1903. Can. Min. 11 (1972), 760. Am. Min. 62 (1977), 89.

Hellyerite, $NiCO_3 \cdot 6H_2O$, G Mon. Vitreous pale blue. Lord Brassy mine, Heazlewood, Tasmania, Australia. H= 2.5, D_m= 1.97, D_c= 1.98, VbC 01. Williams et al., 1959. Am. Min. 44 (1959), 533.

Helmutwinklerite, $PbZn_2(AsO_4)_2 \cdot 2H_2O$, A Tric. Translucent colorless, light blue. Tsumeb, Namibia. H= 4.5, D_m= 5.3, D_c= 5.29, VIIC 17. Süsse & Schnorrer, 1980. Min. Abstr. 80-4913. See also Thometzekite, Tsumcorite.

Helvite, $Be_3Mn_4(SiO_4)_3S$, G Cub. Vitreous/resinous yellow, yellowish/reddish brown, green. Schwarzenberg, Sachsen (Saxony), Germany. H= 6-6.5, D_m= 3.2, D_c= 3.23, VIIIF 08. Werner, 1817. Dana, 6th ed. (1892), 434. Am. Min. 70 (1985), 186. See also Danalite, Genthelvite.

Hematite, α-Fe_2O_3, G Rhom. Metallic/earthy grey; white in reflected light. H= 5-6, D_m= 5.26, D_c= 5.255, IVC 04a. Theophrastus, 315 B.C. Dana, 7th ed.(1944), v.1, 527. Am. Min. 51 (1966), 123. See also Maghemite.

Hematolite, $(Mn,Mg,Al)_{15}(AsO_4)_2(AsO_3)(OH)_{23}$, G Rhom. Vitreous brownish-red, blood-red. Moss mine, Nordmark, Sweden. H= 3.5, D_m= 3.49, D_c= 3.41, VIIB 10c. Igelström, 1884. Dana, 7th ed.(1951), v.2, 777. Am. Min. 63 (1978), 150.

Hematophanite, $Pb_4Fe_3O_8(OH,Cl)$, G Tet. Sub-metallic red-brown. Jakobsberg mine, Finnmossen, Värmland, Sweden. H= 2-3, D_m= 7.70, D_c= 8.11, IVC 01. Johansson, 1928. Min. Mag. 39 (1973), 49. Struct. Repts. 48A (1981), 182.

Hemihedrite, $ZnPb_{10}(CrO_4)_6(SiO_4)_2F_2$, A Tric. Translucent bright orange, brown, blackish. Florence mine, Pinal Co. and Pack Rat Claim, Wickenburg, Maricopa Co., Arizona, USA. H= 3, D_m= 6.42, D_c= 6.50, VIE 02. Williams & Anthony, 1970. Am. Min. 55 (1970), 1088. Am. Min. 55 (1970), 1103. See also Iranite.

Hemimorphite, $Zn_4Si_2O_7(OH)_2 \cdot H_2O$, G Orth. Vitreous/pearly white, bluish, greenish, yellowish, brown. H= 4.5-5, D_m= 3.475, D_c= 3.484, VIIIB 07. Kenngott, 1853. Dana, 6th ed.(1892), 546. Zts. Krist. 146 (1977), 241.

Hemloite, $(Ti,V,Fe,Al)_{12}(As,Sb)_2O_{23}(OH)$, A Tric. Metallic/submetallic black; grey in reflected light. Hemlo deposit, Hemlo, Ontario, Canada. H= 6.5-7, D_m=?, D_c= 4.613, IVC 10. Harris et al., 1989. Can. Min. 27 (1989), 427. Can. Min. 27 (1989), 427.

Hemusite, Cu_6SnMoS_8, A Cub. Metallic violet-grey. Chelopech deposit, Balkin Mts., Bulgaria. H= 4, D_m=?, D_c= 4.494, IIB 03b. Terziev, 1971. Am. Min. 56 (1971), 1847. See also Kiddcreekite.

Hendersonite, $Ca_2V_9O_{24} \cdot 8H_2O$, G Orth. Pearly/sub-adamantine dark greenish-black, black. JJ mine, Paradox Valley, Montrose Co., Colorado, USA and Unspecified locality, New Mexico, USA. H= 2.5, D_m= 2.78, D_c= 2.80, IVF 09. Lindberg et al., 1962. Am. Min. 47 (1962), 1252.

Hendricksite, $K(Zn,Mg,Mn)_3(Si_3Al)O_{10}(OH)_2$, A Mon. Coppery brown, dark reddish-brown, reddish-black. Franklin, Sussex Co., New Jersey, USA. H=?, D_m= 3.4, D_c= 3.28, VIIIE 05b. Frondel & Ito, 1966. Am. Min. 51 (1966), 1107. Min. Abstr. 85M/3798.

Heneuite, $CaMg_5(PO_4)_3(CO_3)(OH)$, [A] Tric. Vitreous pale blue-green. Tingelstadtjern, Buskerud, Modum, Norway. H= 5, D_m= 3.016, D_c= 3.01, VIIB 20. Raade et al., 1986. Am. Min. 73(1988), 440 (Abst.). N.Jb.Min.Mh. (1986), 351.

Henmilite, $Ca_2Cu[B(OH)_4]_2(OH)_4$, [A] Tric. Vitreous bluish-violet. Fuka mine, Okayama, Honshu, Japan. H= 2, D_m=?, D_c= 2.523, Vc 03d. Nakai et al., 1986. Am. Min. 71 (1986), 1234. Am. Min. 71 (1986), 1236.

Henritermierite, $Ca_3(Mn,Al)_2[(SiO_4),(OH)_4]_3$, [A] Tet. Vitreous brown. Tachgagalt mine, Morocco. H=?, D_m= 3.34, D_c= 3.40, VIIIA 06b. Gaudefroy et al., 1969. Am. Min. 54(1969), 1739 (Abst.).

Henryite, $Cu_4Ag_3Te_4$, [A] Cub. Metallic pale blue. Campbell Orebody, Bisbee, Cochise Co., Arizona, USA. H= 3, D_m=?, D_c= 7.86, IIA 13. Criddle et al., 1983. Bull. Min. 106 (1983), 511.

Hentschelite, $CuFe_2(PO_4)_2(OH)_2$, [A] Mon. Translucent dark green. Reichenbach, Bensheim (near), Odenwald, Hessen, Germany. H= 3.5, D_m=?, D_c= 3.79, VIIB 05. Sieber et al., 1987. Am. Min. 72 (1987), 404. Acta Cryst. C43 (1987), 1855.

Hercynite, $FeAl_2O_4$, [G] Cub. Vitreous black. Natschetin, Poběžovice (Ronsberg), Čechy (Bohemia), Czechoslovakia. H= 7.5-8, D_m=?, D_c= 8.119, IVB 01. Zippe, 1847. Dana, 7th ed.(1944), v.1, 689. See also Spinel, Gahnite, Chromite.

Herderite, $CaBePO_4(F,OH)$, [G] Mon. Vitreous colorless, pale yellow, greenish-white. Ehrenfriedersdorf, Sachsen (Saxony), Germany. H= 5-5.5, D_m= 3.01, D_c= 3.00, VIIB 01. Haidinger, 1828. Am. Min. 63 (1978), 913. Struct. Repts. 21 (1957), 383. See also Hydroxylherderite, Bergslagite.

Herschelite, $(Na,Ca,K)(AlSi_2)O_6 \cdot 3H_2O$, [Q] Rhom. Aci Castello, Sicilia (Sicily), Italy. H=?, D_m=?, D_c= 2.05, VIIIF 15. Lévy, 1825. Am. Min. 51 (1966), 909. See also Chabazite.

Herzenbergite, SnS, [G] Orth. Metallic grey. Maria Teresa mine, Huari (near), Bolivia. H= 2, D_m= 5.16, D_c= 5.165, IIB 12. Ramdohr, 1934. Min. Abst. 70-692. Zts. Krist. 148 (1978), 295.

Hessite, Ag_2Te, [G] Mon. Metallic grey. Săcărâmb (Nagyág), Munţii Metaliferi, Judeţul Hunedoara, Romania. H= 2-3, D_m= 8.3, D_c= 8.40, IIA 03. Fröbel, 1843. Am. Min. 36 (1951), 471. Zts. Krist. 112 (1959), 44.

Hetaerolite, $ZnMn_2O_4$, [G] Tet. Sub-metallic black. Sterling Hill, Sussex Co., New Jersey, USA. H= 6, D_m= 5.18, D_c= 5.25, IVB 02a. Moore, 1877. NBS Monogr. 10 (1972), 61. See also Hausmannite.

Heterogenite-2H, CoO(OH), [P] Hex. Metallic black. Midingi, Shaba, Zaïre. H= 5.5, D_m= 4.5, D_c=?, IVF 04. Deliens & Goethals, 1973. Min. Mag. 39 (1973), 152. See also Heterogenite-3R.

Heterogenite-3R, CoO(OH), [G] Rhom. Dull/vitreous black, reddish, blackish-brown. Schneeberg, Sachsen (Saxony), Germany. H= 3-4, D_m= 4.72, D_c= 4.931, IVF 04. Frenzel, 1872. Min. Mag. 39 (1973), 152. See also Heterogenite-2H, Grimaldiite.

Heteromorphite, $Pb_7Sb_8S_{19}$, [G] Mon. Metallic black. Arnsberg, Westphalia, Germany. H= 2.5-3, D_m= 5.73, D_c= 5.81, IID 09a. Zincken & Rammelsberg, 1849. Dana, 7th ed.(1944), v.1, 465. Zts. Krist. 151 (1980), 193.

Heterosite, $(Fe,Mn)PO_4$, [G] Orth. Satiny deep rose, reddish-purple. Limoges (near), Dept. of Haute Vienne, France. H= 4-4.5, D_m= 3.3, D_c= 3.672, VIIA 02. Alluaud, 1825. Dana, 7th ed.(1951), v.2, 675. Am. Min. 57 (1972), 45. See also Purpurite.

Heterotype, [D] A mixture of amphibole and pyroxene. Am. Min. 63 (1978) 1023. Name origin uncertain.

Heulandite, $(Na,K,Ca,Sr,Ba)_5(Al_9Si_{27})O_{72} \cdot 26H_2O$, [G] Mon. Vitreous/pearly white, red, grey, brown. H= 3.5-4, D_m= 2.198, D_c= 2.24, VIIIF 13. Brooke, 1822. Am. Min. 53 (1968), 1120. Struct. Repts. 50A (1983), 324.

Hewettite, $CaV_6O_{16}\cdot 9H_2O$, [G] Mon. Silky red. Minasragra, Junin, Peru. H=?, D_m= 2.62, D_c= 2.59, IVF 09. Hillebrand et al., 1914. Min. Mag. 46 (1982), 503. Can. Min. 27 (1989), 181.

Hexahydrite, $MgSO_4\cdot 6H_2O$, [G] Mon. Pearly/vitreous colorless, white, pale greenish. Lillooet dist., British Columbia, Canada. H=?, D_m= 1.757, D_c= 1.723, VIC 03b. Johnston, 1911. Dana, 7th ed.(1951), v.2, 494. Acta Cryst. 17 (1964), 235.

Hexahydroborite, $Ca[B(OH)_4]_2\cdot 2H_2O$, [A] Mon. Translucent colorless. Solongo deposit, Ural Mts., USSR. H= 2.5, D_m= 1.87, D_c= 1.878, Vc 03a. Simonov et al., 1977. Am. Min. 63(1978), 1283 (Abst.). Acta Cryst. B27 (1971), 1532.

Hexatestibiopanickelite, $(Ni,Pd)_2SbTe$, [Q] Hex. Metallic brownish-grey. China (SW). H= 4, D_m=?, D_c= 8.79, IIB 09d. Am. Min. 61(1976), 182 (Abst.).

Heyite, $Pb_5Fe_2O_4(VO_4)_2$, [A] Mon. Translucent yellow. Betty Jo Claim, Ely (near), White Pine Co., Nevada, USA. H= 4, D_m= 6.3, D_c= 6.284, VIIB 14. Williams, 1973. Min. Mag. 39 (1973), 65.

Heyrovskýite, $AgPb_{10}Bi_5S_{18}$, [A] Orth. Metallic white. Hůrky, Středočeský kraj, Čechy (Bohemia), Czechoslovakia. H= 3.5-4, D_m= 7.17, D_c= 7.18, IID 09d. Klominsky et al., 1971. Am. Min. 57(1972), 325 (Abst.).

Hibonite, $(Ca,Ce)(Al,Ti,Mg)_{12}O_{19}$, [G] Hex. Brownish-black, black. Madagascar. H= 7.5-8, D_m= 3.84, D_c= 3.84, IVC 05a. Curien et al., 1956. Am. Min. 42(1957), 119 (Abst.). See also Magnetoplumbite, Yimengite.

Hibschite, $Ca_3Al_2[(SiO_4),(OH)_4]_3$, [A] Cub. Translucent colorless, pale yellow. Ústí, Čechy (Bohemia), Czechoslovakia. H= 6, D_m= 3.0, D_c= 3.17, VIIIA 06b. Cornu, 1905. Bull. Min. 107 (1984), 605. See also Grossular, Katoite.

Hidalgoite, $PbAl_3(SO_4)(AsO_4)(OH)_6$, [G] Rhom. Dull porcelanous. San Pasquale mine, Zimapan Mining Dist., Hidalgo, Mexico. H= 4.5, D_m= 3.96, D_c= 4.27, VIB 03b. Smith et al., 1953. Am. Min. 38 (1953), 1218.

Hieratite, K_2SiF_6, [G] Cub. Vitreous colorless, white, grey. Vulcano Island, Eolie (Lipari) Islands, Sicilia (Sicily), Italy. H= 2.5, D_m= 2.665, D_c= 2.668, IIIB 02b. Cossa, 1882. Dana, 7th ed.(1951), v.2, 103.

Hilairite, $Na_2ZrSi_3O_9\cdot 3H_2O$, [A] Rhom. Vitreous pale brown. Mont Saint-Hilaire, Rouville Co., Québec, Canada. H= > 4, D_m= 2.724, D_c= 2.739, VIIIC 12. Chao et al., 1974. Min. Abstr. 83M/4219. Min. Abstr. 83M/4219.

Hilgardite-1Tc, $Ca_2B_5O_9Cl\cdot H_2O$, [A] Tric. Vitreous colorless. H= 5, D_m= 2.71, D_c= 3.014, Vc 13. Ghose, 1985. Am. Min. 70 (1985), 636. Min. Abstr. 79-2129. See also Hilgardite-3Tc, Hilgardite-4M, Tyretskite-1Tc.

Hilgardite-3Tc, $Ca_2B_5O_9Cl\cdot H_2O$, [A] Tric. Vitreous colorless. Choctaw salt dome, Iberville Parish, Louisiana, USA. H= 5, D_m= 2.71, D_c= 2.69, Vc 13. Hurlbut, 1938. Am. Min. 70 (1985), 636. Am. Min. 68 (1983), 604. See also Hilgardite-1Tc, Hilgardite-4M.

Hilgardite-4M, $Ca_2B_5O_9Cl\cdot H_2O$, [G] Mon. Vitreous colorless. Choctaw Salt Dome, Iberville Parish, Louisiana, USA. H= 5, D_m= 2.71, D_c= 2.693, Vc 13. Hurlbut & Taylor, 1937. Am. Min. 70 (1985), 636. Am. Min. 64 (1979), 187. See also Hilgardite-1Tc, Hilgardite-3Tc.

Hillebrandite, $Ca_2SiO_4\cdot H_2O$, [G] Mon. Translucent white. Ternanes mine, Velardeña, Durango, Mexico. H= 5.5, D_m= 2.66, D_c= 2.65, VIIIA 10. Wright, 1908. Min. Mag. 30 (1953), 150.

Hingganite-(Ce), $Be(Ce,Y,Fe)SiO_4(O,OH)$, [Q] Mon. Translucent light reddish-brown. Tahara area, Gifu, Japan. H=?, D_m=?, D_c= 4.83, VIIIA 23. Miyawaki et al., 1987. J.Min.Soc.Jap. 18 (1987), 17.

Hingganite-(Y), $Be(Y,Ce)SiO_4(OH)$, [A] Mon. Translucent white, light yellow/green. Heilongjiang, China and/or Tuva, USSR. H=?, D_m= 4.57, D_c= 4.72,

VIIIA 23. Ding et al., 1984. Chem. Abstr. 106 (10), 70456j. Am. Min. 73(1988), 441 (Abst.).

Hingganite-(Yb), Be(Yb,Y)SiO_4(OH), A Mon. Vitreous colorless. Kola Peninsula, USSR. H= 6-7, D_m=?, D_c= 4.83, VIIIA 23. Voloshin et al., 1983. Am. Min. 69(1984), 811 (Abst.). Min. Abstr. 84M/3794.

Hinsdalite, (Pb,Sr)$Al_3(SO_4)(PO_4)(OH)_6$, G Rhom. Vitreous/greasy colorless, greenish. Golden Fleece mine, Lake City (near), Hinsdale Co., Colorado, USA. H= 4.5, D_m= 3.65, D_c= 4.07, VIB 03b. Larsen & Schaller, 1911. JCPDS 16-711.

Hiortdahlite I, $Na_4Ca_8Zr_2(Nb,Mn,Ti,Fe,Mg,Al)_2(Si_2O_7)_4O_3F_5$, G Tric. Vitreous/greasy colorless, yellow, brown. Mittel-Arö island, Langesundfjord, Norway. H= 5-5.6, D_m= 3.256, D_c= 3.25, VIIIB 08. Brögger, 1889. Can. Min. 12 (1974), 241. Min. Abstr. 86M/2906.

Hiortdahlite II, $(Na,Ca)_4Ca_8(Y,Na)_2Zr_2(Si_2O_7)_4O_3F_5$, P Tric. Vitreous/greasy yellow, yellowish-brown. Kipawa River, Villedieu Twp., Temiscaming Co., Québec, Canada. H= 5-5.6, D_m= 3.256, D_c= 3.287, VIIIB 08. Brögger, 1889. Can. Min. 12 (1974), 241. Min. Petrol. 37 (1987), 25.

Hisingerite, $Fe_2Si_2O_5(OH)_4 \cdot 2H_2O$, G Mon. Greasy/vitreous black, brownish-black. H= 3, D_m= 2.67, D_c= 3.23, VIIIE 08d. Berzelius, 1828. JCPDS 26-1140. Min. Abstr. 83M/2626.

Hocartite, $Ag_2(Fe,Zn)SnS_4$, A Tet. Metallic brownish-grey. Tacama, Hocaya, Bolivia. H= 4, D_m=?, D_c= 4.67, IIB 03a. Caye et al., 1968. Am. Min. 54(1969), 573 (Abst.). See also Pirquitasite.

Hochelagaite, $(Ca,Na,Sr)Nb_4O_{11} \cdot 8H_2O$, A Mon. Vitreous white. Francon quarry, St.-Michel Dist., Montreal Island, Québec, Canada. H= 4, D_m= 2.9, D_c= 2.8, IVF 16. Jambor et al., 1986. Can. Min. 24 (1986), 449.

Hodgkinsonite, $Zn_2MnSiO_4(OH)_2$, G Mon. Translucent reddish/violet pink, reddish-brown, orange. Franklin, Sussex Co., New Jersey, USA. H= 4.5-5, D_m= 4.01, D_c= 4.07, VIIIA 13. Palache & Schaller, 1913. Min. Rec. 13 (1982), 229. Zts. Krist. 119 (1963), 117.

Hodrušhite, $Cu_4Bi_6S_{11}$, G Mon. Metallic grey; creamy in reflected light. Banská-Hodruša, Banská-Štiavnica (Schemitz), Středočeský kraj, Slovensko (Slovakia), Czechoslovakia. H= 3.6, D_m= 6.35, D_c= 6.45, IID 06a. Kodera et al., 1970. Min. Mag. 37 (1970), 641. See also Cuprobismutite, Paderaite.

Hoelite, $C_{14}H_8O_2$, G Mon. Translucent yellow. Spitsbergen, Norway. H=?, D_m= 1.42, D_c= 1.43, IXB 01. Oftedal, 1922. JCPDS 28-2002.

Högbomite-4H, $(Mg,Fe)_{1.4}Ti_{0.3}Al_4O_8$, G Hex. Metallic adamantine black; grey in reflected light. Perseus claim, Routevaara, Kvikkjokk (near), Lapland, Sweden. H= 6.5, D_m= 3.81, D_c= 3.72, IVC 05b. Gavelin, 1916. Min. Mag. 33 (1963), 563. Am. Min. 67 (1982), 373.

Högbomite-5H, $(Mg,Fe)_{1.4}Ti_{0.3}Al_4O_8$, P Hex. Metallic adamantine black; grey in reflected light. Routevaara, Kvikkjokk (near), Lapland, Sweden. H=?, D_m=?, D_c= 3.73, IVC 05b. Gavelin, 1916. Min. Mag. 33 (1963), 563.

Högbomite-6H, $(Mg,Fe)_{1.4}Ti_{0.3}Al_4O_8$, P Hex. Metallic adamantine black; grey in reflected light. H=?, D_m=?, D_c= 3.75, IVC 05b. Gavelin, 1916. Min. Mag. 33 (1963), 563.

Högbomite-8H, $(Al,Fe,Mg,Ti,Zn)_{11}O_{15}(OH)$, P Hex. Metallic adamantine black; grey in reflected light. Johannsen's mine, Alice Springs (NNE), Strangways Range, Northern Territory, Australia. H=?, D_m=?, D_c= 4.04, IVC 05b. Gavelin, 1916. Min. Mag. 41 (1977), 337. Am. Min. 67 (1982), 373.

Högbomite-15H, $(Mg,Fe)_{1.4}Ti_{0.3}Al_4O_8$, P Hex. Metallic adamantine black; grey in reflected light. Castor claim, Routevaara, Kvikkjokk (near), Lapland, Sweden. H=?, D_m=?, D_c= 3.73, IVC 05b. Gavelin, 1916. Min. Mag. 33 (1963), 563.

Högbomite-15R, $(Mg,Fe)_{1.4}Ti_{0.3}Al_4O_8$, [P] Rhom. Metallic adamantine black; grey in reflected light. Toombeola, Galway Co., Ireland. H=?, D_m=?, D_c= 3.73, IVC 05b. Gavelin, 1916. Min. Mag. 33 (1963), 563.

Högbomite-18R, $(Mg,Fe)_{1.4}Ti_{0.3}Al_4O_8$, [P] Rhom. Metallic adamantine black; grey in reflected light. Dentz farm, Letaba district, Transvaal, South Africa. H=?, D_m=?, D_c= 3.68, IVC 05b. Gavelin, 1916. Min. Mag. 33 (1963), 563.

Hohmannite, $Fe_2O(SO_4)_2 \cdot 8H_2O$, [G] Tric. Vitreous brown, orange, red. Sierra Gorda, Copiapó (near), Chile. H= 3, D_m= 2.255, D_c= 2.250, VID 02. Frenzel, 1888. Dana, 7th ed.(1951), v.2, 613. Min. Mag. 42 (1978), 144 (M9).

Holdawayite, $Mn_6(CO_3)_2(OH)_7(Cl,OH)$, [A] Mon. Vitreous pink. Kombat mine, Otavi (E), Namibia. H= 3, D_m= 3.19, D_c= 3.24, VbD 01. Peacor et al., 1988. Am. Min. 73 (1988), 632. Am. Min. 73 (1988), 637.

Holdenite, $(Mn,Mg)_6Zn_3(AsO_4)_2(SiO_4)(OH)_8$, [G] Orth. Vitreous pink, yellowish-red, deep red. Franklin, Sussex Co., New Jersey, USA. H= 4, D_m= 4.11, D_c= 4.11, VIIB 10d. Palache, 1921. Am. Min. 34 (1949), 589. Am. Min. 62 (1977), 513.

Hollandite, $(Ba,K)Mn_8O_{16}$, [G] Mon. Metallic silvery-grey, black; white in reflected light. India (central). H= 6, D_m= 4.95, D_c= 4.83, IVD 03b. Fermor, 1906. Am. Min. 64 (1979), 1199. Min. Jour. 13 (1986), 119.

Hollingworthite, $(Rh,Pt,Pd)AsS$, [A] Cub. Metallic grey. Driekop mine, Transvaal, South Africa. H=?, D_m=?, D_c= 7.91, IIC 06a. Stumpfl & Clark, 1965. Am. Min. 50 (1965), 1068. See also Irarsite.

Holmquistite, $Li_2(Mg,Fe)_3Al_2Si_8O_{22}(OH)_2$, [G] Orth. Light/dark blue, violet, violet-black. Utö, Sweden. H= 5-6, D_m= 3.06, D_c= 3.08, VIIID 06. Osann, 1913. Can. Min. 6 (1960), 504. Zts. Krist. 188 (1989), 95. See also Clinoholmquistite, Ferroholmquistite, Magnesioholmquistite.

Holtedahlite, $Mg_{12}(PO_3OH,PO_4)(PO_4)_5(OH,O)_6$, [A] Rhom. Vitreous colorless. Tingelstadtjern deposit, Modum, Norway. H= 4.5-5, D_m= 2.94, D_c= 2.936, VIIB 03c. Raade & Mladeck, 1979. Am. Min. 65(1980), 809 (Abst.). Min. Petrol. 40 (1989), 91. See also Althausite, Satterlyite.

Holtite, $(Al,Ta)_7(Si,Sb)_3BO_{15}(O,OH)_{2.25}$, [A] Orth. Dull/resinous buff, greenish, brown. Greenbushes, Western Australia, Australia. H= 8.5, D_m= 3.90, D_c= 3.88, VIIIA 24. Pryce, 1971. Min. Mag. 38 (1971), 21. Min. Mag. 53 (1989), 457. See also Dumortierite.

Homilite, $Ca_2(Fe,Mg)B_2Si_2O_{10}$, [G] Mon. Resinous/vitreous black, blackish-brown. Stokko, Langesundfjord, Norway. H= 5, D_m= 3.38, D_c= 3.451, VIIIA 23. Paikjull, 1876. Dana, 6th ed. (1892), 505. Acta Cryst. C41 (1985), 13.

Honessite, $(Ni,Fe)_8SO_4(OH)_{16} \cdot nH_2O$, [G] Rhom. Translucent yellow. Linden (near), Iowa Co., Wisconsin, USA. H=?, D_m=?, D_c=?, IVF 03d. Heyl et al., 1959. Min. Mag. 44 (1981), 339. See also Hydrotalcite.

Hongquiite, TiO, [Q] Cub. Metallic white. China. H= 6, D_m=?, D_c= 5.36, IVA 04. Am. Min. 61(1976), 184 (Abst.).

Hongshiite, $PtCu$, [A] Rhom. Metallic grey. Hung dist., China. H= 4, D_m=?, D_c= 15.63, IA 12. Yu et al., 1974. Am. Min. 69(1984), 411 (Abst.). See also Copper.

Hopeite, $Zn_3(PO_4)_2 \cdot 4H_2O$, [G] Orth. Vitreous colorless, greyish-white, pale yellow. Altenberg, Moresnet dist., Belgium (E). H= 3.2, D_m= 3.065, D_c= 3.116, VIIC 06. Spencer, 1908. Min. Jour. 7 (1973), 289. Am. Min. 61 (1976), 987. See also Parahopeite.

Hormites, [D] Unnecessary group name for sepiolite-palygorskite group. Min. Mag. 33 (1962), 261. Mackenzie, 1959.

Hornblende, A group name for amphiboles with the general formula $Ca_2(Mg,Fe,Al)_5(Si,Al)_8O_{22}(OH)_2$. See Magnesio-hornblende, Alumino-ferro-hornblende, Alumino-magnesio-hornblende.

Hörnesite, $Mg_3(AsO_4)_2 \cdot 8H_2O$, G Mon. Pearly white. Oravița (Oravicza), Banat, Romania. H= 1, D_m= 2.6, D_c= 2.602, VIIC 10. Haidinger, 1895. Strunz (1970), 335. See also Erythrite.

Horsfordite, Cu_5Sb, Q Syst=? Metallic white. Mytilene (near), Lesbos, Greece. H= 4-5, D_m= 8.812, D_c=?, IIG 01. Laist & Norton, 1888. Dana, 7th ed.(1944), v.1, 173.

Hotsonite, $Al_{11}(SO_4)_3(PO_4)_2(OH)_{21} \cdot 16H_2O$, A Tric. Dull/silky white. Hotson Sillimanite Quarry, Bushmanland, Pofadder (W), Cape Province, South Africa. H= 2.5, D_m= 2.06, D_c= 2.248, VIID 04. Beukes et al., 1984. Am. Min. 69(1984), 979.

Howardevansite, $NaCuFe_2(VO_4)_3$, A Tric. Metallic black. Izalco Volcano, El Salvador. H=?, D_m=?, D_c= 3.814, VIIA 08. Hughes et al., 1988. Am. Min. 73 (1988), 181. Am. Min. 73 (1988), 181.

Howieite, $Na(Fe, Mg, Al)_{12}(Si_6O_{17})_2(O, OH)_{10}$, A Tric. Dark green, black. Laytonville Quarry, Laytonville (S), Mendocino Co., California, USA. H=?, D_m= 3.378, D_c= 3.35, VIIID 20. Agrell et al., 1964. Min. Mag. 43 (1979), 363. Am. Min. 59 (1974), 86. See also Taneyamalite.

Howlite, $Ca_2B_5SiO_9(OH)_5$, G Mon. Sub-vitreous white. Latonville quarry, Brookville, Windsor (near), Hants Co., Nova Scotia, Canada. H= 3.5, D_m= 2.62, D_c= 2.61, VIIIA 23. Dana, 1868. Am. Min. 55 (1970), 716. Am. Min. 73 (1988), 1138.

Hsianghualite, $Li_2Ca_3Be_3(SiO_4)_3F_2$, G Cub. Vitreous white. Hunan, China. H= 6.5, D_m= 3.0, D_c= 2.956, VIIIF 11. Huang et al., 1958. Am. Min. 44(1959), 1327 (Abst.).

Huanghoite-(Ce), $BaCe(CO_3)_2F$, G Rhom. Greasy yellow, yellowish-green. Bayan-Obo mine, Huangho river, Baotou (Paotow), Inner Mongolia. H= 4.7, D_m= 4.6, D_c= 4.852, VbB 04. Semenov & Chang, 1961. Am. Min. 48(1963), 1179 (Abst.). Min. Abstr. 85M/3834.

Hübnerite, $MnWO_4$, G Mon. Sub-metallic/adamantine yellowish-brown, reddish-brown. Erie and Enterprize veins, Ellsworth, Mammoth dist., Nevada, USA. H= 4-4.5, D_m= 7.12, D_c= 7.265, IVD 08. Riotte, 1865. Dana, 7th ed.(1951), v.2, 1064. Zts. Krist. 144 (1976), 238. See also Ferberite, Sanmartinite.

Huemulite, $Na_4MgV_{10}O_{28} \cdot 24H_2O$, A Tric. Dull/vitreous/sub-adamantine yellowish/reddish orange. Huemul mine, Mendoza, Argentina. H= 2.5-3, D_m= 2.39, D_c= 2.40, VIIC 23. Gordillo et al., 1966. Am. Min. 51 (1966), 1.

Hügelite, $Pb_2(UO_2)_3(AsO_4)_2(OH)_4 \cdot 3H_2O$, G Mon. Translucent brown, orange-yellow. Michael mine, Weiler (near), Schwarzwald, Germany. H=?, D_m=?, D_c= 5.80, VIID 22. Durrfeld, 1913. JCPDS 34-1476. See also Dumontite.

Hulsite, $(Fe, Mg)_2(Fe, Sn, Mg)O_2(BO_3)$, G Mon. Sub-metallic/vitreous black. Brooks Mt., Seward Peninsula, Alaska, USA. H= 3, D_m= 4.57, D_c= 4.62, Vc 01c. Knopf & Schaller, 1908. Dana, 7th ed.(1951), v.2, 326. Am. Min. 61 (1976), 116. See also Magnesiohulsite.

Humberstonite, $K_3Na_7Mg_2(SO_4)_6(NO_3)_2 \cdot 6H_2O$, A Rhom. Vitreous colorless. Oficina Alemania, Antofagasta, Chile. H= 2.5, D_m= 2.252, D_c= 2.252, VID 05. Mrose et al., 1970. Am. Min. 55 (1970), 1518.

Humboldtine, $FeC_2O_4 \cdot 2H_2O$, G Mon. Dull/resinous yellow, amber. Ontario, Canada. H= 1.5-2, D_m= 2.28, D_c= 2.31, IXA 01. Rivero, 1821. NBS Monogr. 10 (1972), 24. Struct. Repts. 21 (1957), 505.

Humite, $(Mg, Fe)_7(SiO_4)_3(F, OH)_2$, G Orth. Translucent yellow, dark orange. Monte Somma, Mt. Vesuvius, Napoli, Campania, Italy. H= 6, D_m= 3.245, D_c= 3.24, VIIIA 16. Bournon, 1813. DHZ, 2nd ed.(1982), v.1A, 380. Am. Min. 56 (1971), 1155.

Hummerite, $KMgV_5O_{14} \cdot 8H_2O$, G Tric. Translucent bright orange. Hummer mine, Paradox Valley, Montrose Co., Colorado, USA. H=?, D_m=?, D_c= 2.663, VIIC 23. Weeks et al., 1950. Am. Min. 40(1955), 314 (Abst.).

Hungchaoite, $MgB_4O_5(OH)_4 \cdot 7H_2O$, G Tric. Translucent white. China. H=?, D_m= 1.706, D_c= 1.700, Vc 07a. Chun et al., 1964. Am. Min. 50(1965), 262 (Abst.). Am. Min. 62 (1977), 1135.

Huntite, $CaMg_3(CO_3)_4$, G Rhom. Chalky white. Current Creek and Gabbs, Nye Co., Nevada, USA. H=?, D_m= 2.696, D_c= 2.875, VbA 03b. Faust, 1953. Am. Min. 38 (1953), 4. Am. Min. 71 (1986), 163. See also Daqingshanite-(Ce).

Huréaulite, $Mn_5(PO_3OH)_2(PO_4)_2 \cdot 4H_2O$, G Mon. Vitreous/greasy orange, red, brown, orange, etc. Huréaux quarry, St. Sylvestre, Haute-Vienne, France. H= 3.5, D_m= 3.17, D_c= 3.24, VIIC 02. Alluaud, 1825. Bull. Min. 99 (1976), 261. Am. Min. 58 (1973), 302.

Hurlbutite, $CaBe_2(PO_4)_2$, G Mon. Vitreous/greasy colorless, greenish-white. Smith mine, Newport, New Hampshire, USA. H= 6, D_m= 2.877, D_c= 2.899, VIIA 01. Mrose, 1951. Am. Min. 37 (1952), 931. Am. Min. 59 (1974), 1267.

Hutchinsonite, $(Tl,Pb)_2As_5S_9$, G Orth. Adamantine red. Lengenbach quarry, Binntal, Valais (Wallis), Switzerland. H= 1.5-2, D_m= 4.6, D_c= 4.58, IID 05c. Solly, 1904. Dana, 7th ed.(1944), v.1, 468. Zts. Krist. 121 (1964), 321.

Huttonite, $ThSiO_4$, G Mon. Translucent colorless, pale cream. Salt Water Creek, South Westland, New Zealand. H=?, D_m= 7.20, D_c= 7.25, VIIIA 08. Pabst, 1951. Min. Mag. 43 (1980), 1031. Acta Cryst. B34 (1978), 1074. See also Thorite.

Hyalophane, $(K,Ba)(Al,Si)_4O_8$, G Mon. Vitreous colorless, white, yellow, red. Lengenbach quarry, Binntal, Valais (Wallis), Switzerland. H= 6-6.5, D_m= 2.805, D_c= 2.88, VIIIF 03. von Waltershausen, 1855. DHZ (1963), v.4, 166. Acta Cryst. B33 (1977), 3073. See also Orthoclase, Celsian.

Hyalotekite, $Ba_2Ca_2Pb_2(B,Si,Al)_2(Si,Be)_{10}O_{28}F$, G Tric. Vitreous/greasy white, grey. Långban mine, Filipstad (near), Värmland, Sweden. H= 5-5.5, D_m= 3.82, D_c= 3.829, VIIIC 11. Nordenskiöld, 1877. Dana, 6th ed. (1892), 422. Am. Min. 67 (1982), 1012.

Hydroamesite, D Inadequate data. Min. Mag. 33 (1962), 261. Erdelyi et al., 1959.

Hydroandradite, $Ca_3Fe_2[SiO_4,(OH)_4]_3$, Q Cub. Translucent light green. Lord Brassey mine, Heazlewood district, Tasmania, Australia. H=?, D_m= 3.660, D_c= 3.804, VIIIA 06b. Ford, 1970. Min. Mag. 37 (1970), 942.

Hydroastrophyllite, $(H_3O,K,Ca)_3(Fe,Mn)_{5-6}Ti_2Si_8(O,OH)_{31}$, Q Tric. Dark brown. Sichuan, China. H=?, D_m= 3.151, D_c= 2.79, VIIID 25. Am. Min. 60(1975), 736 (Abst.).

Hydrobasaluminite, $Al_4SO_4(OH)_{10} \cdot 15H_2O$, G Mon. White. Lodge Pit, Irchester, Wellingsborough (near), Northamptonshire, England. H=?, D_m=?, D_c= 2.277, IVF 02. Bannister & Hollingworth, 1948. Min. Mag. 43 (1980), 931.

Hydrobiotite, $K(Mg,Fe)_6(Si,Al)_8O_{20}(OH)_4 \cdot xH_2O$, G Orth. Black, golden yellow, pinkish. H=?, D_m=?, D_c=?, VIIIE 07a. Schrauf, 1882. JCPDS 13-465. See also Biotite, Vermiculite.

Hydroboracite, $CaMg[B_3O_4(OH)_3]_2 \cdot 3H_2O$, G Mon. Vitreous/silky colorless, white. Lake Inder, Caucasus, USSR. H= 5-6, D_m= 2.15, D_c= 2.170, Vc 06. Hess, 1834. Dana, 7th ed.(1951), v.2, 353. Can. Min. 16 (1978), 75.

Hydrocalcite, D = monohydrocalcite. Min. Mag. 43 (1980), 1055. Marschner, 1969.

Hydrocalumite, $Ca_4Al_2(OH)_{12}(Cl,CO_3,OH,H_2O)_{2.5} \cdot 4H_2O$, G Mon. Vitreous colorless/light green. Scawt Hill, Antrim Co., Northern Ireland. H= 3, D_m= 2.15, D_c= 2.15, IVF 18. Tilley, 1934. N. Jb. Min. Mh. (1980), 322. IMA 1986, Abstr. p. 219.

Hydrocastorite, D A mixture of stilbite and heulandite. Min. Mag. 33 (1962), 262. Grattarola, 1877.

Hydrocatapleiite, D Inadequate data. Min. Mag. 36 (1967), 133. Semenov & Tikhonenkov, 1962.

Hydrocerite, D Probably = karnasurtite. Min. Mag. 33 (1962), 261. Vlasov et al., 1959.

Hydrocerussite, $Pb_3(CO_3)_2(OH)_2$, G Rhom. Adamantine/pearly colorless, white, grey. Långban mine, Filipstad (near), Värmland, Sweden and(?) Moorhead quarry, Mendip Hill, Somerset, England. H= 3.5, D_m= 6.80, D_c= 6.83, VbB 09. Nordenskiold, 1877. TMPM 3 (1953), 298. See also Plumbonacrite.

Hydrochlorborite, $Ca_2B_4O_4(OH)_7Cl \cdot 7H_2O$, G Mon. Vitreous colorless. China. H= 2.5, D_m= 1.852, D_c= 1.876, Vc 05. Chi'en & Chen, 1965. Am. Min. 62 (1977), 147. Am. Min. 63 (1978), 814.

Hydrodelhayelite, $KCa_2(Si_7Al)O_{17}(OH)_2 \cdot 6H_2O$, A Orth. Vitreous greyish-white. Khibina massif, Kola Peninsula, USSR. H= 4, D_m= 2.168, D_c= 2.242, VIIIE 15. Dorfman & Chiragov, 1979. Am. Min. 72(1987), 1024 (Abst.).

Hydrodresserite, $BaAl_2(CO_3)_2(OH)_4 \cdot 3H_2O$, A Tric. Vitreous colorless. Francon quarry, St.-Michel Dist., Montreal Island, Québec, Canada. H= 3-4, D_m= 2.80, D_c= 2.817, VbD 02. Jambor et al., 1977. Can. Min. 15 (1977), 399. Can. Min. 20 (1982), 253.

Hydrogarnet, A group name for hydrous orthosilicates with the garnet structure and with the general formula $Ca_3Al_2(SiO_4)_{3-x}(OH)_{4x}$. See Hibschite, Katoite.

Hydroglauberite, $Na_{10}Ca_3(SO_4)_8 \cdot 6H_2O$, A Syst=? Silky white. Kara-Kalpakii, USSR. H=?, D_m= 1.510, D_c=?, VIC 12. Slyusareva, 1969. Am. Min. 55(1970), 321 (Abst.).

Hydrohalite, $NaCl \cdot 2H_2O$, Q Mon. Hallein, Salzburg, Austria. H=?, D_m= 1.607, D_c= 1.654, IIIC 10b. Hausmann, 1847. Dana, 7th ed.(1944), v.2, 15. Acta Cryst. B30 (1974), 2363.

Hydrohalloysite, D = endellite. Min. Mag. 36 (1967), 133. Sedletzky, 1940.

Hydrohetaerolite, $HZnMn_{2-x}O_4$, G Tet. Creamish-grey in reflected light. Leadville, Lake Co., Colorado, USA. H=?, D_m= 4.64, D_c= 4.77, IVB 02a. Palache, 1929. Am. Min. 41 (1956), 268.

Hydrohonessite, $Ni_6Fe_2SO_4(OH)_{16} \cdot 7H_2O$, A Hex. Translucent yellow. Otter Shoot Nickel mine, Kambalda, Western Australia, Australia. H=?, D_m=?, D_c= 2.636, IVF 03e. Nickel & Wildman, 1981. Min. Mag. 44 (1981), 333.

Hydrokassite, D Inadequate data. Min. Mag. 36 (1968), 1144. Kukharenko et al., 1965.

Hydromagnesite, $Mg_5(CO_3)_4(OH)_2 \cdot 4H_2O$, G Mon. Vitreous/earthy colorless, white. H= 3.5, D_m= 2.25, D_c= 2.25, VbD 01. Wachtmeister, 1827. Dana, 7th ed.(1951), v.2, 271. Acta Cryst. B33 (1977), 1273.

Hydrombobomkulite, $(Ni,Cu)Al_4(NO_3)_2(SO_4)(OH)_{12} \cdot 14H_2O$, A Mon. Blue. Mbobo Mkulu Cave, Transvaal (E), South Africa. H=?, D_m=?, D_c=?, IVF 03e. Martini, 1980. Am. Min. 67(1982), 415 (Abst.).

Hydromolysite, D Inadequately characterized. Min. Mag. 36 (1968), 1144. Povarennykh, 1965.

Hydronaujakasite, D An alteration product of naujakasite. Min. Mag. 38 (1971), 103. Petersen, 1967.

Hydronium jarosite, $(H_3O)Fe_3(SO_4)_2(OH)_6$, A Rhom. Dull/resinous yellow. Staszic mine, Góry Swiętokvzyskia (Holy Cross Mt.), Poland. H= 4-4.5, D_m= 2.7, D_c= 3.17, VIB 03a. Kubisz, 1960. Am. Min. 50 (1965), 1595.

Hydrophilite, $CaCl_2$(?), Q Syst=? Mt. Vesuvius, Napoli, Campania, Italy. H=?, D_m=?, D_c=?, IIIA 05. Hausmann, 1813. Min. Mag. 43 (1980), 682. See also Antarcticite, Sinjarite.

Hydrorinkite, D Inadequate data. Min. Mag. 43 (1980), 1055. Semenov, 1969.

Hydroromarchite, $Sn_3O_2(OH)_2$, [A] Tet. Translucent white. Winnipeg River, Ontario, Canada. H=?, D_m=?, D_c= 4.800, IVF 06. Organ & Mandarino. Am. Min. 58 (1973), 552.

Hydroscarbroite, $Al_{14}(CO_3)_3(OH)_{36} \cdot nH_2O$, [Q] Syst=? White. South Bay, Scarborough, England. H=?, D_m=?, D_c=?, IVF 02. Duffin & Goodyear, 1960. Am. Min. 45(1960), 910 (Abst.).

Hydrosericite, [D] Unnecessary name for hydrous mica. Min. Mag. 36 (1968), 1144. Dimitriev, 1965.

Hydrosodalite, [D] Probably a variety of sodalite. Min. Mag. 33 (1962), 261. Gerasimovskii et al., 1960.

Hydrotalcite, $Mg_4Al_2(OH)_{12}CO_3 \cdot 3H_2O$, [G] Rhom. Waxy colorless. Snarum, Norway. H= 2, D_m= 2.09, D_c= 2.115, IVF 03b. Hochstetter, 1842. Am. Min. 26 (1941), 295. Struct. Repts. 40A (1974), 306. See also Manasseite.

Hydrotungstite, $WO_2(OH)_2 \cdot H_2O$, [G] Mon. Vitreous/dull dark green yellowish-green. Oruro, Bolivia. H= 2, D_m= 4.60, D_c= 4.655, IVF 10. Kerr & Young, 1944. Min. Rec. 12 (1981), 81.

Hydrougrandite, [D] Unnecessary group name for hydrous garnets. Min. Mag. 36 (1967), 133. Tsao, 1964.

Hydroxyapophyllite, $KCa_4Si_8O_{20}(OH,F) \cdot 8H_2O$, [A] Tet. Vitreous/pearly colorless, white. Ore Knob mine, Jefferson, Ashe Co., North Carolina, USA. H= 4.5-5, D_m= 2.37, D_c= 2.36, VIIIE 02. Dunn et al., 1978. Am. Min. 63 (1978), 196. Am. Min. 63 (1978), 196. See also Natroapophyllite, Fluorapophyllite.

Hydroxyl-ascharite, [D] Inadequate data. Min. Mag. 36 (1968), 1144. Grigoriev & Nekrasov, 1966.

Hydroxyl-szajbelyite, [D] A variant of the discredited hydroxyl-ascharite. Min. Mag. 36 (1968), 1144.

Hydroxyl vishnevite, $Na_4(Si,Al)_6O_{12}(OH) \cdot H_2O$, [Q] Hex. Translucent pale blue. Lovozero massif, Kola Peninsula, USSR. H=?, D_m= 2.32, D_c= 2.21, VIIIF 05. Semyonov et al., 1984. Am. Min. 73(1988), 927 (Abst.). See also Carbonate-vishnevite.

Hydroxylapatite, $Ca_5(PO_4)_3(OH)$, [G] Hex. Vitreous/sub-resinous green, grey, blue, red, etc. H= 5, D_m= 3.0, D_c= 3.159, VIIB 16. Schaller, 1912. Dana, 7th ed.(1971), v.2, 879. Am. Min. 74 (1989), 870.

Hydroxyl-bastnäsite-(Ce), $(Ce,La)CO_3(OH,F)$, [G] Hex. Vitreous/greasy waxy-yellow, dark brown. Unspecified locality, USSR. H= 4, D_m= 4.745, D_c= 4.79, VbB 04. Kirillov, 1964. Am. Min. 50(1965), 805 (Abst.). See also Bastnäsite-(Ce).

Hydroxyl-bastnäsite-(La), $LaCO_3(OH,F)$, [Q] Syst=? Hungary and Yugoslavia. H=?, D_m=?, D_c=?, VbB 04. Maksimovic & Panto, 1983. Am. Min. 71(1986), 1277 (Abst.).

Hydroxyl-bastnäsite-(Nd), $(Nd,Ce,La)CO_3(OH,F)$, [A] Hex. Dull whitish. Niksic, Montenegro, Yugoslavia. H=?, D_m=?, D_c= 4.89, VbB 04. Maksimovic & Panto, 1985. Min. Mag. 49 (1985), 717.

Hydroxylellestadite, $Ca_{10}(SiO_4)_3(SO_4)_3(OH,Cl,F)_2$, [A] Mon. Vitreous pale purplish, rose-pink. Chichibu mine, Doshinkubo Orebody, Saitama, Honshu, Japan. H= 4.5, D_m= 3.01, D_c= 3.06, VIIIA 29. Harada et al., 1971. Am. Min. 67 (1982), 90. Acta Cryst. B36 (1980), 1636. See also Fluorellestadite, Apatite.

Hydroxyl-herderite, $CaBePO_4(OH)$, [G] Mon. Vitreous colorless, pale yellow, greenish-white. Paris, Maine, USA. H= 5-5.5, D_m= 2.95, D_c= 2.969, VIIB 01. Penfield, 1894. Am. Min. 63 (1978), 913. Am. Min. 59 (1974), 919. See also Herderite, Bergslagite.

Hydroxyl-pyromorphite, $Pb_5(PO_4)_3(OH)$, [Q] Hex. H=?, D_m=?, D_c= 7.204, VIIB 16. Wondratschek, 1963. Strunz (1970), 327.

Hydrozincite, $Zn_5(CO_3)_2(OH)_6$, G Mon. Earthy/silky white, grey, yellowish, brownish, pinkish, etc. H= 2-2.5, D_m= 4.00, D_c= 4.01, VbB 10. Kenngott, 1853. Can. Min. 8 (1965), 385. Acta Cryst. 17 (1964), 1051.

Hypercinnabar, HgS, A Hex. Adamantine red. Mount Diablo mine, Contra Costa Co., California, USA. H= 3, D_m= 7.43, D_c= 7.54, IIB 01b. Potter & Barnes, 1978. Am. Min. 63, (1978), 1143. See also Cinnabar, Metacinnabar.

Hypersthene, D Unnecessary name for member of enstatite-ferrosilite series. Min. Mag. 52 (1988), 535. Haüy, 1806.

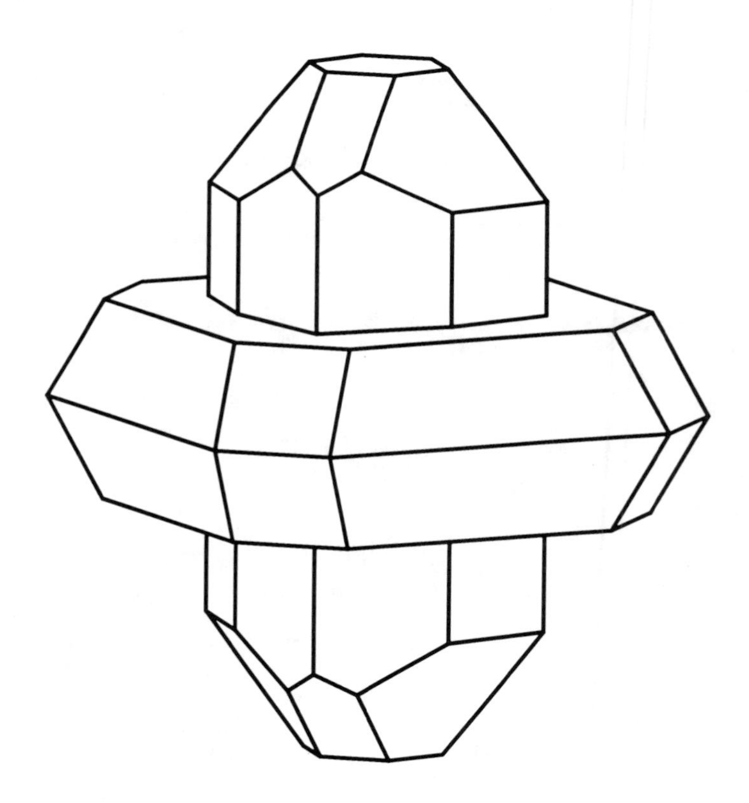

Matlockite surrounding anglesite, from the Mildren mine, Arizona. The forms are {001}, {011} and {111} for the matlockite and {001}, {101}, {111}, {110} and {210} for the anglesite. (After Williams, 1962)

I

Ianthinite, $UO_2 \cdot 5UO_3 \cdot 10H_2O$, G Orth. Sub-metallic violet-black. Shinkolobwe, Shaba, Zaïre. H= 2-3, D_m= 5.16, D_c= 4.99, IVF 11. Schoep, 1926. Am. Min. 44(1959), 1103 (Abst.).

Ice, H_2O, G Hex. Transparent colorless. H= 1.5, D_m=?, D_c= 0.917, IVA 01. Dana, 7th ed.(1944), v.1, 494. Nature, 134 (1934), 900.

Idaite, Cu_5FeS_6, Q Hex. Metallic pale grey/reddish-brown. Ida mine, Khan, Namibia. H=?, D_m= 4.20, D_c= 4.21, IIB 15. Frenzel, 1958. Strunz (1970), 126. See also Nukundamite.

Idrialite, $C_{22}H_{14}$, G Orth. Idrija (Idria), Slovenija (Slovenia), Yugoslavia. H=?, D_m=?, D_c= 1.29, IXB 01. Dumas, 1832. Am. Min. 50(1965), 2109 (Abst.).

Igalikite, D A mixture of analcime and muscovite. Min. Mag. 33 (1962), 262. Bøggild, 1933.

Iimoriite-(Y), $Y_2(SiO_4)(CO_3)$, A Tric. Vitreous/resinous buff/tan. Fukushima, Kawamata, Fusamata, Honshu, Japan. H= 5.5-6, D_m= 4.47, D_c= 4.91, VIIIA 08. Kato & Nagashima, 1970. Am. Min. 69 (1984), 196.

Ikaite, $CaCO_3 \cdot 6H_2O$, A Mon. Chalky white. Ikafjord, Ivigtut, Greenland (SW). H=?, D_m=?, D_c= 1.833, VbC 01. Pauly, 1962. Am. Min. 49(1964), 439. Zts. Krist. 163 (1983), 227.

Ikunolite, $Bi_4(S, Se)_3$, G Rhom. Metallic grey. Ikuno mine, Hyogo, Japan. H= 2, D_m= 7.8, D_c= 7.97, IIA 15. Kato, 1959. Am. Min. 45(1960), 477 (Abst.). Am. Min. 45(1960), 477 (Abst.). See also Laitakarite.

Ilesite, $(Mn, Zn, Fe)SO_4 \cdot 4H_2O$, G Mon. Translucent green. Hall Valley, Park Co., Colorado, USA. H=?, D_m= 2.25, D_c= 2.26, VIC 02. Wünsch, 1881. Dana, 7th ed.(1951), v.2, 486. Acta Cryst. 15 (1962), 815.

Ilímaussite-(Ce), $Na_4Ba_2CeFeNb_2Si_8O_{28} \cdot 5H_2O$, A Hex. Resinous brownish-yellow. Ilímaussaq Intrusion, Greenland (S). H= 4, D_m= 3.6, D_c= 3.669, VIIIA 21. Semenov et al.,1968. Am. Min. 54(1969), 992 (Abst.).

Illite, $(K, H_3O)Al_2(Si_3Al)O_{10}(H_2O, OH)_2$, G Mon. Translucent white. Maquoketa Shale, Gilead (near), Calhoun Co., Illinois, USA. H= 1-2, D_m= 2.8, D_c= 2.61, VIIIE 07a. Grim et al., 1937. Am. Min. 50 (1965), 411. See also Brammallite.

Ilmajokite, $(Na, Ce, Ba)_{10}Ti_5Si_{14}O_{22}(OH)_{44} \cdot nH_2O$, A Mon. Vitreous yellow. Ilmajok River, Lovozero Tundra, Kola Peninsula, USSR. H= 1, D_m= 2.20, D_c= 2.23, VIIIB 09. Bussen et al.,1972. Am. Min. 58(1973), 139 (Abst.).

Ilmenite, $FeTiO_3$, G Rhom. Metallic/sub-metallic black; greyish white in refl. light. Ilmen Mts., Miask, Ural Mts., USSR. H= 5-6, D_m= 4.72, D_c= 4.789, IVC 04b. Kupffer, 1827. Dana, 7th ed.(1944), v.1, 535. Min. Jour. 14 (1989), 179. See also Geikielite, Pyrophanite.

Ilmenorutile, $(Ti, Nb, Fe)O_2$, Q Tet. Black. Miask, Ilmen Mts., USSR. H=?, D_m= 4.20, D_c= 4.97, IVD 02. Koksharov, 1854. JCPDS 31-646. See also Strüverite.

Ilsemannite, $Mo_3O_8 \cdot nH_2O(?)$, Q Syst=? Earthy black, blue-black, blue. Bleiberg, Schwarzenbach, Carinthia, Austria. H=?, D_m=?, D_c=?, IVF 10. Höfer, 1871. Am. Min. 36 (1951), 609.

Ilvaite, $CaFe_3O(Si_2O_7)(OH)$, G Mon. Sub-metallic black, dark greyish-black. Rio Marina, Elba, Toscana, Italy. H= 5.5-6, D_m= 4.0, D_c= 4.448, VIIIB 05. Steffens, 1811. Dana, 6th ed.(1892), 541. Zts. Krist. 163 (1983), 267.

Imandrite, $Na_{12}Ca_3Fe_2Si_{12}O_{36}$, A Orth. Vitreous yellow. Khibina deposits, Lake Imandra (near), Kola Peninsula, USSR. H=?, D_m= 2.93, D_c= 2.92, VIIIC 07. Khomyakov et al., 1979. Am. Min. 65(1980), 810 (Abst.).

Imgreite, NiTe, Q Hex. Metallic grey; pale rose in reflected light. Nittis-Kumuzhya deposit, Monchegorsk, USSR. H= 4, D_m=?, D_c= 8.624, IIB 09a. Yushko-Zakharova, 1964. Powd. Diff. 2 (1987), 257.

Imhofite, $Tl_6As_{15}S_{25}$, G Mon. Translucent red; white in reflected light. Lengenbach quarry, Binntal, Valais (Wallis), Switzerland. H= 2, D_m=?, D_c= 4.39, IID 12. Burri et al., 1965. Am. Min. 51(1966), 531 (Abst.). Zts. Krist. 144 (1976), 323.

Imiterite, Ag_2HgS_2, A Mon. Metallic grey. Imiter mine, Jbel Sarhro, Anti-Atlas, Morocco. H= 2.5, D_m=?, D_c= 7.85, IIA 03. Guillou et al., 1985. Am. Min. 71(1986), 1277 (Abst.). Bull. Min. 108 (1985), 457.

Imogolite, $Al_2SiO_3(OH)_4$, A Syst=? Japan. H=?, D_m=?, D_c=?, VIIIE 10a. Yoshinaga & Aomine, 1962. Min. Mag. 51 (1987), 327. Am. Min. 54 (1969), 50. See also Allophane.

Inaglyite, $Cu_3Pb(Ir,Pt)_8S_{16}$, A Hex. Metallic grey. Inagli massif, Aldan Region, Yakutiya, USSR. H= 4.5-6, D_m=?, D_c=?, IIC 14. Rudashevskii et al., 1984. Am. Min. 71(1986), 228 (Abst.). See also Konderite.

Incaite, $Fe(Pb,Ag)_4Sn_4Sb_2S_{15}$, G Mon. Metallic greyish-white. Poopó, Bolivia. H=?, D_m=?, D_c=?, IID 13. Makovicky, 1974. Am. Min. 60(1975), 486 (Abst.). See also Cylindrite, Potosiite, Incaite.

Inderborite, $CaMg[B_3O_3(OH)_5]_2 \cdot 6H_2O$, G Mon. Vitreous colorless, white. Inder Lake, Kazakhstan (W), USSR. H= 3.5, D_m= 2.00, D_c= 1.919, Vc 05. Bashman & Semenova, 1940. Dana, 7th ed.(1951), v.2, 355. Acta Cryst. 21 (1966), A61.

Inderite, $MgB_3O_3(OH)_5 \cdot 5H_2O$, G Mon. Vitreous/pearly/dull/greasy colorless, white, pink. Inder Lake, Kazakhstan (W), USSR. H= 3, D_m= 1.80, D_c= 1.794, Vc 05. Boldyreva, 1937. Dana, 7th ed.(1951), v.2, 360. Acta Cryst. B32 (1976), 1329. See also Kurnakovite.

Indialite, $Mg_2Al_4Si_5O_{18}$, G Hex. India. H=?, D_m=?, D_c= 2.59, VIIIC 06. Miyashiro & Iiyama, 1954. Am. Min. 40(1955), 787 (Abst.). Can. Min. 15 (1977), 43. See also Cordierite.

Indigirite, $Mg_2Al_2(CO_3)_4(OH)_2 \cdot 15H_2O$, A Syst=? Vitreous/silky white. Indigirka River, Yakutiya (NE), USSR. H= 2, D_m= 1.6, D_c=?, VbD 02. Indolev et al., 1971. Am. Min. 57(1972), 326 (Abstr.).

Indite, $FeIn_2S_4$, G Cub. Metallic white. Little Khingan ridge, Siberia (E), USSR. H= 4.5, D_m=?, D_c= 4.67, IIC 01. Genkin & Murav'eva, 1963. Am. Min. 49(1964), 439 (Abst.). Struct. Repts. 45A (1979), 79.

Indium, In, A Tet. Metallic grey. Unspecified locality, Transbaikal (E), USSR. H= 3, D_m=?, D_c= 7.285, IA 04. Ivanov, 1964. Am. Min. 52(1967), 299 (Abst.).

Inesite, $Ca_2Mn_7Si_{10}O_{28}(OH)_2 \cdot 5H_2O$, G Tric. Vitreous rose/flesh red. Nanzenbach, Dillenburg (near), Germany. H= 6, D_m= 3.03, D_c= 3.03, VIIID 13. Schneider, 1887. Am. Min. 53 (1968), 1614. Am. Min. 63 (1978), 563.

Ingersonite, $Ca_3MnSb_4O_{14}$, A Hex. Vitreous brownish-yellow. Långban mine, Filipstad (near), Värmland, Sweden. H= 6.5, D_m=?, D_c= 5.42, IVC 08. Dunn et al., 1988. Am. Min. 73 (1988), 405.

Ingodite, Bi(Te,S), A Hex. Metallic white. Brandy Gill, Cumberland, England; Ingoda deposit, Transbaikal, USSR. H= 2, D_m=?, D_c= 7.88, IIB 18. Zav'yalov & Begizov, 1981. Am. Min. 70(1985), 220 (Abst.).

Innelite, $Na_2Ba_4CaTi_3O_4(Si_2O_7)_2(SO_4)_2$, G Tric. Vitreous/oily pale yellow, brown. Inagli massif, Aldan Region, Yakutiya, USSR. H= 4.7, D_m= 3.96, D_c= 4.08, VIIIB 12. Kravchenko et al., 1961. Am. Min. 47(1962), 805 (Abst.). Sov.Phys.Cryst. 16(1971), 65.

Insizwaite, $Pt(Bi,Sb)_2$, A Cub. Metallic white. Insizwa deposit, Waterfel Gorge, Transvaal, South Africa. H= 5.3, D_m=?, D_c= 12.8, IIC 05. Cabri & Harris, 1972. Min. Mag. 38 (1972), 794.

Inyoite, $CaB_3O_3(OH)_5 \cdot 4H_2O$, G Mon. Vitreous colorless. Mt. Blanco deposit, Furnace Creek, Death Valley, Inyo Co., California, USA. H= 2, D_m= 1.875, D_c= 1.871, Vc 05. Schaller, 1914. Am. Min. 38 (1953), 912. Struct. Repts. 49A (1982), 239.

Iodargyrite, β–AgI, G Hex. Resinous/adamantine colorless, yellow, green, brown. H= 1.5, D_m= 5.69, D_c= 5.70, IIIA 04. Rammelsberg, 1860. Dana, 7th ed.(1951), v.2, 22.

Iowaite, $Mg_4FeOCl(OH)_8 \cdot 2\text{–}4H_2O$, A Hex. Greasy white. Sioux Co., Iowa, USA. H= 1.5, D_m= 2.11, D_c=?, IVF 03e. Kohls & Rodda, 1967. Am. Min. 52 (1967), 1261.

Iquiqueite, $K_3Na_4Mg(CrO_4)B_{24}O_{39}(OH) \cdot 12H_2O$, A Hex. Vitreous yellow. Iquique, Tarapacá, Chile. H= 2, D_m= 2.05, D_c= 2.05, Vc 16. Ericksen et al., 1986. Am. Min. 71 (1986), 830.

Iranite, $CuPb_{10}(CrO_4)_6(SiO_4)_2(OH)_2$, G Tric. Vitreous yellow. Sebarz mine, Anarak (NE), Iran. H=?, D_m=?, D_c= 6.67, VIE 02. Bariand & Herpin, 1963. Bull. Mineral. 103 (1980), 469. See also Hemihedrite.

Iraqite-(La), $KCa_4(La, Ce, Th)_2Si_{16}O_{40}$, A Tet. Dull/pearly pale greenish-yellow. Shakhi-Rash Mt., Hero, Qala-Diz, Iraq (N). H= 4.5, D_m= 3.27, D_c= 3.28, VIIIE 17. Livingstone et al., 1976. Min Mag. 40 (1976), 441. See also Ekanite, Steacyite.

Irarsite, (Ir, Ru, Rh, Pt)AsS, A Cub. Metallic greyish-white. Onverwacht, Transvaal, South Africa. H= 6.5, D_m=?, D_c=?, IIC 06a. Genkin et al., 1966. Am. Min. 52(1967), 1580 (Abst.). See also Hollingworthite.

Irhtemite, $Ca_4MgH_2(AsO_4)_4 \cdot 4H_2O$, A Mon. Silky white, colorless, pale rose. Irhtem ore deposit, Morocco. H=?, D_m= 3.09, D_c= 3.153, VIIA 09. Pierrot & Schubnel, 1972. Am. Min. 59(1974), 209 (Abst.).

Iridarsenite, $(Ir, Ru)As_2$, A Mon. Metallic grey. New Guinea. H= 5.5, D_m=?, D_c= 10.9, IIC 09. Harris, 1974. Can. Min. 12 (1974), 280.

Iridium, Ir, G Cub. Metallic white. H= 6-6.5, D_m= 22.4, D_c= 22.65, IA 07. Hausmann, 1813. Can. Min. 12 (1973), 104. Acta Cryst. 24A (1968), 469. See also Osmiridium.

Iridosmine, (Os, Ir), A Hex. Metallic white/grey; yellowish-white in reflected light. H= 6-7, D_m= 20, D_c= 23.26, IA 08. Breithaupt, 1839. Dana, 7th ed.(1944), v.1, 111. See also Rutheniridosmine.

Iriginite, $U(MoO_4)_2(OH)_2 \cdot 2H_2O$, G Mon. Vitreous yellow. U–Mo ore deposit, USSR. H= 1-2, D_m= 3.84, D_c=?, VIF 02. Epshtein, 1959. Am. Min. 49 (1964), 408.

Iron, α–Fe, G Cub. Metallic grey; white in reflected light. H= 4, D_m=?, D_c= 7.873, IA 05. Dana, 7th ed.(1944), v.1, 114.

Irtyshite, $Na_2(Ta, Nb)_4O_{11}$, A Hex. Adamantine colorless. Irtysh River Area, Kazakhstan (E), USSR. H= 6.8, D_m=?, D_c= 7.03, IVD 17. Voloshin et al., 1985. Am. Min. 71(1986), 1545 (Abst.).

Ishiganeite, D A mixture of cryptomelane and birnessite. Am. Min. 48 (1963), 952. Kani & Tanaka, 1937.

Ishikawaite, $(U, Fe, Y, Ca)(Nb, Ta)_2O_4$(?), Q Syst=? Waxy black. Ishikawa district, Fukushima, Iwaki, Japan. H= 5-6, D_m= 6.3, D_c=?, IVD 10b. Kimura, 1922. Dana, 7th ed.(1944), v.1, 766. See also Uranpyrochlore.

Isochalcopyrite, (Fe, Cu)S, Q Syst=? Metallic yellow in reflected light. Atlantis II Deep, Red Sea. H=?, D_m=?, D_c=?, IB 02. Missack et al., 1989. Am. Min. 75(1990), 432 (Abst.).

Isoclasite, $Ca_2PO_4(OH) \cdot 2H_2O$, Q Mon. Vitreous/pearly colorless, white. Jáchymov (St. Joachimsthal), Západočeský kraj, Čechy (Bohemia), Czechoslovakia. H= 1.5, D_m= 2.92, D_c=?, VIID 01. Sandberger, 1870. Dana, 7th ed.(1951), v.2, 933.

Isocubanite, $CuFe_2S_3$, A Cub. Metallic bronze. East Pacific Rise, Pacific Ocean; also the Atlantis II Deep, Red Sea. H= 3.5, D_m=?, D_c= 3.928, IIB 08. Caye et al., 1988. Min. Mag. 52 (1988), 509.

Isoferroplatinum, $(Pt,Pd)_3(Fe,Cu)$, A Cub. Metallic. Tulameen River, British Columbia, Canada. H=?, D_m=?, D_c= 18.23, IA 12. Cabri & Feather, 1975. Can. Min. 13 (1975), 117.

Isokite, $CaMg(PO_4)F$, G Mon. Silky colorless, buff, pinkish. Isoka, Zambia. H= 5, D_m= 3.27, D_c= 3.29, VIIB 11a. Deans & McConnell, 1955. Am. Min. 41(1956), 167 (Abst.). See also Panasqueiraite.

Isomertieite, $Pd_{11}(Sb,As)_4$, A Cub. Metallic pale yellow in reflected light. Itabira, Minas Gerais, Brazil. H= 5.5, D_m=?, D_c= 10.33, IIG 04. Clark et al., 1974. Min. Mag. 39 (1974), 528. See also Mertieite, Guanglinite.

Isostannite, D = kësterite/ferrokësterite. Can. Min. 27 (1989), 673. Claringbull & Hey, 1955.

Isowolframite, D Unnecessary name for wolframite series. Min. Mag. 43 (1980), 1055. Senchilo & Mukhlya, 1972.

Itoite, $Pb_3GeO_2(SO_4)_2(OH)_2$, G Orth. Silky white. Tsumeb, Namibia. H=?, D_m=?, D_c= 6.67, VIA 07. Frondel & Strunz, 1960. Am. Min. 45(1960), 1313 (Abst.).

Iwakiite, $Mn(Fe,Mn)_2O_4$, A Tet. Metallic greenish-black; olive-grey in reflected light. Gozaisho mine, Iwaki, Japan. H= 6-6.5, D_m= 4.85, D_c= 4.93, IVB 02b. Matsubara et al., 1979. Am. Min. 65(1980), 406 (Abst.). Zts. Krist. 185 (1988), 605. See also Jacobsite.

Ixiolite, $(Ta,Mn,Nb)O_2$, G Orth. Sub-metallic grey. Skogböle, Kimito, Finland. H= 6-6.5, D_m= 7.1, D_c= 6.93, IVD 09. Nordenskiöld, 1857. IMA 1986, Abst. p. 265. Can. Min. 14 (1976), 540. See also Ashanite.

Izoklakeite, $(Cu,Fe)_2Pb_{27}(Sb,Bi)_{19}S_{57}$, A Orth. Metallic greenish-white in reflected light. Izok Lake, Northwest Territories, Canada. H= 3.5, D_m= 6.47, D_c= 6.505, IID 08. Harris et al., 1986. Can. Min. 24 (1986), 1. N.Jb.Min.Abh. 153 (1986), 121. See also Kobellite.

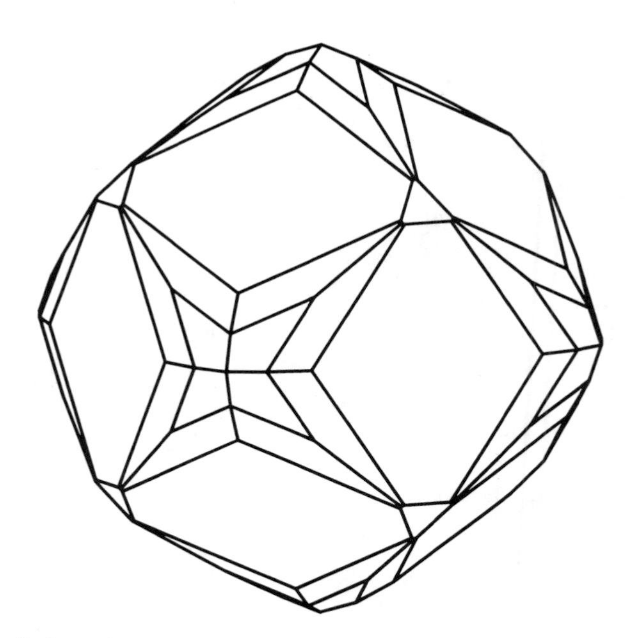

Magnetite from Scalotta showing the forms {311}, {531}, {110} and {111}. (After Cathrein, 1886)

J

Jacobsite, $(Mn,Fe,Mg)(Fe,Mn)_2O_4$, [G] Cub. Metallic/sub-metallic black; grey in reflected light. Jakobsberg mine, Värmland, Sweden. H= 5.5-6.5, D_m= 4.76, D_c= 5.03, IVB 01. Damour, 1869. Dana, 7th ed.(1944), v.1, 698. See also Magnetite, Iwakiite.

Jadeite, $Na(Al,Fe)Si_2O_6$, [G] Mon. Vitreous colorless, white, green, greenish-blue, blue. H= 6, D_m= 3.4, D_c= 3.345, VIIID 01a. Damour, 1863. Am. Min. 51 (1966), 956. MSA Spec. Paper 2 (1969), 31.

Jaffeite, $Ca_6Si_2O_7(OH)_6$, [A] Hex. Vitreous colorless. Kombat mine, Otavi (E), Namibia. H=?, D_m= 2.65, D_c= 2.58, VIIIB 06. Sarp & Peacor, 1989. Am. Min. 74 (1989), 1203.

Jagoite, $(Pb,Na,Ca)_9(Fe,Mg,Mn)_2(Si,Fe,Pb)_{17}O_{41}(Cl,OH)_4$, [G] Hex. Vitreous yellow-green. Långban mine, Filipstad (near), Värmland, Sweden. H= 3, D_m= 5.43, D_c= 5.33, VIIIE 23. Blix et al., 1957. Am. Min. 43(1958), 387 (Abst.). Am. Min. 66 (1981), 852.

Jagowerite, $BaAl_2(PO_4)_2(OH)_2$, [A] Tric. Vitreous light green. Hess River, Yukon Territory, Canada. H= 4.5, D_m= 4.01, D_c= 4.05, VIIB 05. Meagher et al., 1973. Can. Min. 12 (1973), 135. Am. Min. 59 (1974), 291.

Jahnsite-(CaMnFe), $CaMnFe^{2+}_2Fe^{3+}_2(PO_4)_4(OH)_2 \cdot 8H_2O$, [A] Mon. Vitreous/sub-adamantine brown, yellow, greenish-yellow, etc. Tip Top Pegmatite, Custer, Custer Co., South Dakota, USA. H= 4, D_m= 2.86, D_c= 2.88, VIID 24. Moore, 1978. Min. Mag. 42 (1978), 309. See also Jahnsite-(CaMnMg), Jahnsite-(MnMnMn).

Jahnsite-(CaMnMg), $CaMn(Mg,Fe)_2Fe^{3+}_2(PO_4)_4(OH)_2 \cdot 8H_2O$, [A] Mon. Vitreous/sub-adamantine brown, yellow, greenish-yellow, etc. Tip Top Pegmatite, Custer, Custer Co., South Dakota, USA. H= 4, D_m= 2.71, D_c= 2.715, VIID 24. Moore, 1974. Min. Mag. 42 (1978), 309. See also Jahnsite-(CaMnFe).

Jahnsite-(CaMnMn), $CaMnMn_2Fe^{3+}_2(PO_4)_4(OH)_2 \cdot 8H_2O$, [A] Mon. Vitreous brownish-yellow. Mangualde, Beira, Portugal. H= 4, D_m= 2.78, D_c= 2.798, VIID 24. Grice et al., 1990. Am. Min. 75 (1990), 401.

Jahnsite-(MnMnMn), $Mn_4Fe^{3+}_2(PO_4)_4(OH)_2 \cdot 8H_2O$, [Q] Syst=? Stewart mine, Pala, San Diego Co., California, USA. H=?, D_m=?, D_c=?, VIID 24. Moore, 1978. Min. Mag. 42 (1978), 309. See also Jahnsite-(CaMnMg), Jahnsite-(CaMnFe).

Jaipurite, γ–CoS, [Q] Hex. Metallic grey. Khetri mines, Jaipur, Rájputána, India. H=?, D_m= 5.45, D_c= 6.011, IIB 09a. Mallet, 1880. Strunz (1970), 120.

Jalpaite, Ag_3CuS_2, [G] Tet. Metallic grey. Jalpa, Mexico. H= 1.5, D_m= 6.83, D_c= 6.827, IIA 04. Breithaupt, 1858. Am. Min. 53 (1968), 1530.

Jamborite, $(Ni,Fe)_8SO_4(OH)_{16} \cdot nH_2O$, [A] Hex. Translucent green. Bologna and Modena, Italy. H=?, D_m= 2.67, D_c= 2.69, IVF 03b. Morandi & Dalrio, 1973. Am. Min. 58 (1973), 835.

Jamesite, $Pb_2Zn_2Fe_5O_4(AsO_4)_5$, [A] Tric. Sub-adamantine reddish-brown. Tsumeb, Namibia. H= 3, D_m=?, D_c= 5.10, VIIB 06. Keller et al., 1981. Am. Min. 66(1981), 1275 (Abst.).

Jamesonite, $Pb_4FeSb_6S_{14}$, [G] Mon. Metallic grey-black; white in reflected light. H= 2.5, D_m= 5.63, D_c= 5.71, IID 03. Haidinger, 1825. Can. Min. 25 (1987), 667. Zts. Krist. 109 (1957), 161. See also Parajamesonite, Benavidesite.

Janggunite, $(Mn,Fe)_6O_8(OH)_6$, [A] Orth. Dull black. Janggun mine, Bonghwa, Korea. H= 2-3, D_m= 3.59, D_c= 3.58, IVF 05a. Kim, 1977. Min. Mag. 41 (1977), 519.

Janhaugite, $Na_3Mn_3(Ti,Zr,Nb)_2(Si_2O_7)_2O_2(OH,F)_2$, [A] Mon. Vitreous reddish-brown. Lake Gjerdingen (ESE), Nordmarka, Oslo, Norway. H= 5, D_m= 3.60, D_c= 3.71, VIIIB 08. Raade & Mladeck, 1983. Am. Min. 68 (1983), 1216. Min. Abstr. 87M/2103.

Jarlite, $Na_2(Sr,Na)_{14}(Mg)_2Al_{12}F_{64}(OH,H_2O)_4$, [G] Mon. Vitreous colorless, white, grey. Ivigtut, Greenland (SW). H= 4-4.5, D_m= 3.8, D_c= 4.06, IIIC 08. Bøgvad, 1933. Am. Min. 34 (1949), 383. Can. Min. 21 (1983), 553. See also Calcjarlite.

Jarosewichite, $Mn_4AsO_4(OH)_6$, [A] Orth. Sub-vitreous dark red. Franklin, Sussex Co., New Jersey, USA. H= 4, D_m= 3.66, D_c= 3.70, VIIB 10a. Dunn et al., 1982. Am. Min. 67 (1982), 1043. See also Chlorophoenicite, Magnesium-chlorophoenicite.

Jarosite, $KFe_3(SO_4)_2(OH)_6$, [G] Rhom. Sub-adamantine/vitreous yellow, brown. Jaroso ravine, Sierra Almagrera, Spain. H= 2.5-3.5, D_m= 3.1, D_c= 3.127, VIB 03a. Breithaupt, 1852. Am. Min. 72 (1987), 178. Min. Jour. 8 (1977), 419.

Jaskólskiite, $Cu_{0.2}Pb_{2.2}(Sb,Bi)_{1.8}S_5$, [A] Orth. Metallic grey. Vena Cu-Co mine, Askersund, Örebro, Sweden; Izok Lake mine, Mackenzie, Northwest Territories, Canada. H= 3.5, D_m= 6.551, D_c= 6.50, IID 08. Zakrzewski, 1984. Can. Min. 22 (1984), 481. Zts. Krist. 171 (1985), 179.

Jasmundite, $Ca_{11}O_2(SiO_4)_4S$, [A] Tet. Resinous dark brown, greenish-brown, brownish-green. Ettringer Bellerberg, Mayen, Eifel, Rheinland-Pfalz, Germany. H= 5, D_m= 3.03, D_c= 3.23, VIIIA 10. Hentschel et al., 1983. Am. Min. 69(1984), 566 (Abst.). Acta Cryst. B37 (1981), 803.

Jeanbandyite, $(Fe,Mn)Sn(OH)_6$, [A] Tet. Translucent brown-orange. Contacto Vein, Llallagua, Potosí, Bolivia. H= 3.5, D_m= 3.81, D_c= 3.81, IVF 06. Kampf, 1982. Am. Min. 68(1983), 471 (Abst.).

Jeffreyite, $(Ca,Na)_2(Be,Al)Si_2(O,OH)_7$, [A] Orth. Clear, colorless. Jeffrey pit, Asbestos, Shipton Township, Richmond Co., Québec, Canada. H= 5, D_m= 2.99, D_c= 2.98, VIIIB 02b. Grice & Robinson, 1984. Can. Min. 22 (1984), 443. See also Melilite, Gugiaite.

Jennite, $Ca_9Si_6O_{16}(OH)_{10} \cdot 6H_2O$, [A] Tric. Vitreous white. Crestmore quarry, Riverside Co., California, USA. H= 3.5, D_m= 2.33, D_c= 2.319, VIIIB 28. Carpenter et al., 1966. Min. Jour. 14 (1989), 279.

Jeppeite, $(K,Ba)_2(Ti,Fe)_6O_{13}$, [A] Mon. Sub-metallic black, brown. Walgidee Hills, Fitzroy Basin, Kimberley, Western Australia, Australia. H= 5-6, D_m= 3.94, D_c= 3.98, IVD 03b. Pryce et al., 1984. Min. Mag. 48 (1984), 263. Am. Min. 63(1978), 795 (Abstr.).

Jeremejevite, $Al_6(BO_3)_5(F,OH)_3$, [G] Hex. Vitreous colorless, pale yellowish-brown. Mt. Soktuĭ, Dauria, Nerchinsk dist., Siberia, USSR. H= 7.5, D_m= 3.28, D_c= 3.287, Vc 01a. Damour, 1883. Can. Min. 19 (1981), 303. Zts. Krist. 165 (1983), 255.

Jeromite, $As(S,Se)_2$(?), [Q] Amor. Black, cherry-red. United Verde mine, Jerome, Yavapai Co., Arizona, USA. H=?, D_m=?, D_c=?, IIB 19. Lausen, 1928. Dana, 7th ed.(1944), v.1, 144.

Jerrygibbsite, $Mn_9(SiO_4)_4(OH)_2$, [A] Orth. Vitreous violet-pink. Sussex Co., Franklin, New Jersey, USA. H= 5.5, D_m= 4.00, D_c= 4.05, VIIIA 16. Dunn et al., 1984. Am. Min. 69(1984), 546. N.Jb.Min.Mh. (1989), 410. See also Sonolite.

Jervisite, $(Na,Ca,Fe)(Sc,Mg,Fe)Si_2O_6$, [A] Mon. Vitreous light green. Cava Diverio, Baveno, Italy. H=?, D_m=?, D_c= 3.29, VIIID 01a. Mellini et al., 1982. Am. Min. 67 (1982), 599. Acta Cryst. B29 (1973), 2615.

Jimboite, $Mn_3(BO_3)_2$, [G] Orth. Vitreous light purplish-brown. Kaso mine, Kanuma city, Tochigi, Japan. H= 5.5, D_m= 3.98, D_c= 4.1, Vc 01b. Watanabe et al., 1963. Am. Min. 48(1963), 1416 (Abst.). Sov.Phys.Cryst. 23(1979), 272. See also Kotoite.

Jimthompsonite, $(Mg,Fe)_5Si_6O_{16}OH)_2$, [A] Orth. Transparent colorless, pale pinkish-brown. Carlton mine, Chester, Windham Co., Vermont, USA. H=?, D_m=?, D_c= 3.05, VIIID 12. Veblen & Burnham, 1978. Am. Min. 63 (1978), 1000. Am. Min. 63 (1978), 1053. See also Clinojimthompsonite.

Jiningite, [D] Unnecessary name for variety of thorite or thorogummite. Min. Mag. 33 (1962), 261. Kuo, 1959.

Jinshajiangite, $Na_5Ba_4(Fe,Mn)_{15}Ti_8Si_{15}O_{64}(F,OH)_6$, [A] Mon. Vitreous blackish/brownish/golden-red. Jinshajiang River (near), Sichuan, China (SW). H= 5, D_m= 3.61, D_c= 3.56, VIIID 25. Hong & Fu, 1982. Am. Min. 69(1984), 567 (Abst.).

Jixianite, $Pb(W,Fe)_2(O,OH)_7$, [Q] Cub. Resinous red, brownish-red. Jixian, Ji Co., Hebei, China. H= 3, D_m= 6.04, D_c= 7.89, IVC 09a. Liu, 1979. Am. Min. 64(1979), 1330 (Abst.). See also Pyrochlore, Stibiconite.

Joaquinite-(Ce), $NaBa_2FeTi_2Ce_2(SiO_3)_8O_2(OH)\cdot H_2O$, [G] Mon. Silky brown, yellow. Joaquin ridge, Mt. Diablo Range, Contra Costa Co., California, USA. H= 5, D_m= 3.98, D_c= 4.11, VIIIC 05. Louderback, 1909. Am. Min. 57 (1972), 85. Am. Min. 60 (1975), 872. See also Orthojoaquinite-(Ce), Bario-orthojoaquinite.

Joesmithite, $(Ca,Pb)_3(Mg,Fe)_5Si_6Be_2O_{22}(OH)_2$, [A] Mon. Translucent brown. Långban mine, Filipstad (near), Värmland, Sweden. H= 5.5, D_m= 3.83, D_c= 4.02, VIIID 01b. Moore, 1968. Am. Min. 54(1969), 577 (Abst.). MSA Spec. Paper 2 (1969), 111.

Johachidolite, $CaAlB_3O_7$, [A] Orth. Vitreous colorless. Johachido, Kenkyohokudo, North Korea. H= 7.5, D_m= 3.37, D_c= 3.44, Vc 15. Iwase & Saito, 1942. Am. Min. 62 (1977), 327. Nature Phys.Sci. 240(1972), 63.

Johannite, $Cu(UO_2)_2(SO_4)_2(OH)_2\cdot 8H_2O$, [G] Tric. Vitreous green. Jáchymov (St. Joachimsthal), Západočeský kraj, Čechy (Bohemia), Czechoslovakia. H= 2-2.5, D_m= 3.32, D_c= 3.44, VID 08. Haidinger, 1830. N.Jb.Min.Abh. 159 (1988), 297. Min. Abstr. 84M/3844.

Johannsenite, $CaMnSi_2O_6$, [G] Mon. Vitreous brown, greyish, green, colorless, blue. Bohemia Mining Dist., Lane Co., Oregon, USA; Campiglia Marittima (near), Monte Valerio, Toscana, Italy; Franklin, Sussex Co., New Jersey, USA; Tetela del Oro, Puebla, Mexico; Espiritu Santo mine, Pachuca, Real del Monte, Hidalgo, Mexico; Schio-Vincenti mine, Schio (near), Venezia, Italy. H= 6, D_m= 3.4, D_c= 3.522, VIIID 01a. Schaller, 1932. DHZ, 2nd ed.(1978), v.2A, 415. Am. Min. 52 (1967), 709. See also Diopside, Hedenbergite.

Johillerite, $NaCu(Mg,Zn)_3(AsO_4)_3$, [A] Mon. Transparent violet. Tsumeb, Namibia. H= 3, D_m= 4.15, D_c= 4.17, VIIA 07. Keller et al., 1982. Am. Min. 67(1982), 1075 (Abst.). N.Jb.Min.Mh. (1988), 395. See also O'Danielite.

Johnbaumite, $Ca_5(AsO_4)_3(OH)$, [A] Hex. Vitreous colorless. Palmer Shaft (S), Franklin, Sussex Co., New Jersey, USA. H= 4.5, D_m= 3.68, D_c= 3.73, VIIB 16. Dunn et al., 1980. Am. Min. 65 (1980), 1143.

Johninnesite, $Na_2Mg_4Mn_{12}As_2Si_{12}O_{43}(OH)_6$, [A] Tric. Vitreous light yellowish-brown. Kombat mine, Otavi (E), Namibia. H=?, D_m= 3.48, D_c= 3.51, VIIIB 25. Dunn et al., 1986. Min. Mag. 50 (1986), 667.

Johnsomervilleite, $Na_2Ca(Fe,Mg,Mn)_7(PO_4)_6$, [A] Rhom. Vitreous dark brown. Glen Cosaidh, Loch Quoich, Scotland. H= 4.5, D_m= 3.35, D_c= 3.41, VIIA 06. Livingstone, 1980. Min. Mag. 43 (1980), 833. See also Fillowite.

Johnwalkite, $K(Mn,Fe)_2(Nb,Ta)O_2(PO_4)_2\cdot 2(H_2O,OH)$, [A] Orth. Vitreous dark reddish-brown. Champion mine, Keystone, Pennington Co., South Dakota, USA. H= 4, D_m= 3.40, D_c= 3.44, VIID 34. Dunn et al., 1986. Am. Min. 72(1987), 223 (Abst.). See also Olmsteadite.

Jokokuite, $MnSO_4\cdot 5H_2O$, [A] Tric. Vitreous pale pink. Jokoku mine, Hiyama, Hokkaido, Japan. H= 2.5, D_m= 2.03, D_c= 2.094, VIC 03a. Nambu et al., 1978. Am. Min. 64(1979), 655 (Abst.).

Joliotite, $(UO_2)CO_3\cdot 2H_2O$, [G] Orth. Translucent yellow. Menzenschwand, Schwarzwald, Germany. H=?, D_m= 4.04, D_c= 4.55, VbD 04. Walenta, 1976. SMPM 56 (1976), 167.

Jonesite, $(K,Na)_2Ba_4Ti_4Al_2Si_{10}O_{36}\cdot 6H_2O$, [A] Orth. Translucent colorless. Benito Gem mine, San Benito Co., California, USA. H= 3-4, D_m= 3.25, D_c= 3.239, VIIIC 07. Wise & Pabst, 1977. Min. Rec. 8 (1977), 453.

Jordanite, $Pb_{14}(As,Sb)_6S_{23}$, [G] Mon. Metallic lead-grey. Lengenbach quarry, Binntal, Valais (Wallis), Switzerland. H= 3, D_m= 6.4, D_c= 6.34, IID 01f. vom Rath, 1864. Can. Min. 9 (1968), 505. Zts. Krist. 139 (1974), 161. See also Geocronite.

Jordisite, MoS_2, [Q] Amor. Black. Himmelsfürst mine, Freiberg, Sachsen (Saxony), Germany. H=?, D_m=?, D_c=?, IIC 10. Cornu, 1909. Am. Min. 56 (1971), 1832. See also Molybdenite, Molybdenite-3R.

Joséite-A, Bi_4TeS_2, [G] Rhom. Metallic grey. San José, Mariana (near), Minas Gerais, Brazil and/or Hudson Bay mine, Hazelton (N), British Columbia, Canada. H= 2, D_m= 8.10, D_c= 8.26, IIA 15. Kenngott, 1853. Am. Min. 34 (1949), 342. See also Protojoséite.

Joséite-B, Bi_4Te_2S, [G] Rhom. Metallic grey. San José, Mariana (near), Minas Gerais, Brazil. H= 2, D_m= 8.3, D_c= 8.40, IIA 15. Thompson, 1949. Am. Min. 34 (1949), 342. See also Joséite-A, Joséite-C.

Joséite-C, $Bi_{16}Te_3S_9$, [Q] Syst=? Metallic. Sokhondo deposit, Transbaikal (E), USSR. H= 2.5, D_m=?, D_c=?, IIA 15. Godovikov et al., 1970. Am. Min. 56(1971), 1839 (Abst.). See also Joséite-B, Joséite-A.

Jouravskite, $Ca_3Mn(SO_4)(CO_3)(OH)_6\cdot 12H_2O$, [A] Hex. Translucent greenish-yellow, greenish-orange. Anti-Atlas, Morocco. H= 2.5, D_m= 1.95, D_c= 1.93, VID 07. Gaudefroy & Permingeat, 1965. Am. Min. 50(1965), 2102 (Abst.). Acta Cryst. B25 (1969), 1943.

Juanite, $Ca_{10}Mg_4Al_2Si_{11}O_{39}\cdot 4H_2O(?)$, [Q] Syst=? Translucent white. Iron Hill, Gunnison Co., Colorado, USA. H= 5.5, D_m= 3.015, D_c=?, VIIIG 01. Larsen & Goranson, 1932. Min. Abstr. 5 (1932), 146.

Julgoldite-(Fe^{2+}), $(Ca,K)_2Fe^{2+}Fe^{3+}Si_3(O,OH)_{14}$, [A] Mon. Sub-metallic black, green. Långban mine, Filipstad (near), Värmland, Sweden. H= 4.5, D_m= 3.602, D_c= 3.56, VIIIB 26. Moore, 1971. Can. Min. 12 (1973), 219. Min. Mag. 39 (1973), 271. See also Pumpellyite, Julgoldite-(Fe^{3+}), Julgoldite-(Mg).

Julgoldite-(Fe^{3+}), $(Ca,K)_2Fe^{3+}Fe^{3+}Si_3(O,OH)_{14}$, [A] Mon. Sub-metallic black, green. Långban mine, Filipstad (near), Värmland, Sweden. H=?, D_m=?, D_c= 3.53, VIIIB 26. Passaglia & Gottardi, 1973. Can. Min. 12 (1973), 219. See also Pumpellyite, Julgoldite-(Fe^{2+}), Julgoldite-(Mg).

Julgoldite-(Mg), $(Ca,K)_2MgFe^{3+}Si_3(O,OH)_{14}$, [Q] Mon. Sub-metallic black, green. Toba, Sinsen-mura, Japan. H=?, D_m=?, D_c=?, VIIIB 26. Passaglia & Gottardi, 1973. Can. Min. 12 (1973), 219. See also Pumpellyite, Julgoldite-(Fe^{2+}), Julgoldite-(Fe^{3+}).

Juliënite, $Na_2Co(SCN)_4\cdot 8H_2O$, [G] Mon. Translucent blue. Chamitumba, Shaba, Zaïre. H=?, D_m= 1.648, D_c= 1.610, IXA 04. Schoep, 1928. Dana, 7th ed.(1951), v.2, 1106. Acta Cryst. B38 (1982), 1084.

Jungite, $Ca_2Zn_4Fe_8(PO_4)_9(OH)_9\cdot 16H_2O$, [A] Orth. Silky/vitreous dark green. Hagendorf (S) pegmatite, Waidhaus, Oberpfalz, Bayern (Bavaria), Germany. H= 1, D_m= 2.843, D_c= 2.849, VIID 12. Moore & Ito, 1980. Am. Min. 65(1980), 1067 (Abst.).

Junitoite, $CaZn_2Si_2O_7\cdot H_2O$, [A] Orth. Vitreous colorless. Christmas mine, Christmas, Gila Co., Arizona, USA. H= 4.5, D_m= 3.5, D_c= 3.516, VIIIB 07. Williams, 1976. Am. Min. 61 (1976), 1255. Min. Mag. 49 (1985), 91.

Junoite, $Cu_2Pb_3Bi_8(S,Se)_{16}$, [A] Mon. Creamy white. Juno mine, Tennant Creek, Northern Territory, Australia. H= 3-4, D_m=?, D_c= 6.77, IID 08. Mumme, 1975. Econ. Geol. 70 (1975), 369. Am. Min. 60 (1975), 548.

Jurbanite, $AlSO_4(OH) \cdot 5H_2O$, [A] Mon. Translucent colorless. San Manuel mine, Pinal Co., Arizona, USA. H= 2.5, D_m= 1.786, D_c= 1.828, VID 02. Anthony & McLean, 1976. Am. Min. 61 (1976), 1. Zts. Krist. 173 (1985), 33. See also Rostite.

Jusite, $(Ca, Na, K, H)(Si, Al)O_3$, [Q] Syst=? Translucent white. Jus, Schwäbische Alb, Württemberg, Germany. H=?, D_m=?, D_c=?, VIIID 08. Gramling-Mende & Leopold, 1943. Min. Abstr. 9 (1944), 37.

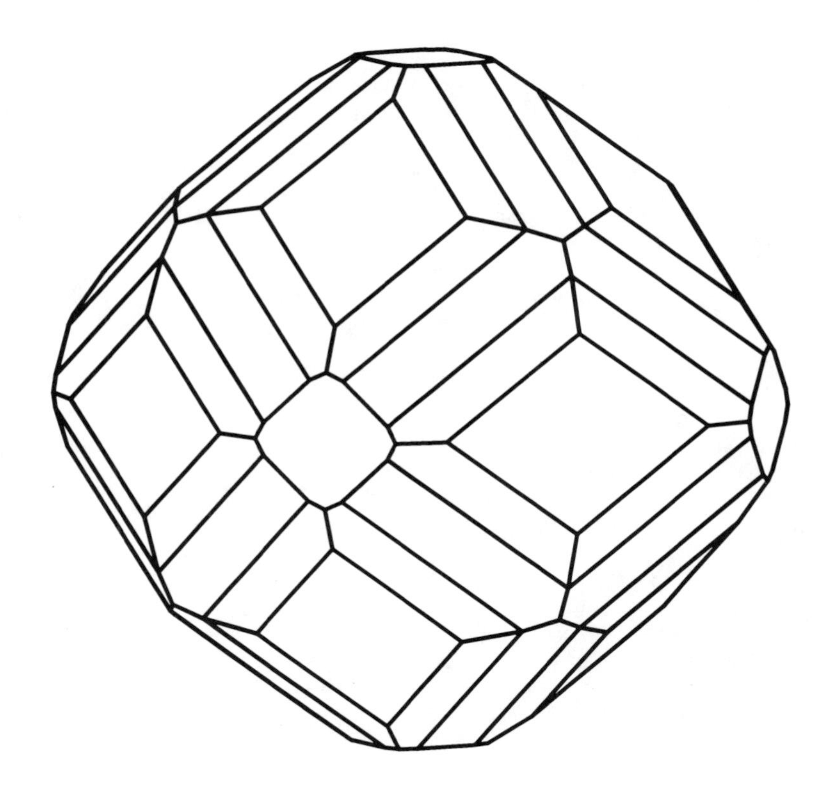

Eglestonite from Terlingua, Texas showing the forms {001}, {112}, {123} and {110}. (After Hillebrand and Schaller, 1909)

K

Kaatialaite, $FeAs_3O_9 \cdot 6\text{–}8H_2O$, [A] Mon. Vitreous greenish-blue. Kaatiala pegmatite, Kuortane, Finland (W). H=?, D_m= 2.64, D_c= 2.62, VIIC 02. Raade et al., 1984. Amer. Mineral. 69 (1984) 383.

Kadyrelite, $Hg_4(Br, Cl)_2O$, [A] Cub. Vitreous/ adamantine orange. Kadyrel deposit, Tuva, Siberia (S), USSR. H= 2.5-3, D_m=?, D_c= 8.79, IIIC 03. Vasil'ev, 1987. Am. Min. 74(1989), 503 (Abst.).

Kaersutite, $NaCa_2(Mg, Fe)_4Ti(Si_6Al_2)O_{22}(OH)_2$, [G] Mon. Dark brown, black. Kaersut, Umanaq dist., Greenland (N). H= 5-6, D_m= 3.2, D_c= 2.94, VIIID 05b. Lorenzen, 1884. Min. Mag. 39 (1973), 390. N.Jb.Min.Mh. (1989), 137. See also Ferrokaersutite.

Kafehydrocyanite, $K_4Fe(CN)_6 \cdot 3H_2O$, [Q] Tet. Translucent yellow. Medvezhiĭ Log mine, Olkhovsk, Sayan (E), USSR. H= 2-2.5, D_m= 1.98, D_c= 1.89, IXA 04. Povarennykh & Rusakova, 1973. Am. Min. 59(1974), 209 (Abst.).

Kahlerite, $Fe(UO_2)_2(AsO_4)_2 \cdot 12H_2O$, [G] Tet. Translucent yellow. Hüttenberg, Carinthia, Austria. H=?, D_m=?, D_c= 3.222, VIID 20a. Meixner, 1953. Am. Min. 39(1954), 1038 (Abst.).

Kainite, $KMg(SO_4)Cl \cdot 3H_2O$, [G] Mon. Vitreous colorless, grey, blue, violet, yellowish, etc. Stassfurt deposit, Sachsen (Saxony), Germany. H= 2.5-3, D_m= 2.15, D_c= 2.176, VID 05. Zincken, 1865. Dana, 7th ed.(1951), v.2, 594. Am. Min. 57 (1972), 1325.

Kainosite-(Y), $Ca_2(Y, Ce)_2(SiO_3)_4(CO_3) \cdot H_2O$, [G] Orth. Greasy yellowish-brown. Igeltjern, Hitterö, Norway. H= 5.5, D_m= 3.61, D_c= 3.562, VIIIC 04. Nordenskiöld, 1886. Can. Min. 8 (1964), 1. N.Jb.Min.Mh. (1989), 153.

Kalborsite, $K_6Al_4BSi_6O_{20}(OH)_4Cl$, [A] Tet. Vitreous/pearly colorless (rose-brown tint). Mt. Rasvumchorr, Khibina massif, Kola Peninsula, USSR. H= 6, D_m= 2.5, D_c= 2.48, VIIIF 22. Khomyakov et al., 1980. Am. Min. 66(1981), 879 (Abst.). Am. Min. 66(1981), 879 (Abst.).

Kaliborite, $HKMg_2B_{12}O_{16}(OH)_{10} \cdot 4H_2O$, [G] Mon. Vitreous colorless, white, reddish-brown. Schmidtmannshall, Aschersleben (near), Germany. H= 4-4.5, D_m= 2.116, D_c= 2.110, Vc 12a. Feit, 1889. Am. Min. 50 (1965), 1079. Struct. Repts. 31A (1966), 172.

Kalicinite, $KHCO_3$, [G] Mon. Transparent colorless, white, yellowish. Chippis, Valais (Wallis), Switzerland. H=?, D_m= 2.17, D_c= 2.169, VbA 01. Pisani, 1865. Dana, 7th ed.(1951), v.2, 136. Acta Cryst. 30B (1974), 1155.

Kalininite, $ZnCr_2S_4$, [A] Cub. Metallic creamy. Slyudyanka complex, Lake Baikal region (S), USSR. H= 5, D_m=?, D_c= 4.04, IIC 01. Reznitskii et al., 1985. Am. Min. 72(1987), 223 (Abst.).

Kalinite, $KAl(SO_4)_2 \cdot 11H_2O$(?), [Q] Syst=? H=?, D_m=?, D_c=?, VIC 07. Dana, 1868. Dana, 7th ed.(1951), v.2, 471.

Kaliophilite, $KAlSiO_4$, [G] Hex. Silky colorless. Monte Somma, Mt. Vesuvius, Napoli, Campania, Italy. H= 6, D_m= 2.49, D_c= 2.65, VIIIF 01. Mierisch, 1886. Am. J. Sci. 255 (1957), 282. See also Kalsilite, Panunzite, Trikalsilite.

Kalipyrochlore, $(K, H_2O)_2Nb_2(O, OH)_7$, [A] Cub. Translucent greenish. Goma (150 km N of), Lueshe, Kivu, Zaïre. H=?, D_m= 3.44, D_c= 3.42, IVC 09a. Van Wambeke, 1978. Am. Min. 63 (1978), 528.

Kalistrontite, $K_2Sr(SO_4)_2$, [G] Rhom. Vitreous colorless. Alshtan, Bashkiria, USSR. H= 2, D_m= 3.20, D_c= 3.31, VIA 06. Voronova, 1962. JCPDS 29-1049. See also Palmierite.

Kalsilite, $KAlSiO_4$, [G] Hex. Translucent colorless, white, grey. Lake Mafuru, Uganda. H= 6, D_m= 2.61, D_c= 2.619, VIIIF 01. Bannister, 1942. DHZ (1963), v.4, 231. Min. Mag. 35 (1965), 588. See also Kaliophilite, Panunzite, Trikalsilit.

Kamacite, α–(Fe, Ni), [G] Cub. Metallic grey. H=?, D_m=?, D_c= 7.85, IA 05. Reichenbach, 1861. Am. Min. 51 (1966), 37. See also Taenite, Tetrataenite.

Kamaishilite, $Ca_2(SiAl_2)O_6(OH)_2$, [A] Tet. Transparent colorless. Kamaishi mine, Japan. H=?, D_m=?, D_c= 2.825, VIIIF 07. Uchida & Iiyama, 1981. Am. Min. 67(1982), 855 (Abst.). See also Bicchulite.

Kambaldaite, $NaNi_4(CO_3)_3(OH)_3 \cdot 3H_2O$, [A] Hex. Translucent green. Otter Shoot, Kambalda, Western Australia, Australia. H= 3, D_m= 3.18, D_c= 3.19, VbD 01. Nickel & Robinson, 1985. Am. Min. 70 (1985), 419. Am. Min. 70 (1985), 423.

Kamchatkite, $KCu_3O(SO_4)_2Cl$, [A] Orth. Vitreous greenish-yellowish brown. Tolbachik volcano, Kamchatka, USSR. H= 3.5, D_m= 3.48, D_c= 3.58, VIB 02. Vergasova et al., 1988. ZVMO 117 (1988), 459.

Kamiokite, $Fe_2Mo_3O_8$, [A] Hex. Metallic black; grey in reflected light. Kamioka mine, Kifu, Japan. H= 4.5, D_m= 5.96, D_c= 6.02, IVC 13. Sasaki et al., 1985. Min. Jour. 12 (1985), 393. Acta Cryst. C42 (1986), 9.

Kamitugaite, $PbAl(UO_2)_5[(P, As)O_4]_2(OH)_9 \cdot 9.5H_2O$, [A] Tric. Translucent yellow. Kobokobo, Kivu, Zaïre. H=?, D_m= 4.03, D_c= 4.47, VIID 22. Deliens & Piret, 1984. Am. Min. 70(1985), 437 (Abst.).

Kamotoite-(Y), $Y_2O_4(UO_2)_4(CO_3)_3 \cdot 14H_2O$, [A] Mon. Vitreous yellow. Kamoto Copper deposit, Shaba (Southern), Zaïre. H=?, D_m= 3.93, D_c= 3.94, VbD 04. Piret & Deliens. Bull. Min. 109 (1986), 643.

Kanaekanite, [D] Unnecessary name for variety of ekanite. Min. Mag. 46 (1982), 514. Povarennykh & Dusmatov, 1970.

Kanemite, $HNaSi_2O_4(OH)_2 \cdot 2H_2O$, [A] Orth. Translucent colorless, white. Lake Chad (NE edge of), Kanem Region, Andaija, Chad. H= 4, D_m= 1.926, D_c= 1.933, VIIIE 16. Johan & Maglione, 1972. Am. Min. 62 (1977), 763.

Kankite, $FeAsO_4 \cdot 3.5H_2O$, [A] Mon. Dull/vitreous yellowish-green. Kaňk, Kutná Hora (Near), Středočeský kraj, Čechy (Bohemia), Czechoslovakiaa. H= 2-3, D_m= 2.70, D_c= 2.732, VIIC 04. Čech et al., 1976. Am. Min. 70(1985), 220 (Abst.).

Kanoite, $(Mn, Mg)SiO_3$, [A] Mon. Vitreous light pinkish-brown. Tatehira mine, Oshima peninsula, Hokkaido, Japan. H= 6, D_m= 3.66, D_c= 3.60, VIIID 01a. Kobayashi, 1977. Am. Min. 63(1978), 598 (Abst.). See also Clinoenstatite, Clinoferrosilite.

Kanonaite, $(Mn, Al)(Al, Mn)SiO_5$, [A] Orth. Vitreous greenish-black. Kanona (N), Serenje, Zambia. H= 6.5, D_m=?, D_c= 3.39, VIIIA 14. Vrana et al., 1978. Am. Min. 64(1979), 655 (Abst.). Am. Min. 66 (1981), 561. See also Andalusite.

Kaolinite, $Al_2Si_2O_5(OH)_4$, [G] Tric. Pearly/earthy white, greyish, yellowish, brownish, etc. H= 2-2.5, D_m= 2.62, D_c= 2.614, VIIIE 10a. Johnson, 1867. Min. Mag. 32 (1961), 902. Clays Cl. Mins. 36 (1988), 225. See also Dickite, Halloysite, Nacrite.

Karelianite, V_2O_3, [G] Rhom. Black; brownish olive-grey in reflected light. Outokumpu, Finland. H= 8-9, D_m= 4.87, D_c= 5.02, IVC 04a. Long et al., 1963. Am. Min. 48 (1963), 33. Zts. Krist. 117 (1962), 235.

Karibibite, $Fe_2As_4(O, OH)_9$, [A] Orth. Translucent brownish-yellow. Karibib pegmatite, Namibia. H=?, D_m= 4.07, D_c= 4.04, VIIA 15. v. Knorring et al., 1973. Am. Min. 59(1974), 382 (Abst.).

Karlite, $Mg_7(BO_3)_3(OH)_4Cl_{0.4}$, [A] Orth. Silky white, light green. Schlegeistal, Furtschaglhaus, Zillertal Alps, Austria. H= 5.5, D_m= 2.83, D_c= 2.895, Vc 01c. Franz et al., 1981. Am. Min. 66 (1981), 872. N.Jb.Min.Mh. (1986), 253.

Karnasurtite-(Ce), $(Ce, La, Th)(Ti, Nb)AlSi_2O_7(OH)_4 \cdot 3H_2O$, [Q] Amor. Greasy yellow. Mt. Karnasurt, Lovozero, Kola Peninsula, USSR. H= 2, D_m= 2.9, D_c=?, VIIIB 16. Kuz'menko & Kozhanov, 1959. Am. Min. 45(1960), 1133 (Abst.).

Karpatite, $C_{24}H_{12}$, [G] Mon. Transcarpathia, USSR. H= 1, D_m= 1.40, D_c= 1.416, IXB 01. Piotrovskii, 1955. Am. Min. 42(1957), 120 (Abst.). Am. Min. 54(1969), 329 (Abst.).

Karpinskite, $(Mg,Ni)_2Si_2O_5(OH)_2$(?), Q Syst=? Dull colorless, light blue, greenish-blue. Ural Mts., USSR. H= 2.5-3, D_m= 2.6, D_c=?, VIIIE 08b. Rukavishnikova, 1956. Am. Min. 42(1957), 584 (Abst.).

Karpinskyite, D A mixture of leifite and Zn-bearing montmorillonite. Am. Min. 57 (1972), 1006 (Abst.). Shilin, 1956.

Kashinite, $(Ir,Rh)_2S_3$, A Orth. Metallic light grey. Nizhniĭ Tailsk massif, Ural Mts., USSR. H= 7.5, D_m=?, D_c= 9.10, IIC 15. Begizov et al., 1985. Am. Min. 72(1987), 223 (Abst.).

Kasolite, $Pb(UO_2)SiO_4 \cdot H_2O$, G Mon. Resinous/greasy yellow, brown. Kasolo, Shaba, Zaïre. H= 4-5, D_m= 5.96, D_c= 6.534, VIIIA 25. Schoep, 1921. Am. Min. 66 (1981), 610. Struct. Repts. 43A (1977), 316.

Kassite, $CaTi_2O_4(OH)_2$, G Orth. Adamantine yellowish, pale yellow. Afrikanda massif, Kola Peninsula, USSR. H= 5, D_m= 3.28, D_c= 3.28, IVF 20. Kukharenko et al., 1965. Am. Min. 71 (1986), 1045.

Katayamalite, $(K,Na)Li_3Ca_7Ti_2(SiO_3)_{12}(OH,F)_2$, A Tric. Vitreous white. Iwagi Islet, Ehima, Japan (SW). H= 3.5-4, D_m= 2.91, D_c= 2.899, VIIIC 05. Murakami et al., 1983. Am. Min. 69(1984), 811 (Abst.). Min. Jour. 12 (1985), 206.

Katoite, $Ca_3Al_2[SiO_4,(OH)_4]_3$, A Cub. Translucent colorless, white. Cava Campomorto, Montalto di Castro, Viterbo, Lazio, Italy. H=?, D_m=?, D_c= 2.76, VIIIA 06b. Passaglia & Rinaldi, 1984. Am. Min. 70(1985), 873 (Abst.). See also Grossular, Hibschite.

Katophorite, A group name for amphiboles with the general formula $Na_2Ca(Mg,Fe)_5(Si,Al)_8O_{22}(OH)_2$. See Alumino-katophorite, Ferri-katophorite, Magnesio-alumino-katophorite.

Katoptrite, $Mn_{13}Al_4Sb_2Si_2O_{28}$, G Mon. Black. Nordmark, Värmland, Sweden. H= 5.5, D_m= 4.56, D_c= 4.530, VIIIA 18. Flink, 1917. Am. Min. 51 (1966), 1494. Min. Abstr. 76-3602.

Kawazulite, Bi_2Te_2Se, A Rhom. Metallic white. Kawazu mine, Shizuoka, Honshu, Japan. H= 1.5, D_m=?, D_c= 8.08, IIC 03. Kato, 1970. Am. Min. 57(1972), 1312 (Abst.).

Kazakhstanite, $Fe_5V_{15}O_{39}(OH)_9 \cdot 8.5H_2O$, A Mon. Adamantine/dull black. Kara Tau dist., Kazakhstan (S), USSR. H= 2.5, D_m= 3.5, D_c= 3.52, VIID 38. Ankinovich et al., 1989. ZVMO 118 (1989), 95.

Kazakovite, $Na_6MnTiSi_6O_{18}$, A Rhom. Vitreous/greasy pale yellow. Mt. Karnasurt, Lovozero massif, Kola Peninsula, USSR. H= 4, D_m= 2.84, D_c= 2.97, VIIIC 07. Khomyakov et al., 1974. Am. Min. 60(1975), 161 (Abst.). Min. Abstr. 80-2849.

Keckite, $(Ca,Mg)(Mn,Zn)_2Fe_3(PO_4)_4(OH)_3 \cdot 2H_2O$, A Mon. Dull brown, yellow-brown, greyish-brown. Hagendorf pegmatite, Oberpfalz, Bayern (Bavaria), Germany. H= 4.5, D_m= 2.6, D_c= 2.682, VIID 24. Mücke, 1979. Am. Min. 64(1979), 1330 (Abst.).

Kegelite, $Pb_8Al_4Si_8O_{20}(SO_4)_2(CO_3)_4(OH)_8$, A Mon. Pearly white to transparent, porous. Tsumeb, Namibia. H=?, D_m= 4.5, D_c= 4.76, VIIIE 23. Medenbach & Schmetzer, 1975. Am. Min. 75 (1990) 702.

Kehoeite, $(Zn,Ca)Al_2(PO_4)_2(OH)_2 \cdot 5H_2O$, G Cub. Chalky white. Merritt mine, Galena, South Dakota, USA. H=?, D_m= 2.34, D_c= 2.32, VIID 13. Headden, 1893. Min. Mag. 33 (1964), 799. Can. Min. 12 (1974), 352.

Keithconnite, $Pd_{20}Te_7$, A Rhom. Metallic cream. Stillwater Complex, Montana, USA. H= 4.5-5, D_m=?, D_c= 11.17, IIA 14. Cabri et al., 1979. J.Less Comm.Met. 51(1977), 35.

Keiviite-(Y), $(Y,Yb)_2Si_2O_7$, A Mon. Vitreous colorless, white. Kola Peninsula, USSR. H= 4-5, D_m= 4.45, D_c= 4.48, VIIIB 01. Voloshin et al., 1985. Am. Min. 73(1988), 191 (Abst.). See also Keiviite-(Yb), Thortveitite.

Keiviite-(Yb), $(Yb,Y)_2Si_2O_7$, [A] Mon. Vitreous colorless. Keiva, Kola Peninsula, USSR. H=?, D_m= 5.95, D_c= 5.99, VIIIB 01. Voloshin et al., 1983. Am. Min. 69(1984), 1191 (Abst.). See also Keiviite-(Y), Thortveitite.

Keldyshite, $NaZrSi_2O_6(OH)$, [G] Tric. Vitreous/greasy white. Lovozero massif, Kola Peninsula, USSR. H= 4, D_m= 3.22, D_c= 3.26, VIIIB 22. Gerasimovskii, 1962. Am. Min. 47(1962), 1216 (Abst.). Min. Abstr. 79-2104. See also Parakeldyshite.

Kellyite, $(Mn,Mg,Al)_3(Si,Al)_2O_5(OH)_4$, [A] Hex. Translucent yellow. Bald Knob, Sparta (near), North Carolina , USA. H=?, D_m= 3.07, D_c= 3.11, VIIIE 10b. Peacor et al., 1974. Am. Min. 59 (1974), 1153.

Kelyanite, $Hg_{36}Sb_3O_{28}(Cl,Br)_9$, [A] Mon. Translucent reddish-brown. Kelyan deposit, Buryat, Ural Mts., USSR. H=?, D_m= 8.57, D_c= 8.51, IIIC 03. Vasil'ev et al., 1982. Am. Min. 68(1983), 1248 (Abst.).

Kemmlitzite, $(Sr,Ce)Al_3(SO_4)(AsO_4)(OH)_6$, [A] Rhom. Translucent light greyish-brown, colorless. Kemmlitz deposit, Sachsen (Saxony), Germany. H= 5.5, D_m= 3.63, D_c= 3.601, VIB 03b. Hak et al., 1969. Am. Min. 55(1970), 320 (Abst.).

Kempite, $Mn_2Cl(OH)_3$, [G] Orth. Translucent green. Alum Rock Park, San Jose (near), Santa Clara Co., California, USA. H= 3.5, D_m= 2.94, D_c= 2.96, IIIC 01. Rogers, 1924. Dana, 7th ed.(1951), v.2, 73.

Kennedyite, [D] Unnecessary name for armalcolite. Am. Min. 73 (1988), 1377. von Knorring, 1961.

Kentrolite, $Pb_2Mn_2(Si_2O_7)O_2$, [G] Orth. Vitreous/sub-metallic/dull dark reddish-brown. Chile (S) and/or Långban mine, Filipstad (near), Värmland, Sweden. H= 5, D_m= 6.19, D_c= 6.21, VIIIB 14. Damour & vom Rath, 1880. Am. Min. 52 (1967), 1085. Struct. Repts. 27 (1962), 714. See also Melanotekite.

Kenyaite, $Na_2Si_{22}O_{41}(OH)_8 \cdot 6H_2O$, [A] Mon. Translucent white. Lake Magadi, Rift Valley, Kenya. H=?, D_m=?, D_c=?, VIIIE 28. Eugster, 1967. Am. Min. 53 (1968), 2061.

Kermesite, Sb_2OS_2, [G] Tric. Adamantine red. H= 1-1.5, D_m= 4.68, D_c= 4.88, IIE 02. Chapman, 1843. Dana, 7th ed.(1944), v.1, 279. N.Jb.Min.Mh. (1987), 419.

Kernite, $Na_2B_4O_6(OH)_2 \cdot 3H_2O$, [G] Mon. Vitreous colorless, white. Kramer deposit, Boron, Kern Co., California, USA. H= 2.5, D_m= 1.908, D_c= 1.904, Vc 07a. Schaller, 1927. Dana, 7th ed.(1951), v.2, 335. Am. Min. 58 (1973), 21.

Kerolite, $Mg_3Si_4O_{10}(OH)_2 \cdot H_2O$, [G] Syst=? Translucent green. Frankenstein, Silesia, Poland. H=?, D_m=?, D_c=?, VIIIE 04. Breithaupt, 1823. Am. Min. 64 (1979), 615.

Kerstenite, $PbSeO_4$, [Q] Orth. Greasy/vitreous yellow. Friedrichsglück mine, Hildburghausen (near), Thüringwald, Germany. H= 3-4, D_m=?, D_c=?, VIA 07. Dana, 1868. Dana, 7th ed.(1951), v.2, 640.

Kësterite, $Cu_2(Zn,Fe)SnS_4$, [G] Tet. Metallic olive brown. Këster deposit, Yakutiya, USSR. H=?, D_m= 4.54, D_c= 4.57, IIB 03a. Orlova, 1956. Can. Min. 17 (1979), 125. Can. Min. 16 (1978), 131. See also Stannite.

Kettnerite, $CaBiO(CO_3)F$, [G] Tet. Translucent brown, yellow. Krupka, Srředočeský kraj, Čechy (Bohemia), Czechoslovakia. H=?, D_m= 5.80, D_c= 5.85, VbB 05. Zak & Syneček, 1957. Am. Min. 43(1958), 385 (Abst.).

Keyite, $(Cu,Zn,Cd)_3(AsO_4)_2$, [A] Mon. Translucent blue. Tsumeb, Namibia. H= 3.5-4, D_m=?, D_c= 4.95, VIIA 08. Embrey et al., 1977. Min. Record 8 (1977), 87.

Keystoneite, $H_{0.8}Mg_{0.8}(Ni,Fe,Mn)_2(TeO_3)_3 \cdot 5H_2O$, [A] Hex. Adamantine yellow. Keystone mine, Magnolia Dist., Boulder Co., Colorado, USA. H=?, D_m=?, D_c= 4.40, VIG 04. Back et al., 1988. GAC-MAC 13 (1988), A4 (Abst.).

Khademite, $Al(SO_4)F \cdot 5H_2O$, [A] Orth. Translucent colorless. Saghand, Iran. H=?, D_m= 1.925, D_c= 1.942, VID 02. Bariand et al., 1973. Min. Mag. 52 (1988), 133. Bull. Min. 104 (1981), 19. See also Rostite.

Khamrabaevite, (Ti, V, Fe)C, [A] Cub. Metallic dark grey. Unspecified locality, Arashan Mts., Chatkal Ranges, USSR. H= > 9, D_m=?, D_c= 10.01, IC 01. Novgorodova et al., 1984. ZVMO 113 (1984), 697.

Khanneshite, $(Na,Ca)_3(Ba,Sr,Ce,Ca)_3(CO_3)_5$, [A] Hex. Translucent pale yellowish. Khanneshin, Afghanistan. H=?, D_m= 3.9, D_c= 3.94, VbA 05. Eremenko & Bel'ko, 1982. Am. Min. 68(1983), 1249 (Abst.). See also Burbankite.

Kharaelakhite, $(Cu,Pt,Pb,Fe,Ni)_9S_8$, [A] Orth. Metallic greyish with brownish-lilac tint. Talnakh, Noril'sk (near), Siberia (N), USSR. H=?, D_m=?, D_c= 15.25, IIA 07. Genkin et al., 1985. Min. Zhurn. 7(1) (1985), 78.

Khatyrkite, $(Cu,Zn)Al_2$, [A] Tet. Metallic steel grey-yellow. Listvenitovȳí, Koriakskhiye Mts., USSR (E). H= 5, D_m=?, D_c= 4.42, IA 02. Razin et al., 1985. ZVMO 114 (1985), 90.

Khibinskite, $K_2ZrSi_2O_7$, [A] Mon. Dull/greasy pale yellowish. Gakman Valley, Khibina massif, Kola Peninsula, USSR. H=?, D_m= 3.3, D_c= 3.368, VIIIB 22. Khomyakov et al., 1974. Am. Min. 59(1974), 1140 (Abst.). Min. Abstr. 78-2697.

Khinite, $Cu_3PbTeO_4(OH)_6$, [A] Orth. Translucent dark green. Old Guard mine, Tombstone, Cochise Co., Arizona, USA. H= 3.5, D_m=?, D_c= 6.69, VIG 09. Williams, 1978. Am. Min. 63 (1978), 1016. See also Parakhinite.

Kiddcreekite, Cu_6WSnS_8, [A] Cub. Metallic grey. Kidd Creek mine, Timmins, Ontario, Canada. H= 3.5, D_m=?, D_c= 4.88, IIB 03b. Harris et al., 1984. Can. Min. 22 (1984), 227. See also Hemusite.

Kidwellite, $NaFe_9(PO_4)_6(OH)_{10}\cdot5H_2O$, [A] Mon. Translucent yellow, greenish-yellow. Coon Creek mine, Shady, Polk Co., Arkansas, USA. H= 3, D_m= 3.3, D_c= 3.34, VIID 15c. Moore & Ito, 1978. Min. Mag. 42 (1978), 137.

Kieserite, $MgSO_4\cdot H_2O$, [G] Mon. Vitreous colorless, greyish-white, yellowish. Stassfurt deposit, Sachsen (Saxony), Germany. H= 3.5, D_m= 2.571, D_c= 2.584, VIC 01. Reichardt, 1861. Dana, 7th ed.(1951), v.2, 477. N.Jb.Min.Abh. 157 (1987), 121.

Kilchoanite, $Ca_3Si_2O_7$, [G] Orth. Translucent colorless. Kilchoan, Ardnamurchan, Scotland. H=?, D_m=?, D_c= 3.002, VIIIB 04. Agrell & Gay, 1961. Am. Min. 46(1961), 1203 (Abst.). Min. Mag. 38 (1971), 26. See also Rankinite.

Killalaite, $Ca_3Si_2O_7\cdot H_2O$, [A] Mon. Translucent colorless. Inishchrone, Killala Bay, Sligo Co., Ireland. H=?, D_m=?, D_c= 2.94, VIIIB 06. Nawaz, 1974. Min. Mag. 39 (1974), 544. Min. Mag. 41 (1977), 363.

Kimrobinsonite, $Ta(OH)_3(O,CO_3)$, [A] Cub. Chalky white. Mt. Holland, Western Australia, Australia. H= 2.5, D_m=?, D_c= 6.87, IVF 17. Nickel & Robinson, 1985. Can. Min. 23 (1985), 573.

Kimuraite-(Y), $CaY_2(CO_3)_4\cdot6H_2O$, [A] Orth. Vitreous/silky light purplish/pinkish white. Saga, Matsura, Higashi, Hizen-cho, Kirigo, Kyushu, Japan. H= 2.5, D_m= 2.6, D_c= 2.98, VbC 03b. Nagashima et al., 1986. Am. Min. 71 (1986), 1028.

Kimzeyite, $Ca_3(Zr,Ti)_2(Si,Al,Fe^{3+})_3O_{12}$, [G] Cub. Translucent brown. Magnet Cove, Hot Spring Co., Arkansas, USA. H= 7, D_m= 4.0, D_c= 3.85, VIIIA 06a. Milton et al., 1961. Am. Min. 46 (1961), 533. Am. Min. 65 (1980), 188.

Kingite, $Al_3(PO_4)_2(OH,F)_3\cdot9H_2O$, [G] Tric. Translucent white. Robertstown, South Australia, Australia. H=?, D_m= 2.2, D_c= 2.465, VIID 06. Norrish et al., 1957. Am. Min. 55 (1970), 515.

Kingsmountite, $(Ca,Mn)_4FeAl_4(PO_4)_6(OH)_4\cdot12H_2O$, [A] Mon. Pearly white, light brown. Foote Mineral Company spodumene mine, Kings Mt. (near), Cleveland Co., North Carolina, USA. H= 2.5, D_m= 2.51, D_c= 2.58, VIID 35. Dunn et al., 1979. Can. Min. 17 (1979), 579. See also Montgomeryite.

Kinichilite, $(H,Na)_2(Fe,Mg,Zn)_2(TeO_3)_3\cdot3H_2O$, [A] Hex. Sub-adamantine dark brown. Kawazu mine, Shimoda City, Izu Peninsula, Japan. H=?, D_m=?, D_c= 3.96, VIG 04. Hori et al., 1981. Min. Abstr. 84M/1932. See also Zemannite.

Kinoite, $Ca_2Cu_2Si_3O_{10} \cdot 2H_2O$, [A] Mon. Transparent deep blue. Santa Rita Mts., Pima Co., Arizona, USA. H= 2-5, D_m= 3.16, D_c= 3.193, VIIIA 12. Anthony & Laughon, 1970. Am. Min. 55 (1970), 709. Am. Min. 56 (1971), 193.

Kinoshitalite, $(Ba,K)(Mg,Mn,Al)_3(Si,Al)_4O_{10}(OH,F)_2$, [A] Mon. Vitreous yellow-brown. Misago ore body, Noda-Tamagawa mine, Iwate, Japan. H= 2.5-3, D_m= 3.30, D_c= 3.33, VIIIE 06. Yoshii et al., 1973. Am. Min. 60(1975), 486 (Abst.). Min. Jour. 9 (1979), 392. See also Anandite.

Kipushite, $(Cu,Zn)_6(PO_4)_2(OH)_6 \cdot H_2O$, [A] Mon. Vitreous green. Kipushi, Shaba, Zaïre. H= 4, D_m= 3.8, D_c= 3.904, VIID 09. Piret et al., 1985. Can. Min. 23 (1985), 35. Can. Min. 23 (1985), 35. See also Philipsburgite.

Kirchheimerite, $Co(UO_2)_2(AsO_4)_2 \cdot 12H_2O$, [Q] Tet. Translucent pale pink. Schwarzwald (Black Forest), Germany. H=?, D_m=?, D_c= 3.243, VIID 20a. Walenta, 1964. Strunz (1970), 352.

Kirkiite, $Pb_{10}Bi_3As_3S_{19}$, [A] Hex. Metallic white. Aghios Philippos, Kirki, Thrace, Greece. H= 3.3, D_m=?, D_c= 6.82, IID 01f. Moëlo et al., 1985. Am. Min. 71(1986), 1278 (Abst.).

Kirschsteinite, $CaFeSiO_4$, [G] Orth. Translucent greenish. Kivu, Zaïre. H=?, D_m= 3.434, D_c= 3.56, VIIIA 04. Sahama & Hytönen, 1957. NBS Monogr. 20 (1983), 28. See also Monticellite, Glaucochroite.

Kitaibelite, $Ag_{10}PbBi_{30}S_{51}$, [Q] Syst=? Nagyborzsony, Hungary (N). H=?, D_m=?, D_c=?, IID 09c. Nagy, 1983. Am. Min. 72(1987), 1027 (Abst.).

Kitkaite, NiTeSe, [G] Rhom. Metallic grey/pink. Kitka river, Kuusamo, Oulu, Finland. H= 3, D_m= 7.22, D_c= 7.19, IIC 11. Häkli et al., 1965. Am. Min. 50 (1965), 581.

Kittatinnyite, $Ca_2Mn_3Si_2O_8(OH)_4 \cdot 9H_2O$, [A] Hex. Vitreous yellow. Franklin, Sussex Co., New Jersey, USA. H= 4, D_m= 2.61, D_c= 2.62, VIIIB 19. Dunn & Peacor, 1983. Am. Min. 68 (1983), 1029. See also Walkilldellite.

Kittlite, [Hg, Ag, Cu, S, Se](?), [Q] Cub. Metallic grey. Jaguel, Llantenes region, La Rioja Province, Argentina. H= 5-5.5, D_m= 5.4, D_c=?, IIB 01a. Rivas, 1970. Am. Min. 57(1972), 1313 (Abst.).

Kivuite, [D] Analysis unsatisfactory. Min. Mag. 33 (1962), 261. van Wambeke, 1959.

Kladnoite, $C_6H_4(CO)_2NH$, [G] Mon. Kladno, Praha (Prague), Stredočský kraj, Čechy (Bohemia), Czechoslovakia. H=?, D_m= 1.47, D_c= 1.469, IXB 01. Rost, 1942. Am. Min. 31 (1946), 605. Acta Cryst. B28 (1972), 415.

Klebelsbergite, $Sb_4O_4SO_4(OH)_2$, [G] Orth. Vitreous pale yellow, orange-yellow. Baia Sprie (Felsöbánya), Romania. H= 3.7, D_m= 4.62, D_c= 4.67, VID 11. Zsivny, 1929. Am. Min. 65 (1980), 499. Am. Min. 65 (1980), 931.

Kleberite, [D] Inadequate description. Am. Min. 64(1979), 655 (Abst.). Bautsch et al., 1978.

Kleemanite, $ZnAl_2(PO_4)_2(OH)_2 \cdot 3H_2O$, [A] Mon. Ochreous yellow, colorless. Iron Monarch Quarry, Iron Knob, South Australia, Australia. H=?, D_m=?, D_c= 2.76, VIID 27. Pilkington et al., 1979. Min. Mag. 43 (1979), 93.

Kleinite, $Hg_2N(Cl,SO_4) \cdot nH_2O$, [G] Hex. Adamantine/greasy yellow/orange. Terlingua, Brewster Co., Texas, USA. H= 3.5-4, D_m= 8.00, D_c= 7.87, VIB 10. Sachs, 1905. Dana, 7th ed.(1951), v.2, 87.

Kliachite, $Al_2O_3 \cdot nH_2O$, [Q] Syst=? Vitreous colorless, pinkish-orange, greyish-pink, creamy. H= < 3, D_m=?, D_c=?, IVF 04. Cornu, 1909. Am. Min. 75 (1990), 431 (Abst.).

Klockmannite, CuSe, [G] Hex. Metallic grey. Sierra de Umango, Argentina; Harz, Germany. H= 3, D_m= 5.99, D_c= 6.12, IIB 15. Ramdohr, 1928. Am. Min. 34 (1939), 435. Min. Abstr. 82M-0177. See also Covellite.

Klyuchevskite, $K_3Cu_3FeO_2(SO_4)_4$, [A] Mon. Vitreous green. Tolbachik volcano, Kamchatka, USSR. H= 3.5, D_m= 3.1, D_c= 3.02, VIB 07. Vergasova et al., 1989. ZVMO 118 (1989), 70.

Kmaite, [D] Unnecessary name for member of celadonite group. Min. Mag. 36 (1967), 133. Illarionov, 1961.

Knorringite, $Mg_3Cr_2(SiO_4)_3$, [A] Cub. Translucent bluish-green. Kao kimberlite pipe, Lesotho. H=?, D_m= 3.756, D_c= 3.852, VIIIA 06a. Nixon & Hornung, 1968. Am. Min. 53 (1968), 1833. See also Pyrope.

Koashvite, $Na_6(Ca, Mn)(Fe, Ti)Si_6O_{18}$, [A] Orth. Vitreous pale yellow. Mt. Koashva, Khibina massif, Kola Peninsula, USSR. H= 6, D_m= 3.00, D_c= 2.99, VIIIC 07. Kapustin et al., 1974. Dokl.Earth Sci. 237(1977), 208. Min. Abstr. 81-2386.

Kobeite-(Y), $(Y, U)(Ti, Nb)_2(O, OH)_6$(?), [Q] Orth. Resinous brown. Kobe-mura, Nakagun, Kyoto, Japan. H=?, D_m= 5.0, D_c=?, IVD 10b. Takubo et al., 1950. JCPDS 11-259. See also Polycrase-(Y).

Kobellite, $(Cu, Fe)_2Pb_{12}(Bi, Sb)_{14}S_{35}$, [G] Orth. Metallic grey; white in reflected light. Vena mine, Askersund (near), Sweden. H= 2.5-3, D_m= 6.33, D_c= 6.53, IID 08. Setterberg, 1839. Can. Min. 9 (1968), 371. Nature,Phys.Sci. 231(1971),133. See also Tintinaite, Izoklakeite.

Koechlinite, Bi_2MoO_6, [G] Orth. Transparent greenish-yellow. Daniel mine, Schneeberg, Sachsen (Saxony), Germany. H=?, D_m=?, D_c= 8.28, IVD 14b. Schaller, 1914. Dana, 7th ed.(1951), v.2, 1092. Acta Cryst. C40 (1984), 2001.

Koenenite, $Na_4Mg_9Al_4Cl_{12}(OH)_{22}$, [G] Hex. Pearly colorless, pale yellow. Volpriehausen, Hannover, Germany. H= 1.5, D_m= 1.98, D_c= 2.162, IVF 03e. Rinne, 1902. Dana, 7th ed.(1951), v.2, 86. Zts. Krist. 126 (1968), 7.

Kogarkoite, $Na_3(SO_4)F$, [A] Mon. Translucent colorless, pale blue. Hortense Hot Spring and Wright's Well, Princeton, Chaffee Co., Colorado, USA. H= 3.5, D_m= 2.667, D_c= 2.676, VIB 04. Pabst & Sharp, 1973. Am. Min. 58 (1973), 116. Min. Mag. 43 (1980), 753.

Koktaite, $(NH_4)_2Ca(SO_4)_2 \cdot H_2O$, [G] Mon. Translucent colorless, white. Zeravice, Kyjov (near), Moravia, Czechoslovakia. H=?, D_m= 2.09, D_c= 2.140, VIC 15. Sekanina, 1948. Per. Mineral. 54 (1985), 29. See also Syngenite.

Kolarite, $PbTeCl_2$, [A] Orth. Grey in reflected light. Champion reef lode, Kolar deposits, India. H=?, D_m=?, D_c= 9.14, IIIB 09. Genkin et al., 1985. Can. Min. 23 (1985), 501.

Kolbeckite, $ScPO_4 \cdot 2H_2O$, [A] Mon. Vitreous/pearly blue, blue-grey. Schmiedeberg, Sachsen (Saxony), Germany. H= 3.5-4, D_m= 2.35, D_c= 2.344, VIIC 05. Edelmann, 1926. Min. Mag. 46 (1982), 493. See also Metavariscite, Phosphosiderite.

Kolfanite, $Ca_2Fe_3O_2(AsO_4)_3 \cdot 2H_2O$, [A] Mon. Adamantine red, orange. Unspecified locality, Kola Peninsula, USSR. H= 2.5, D_m= 3.3, D_c= 3.75, VIID 15a. Voloshin et al., 1983. Am. Min. 68(1983), 280 (Abst.).

Kolicite, $Zn_4Mn_7(AsO_4)_2(SiO_4)_2(OH)_8$, [A] Orth. Vitreous yellow-orange. Sterling Hill mine, Ogdensburg, Sussex Co., New Jersey, USA. H= 4.5, D_m= 4.17, D_c= 4.20, VIIB 10d. Dunn et al., 1979. Am. Min. 64 (1979), 708. Am. Min. 65 (1980), 483.

Kolovratite, [Zn, Ni, VO4](?), [Q] Syst=? Vitreous greenish-yellow. Fergana, Turkestan, USSR. H= 2-3, D_m=?, D_c=?, VIIB 14. Vernadsky, 1922. Can. Min. 7 (1962), 311.

Kolskite, [D] A mixture of lizardite and sepiolite. Am. Min. 59 (1974), 212 (Abst.). Efremov, 1939.

Kolwezite, $(Cu, Co)_2CO_3(OH)_2$, [A] Tric. Black, beige. Kolwezi, Shaba (S), Zaïre. H= 4, D_m= 3.97, D_c= 3.94, VbB 01. Deliens & Piret, 1980. Am. Min. 65(1980), 1067 (Abst.).

Kolymite, Cu_7Hg_6, [A] Cub. Metallic white. Krokhalin mine, Magadan region, USSR. H= 4, D_m=?, D_c= 13.10, IA 06. Markova et al., 1980. ZVMO 109 (1980), 206.

Komarovite, $(Ca,Mn)Nb_2(Si_2O_7)(O,F)_3 \cdot 3.5H_2O$, [A] Orth. Dull pale rose. Mt. Karnasurt, Lovozero, Kola Peninsula, USSR. H= 1.5-2, D_m= 3.0, D_c= 2.96, VIIIB 08. Portnov et al., 1971. Am. Min. 57(1972), 1315 (Abst.). See also Na-komarovite.

Kombatite, $Pb_{14}O_9(VO_4)_2Cl_4$, [A] Mon. Adamantine yellow. Kombat mine, Otavi (E), Namibia. H= 2-3, D_m=?, D_c= 7.979, IIID 03. Rouse et al., 1986. Am. Min. 73(1988), 928 (Abst.).

Kondërite, $Cu_3Pb(Rh,Pt,Ir)_8S_{16}$, [A] Hex. Metallic grey. Konder massif, Aldan shield, Siberia, USSR (E). H= 4.5-6, D_m=?, D_c=?, IIC 14. Rudashevskii et al., 1984. Am. Min. 71(1986), 229 (Abst.). See also Inaglyite.

Koninckite, $FePO_4 \cdot 3H_2O$, [G] Tet. Vitreous yellowish, whitish. Richelle, Visé (near), Liège, Belgium. H= 3.5, D_m= 2.40, D_c= 2.625, VIIC 08. Cesaro,1884. Bull. Min. 91 (1968), 487.

Konyaite, $Na_2Mg(SO_4)_2 \cdot 5H_2O$, [A] Mon. Translucent white. Great Konya Basin, Turkey. H= 2.5, D_m= 2.088, D_c= 2.097, VIC 11. Van Doesburg et al., 1982. Am. Min. 67 (1982), 1035.

Koritnigite, $Zn(AsO_3OH) \cdot H_2O$, [A] Tric. Transparent colorless. Tsumeb, Namibia. H= 2, D_m= 3.54, D_c= 3.56, VIID 32. Keller et al., 1980. Am. Min. 65(1980), 206 (Abst.). Min. Abstr. 81-0253. See also Cobaltkoritnigite.

Kornelite, $Fe_2(SO_4)_3 \cdot 7H_2O$, [G] Mon. Silky, rose-pink, violet. Smolnik (Szomolnok), Slovensko (Slovakia), Czechoslovakia. H=?, D_m= 2.306, D_c= 2.254, VIC 04. Krenner, 1888. Dana, 7th ed.(1951), v.2, 530. Am. Min. 58 (1973), 535.

Kornerupine, $Mg_4Al_6(Si,Al,B)_5O_{21}(OH)$, [G] Orth. Vitreous white, colorless, yellow. Fiskenäs, Nuuk area, Greenland (SW). H= 6.5, D_m= 3.273, D_c= 3.288, VIIIA 24. Lorenzen, 1884. Contr.Min.Petr. 67 (1978), 247. Am. Min. 74 (1989), 642.

Korshunovskite, $Mg_2Cl(OH)_3 \cdot 3.5\text{–}4H_2O$, [A] Tric. Transparent colorless. Korshunov deposit, Irkutsk, USSR. H= 2, D_m= 1.798, D_c= 1.787, IIIC 01. Malinko et al., 1982. Am. Min. 68(1983), 643 (Abst.).

Korzhinskite, $CaB_2O_4 \cdot H_2O$, [Q] Syst=? Transparent colorless. Ural Mts., USSR. H=?, D_m=?, D_c=?, Vc 01g. Malinko, 1963. Am. Min. 49(1964), 441 (Abst.).

Kosmochlor, $NaCrSi_2O_6$, [A] Mon. Vitreous dark green. Toluca meteorite. H= 6, D_m= 3.60, D_c= 3.603, VIIID 01a. Laspeyres, 1897. DHZ, 2nd ed.(1978), v.2A, 520. MSA Spec. Paper 2 (1969), 31.

Kostovite, $CuAuTe_4$, [A] Orth. Metallic creamy-white. Chelopech deposit, Bulgaria. H= 2-2.5, D_m= 8.43, D_c= 7.94, IIC 04. Terziev. 1966. JCPDS 35-521. See also Sylvanite.

Kostylevite, $K_2ZrSi_3O_9 \cdot H_2O$, [A] Mon. Vitreous colorless. Khibina massif, Vuonnemi river, Kola Peninsula, USSR. H= 5, D_m= 2.74, D_c= 2.75, VIIID 09. Khomyakov et al., 1983. Am. Min. 69(1984), 812 (Abst.). Min. Abstr. 83M/4213. See also Umbite.

Kotoite, $Mg_3(BO_3)_2$, [G] Orth. Vitreous colorless. Hol Kol Gold mine, Wall Rock of "New Ore Body", Suan, Korea. H= 6.5, D_m= 3.04, D_c= 3.10, Vc 01b. Watanabe, 1939. JCPDS 5-648. Min. Abstr. 84M/2549. See also Jimboite.

Köttigite, $Zn_3(AsO_4)_2 \cdot 8H_2O$, [G] Mon. Translucent red. Daniel mine, Schneeberg, Sachsen (Saxony), Germany. H= 2-3, D_m= 3.33, D_c= 3.24, VIIC 10. Dana, 1850. Can. Min. 14 (1976), 437. Am. Min. 64 (1979), 376. See also Parasymplesite.

Kotulskite, $Pd(Te,Bi)$, [G] Hex. Metallic cream. Monchegorsk, USSR. H=?, D_m=?, D_c= 9.18, IIB 09a. Genkin et al., 1963. Am. Min. 48(1963), 1181 (Abst.).

Koutekite, Cu_5As_2, [G] Hex. Metallic bluish-white in reflected light. Černý důl (Schwarzenthal), Krkonoše (Giant Mtns.), Vychodočeský kraj, Čechy (Bohemia),

Czechoslovakia. H= 3.7, D_m=?, D_c= 8.376, IIG 01. Johan, 1958. Bull. Min. 90 (1967), 82.

Kovdorskite, $Mg_2PO_4(OH)\cdot3H_2O$, A Mon. Translucent pale rose, pale blue. Kovdor massif, Kola Peninsula, USSR. H= 4, D_m= 2.60, D_c= 2.271, VIID 02. Kapustin et al., 1980. Am. Min. 66(1981), 437 (Abst.). Min. Abstr. 82M/1161.

Kôzulite, $Na_3(Mn,Mg,Fe,Al)_5Si_8O_{22}(OH)_2$, A Mon. Vitreous reddish-black, black. Tanahota mine, Iwate, Japan. H= 5, D_m= 3.30, D_c= 3.36, VIIID 05d. Nambu et al., 1969. Am. Min. 55(1970), 1815 (Abst.). Acta Cryst. A28 (1972), S71.

Kraisslite, $Zn_3FeMn_{24}As_5O_{18}(SiO_4)_6(OH)_{18}$, A Hex. Sub-metallic pale red-brown, deep coppery-brown. Sterling Hill mine, Ogdensburg, Sussex Co., New Jersey, USA. H= 3-4, D_m= 3.876, D_c= 3.903, VIIB 10c. Moore & Ito, 1978. Am. Min. 65 (1980), 957.

Kratochvílite, $C_{13}H_{10}$, G Orth. Kladno, Praha (Prague), Stredočský kraj, Čechy (Bohemia), Czechoslovakia. H=?, D_m= 1.206, D_c= 1.197, IXB 01. Rost, 1937. Strunz (1970), 496. Struct. Repts. 19 (1955), 583.

Krausite, $KFe(SO_4)_2\cdot H_2O$, G Mon. Vitreous pale lemon-yellow. Sulfur Hole, Calico Hills, Borate, San Bernardino Co., California, USA. H= 2.5, D_m= 2.840, D_c= 2.827, VIC 07. Foshag, 1931. Dana, 7th ed.(1951), v.2, 462. Am. Min. 71 (1986), 202.

Krauskopfite, $BaSi_2O_5\cdot3H_2O$, A Mon. Sub-vitreous/pearly white, colorless. Big Creek and Rush Creek, Fresno Co., California, USA. H= 4, D_m= 3.14, D_c= 3.10, VIIID 15. Alfors et al., 1965. Am. Min. 50 (1965), 314. Struct. Repts. 32A (1967), 459.

Krautite, $Mn(AsO_3OH)\cdot H_2O$, A Mon. Translucent pale rose. Cavnic, Crisana-Maramures, Romania. H= $<$ 4, D_m= 3.30, D_c= 3.274, VIID 32. Fontan et al., 1975. Am. Min. 61(1976), 503 (Abst.). Am. Min. 64 (1979), 1248.

Kremersite, $(NH_4,K)_2FeCl_5\cdot H_2O$, G Orth. Vitreous red, brownish-red. Mt. Vesuvius, Napoli, Campania, Italy. H=?, D_m= 2.00, D_c= 2.20, IIIB 07. Kenngott, 1853. Dana, 7th ed.(1951), v.2, 101. See also Erythrosiderite.

Krennerite, $AuTe_2$, G Orth. Metallic white/yellow; creamy-white in reflected light. Săcărâmb (Nagyág), Transylvania, Romania. H= 2-3, D_m= 8.62, D_c= 8.86, IIC 04. vom Rath, 1877. Dana, 7th ed.(1944), v.1, 333. Min. Abstr. 85M/2407.

Kribergite, $Al_5(PO_4)_3(SO_4)(OH)_4\cdot4H_2O$, G Tric. Chalky white. Kristineberg mine, Västerbotten, Sweden. H=?, D_m= 1.92, D_c= 1.95, VIID 04. du Rietz, 1945. Min. Mag. 53 (1989), 385.

Krinovite, $NaMg_2CrSi_3O_{10}$, A Tric. Translucent green. Canyon Diablo (USA), Wichita Co. (USA) and Youndegin (Australia) meteorites. H= 5.5-7, D_m= 3.38, D_c= 3.332, VIIID 07. Olsen & Fuchs, 1968. Am. Min. 54(1969), 578 (Abst.). Zeits. Krist. 187 (1989), 133.

Kröhnkite, $Na_2Cu(SO_4)_2\cdot2H_2O$, G Mon. Vitreous blue, greenish-blue. Chile. H= 2.5-3, D_m= 2.90, D_c= 2.913, VIC 11. Domeyko, 1876. Dana, 7th ed.(1951), v.2, 444. Acta Cryst. B31 (1975), 1753.

Krupkaite, $CuPbBi_3S_6$, A Orth. Metallic grey; greyish-white in reflected light. Krupka, Sředočeský kraj, Čechy (Bohemia), Czechoslovakia. H= 3.5, D_m=?, D_c= 6.95, IID 05a. Zak et al., 1974. Am. Min. 60(1975), 737 (Abst.). Am. Min. 60 (1975), 300.

Krut̆aite, $CuSe_2$, A Cub. Metallic grey. Petrovice Žd'ar, Jihomoravský jraj, Morava (Moravia), Czechoslovakia. H= 4, D_m=?, D_c= 6.53, IIC 05. Johan et al., 1972. Am. Min. 59(1974), 210 (Abst.).

Krutovite, $NiAs_2$, A Cub. Metallic greyish-white. Potučhky, Čechy (Bohemia), Czechoslovakia. H= 5.5, D_m=?, D_c= 6.93, IIC 06b. Vinogradova et al., 1976. Am. Min. 62(1977), 173 (Abst.). See also Rammelsbergite, Pararammelsbergite.

Kryzhanovskite, $(Fe,Mn)_3(PO_4)_2(OH,H_2O)_3$, G Orth. Vitreous/dull brown, greenish-brown. Kalbinsk pegmatite, USSR. H= 3.5-4, D_m= 3.31, D_c= 3.464, VIIC 04. Ginzburg, 1950. Am. Min. 56 (1971), 1. Min. Mag. 43 (1980), 789. See also Garyansellite.

Ktenasite, $(Cu,Zn)_5(SO_4)_2(OH)_6 \cdot 6H_2O$, G Mon. Vitreous blue-green. Kamariza mine, Lávrion (Laurium), Attikí, Greece. H= 2-2.5, D_m= 2.98, D_c= 2.93, VID 14. Kokkoros, 1950. Am. Min. 36(1951), 381 (Abst.). Zts. Krist. 147 (1978), 129.

Kulanite, $Ba(Fe,Mn,Mg)_2Al_2(PO_4)_3(OH)_3$, A Tric. Translucent green, blue. Rapid Creek (Cross-cut Creek), Big Fish River–Blow River Area, Yukon Territory, Canada. H= 4, D_m= 3.91, D_c= 3.92, VIIB 19. Mandarino & Sturman, 1976. Can. Min. 14 (1976), 127. See also Penikisite.

Kuliokite-(Y), $(Y,Yb)_4Al(SiO_4)_2(OH)_2F_5$, A Tric. Transparent colorless. Kola Peninsula, USSR. H= 4-5, D_m= 4.3, D_c= 4.26, VIIIA 11. Voloshin et al., 1986. Am. Min. 73(1988), 192 (Abst.).

Kulkeite, $Na_{0.3}Mg_8Al(Si,Al)_8O_{20}(OH)_{10}$, A Mon. Pearly colorless. Ksar el Boukhari (W), Derrag, Algeria. H= 2, D_m=?, D_c= 2.70, VIIIE 08c. Abraham et al., 1980. Contr. Min. Pet. 80(1982), 103. See also Talc, Chlorite.

Kullerudite, $NiSe_2$, G Orth. Metallic grey. Kuusamo, Finland. H=?, D_m=?, D_c= 6.72, IIC 07. Vuorelainen et al., 1964. Am. Min. 50 (1965), 519.

Kupletskite, $(K,Na)_3(Mn,Fe)_7(Ti,Nb)_2Si_8(O,OH)_{31}$, G Syst=? Dark brown, black. Kola Peninsula, USSR. H= 3, D_m=?, D_c=?, VIIID 25. Semenov, 1956. Am. Min. 42(1957), 118 (Abst.). See also Astrophyllite, Cesium kupletskite.

Kuramite, Cu_3SnS_4, A Tet. Metallic grey. Unspecified locality, Kuramin mountains, Uzbekistan, USSR. H= 4.5, D_m=?, D_c= 4.56, IIB 03a. Kovalenker et al., 1979. Am. Min. 65(1980), 1067 (Abst.).

Kuranakhite, $PbMnTeO_6$, A Orth. Vitreous brownish-black. Kuranakh gold deposit, Yakutiya (S), USSR. H= 4-5, D_m=?, D_c= 6.7, VIG 10. Yablokova et al., 1975. Am. Min. 61(1976), 339 (Abst.).

Kurchatovite, $Ca(Mg,Mn,Fe)B_2O_5$, A Orth. Vitreous pale grey. Siberia, USSR. H= 4.5, D_m= 3.02, D_c= 3.10, Vc 02. Malinko et al., 1965. Am. Min. 51(1966), 1817 (Abst.). Min. Abstr. 78-260. See also Clinokurchatovite.

Kurnakovite, $MgB_3O_3(OH)_5 \cdot 5H_2O$, G Tric. Vitreous white. Inder, Kazakhstan, USSR. H= 3, D_m= 1.847, D_c= 1.855, Vc 05. Godlevsky, 1940. Dana, 7th ed.(1951), v.2, 360. Acta Cryst. B30 (1974), 2194. See also Inderite.

Kurumsakite, $(Zn,Ni,Cu)_8Al_8V_2Si_5O_{35} \cdot 27H_2O(?)$, Q Syst=? Vitreous/silky yellow. Kara Tau, Dzhambul (near), Kazakhstan, USSR. H=?, D_m= 4.03, D_c=?, VIIIE 08b. Ankinovich, 1954. Am. Min. 42(1957), 583 (Abst.).

Kutínaite, $Ag_6Cu_{14}As_7$, A Cub. Metallic grey; greyish-white/blue-grey in reflected light. Černý důl (Schwarzenthal), Krkonoše (Giant Mtns.), Vychodočeský kraj, Čechy (Bohemia), Czechoslovakia. H= 4.8, D_m= 8.37, D_c= 8.37, IIG 01. Hak et al., 1979. TMPM 34 (1985), 183.

Kutnohorite, $Ca(Mn,Mg,Fe)(CO_3)_2$, G Rhom. Translucent pink. Kaňk, Kutná Hora (Near), Středočeský kraj, Čechy (Bohemia), Czechoslovakiaa. H=?, D_m=?, D_c= 4.68, VbA 03a. Bukovsky, 1901. Powd. Diff. 3 (1988), 172. N.Jb.Min.Mh. (1988), 539. See also Dolomite, Ankerite.

Kuzminite, $Hg(Br,Cl)$, A Tet. Translucent colorless, bluish-grey, white. Kadyrel deposit, Tuva, Siberia (S), USSR. H= 1.7, D_m=?, D_c= 7.64, IIIA 04. Vasil'ev et al., 1986. Am. Min. 73(1988), 192 (Abst.).

Kuznetsovite, $Hg_6As_2O_9Cl_2$, A Cub. Vitreous/adamantine yellow, orange. Arzak, Tuva and Khaidarkan deposit, Kirgizia, USSR. H= 2.5-3, D_m= 8.7, D_c= 8.786, IIIC 03. Vasil'eva & Lavrent'ev, 1980. Am. Min. 66(1981), 1100 (Abst.).

Kvanefjeldite, $Na_4(Ca,Mn)Si_6O_{14}(OH)_2$, A Orth. Vitreous/pearly violet-pink. Kvanefjeld Plateau, Ilímaussaq Intrusion, Greenland (S). H= 5.5-6, D_m= 2.55,

D_c= 2.53, VIIIE 19. Petersen et al., 1984. Can. Min. 22 (1984), 465. N.Jb.Min.Mh. (1983), 505.

Kyanite, Al_2SiO_5, G Tric. Translucent blue, white, grey, green yellow, pink, etc. H= 5.5-7, D_m= 3.6, D_c= 3.666, VIIIA 14. Werner, 1789. DHZ, 2nd ed.(1982), v.1A, 780. Am. Min. 64 (1979), 573. See also Andalusite, Sillimanite.

Kyanophyllite, D A mixture of paragonite and muscovite. Am. Min. 58 (1973), 807 (Abst.). Rao, 1945.

Kyzylkumite, $V_2Ti_3O_9$, A Mon. Vitreous/resinous black. Kyzyl-Kum, Uzbekistan, USSR. H=?, D_m= 3.75, D_c= 3.77, IVD 05. Smyslova et al., 1981. Am. Min. 67(1982), 855 (Abst.). See also Schreyerite.

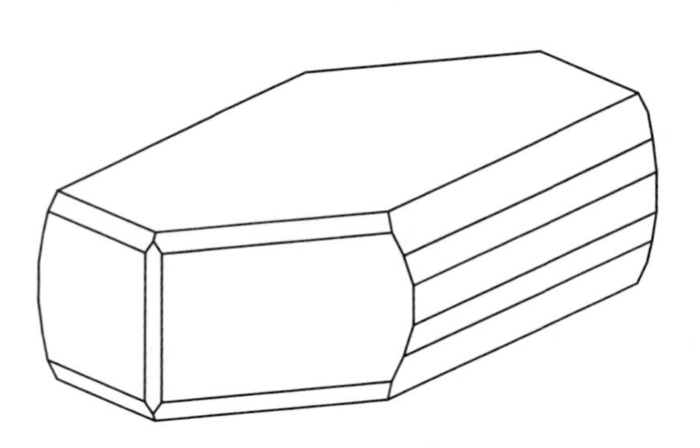

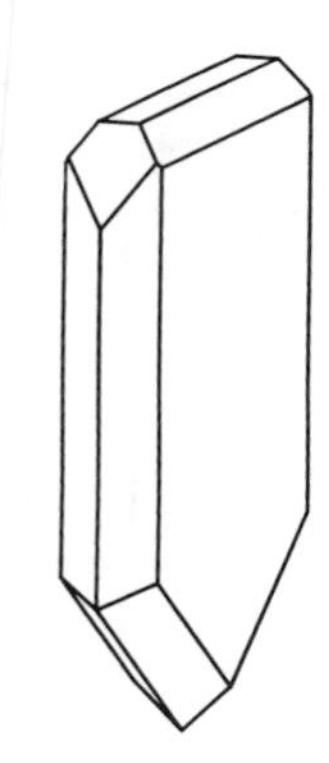

(Left) Cerrusite from Dognáeska, Hungary showing the forms {001}, {100}, {110}, {111}, {021}, {031} and {010} (after Palache, Berman & Frondel, 1951).

(Right) Hemimorphite showing the forms {001}, {301}, {031}, {010}, {110} and $\{12\bar{1}\}$ (after Rose, 1843).

L

Labradorite, (Ca, Na)(Si, Al)$_4$O$_8$, [G] Tric. Pearly/vitreous grey, brown, greenish, colorless, etc. Isle of Paul, Labrador (coast), Newfoundland, Canada. H= 5-6, D_m= 2.71, D_c= 2.70, VIIIF 03. Werner, 1780. DHZ (1963), v.4, 94. Am. Min. 65 (1980), 81. See also Plagioclase.

Labuntsovite, (K, Na)$_8$(Ti, Nb)$_9$(SiO$_3$)$_{16}$(O, OH)$_{10}$•xH$_2$O, [G] Mon. Translucent rose, brownish-yellow. Khibina tundra, Kola Peninsula, USSR. H= 6, D_m= 2.9, D_c= 2.91, VIIID 09. Semenov & Burova, 1955. Am. Min. 41(1956), 163 (Abst.). Sov.Phys.Cryst. 18(1973), 596.

Lacroixite, (Na, Li)AlPO$_4$(F, OH), [G] Mon. Vitreous/resinous grey, colorless, pale yellow/green. Greifenstein, Germany. H= 4.5, D_m= 3.29, D_c= 3.29, VIIB 11a. Slavik, 1914. Can. Min. 27 (1989), 211. Am. Min. 70 (1985), 849. See also Durangite.

Laffittite, AgHgAsS$_3$, [A] Mon. Metallic bluish-white. Jas Roux, Hautes-Alpes, France. H= 3, D_m=?, D_c= 6.071, IID 07. Johan et al., 1974. Bull. Min. 97 (1974), 48. Am. Min. 68 (1983), 235. See also Freieslebenite, Marrite.

Laihunite, Fe$_{1.6}$SiO$_4$, [G] Mon. Sub-metallic/metallic black. Lai-He village, China (NE). H= 6.1, D_m= 3.92, D_c= 4.11, VIIIA 03. Am. Min. 71 (1986), 1455. Min. Jour. 11 (1983), 382.

Laitakarite, Bi$_4$(Se, S)$_3$, [G] Rhom. Metallic. Orijärvi mine, Orijärvi, Finland (SW). H=?, D_m= 7.93, D_c= 8.279, IIA 15. Vorma, 1959. Can. Min. 7 (1963), 677. See also Ikunolite.

Lammerite, Cu$_3$[(As, P)O$_4$]$_2$, [A] Mon. Vitreous/adamantine dark green. Sica Sica, Laurani, Veta Negra, La Paz, Bolivia. H= 3.5-4, D_m= 5.18, D_c= 5.264, VIIA 08. Keller et al., 1981. Am. Min. 67(1982), 415 (Abst.). Am. Min. 71 (1986), 206.

Lamprophyllite, (Na, Ca)(Na, Mn)$_2$(Sr, Ba)$_2$Ti$_3$(Si$_2$O$_7$)$_2$(O, OH, F)$_4$, [G] Mon. Translucent golden-brown, brown. Kola Peninsula, USSR. H= 2-3, D_m= 3.46, D_c= 3.54, VIIIB 10. Ramsay & Hackmann, 1894. Am. Min. 27 (1942), 397. Sov.Phys.Dokl. 28 (1983), 206. See also Barytolamprophyllite.

Lanarkite, Pb$_2$O(SO$_4$), [G] Mon. Translucent grey, greenish-white, pale yellow. Leadhills, Lanarkshire, Scotland. H= 2-2.5, D_m= 6.92, D_c= 7.000, VIE 02. Beudant, 1832. Dana, 7th ed.(1951), v.2, 550. Zts. Krist. 132 (1970), 99. See also Phoenicochroite.

Landauite, NaMnZn$_2$(Ti, Fe)$_6$Ti$_{12}$O$_{38}$, [A] Rhom. Semi-metallic black. North Baikal Region, Burpala Alkaline Complex, USSR. H= 7.5, D_m= 4.42, D_c= 4.46, IVC 06. Portnov et al., 1966. Am. Min. 51(1966), 1546 (Abst.). Can. Min. 16 (1978), 63.

Landesite, (Mn, Mg)$_9$Fe$_3$(PO$_4$)$_8$(OH)$_3$•9H$_2$O, [G] Orth. Translucent brown. Berry Quarry, Poland, Androseoggin Co., Maine, USA. H= 3-3.5, D_m= 3.026, D_c= 3.210, VIIC 04. Berman & Gonyer, 1930. Am. Min. 49 (1964), 1122. Min. Mag. 43 (1980), 789.

Långbanite, (Mn, Ca)$_5$(Mn, Fe, Sb)$_2$SiO$_{11}$, [G] Rhom. Black. Långban mine, Filipstad (near), Värmland, Sweden. H= 6.5, D_m= 4.918, D_c= 4.39, VIIIA 18. Flink, 1877. Am. Min. 55 (1970), 1489. Sov.Phys.Cryst. 18(1973), 320.

Langbeinite, K$_2$Mg$_2$(SO$_4$)$_3$, [G] Cub. Vitreous colorless, yellowish, rose, reddish, greenish, etc. Prussian salt deposits, Holberstadt, Germany. H= 3.5-4, D_m= 2.83, D_c= 2.824, VIA 02. Zuckschwert, 1891. Per. Mineral. 54 (1985), 1. N.Jb.Min.Mh. (1979), H4, 182. See also Manganolangbeinite.

Langisite, (Co, Ni)As, [A] Hex. Metallic pinkish-buff. Langis mine, Harris Twp., Timiskaming Dist, Ontario, Canada. H= 6, D_m=?, D_c= 8.17, IIB 09a. Petruk et al., 1969. Can. Min. 9 (1969), 597.

Langite, Cu$_4$SO$_4$(OH)$_6$•2H$_2$O, [G] Mon. Vitreous blue, greenish-blue. St. Blazey and St. Just, Cornwall, England. H= 2.5-3, D_m= 3.28, D_c= 3.37, VID 01. Maskelyne, 1864. Bull. Min. 81 (1975), 257. Bull. Min. 107 (1984), 641. See also Wroewolfeite.

Lannonite, $HCa_4Mg_2Al_4(SO_4)_8F_9 \cdot 32H_2O$, A Tet. Chalky white. Lone Pine mine, Silver City, Grant Co., New Mexico , USA. H= 2, D_m= 2.22, D_c= 2.322, VID 09. Williams & Cesbron, 1983. Min. Mag. 47(1983), 37.

Lansfordite, $MgCO_3 \cdot 5H_2O$, G Mon. Vitreous colorless, white. Nesquehoning, Lansford (near), Carbon Co., Pennsylvania, USA. H= 2.5, D_m= 1.70, D_c= 1.693, VbC 01. Genth, 1888. Min. Mag. 46 (1982), 453.

Lanthanite-(Ce), $(Ce,La,Nd)_2(CO_3)_3 \cdot 8H_2O$, A Orth. Vitreous colorless. Brittania mine, Snowdonia, Gwynedd, Wales. H= 2.5, D_m= 2.76, D_c= 2.79, VbC 03b. Bevins et al., 1985. Am. Min. 70 (1985), 411.

Lanthanite-(La), $(La,Ce)_2(CO_3)_3 \cdot 8H_2O$, G Orth. Pearly colorless, white, pink, yellowish. Bastnäs, Roddarjuttam, Sweden (?). H= 2.5-3, D_m= 2.7, D_c= 2.78, VbC 03b. Haidinger, 1845. Am. Min. 62 (1977), 142. Am. Min. 62 (1977), 142.

Lanthanite-(Nd), $(Nd,La)_2(CO_3)_3 \cdot 8H_2O$, A Orth. Vitreous/pearly pink. Curitiba, Paraná, Brazil. H= 2.5-3, D_m= 2.81, D_c= 2.82, VbC 03b. Roberts et al., 1980. Am. Min. 66(1981), 637 (Abst.).

Laphamite, $As_2(Se,S)_3$, A Mon. Resinous dark red. Burnside, Northumberland Co., Pennsylvania, USA. H=?, D_m= 4.5, D_c= 4.60, IIB 19. Dunn et al., 1986. Min. Mag. 50 (1986), 279. See also Orpiment.

Lapieite, $CuNiSbS_3$, A Orth. Metallic greenish-grey/grey in reflected light. Lapie River, Yukon Territory, Canada. H= 4.5-5, D_m=?, D_c= 4.966, IID 19. Harris et al., 1984. Can. Min. 22 (1984), 561.

Laplandite-(Ce), $Na_4CeTiPSi_7O_{22} \cdot 5H_2O$, A Orth. Vitreous/greasy light grey, yellowish, bluish. Yubileĭnaya (Jubilejnaja) pegmatite, Mt. Karnasurt, Lovozero massif, Kola Peninsula, USSR. H= 2-3, D_m= 2.83, D_c= 2.71, VIIIB 09. Eskova et al., 1974. ZVMO 103 (1974), 571.

Larderellite, $NH_4B_5O_7(OH)_2 \cdot H_2O$, G Mon. Translucent white. Larderello, Val di Cecina, Piza, Toscana, Italy. H=?, D_m= 1.905, D_c= 1.877, Vc 09. Bechi, 1854. Am. Min. 45 (1960), 1087. Acta Cryst. B25 (1969), 2264.

Larnite, β-Ca_2SiO_4, G Mon. Translucent grey. Larne, Antrim Co., Northern Ireland. H=?, D_m= 3.28, D_c= 3.33, VIIIA 05. Tilley, 1929. Min. Mag. 22 (1929), 77. Struct. Repts. 16 (1952), 335.

Larosite, $(Cu,Ag)_{21}(Pb,Bi)_2S_{13}$, A Orth. Whitish buff in reflected light. Foster mine, Coleman Twp., Cobalt, Ontario, Canada. H= 3, D_m=?, D_c= 6.18, IID 11. Petruk, 1972. Can. Min. 11 (1972), 886.

Larsenite, $ZnPbSiO_4$, G Orth. Adamantine white. Franklin Furnace, Sussex Co., New Jersey, USA. H=?, D_m= 5.90, D_c= 6.12, VIIIA 02. Palache et al., 1928. Am. Min. 13 (1928), 142. Zts. Krist. 124 (1967), 115. See also Esperite.

Latiumite, $KCa_3(Al,Si)_5O_{11}(SO_4,CO_3)$, G Mon. Vitreous white, colorless. Albano, Lazio (Latium), Italy. H= 5.5-6, D_m= 2.93, D_c= 2.90, VIIIE 20. Tilley & Henry, 1953. Min. Mag. 30 (1953), 39. Am. Min. 58 (1973), 466. See also Tuscanite.

Latrappite, $(Ca,Na)(Nb,Ti)O_3$, A Orth. Sub-metallic black. La Trappe, Oka, Québec, Canada. H=?, D_m= 4.40, D_c= 4.457, IVC 07. Nickel, 1964. Can. Min. 7 (1963), 683.

Laubmannite, $Fe_9(PO_4)_4(OH)_{12}$, G Orth. Vitreous/silky greyish-green, greenish-brown, brown. Shady, Polk Co., Arkansas, USA. H= 3.5-4, D_m= 3.33, D_c= 3.258, VIIB 07. Frondel, 1949. Am. Min. 55 (1970), 135.

Laueite, $MnFe_2(PO_4)_2(OH)_2 \cdot 8H_2O$, G Tric. Translucent honey-brown. Hagendorf pegmatite, Oberpfalz, Bayern (Bavaria), Germany. H= 3, D_m= 2.46, D_c= 2.56, VIID 03. Strunz, 1954. Am. Min. 39(1954), 1038 (Abst.). Am. Min. 54 (1969), 1312. See also Stewartite, Strunzite.

Laumontite, $Ca(Al_2Si_4)O_{12} \cdot 4H_2O$, G Mon. Vitreous/pearly white, yellow, grey. Huelgoet mines, Brittany, France. H= 3.5-4, D_m= 2.3, D_c= 2.25, VIIIF 11. Werner, 1805. DHZ (1963), v.4, 401. Sov.Phys.Cryst. 30(1985), 624.

Launayite, $Pb_{22}(Sb,As)_{26}S_{61}$, [A] Mon. Metallic white/grey in reflected light. Taylor Pit, Madoc, Huntingdon Twp., Hasting Co., Ontario, Canada. H= 3.5, D_m=?, D_c= 5.83, IID 10. Jambor, 1967. Can. Min. 9 (1967), 191.

Laurelite, $Pb(F,Cl,OH)_2$, [A] Hex. Silky colorless. Grand Reef mine, Laurel Canyon, Graham Co., Arizona, USA. H= 2, D_m= 6.2, D_c= 6.52, IIIC 16. Kampf et al., 1989. Am. Min. 74 (1989), 927.

Laurionite, $PbCl(OH)$, [G] Orth. Adamantine colorless, white. Lávrion (Laurium), Attikí, Greece. H= 3-3.5, D_m= 6.241, D_c= 6.212, IIIC 05. Köchlin, 1887. Dana, 7th ed.(1951), v.2, 62. Zts. Krist. 141 (1975), 246. See also Paralaurionite.

Laurite, RuS_2, [G] Cub. Metallic black; white in reflected light. Borneo, Malaysia. H= 7, D_m= 6.43, D_c= 6.39, IIC 05. Wöhler, 1866. Min. Mag. 47 (1983), 465. See also Erlichmanite.

Lausenite, $Fe_2(SO_4)_3 \cdot 6H_2O$, [G] Mon. Silky white. United Verde mine, Jerome, Arizona, USA. H=?, D_m=?, D_c=?, VIC 04. Butler, 1928. Dana, 7th ed., vol. 2, 530.

Lautarite, $Ca(IO_3)_2$, [G] Mon. Transparent colorless, yellowish. Pampa del Pique III, Oficina Lautaro, Atacama desert, Antofagasta, Chile. H= 3.5-4, D_m= 4.50, D_c= 4.48, IVG 01. Dietze, 1891. Dana, 7th ed.(1951), v.2, 312. Acta Cryst. B34 (1978), 84.

Lautite, $CuAsS$, [G] Orth. Metallic black/grey; greyish-white in reflected light. Lauta, Marienberg (near), Sachsen (Saxony), Germany. H= 3-3.5, D_m= 4.91, D_c= 4.88, IIC 08. Frenzel, 1881. Dana, 7th ed.(1944), v.1, 327. Acta Cryst. 19 (1965), 543.

Lavendulan, $NaCaCu_5(AsO_4)_4Cl \cdot 5H_2O$, [G] Orth. Vitreous/waxy/satiny lavender-blue. Jáchymov (St. Joachimsthal), Západočeský kraj, Čechy (Bohemia), Czechoslovakia. H= 2.5-3, D_m= 3.54, D_c= 3.64, VIID 16. Breithaupt, 1837. Bull. Min. 79 (1956), 7. See also Sampleite.

Låvenite, $Na_2(Mn,Ca,Fe)(Zr,Nb)(Si_2O_7)(O,F)_2$, [G] Mon. Vitreous light yellow. Låven island, Langesundfjord, Norway. H= 6, D_m= 3.53, D_c= 3.41, VIIIB 08. Brögger, 1885. Dana, 6th ed. (1892), 375. Min. Abstr. 83M/1191.

Lavrentievite, $Hg_3S_2(Cl,Br)_2$, [A] Mon. Adamantine colorless/light olive-green. Arzak and Kadyrel deposits, Tuva, USSR. H= 2-2.5, D_m= 7.4, D_c= 7.26, IIF 01. Vasil'ev et al., 1984. Am. Min. 70(1985), 873 (Abst.). See also Arzakite, Corderoite.

Lawrencite, $(Fe,Ni)Cl_2$, [G] Rhom. Translucent white, green, brown. H=?, D_m= 3.16, D_c= 3.22, IIIA 06. Daubrée, 1877. Dana, 7th ed.(1951), v.2, 40.

Lawsonbauerite, $(Mn,Mg)_9Zn_4(SO_4)_2(OH)_{22} \cdot 8H_2O$, [A] Mon. Dull/vitreous colorless, white. Sterling Hill mine, Ogdensburg, Sussex Co., New Jersey, USA. H= 4.5, D_m= 2.87, D_c= 2.92, IVF 03e. Dunn et al., 1979. Am. Min. 64 (1979), 949. Am. Min. 67 (1982), 1029. See also Mooreite, Shigaite, Torreyite.

Lawsonite, $CaAl_2Si_2O_7(OH)_2 \cdot H_2O$, [G] Orth. Translucent colorless, white, bluish. Tiburon Peninsula, Marin Co., California, USA. H= 6, D_m= 3.1, D_c= 3.088, VIIIB 05. Ransome, 1895. DHZ, 2nd ed.(1986), v.1B, 180. Am. Min. 63 (1978), 311. See also Partheite.

Lazarenkoite, $(Ca,Fe)FeAs_3O_7 \cdot 3H_2O$, [A] Orth. Resinous/silky bright orange. Siberia (W), USSR. H= 1, D_m= 3.45, D_c= 3.59, VIIC 12. Yakhontova & Plosina, 1981. Am. Min. 67(1982), 415 (Abst.).

Lazarevićite, [D] = arsenosulvanite. Min. Mag. 33 (1962), 261. Sclar & Drovenik, 1960.

Lazulite, $(Mg,Fe)Al_2(PO_4)_2(OH)_2$, [G] Mon. Vitreous blue, bluish-green. H= 5.5-6, D_m= 3.08, D_c= 3.09, VIIB 05. Klaproth, 1795. Am. Min. 67 (1982), 610. Min. Abstr. 85M/0190. See also Scorzalite, Barbosalite.

Lazurite, $Na_3Ca(Si_3Al_3)O_{12}S$, [G] Cub. Vitreous blue, violet-blue, greenish-blue. Badakhstan, Afghanistan (NE) (?). H= 5-5.5, D_m= 2.4, D_c= 2.39, VIIIF 07. Brögger, 1890. Lithos 9 (1976), 39. Acta Cryst. C41 (1985), 827.

Lead, Pb, G Cub. Metallic grey; greyish-white in reflected light. H= 1.5, D_m= 11.347, D_c= 11.339, IA 01a. Dana, 7th ed.(1944), v.1, 102.

Leadamalgam, $Pb_{0.7}Hg_{0.3}$, A Tet. Metallic white. Shiaonanshan, Inner Mongolia, China. H= 1.6, D_m=?, D_c= 11.96, IA 06. Chen et al., 1981. Am. Min. 70(1985), 216 (Abst.).

Leadhillite, $Pb_4(SO_4)(CO_3)_2(OH)_2$, G Mon. Resinous/adamantine colorless, white, grey, pale green, etc. Leadhills, Lanarkshire, Scotland. H= 2.5-3, D_m= 6.55, D_c= 6.576, VbB 03. Beudant, 1832. Min. Mag. 49 (1985), 759. See also Macphersonite, Susannite.

Lechatelierite, SiO_2, G Amor. Vitreous colorless. H=?, D_m=?, D_c=?, IVD 01. Lacroix, 1915. Dana, 7th ed.(1962), v.3, 325.

Lecontite, $(NH_4,K)Na(SO_4)\cdot 2H_2O$, G Orth. Vitreous colorless. Las Piedras (cave), Comayagua (near), Honduras. H= 2-2.5, D_m= 1.745, D_c= 1.747, VIC 13. Taylor, 1858. Am. Min. 48 (1963), 180. Acta Cryst. 22 (1967), 683.

Legrandite, $Zn_2AsO_4(OH)\cdot H_2O$, G Mon. Translucent colorless, yellow. Flor de Pena mine, Lampazos, Nueva Leon, Mexico. H= 5, D_m= 3.98, D_c= 4.01, VIID 02. Drugman & Hey, 1932. Am. Min. 48 (1963), 1258. Am. Min. 56 (1971), 1147.

Lehiite, D Discredited as crandallite. Am. Min. 71 (1986), 1515. Larsen & Shannon, 1930.

Lehnerite, $Mn(UO_2)_2(PO_4)_2\cdot 8H_2O$, A Mon. Bronze-yellow, yellow. Hagendorf pegmatite, Oberpfalz, Bayern (Bavaria), Germany. H=?, D_m=?, D_c= 3.67, VIID 20c. Mücke, 1988. Min. Mag. 52(1988), 727 (Abst.).

Leifite, $Na_6Be_2Al_2Si_{16}O_{39}(OH)_2\cdot 1.5H_2O$, G Rhom. Translucent colorless. Narsarsuk, Greenland (S). H= 6, D_m= 2.57, D_c= 2.59, VIIIF 06. Bøggild, 1915. Am. Min. 57(1972), 1006 (Abst.). Acta Cryst. B30 (1974), 396.

Leightonite, $K_2Ca_2Cu(SO_4)_4\cdot 2H_2O$, G Orth. Vitreous pale blue, greenish-blue. Chuquicamata, Antofagasta, Chile. H= 3, D_m= 2.95, D_c= 2.953, VIC 12. Palache, 1938. Can. Min. 7 (1962), 272. See also Polyhalite.

Leiteite, $ZnAs_2O_4$, A Mon. Pearly colorless, brown. Tsumeb, Namibia. H= 1.5-2, D_m= 4.3, D_c= 4.619, VIIA 15. Cesbron et al., 1977. Am. Min. 62(1977), 1259 (Abst.). Am. Min. 72 (1987), 629.

Lemoynite, $(Na,K)_2CaZr_2Si_{10}O_{26}\cdot 5\text{–}6H_2O$, A Mon. Translucent yellowish-white. Mont Saint-Hilaire, Rouville Co., Québec, Canada. H= 4, D_m= 2.29, D_c= 2.38, VIIID 18. Perrault et al., 1967. Can. Min. 9 (1967), 585. Can. Min. 14 (1976), 132.

Lengenbachite, $(Ag,Cu)_2Pb_6As_4S_{13}$, G Tric. Metallic grey. Lengenbach, Binntal, Valais (Wallis), Switzerland. H= 3, D_m= 5.8, D_c= 5.64, IID 08. Solly, 1904. Dana, 7th ed.(1944), v.1, 398. Am. Min. 73 (1988), 1426.

Lennilenapeite, $K_7Mg_{48}(Si,Al)_{72}(O,OH)_{216}\cdot 16H_2O$, A Tric. Vitreous/resinous black. Franklin, Sussex Co., New Jersey, USA. H= 3, D_m= 2.72, D_c=?, VIIIE 07c. Dunn et al., 1984. Can. Min. 22 (1984), 259. See also Stilpnomelane.

Lenoblite, $V_2O_4\cdot 2H_2O$, A Syst=? Blue, altering to greenish. Mounana mine, Franceville, Haut-Ogooué, Gabon. H=?, D_m=?, D_c=?, IVF 07. Cesbron & Vachey, 1970. Am. Min. 56(1971), 635 (Abst.).

Leonite, $K_2Mg(SO_4)_2\cdot 4H_2O$, G Mon. Waxy/vitreous colorless, yellowish. Westregeln and Leopoldshall, Prussia, Germany. H= 2.5-3, D_m= 2.201, D_c= 2.203, VIC 10. Tenne, 1896. Dana, 7th ed.(1951), v.2, 450. Zts. Krist. 173 (1985), 75.

Lepersonnite-(Gd), $Ca(Gd,Dy)_2(UO_2)_{24}(CO_3)_8Si_4O_{12}\cdot 60H_2O$, A Orth. Translucent yellow. Shinkolobwe, Shaba, Zaïre. H=?, D_m= 3.97, D_c= 4.01, VbD 04. Deliens & Piret, 1982. Can. Min. 20 (1982), 231.

Lepidocrocite, γ–FeO(OH), G Orth. Sub-metallic red, reddish-brown. Zlattahora, Czechoslovakia. H= 5, D_m= 4.09, D_c= 3.960, IVF 04. Ullmann, 1813. Dana, 7th ed.(1944), v.1, 642. Struct. Repts. 44A (1978), 224. See also Akaganéite, Feroxyhyte, Goethite.

Lepidolite-1M, $K(Li,Al)_3(Si,Al)_4O_{10}(F,OH)_2$, G Mon. Translucent colorless, pink, purple. H= 2.5-4, D_m= 2.85, D_c= 2.83, VIIIE 05b. Klaproth, 1792. DHZ (1962), v.3, 85. Am. Min. 66 (1981), 1221.

Lepidolite-2M₂, $K(Li,Al)_3(Si,Al)_4O_{10}(F,OH)_2$, P Mon. Colorless, pink, purple. H= 2.5-4, D_m= 2.85, D_c= 2.84, VIIIE 05b. Klaproth, 1792. DHZ (1962), v.3, 85. Am. Min. 66 (1981), 1221.

Lepidolite-2M₁, $K(Li,Al)_3(Si,Al)_4O_{10}(F,OH)_2$, P Mon. Colorless, pink, purple. H= 2.5-4, D_m= 2.85, D_c=?, VIIIE 05b. Klaproth, 1792. DHZ (1962), v.3, 85. Min. Abstr. 78-1488.

Lermontovite, $UPO_4(OH)\cdot H_2O$(?), G Orth. Dull/silky greyish-green. Kola Peninsula, USSR. H=?, D_m= 4.50, D_c= 3.92, VIID 19. Melkov, 1952. Am. Min. 69(1984). 214 (Abst.).

Lessingite-(Ce), $(Ce,Ca)_5(SiO_4)_3F$, Q Mon. Vitreous greenish/reddish yellow. Kyshtym district, Ural Mts., USSR. H= 4.5, D_m= 4.877, D_c= 4.918, VIIIA 29. Zilbermintz, 1929. Min. Mag. 31 (1957), 455. Acta Cryst. A37 (1981), C-188.

Letovicite, $(NH_4)_3H(SO_4)_2$, G Tric. Transparent colorless, white. Letovice, Moravia, Czechoslovakia. H=?, D_m= 1.83, D_c= 1.82, VIA 04. Sekanina, 1932. Dana, 7th ed.(1951), v.2, 397. J. Appl. Cryst. 17 (1984), 331.

Leucite, $K(AlSi_2)O_6$, G Tet. Vitreous white, grey. Mt. Vesuvius, Napoli, Campania, Italy. H= 5.5-6, D_m= 2.50, D_c= 2.461, VIIIF 11. Werner, 1791. DHZ (1963), v.4, 276. Am. Min. 61 (1976), 108. See also Ammonioleucite.

Leucophane, $NaCaBeSi_2O_6F$, G Orth. Vitreous whitish-green, greenish-white, yellowish-green, etc. Låven, Langesundfjord, Norway. H= 4, D_m= 2.959, D_c= 2.948, VIIIB 02b. Esmark, 1840. Dana, 6th ed.(1892), 417. Can. Min. 27 (1989), 193.

Leucophoenicite, $Mn_7(SiO_4)_3(OH)_2$, G Mon. Translucent light purplish-red. Franklin, Sussex Co., New Jersey, USA. H= 5.5-6, D_m= 3.8, D_c= 3.848, VIIIA 16. Penfield & Warren, 1899. Dana/Ford (1932), 631. Am. Min. 55 (1970), 1146.

Leucophosphite, $K(Fe,Al)_2(PO_4)_2(OH)\cdot 2H_2O$, G Mon. Chalky buff, white, greenish. Ninghanboun Hills, Western Australia, Australia. H=?, D_m= 2.948, D_c= 2.911, VIID 14a. Simpson, 1932. Am. Min. 42 (1957), 214. Am. Min. 57 (1972), 397. See also Spheniscidite, Tinsleyite.

Leucosphenite, $Na_4BaTi_2B_2Si_{10}O_{30}$, G Mon. Translucent white. Narsarsuk, Greenland (S). H= 6.5, D_m= 3.0, D_c= 3.090, VIIID 19. Flink, 1901. Am. Min. 57 (1972), 1801. Min. Abstr. 82M/3923.

Levyne, $NaCa_{2.5}(Al_6Si_{12})O_{36}\cdot 18H_2O$, G Rhom. Vitreous white, grey, greenish, reddish, yellowish. Dalsnipa, Sandoy, Faröe Islands. H= 4-4.5, D_m= 2.1, D_c= 2.145, VIIIF 15. Brewster, 1825. Natural Zeolites (1985), 192. Struct. Repts. 23 (1959), 491.

Lewisite, $(Ca,Fe,Na)_2(Sb,Ti)_2(O,OH)_7$, Q Cub. Tripuhy, Ouro Preto, Minas Gerais, Brazil. H=?, D_m= 4.95, D_c= 5.31, IVC 08. Hussak & Prior, 1895. JCPDS 7-66. See also Romeite.

Liandratite, $U(Nb,Ta)_2O_8$, A Rhom. Translucent yellow, yellow-brown. Antsakoa I pegmatite, Berere region, Tsaratanana (near), Madagascar. H= 3.5, D_m= 6.8, D_c= 6.87, IVD 18. Mücke & Strunz, 1978. Am. Min. 63 (1978), 941.

Liberite, Li_2BeSiO_4, G Mon. Vitreous pale yellow, brown. Nan Ling Range, China (S). H= 7, D_m= 2.69, D_c= 2.69, VIIIA 01. Chao, 1964. Am. Min. 50(1965), 519 (Abst.). Struct. Repts. 31A (1966), 230.

Libethenite, $Cu_2PO_4(OH)$, G Orth. Vitreous light/dark olive-green, deep green, blackish-green. Lubietová (Libethen), Banská Bystrica (Neusohl), Středoslovenský kraj, Slovensko (Slovakia), Czechoslovakia. H= 4, D_m= 3.97, D_c= 3.972, VIIB 04. Breithaupt, 1823. N.Jb.Min.Abh. 134 (1979), 147. Can. Min. 16 (1978), 153. See also Olivenite.

Liddicoatite, $(Li,Al)_3CaAl_6(BO_3)_3Si_6O_{18}(O,OH,F)_4$, [A] Rhom. Vitreous brown, pale pink. Antsirabe, Madagascar. H= 7.5, D_m= 3.02, D_c= 3.05, VIIIC 08. Dunn et al., 1977. Am. Min. 62 (1977), 1121. Min. Abstr. 82M/0143.

Liebenbergite, $(Ni,Mg)_2SiO_4$, [A] Orth. Translucent yellowish-green. Scotia Talc mine, Bon Accord, Barberton Dist., Transvaal, South Africa. H= 6-6.5, D_m=?, D_c= 4.60, VIIIA 03. DeWaal & Calk, 1973. Am. Min. 58 (1973), 733. Am. Min. 66 (1981), 770.

Liebigite, $Ca_2(UO_2)(CO_3)_3 \cdot 11H_2O$, [G] Orth. Vitreous green, yellowish-green. Edirne, Adrianople (near), Turkey. H= 2.5-3, D_m= 2.41, D_c= 2.409, VbD 04. Smith, 1848. Dana, 7th ed.(1951), v.2, 240. Min. Abstr. 84M/3848.

Likasite, $Cu_3NO_3(OH)_5 \cdot 2H_2O$, [G] Orth. Translucent blue. Likasi Copper mine, Shaba, Zaïre. H=?, D_m= 2.97, D_c= 2.894, Va 03. Schoep et al., 1955. Acta Cryst. B33 (1977), 1422. N.Jb.Min.Mh. (1986), 101.

Lillianite, $Pb_3Bi_2S_6$, [G] Orth. Metallic grey; white in reflected light. Lillian Mining Co., Printerboy Hill, Leadville (near), Colorado, USA. H= 2-3, D_m= 7.0, D_c= 7.07, IID 09b. Keller, 1889. Am. Min. 54(1969), 579 (Abst.). Acta Cryst. B28 (1972), 649. See also Gustavite.

Lime, CaO, [G] Cub. Translucent colorless. H= 3.5, D_m= 3.3, D_c= 3.342, IVA 04. NBS Circ. 539, v.1 (1953), 43.

Linarite, $CuPbSO_4(OH)_2$, [G] Mon. Vitreous/sub-adamantine blue. Linares, Jaén, Spain. H= 2.5, D_m= 5.35, D_c= 5.31, VIB 06. Brooke, 1822. Min. Abstr. 87M/3984. Acta Cryst. 14 (1961), 747. See also Schmiederite.

Lindackerite, $H_2Cu_5(AsO_4)_4 \cdot 9H_2O$, [G] Mon. Vitreous green, greenish-white, greenish-blue. Jáchymov (St. Joachimsthal), Západočeský kraj, Čechy (Bohemia), Czechoslovakia. H= 2-3, D_m= 3.27, D_c= 3.46, VIIC 07. Haidinger, 1853. Bull. Min. 79 (1956), 7.

Lindgrenite, $Cu_3(MoO_4)_2(OH)_2$, [G] Mon. Transparent green, yellowish-green. Chuquicamata, Antofagasta, Chile. H= 4.5, D_m= 4.26, D_c= 4.310, VIF 03. Palache, 1935. Can. Min. 6 (1957), 31. N.Jb.Min.Mh. (1985), 234.

Lindsleyite, $(Ba,Sr)(Ti,Cr,Fe)_{21}O_{38}$, [A] Rhom. Metallic black; tan in reflected light. Kimberlite deposits, South Africa. H= 7.3, D_m=?, D_c= 4.63, IVC 06. Haggerty et al., 1983. Am. Min. 68 (1983), 494.

Lindströmite, $Cu_3Pb_3Bi_7S_{15}$, [A] Orth. Metallic grey. Gladhammar, Småland, Kalmar, Sweden. H= 3-3.5, D_m= 7.01, D_c= 7.03, IID 05a. Johansson, 1924. N.Jb.Min.Mh. (1990), 35. Can. Min. 15 (1977), 527.

Linnaeite, Co_3S_4, [G] Cub. Metallic grey; white in reflected light. H= 4.5-5.5, D_m= 4.6, D_c= 4.85, IIC 01. Haidinger, 1845. Am. Min. 49 (1964), 543. Struct. Repts. 11 (1947), 288. See also Polydymite.

Liottite, $(Ca,Na)_8(Si,Al)_{12}O_{24}(SO_4,Cl,OH)_4 \cdot 2H_2O$, [A] Hex. Transparent colorless. Pitigliano, Grosseto, Toscana, Italy. H= 5, D_m= 2.56, D_c= 2.51, VIIIF 05. Merlino & Orlandi, 1977. Am. Min. 62 (1977), 321.

Lipscombite, $Fe_3(PO_4)_2(OH)_2$, [Q] Tet. Sepucaia pegmatite, Coundelheiro Peña, Minas Gerais, Brazil. H=?, D_m= 3.8, D_c= 3.68, VIIB 05. Gheith, 1953. Struct. Repts. 48A (1981), 286. Am. Min. 74 (1989), 456.

Liroconite, $Cu_2AlAsO_4(OH)_4 \cdot 4H_2O$, [G] Mon. Vitreous/resinous blue, green. Cornwall, England. H= 2-2.5, D_m= 3.0, D_c= 3.08, VIID 33. Haidinger, 1825. Am. Min. 36 (1951), 484. Struct. Repts. 33A (1968), 424.

Lisetite, $Na_2CaAl_4(SiO_4)_4$, [A] Orth. Transparent colorless. Liset pod, Selje dist., Western Gneiss region, Norway. H=?, D_m=?, D_c= 2.73, VIIIF 02. Smith et al., 1986. Am. Min. 71 (1986), 1372. Am. Min. 71 (1986), 1378.

Liskeardite, $(Al,Fe)_3AsO_4(OH)_6 \cdot 5H_2O$, [Q] Syst=? Translucent white, greenish, bluish, brownish. Marl Valley mine, Liskeard, Cornwall, England. H=?, D_m=?, D_c=?, VIID 10. Maskelyne, 1878. Dana, 7th ed.(1951),v.2, 924.

Litharge, γ-PbO, G Tet. Greasy transparent red. H= 2, D_m= 9.14, D_c= 9.36, IVA 07. Wherry, 1917. NBS Circ. 539, v.2 (1953), 30. See also Massicot.

Lithiomarsturite, $LiMn_2Ca_2HSi_5O_{15}$, A Tric. Vitreous light pinkish-brown to light yellow. Foote mine, Kings Mt., Cleveland Co., North Carolina, USA. H= 6, D_m= 3.32, D_c= 3.27, VIIID 13. Peacor et al., 1990. Am. Min. 75 (1990), 409.

Lithiophilite, $Li(Mn,Fe)PO_4$, G Orth. Vitreous/sub-resinous brown, yellowish-brown, yellow, salmon. Branchville, Fairfield Co., Connecticut, USA. H= 4-5, D_m= 3.34, D_c= 3.44, VIIA 02. Brush & Dana, 1878. Dana, 7th ed.(1951), v.2, 665. See also Triphylite.

Lithiophorite, $Li_6Al_{14}Mn_{21}O_{42}(OH)_{42}$, G Hex. Dull/metallic bluish-black. Schneeberg, Sachsen (Saxony), Germany. H= 3, D_m= 3.2, D_c= 3.376, IVF 05b. Breithaupt, 1870. Acta Cryst. 5 (1952), 676. Am. Min. 67 (1982), 817.

Lithiophosphate, Li_3PO_4, G Orth. Vitreous colorless, white, light rose. Kola Peninsula, USSR. H= 4, D_m= 2.46, D_c= 2.430, VIIA 01. Mathias & Bondareva, 1957. Am. Min. 42(1957), 585 (Abst.). Sov.Phys.Dokl. 23 (1978), 287.

Lithiotantite, $Li(Ta,Nb)_3O_8$, A Mon. Transparent colorless. Kazakhstan (E), USSR. H= 6-6.5, D_m= 7.0, D_c= 7.08, IVD 17. Voloshin et al., 1983. Am. Min. 69(1984), 1191 (Abst.).

Lithiowodginite, $LiTa_3O_8$, A Mon. Adamantine dark pink to red. Kazakhstan (E), USSR. H= 6.5, D_m= 7.5, D_c= 7.74, IVD 09. Voloshin et al., 1990. Min. Zhurn. 112 (1990), 94.

Lithosite, $K_6Al_4Si_8O_{25}\cdot 2H_2O$, A Mon. Vitreous colorless. Khibina massif, Kola Peninsula, USSR. H= 5.5, D_m= 2.51, D_c= 2.54, VIIIF 14. Khomyakov et al., 1983. Am. Min. 69(1984), 210 (Abst.).

Litidionite, $KNaCuSi_4O_{10}$, G Tric. Earthy white. Mt. Vesuvius, Napoli, Campania, Italy. H= 5-6, D_m= 2.75, D_c= 2.85, VIIID 18. Scacchi, 1880. Bull. Min. 104 (1981), 387. Am. Min. 60 (1975), 471.

Liujinyinite, Ag_3AuS_2, Q Tet. Metallic greenish-grey. Guangdong, China. H= 2.8, D_m=?, D_c= 7.95, IIA 12. Chen et al., 1979. Am. Min. 67(1982), 1081 (Abst.).

Liveingite, $Pb_9As_{13}S_{28}$, G Mon. Bluish-grey; white in reflected light. Lengenbach quarry, Binntal, Valais (Wallis), Switzerland. H= 3, D_m= 5.3, D_c= 5.365, IID 02. Solly, 1902. Am. Min. 54(1969), 1498 (Abst.).

Livingstonite, $HgSb_4S_8$, G Mon. Adamantine/metallic grey; white in reflected light. Huitzuco, Guerrero, Mexico. H= 2, D_m= 5.00, D_c= 4.89, IID 04. Barcena, 1874. Dana, 7th ed.(1944), v.1, 485. Zts. Krist. 141 (1957), 174.

Lizardite-1T, $Mg_3Si_2O_5(OH)_4$, G Rhom. Translucent green, white. Kennack Cove, Lizard, Cornwall, England. H= 2.5, D_m= 2.55, D_c= 2.58, VIIIE 10b. Whittaker & Zussman, 1956. Rev. Min. 19 (1988), 91. Am. Min. 72 (1987), 943. See also Lizardite-$2H_1$, Antigorite, Clinochrysotile, Orthochrysotile, Parachrysotile , Nepouite.

Lizardite-$2H_1$, $Mg_3Si_2O_5(OH)_4$, P Hex. Translucent green, white. Monte dei Tre Abati, Coli, Italy. H= 2.5, D_m= 2.5, D_c= 2.58, VIIIE 10b. Whittaker & Zussman, 1955. DHZ (1962), v.3, 170. Am. Min. 72 (1987), 943. See also Lizardite-1T.

Lokkaite-(Y), $CaY_4(CO_3)_7\cdot 9H_2O$, A Orth. Translucent white. Kangasala, Finland (SW). H=?, D_m=?, D_c= 2.92, VbC 03b. Perttunen, 1970. Am. Min. 71 (1986), 1028.

Löllingite, $FeAs_2$, G Orth. Metallic white. Lölling, Hüttenberg, Carinthia, Austria. H= 5-5.5, D_m= 7.40, D_c= 7.473, IIC 07. Haidinger,, 1845. Am. Min. 53 (1968), 1856. Struct. Repts. 54A (1987), 43.

Lomonosovite, $Na_5Ti_2O_2(Si_2O_7)(PO_4)$, G Tric. Vitreous/adamantine dark brown, black, rose-violet. Lovozero massif, Kola Peninsula, USSR. H= 3-4, D_m=?, D_c= 3.04, VIIIB 11. Gerasimovsky, 1941. Am. Min. 35(1950), 1092 (Abst.). Min. Abstr. 79-2144. See also Beta-Lomonosovite.

Lonecreekite, $NH_4(Fe, Al)(SO_4)_2 \cdot 12H_2O$, [A] Cub. Colorless transparent. Lone Creek Fall Cave, Sabie, Transvaal (E), South Africa. H=?, D_m= 1.693, D_c= 1.691, VIC 08. Martini, 1983. Am. Min. 71(1986), 229 (Abst.). See also Potassium Alum.

Lonsdaleite, C, [A] Hex. Transparent brownish-yellow. Canyon Diablo meteorite, Meteor Crater, Coconino Co., Arizona, USA. H=?, D_m=?, D_c= 3.55, IB 02b. Frondel & Marvin, 1967. Am. Min. 52(1967), 1579 (Abst.). See also Diamond, Chaoite, Graphite-2H, Graphite-3R.

Loparite-(Ce), (Ce, Na, Ca)(Ti, Nb)O3, [G] Cub. Black. Khibina tundra, Kola Peninsula, USSR. H=?, D_m= 4.01, D_c= 5.25, IVC 07. Fersman, 1922. JCPDS 20-272.

Lopezite, $K_2Cr_2O_7$, [G] Tric. Transparent orange-red, red. Huara, Iquique Pampa, Officiana Rosario, Tarapacá, Chile. H= 2.5, D_m= 2.69, D_c= 2.704, VIE 01. Bandy, 1937. Dana, 7th ed.(1951), v.2, 645. Can. Jour. Chem. 46(1968), 933.

Lorandite, $TlAsS_2$, [G] Mon. Metallic adamantine red/grey; bluish gray-white in reflected light. Alšar (Allchar), Rožden (near), Makedonija (Macedonia), Yugoslavia. H= 2-2.5, D_m= 5.53, D_c= 5.56, IID 15. Krenner, 1894. Dana, 7th ed.(1944), v.1, 437. Zts. Krist. 138 (1973), 147.

Loranskite-(Y), $(Y, Ce, Ca)(Zr, Ta)_2O_6$(?), [Q] Syst=? Sub-metallic black. Impilakhti, Ladozhskoye Ozero (Lake Ladoga), USSR. H= 5, D_m= 4.3, D_c=?, IVD 10b. Melnikov, 1899. Dana, 7th ed.(1944), v.1, 767.

Lorenzenite, $Na_2Ti_2O_3(Si_2O_6)$, [G] Orth. Vitreous/sub-metallic dark brown to yellow-brown. Narsarsuk, Greenland (S). H= 6, D_m= 3.45, D_c= 3.44, VIIID 28. Flink, 1897. Am. Min. 32 (1947), 59. Am. Min. 72 (1987), 173.

Lorettoite, [D] Natural occurrence not proven. Am. Min. 64 (1979), 1303. Wells & Larsen, 1916.

Loseyite, $(Mn, Zn)_7(CO_3)_2(OH)_{10}$, [G] Mon. Transparent bluish-white, brownish. Franklin, Sussex Co., New Jersey, USA. H= 3, D_m= 3.27, D_c= 3.341, VbB 11. Bauer & Berman, 1929. Dana, 7th ed.(1951), v.2, 244. Acta Cryst. B37 (1981), 1323.

Lotharmeyerite, $CaZnMn(AsO_3OH)_2(OH)_3$, [A] Mon. Vitreous reddish-orange. Mapimi, Durango, Mexico. H= 3, D_m= 4.23, D_c= 4.29, VIIB 03c. Dunn, 1983. Min. Rec. 15 (1984), 223.

Loudounite, $NaCa_5Zr_4Si_{16}O_{40}(OH)_{11} \cdot 8H_2O$, [A] Syst=? Translucent light green, white. Goose Creek Quarry, Loudoun Co., Virginia, USA. H= 5, D_m= 2.48, D_c=?, VIIIC 03. Dunn & Newbury, 1983. Can. Min. 21 (1983), 37.

Loughlinite, $Na_2Mg_3Si_6O_{16} \cdot 8H_2O$, [G] Syst=? Silky pearly-white. Westvaco Trona mine, Sweetwater Co., Wyoming, USA. H=?, D_m= 2.165, D_c=?, VIIIE 13. Fahey et al., 1960. Am. Min. 45 (1960), 270.

Lourenswalsite, $(K, Ba)_2Ti_4(Si, Al)_6O_{14}(OH)_{12}$, [A] Hex. Pearly/dull silvery-grey, light brownish-orange. Diamond Jo quarry, Magnet Cove, Hot Springs (near), Arkansas, USA. H=?, D_m= 3.17, D_c= 3.199, VIIIC 07. Appleman et al., 1987. Min. Mag. 51 (1987), 417.

Lovdarite, $K_2Na_6Be_4Si_{14}O_{36} \cdot 9H_2O$, [A] Orth. Translucent white, yellowish. Mt. Karnasurt, Lovozero massif, Kola Peninsula, USSR. H= 5-6, D_m= 2.33, D_c= 2.34, VIIIB 02b. Men'shikov et al., 1973. Am. Min. 68(1983), 474 (Abst.). Acta Cryst. A37 (1981), C189.

Loveringite, $(Ca, Ce, La)(Ti, Fe, Cr)_{21}O_{38}$, [A] Rhom. Metallic/sub-metallic black; white in reflected light. Norseman (near), Jimberlana Intrusion, Western Australia, Australia. H= 5, D_m=?, D_c= 4.42, IVC 06. Gatehouse et al., 1978. Can. Min. 17 (1979), 635. Am. Min. 63 (1978), 28.

Lovozerite, $(Na, Ca)_3(Zr, Ti)Si_6(O, OH)_{18}$, [G] Mon. Resinous dark brown, black. Lovozero, Kola Peninsula, USSR. H= 5, D_m= 2.64, D_c= 2.63, VIIIC 07. Gerasimovsky, 1939. Am. Min. 59(1974), 632 (Abst.). Struct. Repts. 24 (1960), 498.

Löweite, $Na_{12}Mg_7(SO_4)_{13} \cdot 15H_2O$, [G] Rhom. Vitreous colorless, reddish-yellow. Ischl salt mine, Oberösterreich (Upper Austria), Austria. H= 2.5-3, D_m= 2.36,

D_c= 2.35, VIC 10. Haidinger, 1846. Dana, 7th ed.(1951), v.2, 446. Am. Min. 55 (1970), 378.

Luanheite, Ag_3Hg, [A] Hex. Metallic white. Luanhe River, Hebei, China. H= 2.5, D_m= 12.5, D_c= 12.57, IA 06. Shao et al., 1984. Am. Min. 73(1988), 192 (Abst.).

Lucasite-(Ce), $(Ce, La)Ti_2(O, OH)_6$, [A] Mon. Resinous brown, grey. Argyle Ak1, Western Australia, Australia. H= 6, D_m=?, D_c= 5.00, IVD 11. Nickel et al., 1987. Am. Min. 72 (1987), 1006. Am. Min. 72 (1987), 1006.

Luddenite, $Cu_2Pb_2Si_5O_{14} \cdot 14H_2O$, [A] Mon. Translucent green. Artillery Peak, Mohave Co., Arizona, USA. H= 4, D_m= 4.45, D_c= 4.98, VIIIB 13. Williams, 1982. Min. Mag. 46 (1982), 363.

Ludjibaite, $Cu_5(PO_4)_2(OH)_4$, [A] Tric. Vitreous blue-green. Ludjiba deposit, Zaïre. H=?, D_m=?, D_c= 4.36, VIIB 08. Piret & Deliens, 1988. Am. Min. 73(1988), 1495 (Abst.).

Ludlamite, $(Fe, Mg, Mn)_3(PO_4)_2 \cdot 4H_2O$, [G] Mon. Vitreous green. Wheal Jane mine, Truro, Cornwall, England. H= 3.5, D_m= 3.165, D_c= 3.176, VIIC 07. Field, 1877. Dana, 7th ed.(1951), v.2, 952. Struct. Repts. 31A (1966), 187.

Ludlockite, $(Fe, Pb)As_2O_6$, [A] Tric. Sub-adamantine red. Tsumeb, Namibia. H= 1.5-2, D_m= 4.40, D_c= 4.35, VIIA 15. Davis et al., 1970. Am. Min. 57(1972), 1003 (Abst.).

Ludwigite, $Mg_2FeO_2(BO_3)$, [G] Orth. Silky black, dark green. Moravicza, Banat, Romania. H= 5, D_m= 3.86, D_c= 3.788, Vc 01c. Tschermak, 1874. Dana, 7th ed.(1951), v.2, 321. N.Jb.Min.Mh. (1989), 69. See also Vonsenite.

Lueshite, $NaNbO_3$, [G] Orth. Black. Lueshe, Goma, Zaïre. H= 5.5, D_m= 4.44, D_c= 4.578, IVC 07. Safiannikoff, 1959. Am. Min. 46(1961), 1004 (Abst.). Zts. Krist. 143 (1976), 444. See also Natroniobite.

Luetheite, $Cu_2Al_2(AsO_4)_2(OH)_4 \cdot H_2O$, [A] Mon. Translucent blue, pale green. Patagonia Mining Dist., Santa Cruz Co., Arizona, USA. H= 3, D_m= 4.28, D_c= 4.40, VIID 33. Williams, 1977. Min. Mag. 41 (1977), 27. See also Chenevixite.

Lüneburgite, $Mg_3[B(OH)_3]_2(PO_4)_2 \cdot 5H_2O$, [G] Mon. Translucent white, brownish-white. Lüneburg, Hannover, Germany. H= 2, D_m= 2.07, D_c= 2.04, Vc 01h. Nollner, 1870. JCPDS 25-1155.

Lun'okite, $(Mg, Fe)(Mn, Ca)Al(PO_4)_2(OH) \cdot 4H_2O$, [A] Orth. Translucent colorless, white. Kola Peninsula, USSR. H= 3-4, D_m= 2.66, D_c= 2.69, VIID 25. Voloshin et al., 1984. Am. Min. 69(1984), 210 (Abst.).

Lusungite, $(Sr, Pb)Fe_3(PO_4)(PO_3OH)(OH)_6$, [G] Rhom. Translucent yellow-brown. Kobokobo, Kivu, Zaïre. H=?, D_m=?, D_c= 4.12, VIIB 15a. Van Wambeke, 1958. Am. Min. 44(1959), 906 (Abst.).

Luzonite, Cu_3AsS_4, [G] Tet. Metallic pinkish-brown. Mancayan, Lepanto district, Luzon island, Philippines. H= 3.5, D_m= 4.38, D_c= 4.438, IIB 03a. Weisbach, 1874. Econ. Geol. 72 (1977), 271. Am. Min. 42 (1957), 766. See also Enargite.

Lyonsite, $Cu_3Fe_4(VO_4)_6$, [A] Orth. Metallic black; creamy-white in reflected light. Izalco Volcano, El Salvador. H=?, D_m=?, D_c= 4.215, VIIA 08. Hughes et al., 1987. Am. Min. 72 (1987), 1000. Am. Min. 72 (1987), 1000.

M

Macaulayite, $(Fe, Al)_{24}Si_4O_{43}(OH)_2$, [A] Mon. Earthy red. Bennachie, Inverurie (near), Aberdeenshire, Scotland. H=?, D_m=?, D_c= 4.41, VIIIE 11. Wilson et al., 1984. Min. Mag. 48 (1984),127.

Macdonaldite, $BaCa_4Si_{16}O_{36}(OH)_2 \cdot 10H_2O$, [A] Orth. Satiny/vitreous white, colorless. Big Creek and Rush Creek, Fresno Co., California, USA. H= 3.5-4, D_m= 2.27, D_c= 2.36, VIIIE 15. Alfors et al., 1965. Am. Min. 50 (1965), 314. Struct. Repts. 33A (1968), 489.

Macedonite, $PbTiO_3$, [A] Tet. Vitreous black; greyish-white in reflected light. Crni Kamen, Prilep (near), Makedonija (Macedonia), Yugoslavia. H= 5.5, D_m= 7.82, D_c= 8.09, IVC 07. Radusinovič & Markov, 1971. Am. Min. 56 (1971), 387.

Macfallite, $Ca_2Mn_3(SiO_4)(Si_2O_7)(OH)_3$, [A] Mon. Silky/sub-adamantine reddish-brown, maroon. Copper Harbor, Lake Manganese (near), Keeweenaw Co., Michigan, USA. H= 5, D_m= 3.43, D_c= 3.53, VIIIB 19. Moore et al., 1979. Min. Mag. 43 (1979), 325. Am. Min. 70 (1985), 171. See also Sursassite.

Machatschkiite, $(Ca, Na)_6(AsO_4)(AsO_3OH)_3PO_4 \cdot 15H_2O$, [A] Rhom. Translucent colorless. Anton mine, Schiltach (near), Heubachtal, Schwarzwald, Germany. H= 2-3, D_m= 2.50, D_c= 2.59, VIIC 16. Walenta, 1977. Am. Min. 62(1977), 1260 (Abst.). Am. Min. 68(1983), 851 (Abst.).

Mackayite, $FeTe_2O_5(OH)$, [G] Tet. Vitreous green, brownish-green. Mowhawk mine, Goldfield, Nevada, USA. H= 4.5, D_m= 4.86, D_c= 5.281, VIG 02. Frondel & Pough, 1944. Am. Min. 55(1970), 1072 (Abst.). N.Jb.Min.Mh. (1977), 145.

Mackinawite, $(Fe, Ni)_9S_8$, [G] Tet. Metallic reddish-grey. Mackinaw mine, Snohomish Co., Washington, USA. H= 5, D_m=?, D_c= 4.12, IIB 15. Evans et al., 1962. Am. Min. 48 (1963), 511.

Macphersonite, $Pb_4(SO_4)(CO_3)_2(OH)_2$, [A] Orth. Resinous/adamantine white. Leadhills, Lanarkshire, Scotland; Aregentolle mine, Saint-Prix, Saône-et-Loire, France. H= 2.5-3, D_m= 6.5, D_c= 6.6, VbB 03. Livingstone & Sarp, 1984. Min. Mag. 48 (1984), 277. See also Leadhillite, Susannite.

Macquartite, $CuPb_3(CrO_4)SiO_3(OH)_4 \cdot 2H_2O$, [A] Mon. Translucent orange. Mammoth mine, Tiger, Pinal Co., Arizona, USA. H= 3.5, D_m= 5.49, D_c= 5.58, VIIIB 13. Williams & Duggan, 1980. Am. Min. 66(1981), 638 (Abst.).

Macrokaolinite, [D] Rock name. Min. Mag. 43 (1980), 1055. Isphording & Lodding, 1968.

Madocite, $Pb_{17}(Sb, As)_{16}S_{41}$, [G] Orth. Metallic grey-black; grey/white in reflected light. Taylor pit, Madoc, Huntingdon Twp., Hastings Co., Ontario, Canada. H= 3.3, D_m=?, D_c= 5.98, IID 02. Jambor, 1967. Can. Min. 9 (1967), 7.

Magadiite, $NaSi_7O_{13}(OH)_3 \cdot 3H_2O$, [A] Mon. Translucent white. Lake Magadi, Rift Valley, Kenya. H=?, D_m=?, D_c=?, VIIIE 28. Eugster, 1967. Am. Min. 54 (1969), 1583.

Magbasite, $KBa(Mg, Fe)_6(Al, Sc)Si_6O_{20}F_2$, [G] Syst=? Vitreous colorless, rose-violet. Unspecified locality, Asia, USSR. H= 5, D_m= 3.41, D_c=?, VIIID 07. Semenov et al., 1965. Am. Min. 51(1966), 530 (Abst.).

Maghagendorfite, $Na(Mg, Fe)_2Mn(PO_4)_3$, [A] Mon. Dyke Lode, Custer, South Dakota, USA. H=?, D_m=?, D_c=?, VIIA 05b. Moore & Ito, 1979. Min. Mag. 43 (1979), 227. See also Alluaudite, Ferro-alluaudite, Hagendorfite.

Maghemite, $Fe_{2.67}O_4$, [G] Cub. Brown; white/greyish-blue in reflected light. Iron Mt. mine, Redding (near), Shasta Co., California, USA. H= 5, D_m=?, D_c= 4.860, IVB 01. Wagner, 1927. Powd. Diff. 3 (1988), 104. See also Hematite.

Magnesio-alumino-katophorite, $Na_2CaMg_4Al(Si_7Al)O_{22}(OH)_2$, [A] Mon. H=?, D_m=?, D_c=?, VIIID 05c. Leake et al., 1978. Am. Min. 63 (1978), 1023. See also Aluminokatophorite.

Magnesio-alumino-taramite, $Na_2CaMg_3Al_2(Si_6Al_2)O_{22}(OH)_2$, [A] Mon. H=?, D_m=?, D_c=?, VIIID 05c. Leake et al., 1978. Am. Min. 63 (1978), 1023. See also Taramite.

Magnesio-anthophyllite, $(Mg,Fe)_7Si_8O_{22}(OH)_2$, [A] Orth. H= 5.5-6, D_m=?, D_c= 2.95, VIIID 06. Shannon, 1921. Jour. Petrol. 4 (1963), 317. See also Anthophyllite, Ferro-anthophyllite.

Magnesio-arfvedsonite, $Na_3(Mg,Fe)_4Fe^{3+}Si_8O_{22}(OH)_2$, [A] Mon. Vitreous green, greenish-black, black. Japan. H=?, D_m= 3.17, D_c= 3.17, VIIID 05d. Miyashiro, 1957. JCPDS 31-1283. See also Arfvedsonite.

Magnesioaubertite, $(Mg,Cu)Al(SO_4)_2Cl \cdot 14H_2O$, [A] Tric. Transparent blue. Grotta de Faraglione, Vulcano Island, Lipari Islands, Italy. H=?, D_m= 1.80, D_c= 1.75, VID 09. Gebhard et al., 1988. Min. Abstr. 89M/0935.

Magnesio-axinite, $Ca_2MgAl_2(BO_3OH)(SiO_3)_4$, [A] Tric. Translucent pale blue. Tanzania. H= 6-7, D_m= 3.178, D_c= 3.183, VIIIC 04. Jobbins et al., 1976. DHZ, 2nd ed.(1986), v.1B, 603.

Magnesiocarpholite, $(Mg,Fe)(Al,Fe)_2Si_2O_6(OH)_4$, [A] Orth. Translucent white. Vanoise, French Alps, France. H=?, D_m=?, D_c= 2.82, VIIID 03. Goffe et al., 1979. Am. Min. 65(1980), 406 (Abst.). See also Ferrocarpholite, Carpholite.

Magnesiochloritoid, $(Mg,Fe)Al_2SiO_5(OH)_2$, [Q] Mon. Alps (numerous localities), Switzerland and Italy. H=?, D_m=?, D_c= 3.332, VIIIA 22. Chopin, 1983. Am. Min. 70(1985), 216 (Abst.). Am. Min. 73 (1988), 358. See also Chloritoid, Ottrelite.

Magnesiochromite, $MgCr_2O_4$, [G] Cub. Metallic black; grey in reflected light. H= 5.5, D_m= 4.2, D_c= 4.43, IVB 01. Bock, 1868. Dana, 7th ed.(1944), v.1, 709. See also Chromite, Spinel.

Magnesioclinoholmquistite, $Li_2(Mg,Fe)_3Al_2Si_8O_{22}(OH)_2$, [A] Mon. Translucent light blue, violet. Unspecified locality, Siberia, USSR. H=?, D_m= 3.00, D_c= 3.02, VIIID 05a. Ginzburg, 1965. Am. Min. 52(1967), 1585 (Abst.). See also Magnesioholmquistite, Clinoholmquistite, Ferroclinoholmquistite.

Magnesiocopiapite, $MgFe_4(SO_4)_6(OH)_2 \cdot 20H_2O$, [G] Tric. Pearly yellow, orange, greenish-yellow. Blythe, Riverside Co., California, USA; Las Vegas (near), San Miguel Co., New Mexico, USA. H= 2.5-3, D_m= 2.04, D_c= 2.05, VID 04b. Berry, 1938. Can. Min. 23 (1985), 53. Zts. Krist. 135 (1972), 34.

Magnesio-cummingtonite, $(Mg,Fe)_7Si_8O_{22}(OH)_2$, [A] Mon. Translucent green, brown. Kongsberg, Buskerud, Norway (S). H= 5-6, D_m= 3.13, D_c= 3.18, VIIID 05a. Dewey, 1824. Am. Min. 49 (1964), 963. Acta Cryst. 14 (1961), 622. See also Cummingtonite, Grünerite.

Magnesio-ferri-katophorite, $Na_2CaMg_4Fe^{3+}(Si_7Al)O_{22}(OH)_2$, [A] Mon. H=?, D_m=?, D_c=?, VIIID 05c. Leake et al, 1978. Am. Min. 63 (1978), 1023.

Magnesio-ferri-taramite, $Na_2CaMg_3Fe_2^{3+}(Si_6Al_2)O_{22}(OH)_2$, [A] Mon. H=?, D_m=?, D_c=?, VIIID 05c. Leake et al., 1978. Am. Min. 63 (1978), 1023. See also Taramite.

Magnesioferrite, $MgFe_2O_4$, [G] Cub. Metallic/sub-metallic black; grey in reflected light. Mt. Vesuvius, Napoli, Campania, Italy. H= 5.5-6.5, D_m= 4.6, D_c= 4.51, IVB 01. Rammelsberg, 1859. Dana, 7th ed.(1944), v.1, 698. See also Magnetite.

Magnesio-gedrite, $(Mg,Fe)_5Al_2(Si_6Al_2)O_{22}(OH)_2$, [A] Orth. H=?, D_m=?, D_c= 3.18, VIIID 06. Leake et al., 1978. Am. Min. 63 (1978), 1023. Am. Min. 55 (1970), 1945. See also Gedrite, Ferrogedrite.

Magnesio-hastingsite, $NaCa_2(Mg,Fe)_4Fe(Si_6Al_2)O_{22}(OH)_2$, [G] Mon. Translucent green. H= 5-6, D_m= 3.225, D_c= 3.243, VIIID 05b. Billings, 1928. Am. Min. 63 (1978), 1023. Zts. Krist. 156 (1981), 197. See also Hastingsite.

Magnesio-holmquistite, $Li_2(Mg,Fe)_3Al_2Si_8O_{22}(OH)_2$, [A] Orth. Translucent blue, violet. Utö, Sweden. H= 5-6, D_m= 3.13, D_c= 2.92, VIIID 06. Leake et al., 1978. Can. Min. 6 (1960), 504. Acta Cryst. B25 (1969), 394. See also Magnesioclinoholmquistite, Holmquistite, Ferroholmquistite.

Magnesio-hornblende, $Ca_2(Mg,Fe)_4Al(Si_7Al)O_{22}(OH,F)_2$, [A] Mon. Vitreous green. H= 5-6, D_m= 3.0, D_c= 3.11, VIIID 05b. Leake, 1978. Am. Min. 63 (1978), 1023. Struct. Repts. 30A (1965), 433. See also Alumino-magnesio-hornblende.

Magnesiohulsite, $(Mg,Fe)_2(Fe,Sn,Mg)O_2(BO_3)$, [A] Mon. Sub-metallic black; bluish-grey, greyish-white in refl.light. Qiliping, Changning Co., Hunan, China. H= 6, D_m= 4.18, D_c= 4.15, Vc 01c. Yang et al., 1985. Am. Min. 73(1988), 929 (Abst.). See also Hulsite.

Magnesiolaumontite, [D] Unnecessary name for magnesian laumontite. Min. Mag. 36 (1967), 133. Borcos, 1962.

Magnesio-riebeckite, $Na_2(Mg,Fe)_3Fe^{3+}_2Si_8O_{22}(OH)_2$, [A] Mon. Japan. H=?, D_m=?, D_c= 3.11, VIIID 05d. Miyashiro, 1957. JCPDS 29-1237. See also Riebeckite.

Magnesiosadanagaite, $(K,Na)Ca_2(Mg,Fe,Al)_5(Si,Al)_8O_{22}(OH)_2$, [A] Mon. Vitreous dark brown, black. Yuge Island and Myojin Island, Japan. H= 6, D_m=?, D_c= 3.27, VIIID 05b. Shimazaki et al., 1984. Am. Min. 69 (1984), 465. See also Sadanagaite.

Magnesite, $MgCO_3$, [G] Rhom. Vitreous colorless, white, greyish-white, yellowish, brown. Magnesia, Greece and/or Baldissero Canavese, Piemonte, Italy. H= 4, D_m= 3.00, D_c= 3.010, VbA 02. Karsten, 1808. Rev. Min. 11 (1983). Zts. Krist. 156 (1981), 233. See also Gaspéite, Siderite.

Magnesium szomolnokite, [D] Unnecessary name for mangesian szomolnokite. Min. Mag. 33 (1962), 261. Kubisz, 1960.

Magnesium astrophyllite, $K_2Na_2Mg_2Fe_5Ti_2Si_8O_{24}(O,OH,F)_7$, [G] Mon. Vitreous yellow. Unspecified locality, Khibina, Kola Peninsula, USSR. H=?, D_m=?, D_c= 3.32, VIIID 25. Peng & Ma, 1963. Am. Min. 60(1975), 737 (Abst.).

Magnesium-chlorophoenicite, $(Mg,Mn)_3Zn_2AsO_4(OH,O)_6$, [A] Mon. Translucent colorless, white. Franklin, Sussex Co., New Jersey, USA. H=?, D_m= 3.45, D_c= 3.18, VIIB 10a. Palache, 1935. Powd. Diff. 2 (1987), 225. See also Chlorophoenicite, Jarosewichite.

Magnesium-zippeite, $Mg_2(UO_2)_6(SO_4)_3(OH)_{10} \cdot 16H_2O$, [A] Syst=? Translucent yellow. Lucky Strike #2 mine, Emery Co., Utah, USA. H= 2, D_m=?, D_c=?, VID 08. Frondel et al., 1976. Can. Min. 14 (1976), 429. See also Zippeite, Sodium-zippeite, Nickel-zippeite, Zinc-zippeite, Cobalt-zippeite.

Magnetite, Fe_3O_4, [G] Cub. Metallic black; brownish-grey in reflected light. Magnesia, Thessalia, Greece. H= 5.5-6.5, D_m= 5.175, D_c= 5.202, IVB 01. Haidinger, 1845. Dana, 7th ed.(1944), v.1, 698. Acta Cryst. C40 (1984), 1491. See also Jacobsite, Magnesioferrite.

Magnetoplumbite, $Pb(Fe,Mn)_{12}O_{19}$, [G] Hex. Grey-black. Långban mine, Filipstad (near), Värmland, Sweden. H= 6, D_m= 5.52, D_c= 5.71, IVC 05a. Aminoff, 1925. Am. Min. 36 (1951), 512. Am. Min. 74 (1989), 1194. See also Hibonite, Yimengite.

Magniotriplite, $(Mg,Fe,Mn)_2PO_4(F,OH)$, [G] Mon. Vitreous reddish-brown. Turkestan ridge, USSR. H= 4, D_m= 3.59, D_c= 3.68, VIIB 03a. Ginsburg et al., 1951. Bull. Min. 104 (1981), 672. Bull. Min. 104 (1981), 677. See also Triplite, Zwieselite.

Magnocolumbite, $(Mg,Fe,Mn)(Nb,Ta)_2O_6$, [G] Orth. Semi-metallic black, brownish-black. Kugi-Lyal, Pamir (SW), USSR. H=?, D_m= 5.17, D_c= 4.99, IVD 10a. Mathias et al., 1963. Am. Min. 48(1963), 1182(Abst.). See also Ferrocolumbite, Manganocolumbite.

Magnodravite, [D] Unnecessary name for magnesian dravite. Min. Mag. 36 (1968), 1144. Wang & Hsu, 1966.

Magnolite, Hg_2TeO_3, G Orth. Silky/adamantine colorless, creamy, brown, yellow-green. Keystone mine, Magnolia Dist., Boulder Co., Colorado, USA. H=?, D_m=?, D_c= 8.12, VIG 11. Genth, 1877. Can. Min. 27 (1989), 129. Can. Min. 27 (1989), 133.

Magnussonite, $Mn_5As_3O_9(OH,Cl)$, G Cub. Vitreous green, blue-green. Långban mine, Filipstad (near), Värmland, Sweden. H= 3.5-4, D_m= 4.30, D_c= 4.55, VIID 31. Gabrielson, 1956. Am. Min. 69 (1984), 800. Am. Min. 64 (1979), 390.

Maigruen, D Inadequately characterized. Min. Mag. 43 (1980), 1055. Geier & Ottemann, 1970.

Majakite, PdNiAs, A Hex. Metallic greyish-white in reflected light. Majak mine, Talnakh, Noril'sk (near), Siberia (N), USSR. H= 5, D_m= 9.33, D_c= 10.5, IIG 04. Genkin et al., 1976. Am. Min. 62(1977), 1260 (Abst.).

Majorite, $Mg_3(Fe,Al,Si)_2(SiO_4)_3$, A Cub. Translucent purple, pale yellowish-brown. Rawlina (near), Coorara Meteorite Crater, Western Australia, Australia. H=?, D_m=?, D_c= 3.76, VIIIA 06a. Smith & Mason, 1970. Can. Min. 15 (1977), 97.

Makatite, $Na_2Si_4O_8(OH)_2 \cdot 4H_2O$, A Mon. Translucent white. Lake Magadi, Rift Valley, Kenya. H=?, D_m= 2.03, D_c= 2.04, VIIIE 16. Sheppard & Gude, 1970. Am. Min. 55 (1970), 358. Zts. Krist. 159 (1982), 203.

Mäkinenite, γ–NiSe, G Rhom. Metallic orange-yellow. Kuusamo, Finland (NE). H= 3, D_m=?, D_c= 7.22, IIB 10. Vuorelainen et al., 1964. Am. Min. 50(1965), 519 (Abst.). See also Millerite.

Malachite, $Cu_2CO_3(OH)_2$, G Mon. Adamantine/vitreous/silky/earthy green. H= 3.5-4, D_m= 4.05, D_c= 4.030, VbB 01. Pliny, 77. Dana, 7th ed.(1951), v.2, 253. Zts. Krist. 145 (1977), 412. See also Pokrovskite.

Malanite, $Cu(Pt,Ir)_2S_4$, Q Cub. Metallic white. Unspecified locality, China. H=?, D_m=?, D_c= 5.83, IIC 01. Am. Min. 67(1982), 1081 (Abst.). See also Cuproiridsite, Cuprorhodsite, Dayingite.

Malayaite, $CaSnSiO_5$, G Mon. Pale creamy-yellow, deep orange-yellow. Sungei Lok, Chenderiang, Perak, Malaysia (Malaya). H=?, D_m= 4.55, D_c= 4.546, VIIIA 20. Alexander & Flinter, 1965. Min. Mag. 35 (1965), 622. Am. Min. 62 (1977), 801. See also Titanite.

Maldonite, Au_2Bi, G Cub. Metallic pinkish-white. Nuggety Reef, Maldon, Victoria, Australia. H= 4, D_m= 15.46, D_c= 15.70, IIG 02. Ulrich, 1869. ZVMO 105 (1976), 453.

Malladrite, Na_2SiF_6, G Rhom. Translucent pale rose, white. Mt. Vesuvius, Napoli, Campania, Italy. H=?, D_m= 2.755, D_c= 2.74, IIIB 02a. Zambonini & Carobbi, 1926. Dana, 7th ed.(1951), v.2, 105. Acta Cryst. 17 (1964), 1408.

Mallardite, $MnSO_4 \cdot 7H_2O$, G Mon. Vitreous colorless, white, pale rose. Lucky Boy mine, Butterfield Canyon, Salt Lake Co., Utah, USA. H= 2, D_m= 1.846, D_c= 1.838, VIC 03c. Carnot, 1879. Min. Abstr. 82M/4639.

Mammothite, $Cu_4Pb_6AlSbO_2(SO_4)_2Cl_4(OH)_{16}$, A Mon. Vitreous blue. Mammoth mine, Tiger, Pinal Co., Arizona, USA. H= 3, D_m=?, D_c= 5.25, IIID 04. Peacor et al., 1985. Min. Rec. 16 (1985), 117. Am. Min. 71(1986), 1548 (Abst.).

Manandonite, $LiAl_4(Si_2AlB)O_{10}(OH)_8$, G Orth. Pearly white. Antandrokomby pegmatite, Manandona River, Sahatany Valley, Madagascar. H=?, D_m= 2.78, D_c= 2.79, VIIIE 09b. Lacroix, 1912. Europ.Jour.Min. 1 (1989), 633.

Manasseite, $Mg_6Al_2CO_3(OH)_{16} \cdot 4H_2O$, G Hex. Pearly white, bluish-white. Snarum and Kongsberg, Norway; Amity, New York, USA. H= 2, D_m= 2.05, D_c= 2.00, IVF 03c. Frondel, 1941. Am. Min. 26 (1941), 295. See also Hydrotalcite.

Mandarinoite, $Fe_2(SeO_3)_3 \cdot 6H_2O$, A Mon. Vitreous/greasy light green. Pacajake mine, Colquechaca (Near), Potosí, Bolivia. H= 2.5, D_m= 2.93, D_c= 3.037, VIG 04. Dunn et al., 1978. Can. Min. 16 (1978), 605. Can. Min. 22 (1984), 475.

Manganarsite, $Mn_3As_2O_4(OH)_4$, [A] Hex. Vitreous light pinkish-brown. Långban mine, Filipstad (near), Värmland, Sweden. H= 3, D_m= 3.64, D_c= 3.60, VIID 31. Peacor et al., 1986. Am. Min. 71 (1986), 1517.

Manganaxinite, $Ca_2MnAl_2(BO_3OH)(SiO_3)_4$, [A] Tric. Translucent yellowish, dark greyish-brown. Harz, Germany. H= 6.5-7, D_m=?, D_c= 3.317, VIIIC 04. Fromme, 1909. DHZ, 2nd ed.(1986), v.1B, 603. See also Ferroaxinite, Tinzenite.

Manganbabingtonite, $Ca_2(Mn,Fe)FeSi_5O_{14}(OH)$, [G] Tric. Rudnyi Kaskad deposit, Sayan (E), USSR. H=?, D_m= 3.452, D_c= 3.59, VIIID 13. Vinogradova & Plyusinna, 1967. Am. Min. 53(1968), 1064 (Abst.). See also Babingtonite.

Manganbelyankinite, $(Mn,Ca)(Ti,Nb)_5O_{12}{\bullet}9H_2O$, [G] Amor. Resinous brownish-black. Lovozero massif, Kola Peninsula, USSR. H=?, D_m=?, D_c=?, IVF 17. Semenov, 1957. Am. Min. 43(1958), 1220 (Abst.). See also Belyankinite, Gerasimovskite.

Manganberzeliite, $(Ca,Na)_3(Mn,Mg)_2(AsO_4)_3$, [G] Cub. Resinous yellow-red. Långban mine, Filipstad (near), Värmland, Sweden. H= 4.5-5, D_m= 4.21, D_c= 4.379, VIIA 07. Igelström, 1894. Am. Min. 53 (1968), 316. See also Berzeliite, Garnet.

Manganese-hörnesite, $(Mn,Mg)_3(AsO_4)_2{\bullet}8H_2O$, [G] Mon. Långban mine, Filipstad (near), Värmland, Sweden. H=?, D_m=?, D_c= 2.76, VIIC 10. Gabrielson, 1951. Am. Min. 39(1954), 159 (Abst.). See also Bobierrite, Vivianite, Hörnesite.

Manganese-shadlunite, $(Fe,Cu)_8(Mn,Pb)S_8$, [A] Cub. Metallic greyish-yellow. Oktyabr deposit, Talnakh, Noril'sk (near), Siberia (N), USSR. H= 3.5, D_m=?, D_c= 4.44, IIA 07. Evstigneeva et al., 1973. ZVMO 102 (1973), 63. See also Shadlunite.

Manganhumite, $(Mn,Mg)_7(SiO_4)_3(OH)_2$, [A] Orth. Sub-adamantine pale/deep brownish-orange. Brattfors mine and Långban mine, Filipstad (near), Värmland, Sweden. H= 4, D_m= 3.83, D_c= 3.83, VIIIA 16. Moore, 1978. Min. Mag. 42 (1978), 133. Am. Min. 63 (1978), 874.

Manganite, γ-MnO(OH), [G] Mon. Sub-metallic grey, black. Ilfeld, Harz, Germany. H= 4, D_m= 4.33, D_c= 4.38, IVF 04. Haidinger, 1827. Dana, 7th ed.(1944), v.1, 646. Zts. Krist. 118 (1963), 303. See also Feitknechtite, Groutite.

Mangan-neptunite, $KNa_2Li(Mn,Fe)_2Ti_2Si_8O_{24}$, [G] Mon. Black. Kola Peninsula, USSR. H=?, D_m= 3.23, D_c= 3.26, VIIID 22. Kurbatov, 1923. JCPDS 29-823. See also Neptunite.

Manganochromite, $(Mn,Fe)(Cr,V)_2O_4$, [A] Cub. Brownish-grey in reflected light. Nairne deposit, Brukunga, Adelaide (E), South Australia, Australia. H= 6.5, D_m=?, D_c= 4.88, IVB 01. Graham, 1978. Am. Min. 63 (1978), 1166.

Manganocolumbite, $(Mn,Fe)(Nb,Ta)_2O_6$, [G] Orth. Black, brownish-black. H= 6, D_m= 5.20, D_c=?, IVD 10a. Dana, 1892. N. Jb. Min., Mh. (1985), 372. See also Manganotantalite, Ferrocolumbite, Magnocolumbite.

Manganolangbeinite, $K_2Mn_2(SO_4)_3$, [G] Cub. Translucent rose-red. Mt. Vesuvius, Napoli, Campania, Italy. H=?, D_m= 3.02, D_c= 3.06, VIA 02. Zambonini & Carobbi, 1924. NBS Monogr. 6 (1968), 43. See also Langbeinite.

Manganosite, MnO, [G] Cub. Vitreous green. Långban mine, Filipstad (near), Värmland, Sweden. H= 5.5, D_m= 5.364, D_c= 5.361, IVA 04. Blomstrand, 1874. Dana, 7th ed.(1944), v.1, 501. Acta Cryst. A36 (1980), 904.

Manganosteenstrupine, [D] Unnecessary name for steenstrupine-(Ce). Min. Mag. 33 (1962), 261. Vlasov et al., 1959.

Manganostibite, $(Mn,Fe)_7Sb(As,Si)O_{12}$, [G] Orth. Greasy black. Brattfors mine, Nordmark, Vårmland, Sweden. H=?, D_m= 4.949, D_c= 5.00, VIIA 15. Igelström, 1884. Am. Min. 73(1988), 667 (Abst.). Am. Min. 55 (1970), 1489.

Manganotantalite, $(Mn,Fe)(Ta,Nb)_2O_6$, [G] Orth. Black to brownish-black. Utö, Sweden. H= 6-6.5, D_m= 6.76, D_c= 7.073, IVD 10a. Nordenskiöld, 1877.

N. Jb. Min., Mh. (1985), 372. Can. Min. 14 (1976), 540. See also Manganotapiolite, Manganocolumbite, Ferrotantalite.

Manganotapiolite, $(Mn,Fe)(Ta,Nb)_2O_6$, A Tet. Brown; grey in reflected light. Viitaniemi, Eräjärvi, Orivesi, Finland. H= 5.8, D_m=?, D_c= 7.72, IVD 04. Lahti et al., 1983. Am. Min. 70(1985), 217 (Abst.). See also Manganotantalite, Staringite, Tapiolite.

Manganpyrosmalite, $(Mn,Fe)_8Si_6O_{15}(OH,Cl)_{10}$, G Rhom. Translucent brown. Sterling Hill mine, Ogdensburg, Sussex Co., New Jersey, USA. H= 4.5, D_m= 3.13, D_c= 3.03, VIIIE 12. Frondel & Bauer, 1953. Am. Min. 38 (1953), 755. Can. Min. 21 (1983), 1. See also Pyrosmalite, Brokenhillite.

Manganseverginite, D = manganaxinite. Min. Mag. 38 (1971), 103. Kurshakova, 1967.

Manjiroite, $(Na,K)Mn_8O_{16} \cdot nH_2O$, A Tet. Dull brownish-grey. Kohare mine, Iwate, Japan. H= 3.5, D_m= 4.29, D_c= 4.41, IVD 03b. Nambu & Tanida, 1967. Am. Min. 53(1968), 2103 (Abst.).

Mannardite, $BaTi_6(V,Cr)_2O_{16} \cdot H_2O$, A Tet. Adamantine black. Rough Claims, Sifton Pass (N of), Kechica River, British Columbia, Canada. H= 7, D_m= 4.12, D_c= 4.28, IVD 03b. Scott & Peatfield, 1986. Can. Min. 24 (1986), 55. Can. Min. 24 (1986), 67.

Mansfieldite, $AlAsO_4 \cdot 2H_2O$, G Orth. Vitreous/sub-resinous white, pale grey. Hobart Butte, Lane Co., Oregon, USA. H= 3.5-4, D_m= 3.02, D_c= 3.08, VIIC 05. Allen et al., 1948. JCPDS 23-123. See also Scorodite.

Mantienneite, $KMg_2Al_2Ti(PO_4)_4(OH)_3 \cdot 15H_2O$, A Orth. Bright yellow-brown. Anloua, Cameroun. H=?, D_m= 2.31, D_c= 2.25, VIID 29. Fransolet et al., 1984. Am. Min. 70(1985), 1330 (Abst.). See also Paulkerrite.

Mapimite, $Zn_2Fe_3(AsO_4)_3(OH)_4 \cdot 10H_2O$, A Mon. Vitreous blue, bluish-green, green. Ojuela mine, Mapimi, Mexico. H= 3, D_m= 2.95, D_c= 3.02, VIID 27. Cesbron et al., 1981. Am. Min. 67(1982), 623 (Abst.). Acta Cryst. B37 (1981), 1040.

Marcasite, FeS_2, G Orth. Metallic white/bronze-yellow. H= 6-6.5, D_m= 4.887, D_c= 4.875, IIC 07. Haidinger, 1845. Dana, 7th ed.(1944), v.1, 311. Struct. Repts. 35A (1970), 75. See also Pyrite.

Margaritasite, $(Cs,H_3O,K)_2(UO_2)_2(VO_4)_2 \cdot H_2O$, A Mon. Translucent yellow. Margarita's deposit, Chihuahua (near), Pena Blanca Dist., Chihuahua, Mexico. H=?, D_m=?, D_c= 5.40, VIID 23. Wenrich et al., 1982. Am. Min. 67 (1982), 1273. See also Carnotite, Tyuyamunite.

Margarite, $CaAl_2(Si_2Al_2)O_{10}(OH)_2$, G Mon. Translucent greyish-pink, pale yellow/green. Tyrol, Austria. H= 3.5-4.5, D_m= 3.0, D_c= 3.079, VIIIE 06. Fuchs, pre-1823. Am. Min. 60 (1975), 1023. Am. Min. 63 (1978), 186.

Margarosanite, $Ca_2PbSi_3O_9$, G Tric. Pearly colorless. Franklin, Sussex Co., New Jersey, USA. H= 2.5-3, D_m= 4.33, D_c= 4.359, VIIIC 02. Ford & Bradley, 1916. Dana/Ford (1932), 584. Zts. Krist. 128 (1969), 213.

Marialite, $Na_4(Si,Al)_{12}O_{24}Cl$, A Tet. Vitreous colorless, white. Pianura, Napoli (Naples), Italy. H= 5.5-6, D_m= 2.566, D_c= 2.543, VIIIF 09. vom Rath, 1866. Min. Mag. 51 (1987), 176. Acta Cryst. B29 (1973), 1272. See also Meionite.

Marićite, $NaFePO_4$, A Orth. Vitreous colorless, grey, pale brown. Big Fish River, Yukon Territory, Canada. H= 4-4.5, D_m= 3.66, D_c= 3.69, VIIA 02. Sturman et al., 1977. Can. Min. 15 (1977), 396. Can. Min. 15 (1977), 518.

Maricopaite, $Ca_2Pb_7(Si,Al)_{48}O_{100} \cdot 32H_2O$, A Orth. Silky/vitreous white, colorless. Moon Anchor mine, Tonopah (near), Maricopa Co., Arizona, USA. H=?, D_m= 2.94, D_c= 2.96, VIIIF 12. Peacor et al., 1988. Can. Min. 26 (1988), 309.

Marokite, $CaMn_2O_4$, A Orth. Black; brownish-grey in reflected light. Tachgagalt, Ouarzazate, Morocco. H=?, D_m= 4.64, D_c= 4.63, IVB 02a. Gaudefroy et al., 1963. Am. Min. 49(1964), 817 (Abst.). Bull. Min. 89 (1966), 318.

Marrite, $AgPbAsS_3$, [G] Mon. Metallic grey; white in reflected light. Lengenbach quarry, Binntal, Valais (Wallis), Switzerland. H= 3, D_m=?, D_c= 5.876, IID 07. Solly, 1904. Am. Min. 50(1965), 812 (Abst.). Zts. Krist. 125 (1967), 459. See also Freieslebenite, Laffittite.

Marshite, CuI, [G] Cub. Adamantine colorless, yellow. Broken Hill, New South Wales, Australia. H= 2.5, D_m= 5.68, D_c= 5.686, IIIA 01. Liversidge, 1892. Dana, 7th ed.(1951), v.2, 20. See also Miersite, Nantokite.

Marsturite, $NaCaMn_3Si_5O_{14}(OH)$, [A] Tric. Translucent white, light pink. Franklin, Sussex Co., New Jersey, USA. H= 6, D_m= 3.46, D_c= 3.465, VIIID 13. Peacor et al., 1978. Am. Min. 63 (1978), 1187.

Marthozite, $Cu(UO_2)_3(SeO_3)_3(OH)_2 \cdot 7H_2O$, [A] Orth. Translucent yellowish-green, greenish-brown. Musunoi, Shaba, Zaïre. H=?, D_m= 4.4, D_c= 4.7, VIG 03. Cesbron et al., 1969. Bull. Min. 92 (1969), 278.

Mascagnite, $(NH_4)_2SO_4$, [G] Orth. Vitreous/dull colorless, grey, yellowish-grey, yellow. Mt. Vesuvius and Mt. Ettna, Napoli, Campania, Italy. H= 2-2.5, D_m= 1.768, D_c= 1.77, VIA 05. Karsten, 1800. Per. Mineral. 54 (1985), 32. See also Arcanite.

Maslovite, $PtBiTe$, [A] Cub. Metallic light grey. Oktyabr deposit, Talnakh, Noril'sk (near), Siberia (N), USSR. H= 4-5, D_m=?, D_c= 11.81, IIC 06b. Kovalenker et al., 1979. Am. Min. 74 (1989), 1168. See also Michenerite, Testibiopalladite.

Massicot, β-PbO, [G] Orth. Greasy/dull yellow. H= 2, D_m= 9.56, D_c= 9.640, IVA 07. Huot, 1841. Dana, 7th ed.(1944), v.1, 516. Acta Cryst. C41 (1985), 1281. See also Litharge.

Masutomilite, $K(Li, Al, Mn)_3(Si, Al)_4O_{10}(F, OH)_2$, [A] Mon. Translucent pale purplish-pink, purple. Tanakamiyama, Gifu, Japan. H= 2.5-3, D_m= 2.92, D_c= 2.94, VIIIE 05b. Harada et al., 1976. Am. Min. 62(1977), 594 (Abst.). Min. Jour. 13 (1986), 13.

Masuyite, $UO_3 \cdot 2H_2O$, [G] Orth. Translucent reddish/brownish orange. Shinkolobwe, Shaba, Zaïre. H=?, D_m=?, D_c= 5.338, IVF 11. Vaes, 1947. Am. Min. 45 (1960), 1026.

Mathewrogersite, $Pb_7(Fe, Cu)Al_3GeSi_{12}O_{36}(OH, H_2O)_6$, [A] Rhom. Translucent/pearly colorless, white, pale greenish-yellow. Tsumeb, Namibia. H= 2, D_m= 4.7, D_c= 4.76, VIIIE 23. Keller & Dunn, 1986. Am. Min. 72(1987), 1025 (Abst.).

Mathiasite, $(K, Ca, Sr)(Ti, Cr, Fe, Mg)_{21}O_{38}$, [A] Rhom. Metallic black; tan in reflected light. Jagesfontein mine, Kimberlite deposits, Orange Free State, South Africa. H= 7.3, D_m=?, D_c= 4.39, IVC 06. Haggerty et al., 1983. Am. Min. 68 (1983), 494. Acta Cryst. C39 (1983), 421.

Matildite, $AgBiS_2$, [G] Hex. Metallic black/grey; white in reflected light. Matilda mine, Morococha (near), Peru. H= 2.5, D_m= 6.9, D_c= 6.99, IID 07. D'Achiardi, 1883. Can. Min. 9 (1969), 655. See also Bohdanowiczite, Volynskite, Schapbachite.

Matlockite, $PbClF$, [G] Tet. Adamantine colorless, yellow, pale amber, greenish. Cromford, Matlock (near), Derbyshire, England. H= 2.5-3, D_m= 7.12, D_c= 7.16, IIIC 16. Greg, 1851. Dana, 7th ed.(1951), v.2, 59.

Matorolite, [D] = chalcedonic quartz. Min. Mag. 38 (1971), 103.

Matraite, ZnS, [P] Rhom. Transparent. Matra Mts., Gyöngyösoroszi, Hungary. H=?, D_m=?, D_c= 4.13, IIB 06. Koch, 1958. Am. Min. 45(1960), 1131 (Abst.). See also Sphalerite, Wurtzite.

Mattagamite, $(Co, Fe)Te_2$, [A] Orth. Metallic violet-grey. Mattagami Lake mine, Galinée Twp., Abitibi Co., Québec, Canada. H= 4.5-5.5, D_m=?, D_c= 7.92, IIC 07. Thorpe & Harris, 1973. Can. Min. 12 (1973), 55. Struct. Repts. 35A (1970), 75. See also Frohbergite.

Matteuccite, $NaH(SO_4) \cdot H_2O$, [G] Mon. Mt. Vesuvius, Napoli, Campania, Italy. H=?, D_m= 2.118, D_c= 2.108, VIC 13. Carobbi & Cipriani, 1952. Am. Min. 39(1954), 848 (Abst.). Struct. Repts. 43A (1977), 277.

Mattheddleite, $Pb_5(SiO_4, SO_4)_3Cl$, [A] Hex. Adamantine creamy-white. Leadhills, Lanarkshire, Scotland. H=?, D_m=?, D_c= 6.96, VIIIA 29. Livingstone et al., 1987. Am. Min. 73(1988), 929 (Abst.).

Matulaite, $CaAl_{18}(PO_4)_{12}(OH)_{20} \cdot 28H_2O$, [A] Mon. Pearly colorless, white. Bachman mine, Hellertown, Northampton Co., Pennsylvania, USA. H= 1, D_m= 2.330, D_c=?, VIID 06. Moore & Ito, 1980. Am. Min. 65(1980), 1067 (Abst.).

Maucherite, $Ni_{11}As_8$, [G] Tet. Metallic grey; pinkish grey in reflected light. Kupferschiefer of Eisleben, Thüringen, Germany. H= 5, D_m= 8.00, D_c= 8.04, IIG 03. Grünling, 1913. Dana, 7th ed.(1944), v.1, 192. Am. Min. 58 (1973), 203.

Maufite, $(Mg, Ni)Al_4Si_3O_{13} \cdot 4H_2O$(?), [Q] Syst=? Translucent green. Lomagundi Dist., Ruorka Ranch, Zimbabwe. H= 3, D_m= 2.27, D_c=?, VIIIE 10b. Keep, 1930. Min. Abstr. 4 (1930), 248.

Mawbyite, $(Fe, Zn)_2Pb(AsO_4)_2(OH, H_2O)_2$, [A] Mon. Adamantine pale brown, reddish brown. Kintore open cut, Broken Hill, New South Wales, Australia. H= 4, D_m=?, D_c= 5.53, VIIC 17. Pring et al., 1989. Am. Min. 74 (1989), 1377.

Mawsonite, $Cu_6Fe_2SnS_8$, [A] Tet. Metallic orange. Mt. Lyell, Tasmania and Tingha, New South Wales, Australia. H= 3.5, D_m=?, D_c= 4.65, IIB 03b. Markham & Lawrence, 1965. Am. Min. 50 (1965), 900. Can. Min. 14 (1976), 529. See also Chatkalite.

Mayenite, $Ca_{12}Al_{14}O_{33}$, [A] Cub. Clear colorless. Ettringer Bellerberg, Mayen, Eifel, Rheinland-Pfalz, Germany. H=?, D_m= 2.85, D_c= 2.68, IVA 08. Hentschel, 1964. Am. Min. 50(1965), 2106 (Abst.).

Mazzite, $K_2CaMg_2(Si, Al)_{36}O_{72} \cdot 28H_2O$, [A] Hex. Mont Semiol (Mt. Semiouse), Montbrison, Loire, France. H=?, D_m=?, D_c= 2.108, VIIIF 14. Galli et al., 1974. Am. Min. 60(1975), 340 (Abst.). Min. Abstr. 76-3282.

Mbobomkulite, $(Ni, Cu)Al_4(NO_3, SO_4)_2(OH)_{12} \cdot 3H_2O$, [A] Mon. Powdery sky-blue. Mbobo Mkulu Cave, Transvaal (E), South Africa. H=?, D_m= 2.30, D_c= 2.344, IVF 03e. Martini, 1980. Am. Min. 67(1982), 415 (Abst.). See also Chalcoalumite, Nickelalumite.

Mcallisterite, $Mg_2[B_6O_7(OH)_6]_2 \cdot 9H_2O$, [A] Rhom. Translucent white. Furnace Creek Wash, Death Valley, Inyo Co., California, USA. H= 2.5, D_m= 1.868, D_c= 1.864, Vc 11. Schaller et al., 1965. Am. Min. 50 (1965), 629. Struct. Repts. 41A (1975), 421. See also Admontite.

Mcauslanite, $Fe_3Al_2(PO_4)_3(PO_3OH)F \cdot 18H_2O$, [A] Tric. Vitreous yellowish-white. East Kemptville Tin mine, Yarmouth Co., Nova Scotia, Canada. H= 3.5, D_m= 2.22, D_c= 2.17, VIID 06. Richardson et al., 1988. Can. Min. 26 (1988), 917.

Mcbirneyite, $Cu_3(VO_4)_2$, [A] Tric. Metallic dark grey. Izalco volcano, El Salvador. H=?, D_m=?, D_c= 4.50, VIIA 08. Hughes et al., 1987. Am. Min. 73(1988), 1495 (Abst.).

Mcconnellite, $CuCrO_2$, [A] Rhom. Translucent red. Merume River, Mazaruni district, Guayana. H=?, D_m= 5.49, D_c= 5.61, IVA 06. Milton et al., 1976. Am. Min. 62(1977), 593 (Abst.). See also Delafossite.

Mcgillite, $(Mn, Fe)_8Si_6O_{15}(OH)_8Cl_2$, [A] Mon. Pearly light to dark pink. Sullivan mine, Kimberley, British Columbia, Canada. H= 4.2, D_m= 2.98, D_c= 3.08, VIIIE 12. Donnay et al., 1980. Can. Min. 18 (1980), 31. Can. Min. 21 (1983), 7. See also Friedelite.

Mcgovernite, $Mg_4Mn_9Zn_2As_2Si_2O_{17}(OH)_{14}$, [G] Rhom. Translucent reddish/bronze. Stirling Hill mine, Ogdensburg, Sussex Co., New Jersey, USA. H=?, D_m= 3.719, D_c= 3.575, VIIB 10c. Palache & Bauer, 1927. Am. Min. 45 (1960), 937.

Mcguinnessite, $(Mg,Cu)_2CO_3(OH)_2$, [A] Mon. Vitreous/silky blue-green/white. Red Mountain (locality #2), Mendocino Co., California, USA. H= 2.5, D_m= 3.02, D_c= 3.076, VbB 01. Erd et al., 1981. Am. Min. 66(1981), 1276 (Abst.). See also Pokrovskite.

Mckelveyite-(Y), $NaBa_3(Ca,U)Y(CO_3)_6 \cdot 3H_2O$, [G] Tric. Translucent green, black. Green River Formation, Uintah Co., Wyoming, USA. H=?, D_m= 3.62, D_c= 3.5, VbC 03a. Milton et al., 1965. Can. Min. 16 (1978), 335. See also Donnayite-(Y), Weloganite.

Mckinstryite, $(Ag,Cu)_2S$, [G] Orth. Metallic grey. Foster mine, Coleman Twp., Cobalt, Ontario, Canada. H=?, D_m=?, D_c= 6.61, IIA 04. Skinner et al., 1966. Econ. Geol. 61 (1966), 1383.

Mcnearite, $NaCa_5(AsO_4)(AsO_3OH)_4 \cdot 4H_2O$, [A] Tric. Pearly white. Ste.-Marie-aux-Mines, Vosges Mtns., Haut-Rhin, Alsace, France. H=?, D_m= 2.60, D_c= 2.85, VIIC 11. Sarp et al., 1981. Am. Min. 67(1982), 856 (Abst.).

Medaite, $(Mn,Ca)_6(V,As)Si_5O_{18}(OH)$, [A] Mon. Sub-adamantine brownish-red. Molinello mine, Val Graveglia, Liguria, Italy. H=?, D_m= 3.70, D_c= 3.727, VIIIB 23. Gramaccioli et al., 1982. Am. Min. 67 (1982), 85. Acta Cryst. B37 (1981), 1972.

Medmontite, [D] A mixture of chrysocolla and mica. Am. Min. 54 (1969), 994 (Abst.). Chukhrov & Anosov, 1951.

Meionite, $Ca_4(Si,Al)_{12}O_{24}(CO_3)$, [A] Tet. Vitreous colorless, white. Monte Somma, Mt. Vesuvius, Napoli, Campania, Italy. H= 5.5-6, D_m= 2.72, D_c= 2.76, VIIIF 09. Haüy, 1801. Min. Mag. 51 (1987), 176. N.Jb.Min.Abh. 149 (1984), 309. See also Marialite.

Meixnerite, $Mg_6Al_2(OH)_{18} \cdot 4H_2O$, [A] Rhom. Transparent colorless. Ybbs-Persenberg, Austria. H=?, D_m=?, D_c= 1.953, IVF 03b. Koritnig & Süsse, 1975. Am. Min. 61(1976), 176 (Abst.). See also Hydrotalcite.

Melanocerite-(Ce), $(Ce,Ca)_5(Si,B)_3O_{12}(OH,F) \cdot nH_2O(?)$, [G] Amor. Greasy/vitreous brown, black. Kjeö, Barkevik (near), Langesundfjord, Norway (S). H= 5-6, D_m= 4.13, D_c=?, VIIIA 23. Brögger, 1887. Dokl. Earth Sci. 185(1969),107.

Melanophlogite, $C_2H_{17}O_5 \cdot Si_{46}O_{92}$, [G] Tet. Transparent colorless, white. Giona mine, Girgenti (Recalmuto), Sicilia (Sicily), Italy. H= 5.7, D_m= 2.00, D_c= 2.30, IVD 01. Lasaulx, 1876. Am. Min. 57 (1972), 779.

Melanostibite, $Mn(Sb,Fe)O_3$, [G] Rhom. Black. Sjögruvan, Grythyttan (near), Örebro, Sweden. H= 4, D_m=?, D_c= 5.63, IVC 04b. Igelström, 1892. Am. Min. 55 (1970), 1489. Am. Min. 53 (1968), 1104.

Melanotekite, $Pb_2Fe_2O_2(Si_2O_7)$, [G] Orth. Metallic/greasy black, blackish-grey. Långban mine, Filipstad (near), Värmland, Sweden. H= 6.5, D_m= 6.19, D_c= 6.21, VIIIB 14. Lindström, 1880. Am. Min. 52 (1967), 1085. Struct. Repts. 27 (1962), 714. See also Kentrolite.

Melanothallite, Cu_2OCl_2, [G] Orth. Vitreous black/bluish-black. Mt. Vesuvius, Napoli, Campania, Italy. H=?, D_m=?, D_c= 4.08, IIIC 01. Scacchi, 1870. Am. Min. 68(1983), 852 (Abst.).

Melanovanadite, $CaV_4O_{10} \cdot 5H_2O$, [G] Tric. Sub-metallic black. Minasragra, Junin, Peru. H= 2.5, D_m= 2.55, D_c= 2.53, IVF 09. Lindgren, 1921. Am. Min. 70 (1985), 644. Am. Min. 72 (1987), 637.

Melanterite, $FeSO_4 \cdot 7H_2O$, [G] Mon. Vitreous green, greenish-blue, blue. H= 2, D_m= 1.895, D_c= 1.897, VIC 03c. Haidinger, 1850. Dana, 7th ed.(1951), v.2, 499. Acta Cryst. 17 (1964), 1167.

Melilite, $(Ca,Na)_2(Al,Mg)(Si,Al)_2O_7$, [G] Tet. Translucent white, yellow, brown, greenish-brown. Capo di Bove, Roma (near), Italy. H= 5-6, D_m= 3.0, D_c= 3.09, VIIIB 02a. Delamétherie, 1796. DHZ, 2nd ed.(1986), v.1B, 285. Am. Min. 38 (1953), 643. See also Akermanite, Gehlenite.

Meliphanite, $Na(Na,Ca)BeSi_2O_6F$, [G] Tet. Vitreous yellow, red. Frediksvärn (Near), Norway. H= 5-5.5, D_m= 3.012, D_c= 3.024, VIIIB 02b. Scheerer, 1852. Dana, 6th ed.(1892), 418. Acta Cryst. 23 (1967), 260. See also Melilite.

Melkovite, $CaFe_2Mo_5O_{10}(PO_4)_2(OH)_{12}\cdot 8H_2O$, [A] Mon. Dull/waxy yellow, brownish-yellow. Shunak Mts., Kazakhstan, USSR. H= 3, D_m= 2.971, D_c= 2.512, VIID 15b. Egorov et al., 1969. Am. Min. 55(1970), 320 (Abst.). Inorg. Chem. 16 (1977), 1096.

Mellite, $Al_2C_6(COO)_6\cdot 16H_2O$, [G] Tet. Resinous/vitreous yellow, reddish, brownish, white. Arten, Thüringen, Germany. H= 2-2.5, D_m= 1.64, D_c= 1.605, IXA 02. Gmelin, 1793. Powd. Diff. 4 (1989), 172. Acta Cryst. B29 (1973), 26.

Melonite, $NiTe_2$, [G] Rhom. Metallic light pink. Stanislaus mine, Carson Hill, Calaveras Co., California, USA. H= 2, D_m= 7.72, D_c= 7.74, IIC 11. Genth, 1868. Am. Min. 34 (1949), 342.

Melonjosephite, $CaFe_2(PO_4)_2(OH)$, [A] Orth. Brilliant/resinous dark green. Pegmatite D'Angarf-sud, Anti-atlas, Plaine des Zenaga, Morocco. H= < 5, D_m= 3.65, D_c= 3.615, VIIB 03c. Fransolet, 1973. Am. Min. 60(1975), 946 (Abst.). Am. Min. 62 (1977), 60.

Mendipite, $Pb_3O_2Cl_2$, [G] Orth. Resinous/adamantine colorless, white, grey. Mendip Hills, Somersetshire, England. H= 2.5, D_m= 7.24, D_c= 7.3, IIIC 07. Glocker, 1839. Dana, 7th ed.(1951), v.2, 56. Bull. Min. 94 (1971), 323.

Mendozavilite, $NaCa_2Fe_6(PO_4)_2(PMo_{11}O_{39})(OH,Cl)_{10}\cdot 33H_2O$, [A] Syst=? Vitreous yellow, orange. Cumobabi, Sonora, Mexico. H= 1.5, D_m= 3.85, D_c=?, VIID 15b. Williams, 1986. Am. Min. 73(1988), 193 (Abst.).

Mendozite, $NaAl(SO_4)_2\cdot 11H_2O$, [G] Mon. Transparent white. St. Juan, Mendoza (near), Argentina. H= 3, D_m= 1.765, D_c= 1.781, VIC 07. Dana, 1868. Dana, 7th ed.(1951), v.2, 469. Am. Min. 57 (1972), 1081.

Meneghinite, $CuPb_{13}Sb_7S_{24}$, [G] Orth. Metallic grey; white in reflected light. Bottino mine, Serravezza, Alpe Apuane, Toscana, Italy. H= 2.5, D_m= 6.36, D_c= 6.44, IID 08. Bechi, 1852. Dana, 7th ed.(1944), v.1, 402. Zts. Krist. 113 (1960), 345.

Mengxianminite, $(Ca,Na)_3Mg_2(Fe,Mn)_2(Sn,Zn)_5Al_8O_{29}$, [Q] Orth. Vitreous brownish-green. Unspecified locality, China. H= 6, D_m= 3.85, D_c=?, IVC 15. Huang et al., 1986. IMA 1986, Abstr. p. 130.

Mercallite, $KHSO_4$, [G] Orth. Vitreous colorless, blue. Mt. Vesuvius, Napoli, Campania, Italy. H=?, D_m= 2.329, D_c= 2.322, VIA 04. Carobbi, 1935. Dana, 7th ed.(1951), v.2, 395. Acta Cryst. B31 (1975), 302.

Mercury, Hg, [G] Rhom. Metallic grey liquid. H=?, D_m= 13.60, D_c= 14.26, IA 06. Dana, 7th ed., v.1 (1944), 110. Acta Cryst. 19 (1965), 807.

Merenskyite, $(Pd,Pt)(Te,Bi)_2$, [A] Rhom. Metallic white. Rustenburg mine, Pretoria, Transvaal, South Africa. H= 3.5-4, D_m=?, D_c= 8.29, IIC 11. Kingston, 1966. Min. Mag. 35 (1966), 815.

Merlinoite, $(K,Na)_5(Ba,Ca)_2(Si_{23}Al_9)O_{64}\cdot 24H_2O$, [A] Orth. Translucent white. Rieti, Santa Rufina (near), Cupaello Quarry, Lazio (Latium), Italy. H=?, D_m= 2.14, D_c= 2.19, VIIIF 14. Pasaglia et al., 1977. Natural Zeolites (1985), 155. N.Jb.Min.Mh. (1979), 1.

Merrihueite, $(K,Na)_2(Fe,Mg)_5Si_{12}O_{30}$, [A] Hex. Translucent greenish-blue. Mozö-Madaras meteorite. H=?, D_m=?, D_c= 2.636, VIIIC 10. Dodd et al., 1965. Am. Min. 50 (1965), 2096. Acta Cryst. B28 (1972), 267.

Mertieite-I, $Pd_{11}(Sb,As)_4$, [A] Hex. Metallic brassy yellow. Goodnews Bay, Alaska, USA. H= 5.5, D_m=?, D_c=?, IIG 04. Desborough et al., 1973. Am. Min. 58 (1973), 1. See also Isomertieite.

Mertieite-II, $Pd_8(Sb,As)_3$, [P] Rhom. Metallic brassy yellow. Goodnews Bay, Alaska, USA. H=?, D_m=?, D_c= 11.2, IIG 04. Desborough et al, 1973. Can. Min. 13 (1975), 321.

Merwinite, $Ca_3Mg(SiO_4)_2$, [G] Mon. Translucent light green. Wet Weather Quarry, Crestmore, Riverside (near), Riverside Co., California, USA. H= 6, D_m= 3.15, D_c= 3.32, VIIIA 04. Larsen & Foshag, 1921. Am. Min. 6 (1921), 143. Am. Min. 57 (1972), 1355.

Mesolite, $Na_2Ca_2(Al_6Si_9)O_{30} \cdot 8H_2O$, [G] Orth. Vitreous/silky white, colorless, greyish, yellowish. Nova Scotia, Canada. H= 5, D_m= 2.3, D_c= 2.27, VIIIF 10. Fuchs & Gehlen, 1816. DHZ (1963), v.4, 358. Acta Cryst. C42 (1986), 937.

Messelite, $Ca_2(Fe,Mn)(PO_4)_2 \cdot 2H_2O$, [Q] Syst=? Vitreous greenish white/grey. Messel (near), Hessen, Germany. H= 3.5, D_m= 3.16, D_c=?, VIIC 12. Muthmann, 1890. Am. Min. 44(1959), 469 (Abst.).

Meta-aluminite, $Al_2SO_4(OH)_4 \cdot 5H_2O$, [A] Mon. Silky white. Fuemrole mine, Temple Mt., Emery Co., Utah, USA. H=?, D_m= 2.18, D_c= 2.17, VID 03. Frondel, 1968. Zts. Krist. 151 (1980), 141. See also Aluminite.

Meta-alunogen, $Al_2(SO_4)_3 \cdot 14H_2O$, [G] Orth. Waxy/pearly colorless. Francisco de Vergara, Chile. H=?, D_m=?, D_c=?, VIC 04. Gordon, 1942. JCPDS 22-23. See also Alunogen.

Meta-ankoleite, $K_2(UO_2)_2(PO_4)_2 \cdot 6H_2O$, [A] Tet. Translucent yellow. Mungenyi pegmatite, Ankole district, Uganda (SW). H=?, D_m=?, D_c= 3.54, VIID 20b. Gallather & Atkin, 1966. Am. Min. 52(1967), 560 (Abst.).

Meta-autunite, $Ca(UO_2)_2(PO_4)_2 \cdot 6H_2O$, [G] Tet. Vitreous light/dark green, yellow, greenish-yellow. H= 2-2.5, D_m= 3.44, D_c= 3.58, VIID 20b. Gaubert, 1904. Am. Min. 48 (1963), 1389. Struct. Repts. 24 (1960). 412. See also Autunite.

Metaberyllite, $Be_3SiO_4(OH)_2 \cdot H_2O$, [Q] Syst=? Lovozero massif, Kola Peninsula, USSR. H=?, D_m=?, D_c=?, VIIIA 13. Semenov, 1972. Min. Mag. 39(1974), 920 (Abst.). See also Beryllite.

Metaborite, γ–HBO_2, [G] Cub. Vitreous colorless, brownish. Unspecified locality, Kazakhstan(?), USSR. H= 5, D_m= 2.47, D_c= 2.487, Vc 04a. Lobanova & Avrova, 1964. Am. Min. 50(1965), 261 (Abst.). Acta Cryst. 16 (1963), 380.

Metacalciouranoite, $(Ca,Na,Ba)U_2O_7 \cdot 2H_2O$, [A] Syst=? Translucent orange. Unspecified locality, USSR. H=?, D_m= 4.90, D_c=?, IVF 12. Rogova et al., 1973. Am. Min. 58(1973), 1111 (Abst.).

Metacinnabar, HgS, [G] Cub. Metallic greyish-black; greyish-white in reflected light. Reddington mine, Lake Co., California, USA. H= 3, D_m= 7.65, D_c= 7.632, IIB 01a. Moore, 1870. Dana, 7th ed.(1944), v.1, 215. Struct. Repts. 29 (1964), 67. See also Cinnabar, Hypercinnabar, Tiemannite.

Metadelrioite, $SrCa(VO_3OH)_2$, [A] Tric. Translucent pale yellow-green. Jo Dandy mine, Montrose Co., Colorado, USA. H= 2, D_m= 4.3, D_c= 4.21, VIID 32. Smith, 1970. Am. Min. 55 (1970), 185.

Metadomeykite, β–Cu_3As, [Q] Hex. Metallic white/grey. H=?, D_m= 7.20, D_c= 8.13, IIG 01. Mikheyen, 1949. JCPDS 14-454. See also Domeykite.

Meta-haiweeite, $Ca(UO_2)_2Si_6O_{15} \cdot nH_2O$, [Q] Syst=? Haiwee Reservoir, Coso Mts., Inyo Co., California, USA. H=?, D_m=?, D_c=?, VIIIA 25. McBurney & Murdoch, 1959. Am. Min. 44 (1959), 839. See also Haiweeite.

Metaheinrichite, $Ba(UO_2)_2(AsO_4)_2 \cdot 8H_2O$, [G] Tet. Vitreous/pearly yellow, green. White King mine, Lakeview (near), Lake Co., Oregon, USA. H= 2.5, D_m= 4.04, D_c= 4.09, VIID 20b. Gross et al., 1958. Am. Min. 43 (1958), 1134. See also Heinrichite.

Metahewettite, $CaV_6O_{16} \cdot 3H_2O$, [G] Mon. Dull/silky red. Jo Dandy Claim, East Paradox Valley, Montrose Co., Colorado, USA; Henry Mountains, Utah, USA; Thompsons (SE), Yellow Cat Creek, Utah, USA. H=?, D_m= 2.94, D_c= 3.05, IVF 09. Hillebrand et al., 1914. Min. Mag. 43 (1979), 550.

Metahohmannite, $Fe_2(SO_4)_2(OH)_2 \cdot 3H_2O$, [Q] Syst=? Pulverulent orange. Chile. H=?, D_m=?, D_c=?, VID 02. Bandy, 1938. Dana, 7th ed.(1951), v.2, 608.

Metajennite, [D] An artificial product. Min. Mag. 36 (1968), 1144. Carpenter et al., 1966.

Metakahlerite, $Fe(UO_2)_2(AsO_4)_2 \cdot 8H_2O$, [G] Tet. Translucent yellow. Sophia mine, Wittichen, Schwarzwald, Germany. H=?, D_m= 3.73, D_c= 3.77, VIID 20b. Walenta, 1958. Am. Min. 71 (1986), 1037.

Metakirchheimerite, $Co(UO_2)_2(AsO_4)_2 \cdot 8H_2O$, [G] Tet. Translucent pale rose. Sophia mine, Wittichen, Schwarzwald, Germany. H= 2-2.5, D_m=?, D_c= 3.97, VIID 20b. Walenta, 1958. Strunz (1970), 353.

Metaköttigite, $(Zn,Fe)_3(AsO_4)_2 \cdot 8(H_2O,OH)$, [A] Tric. Translucent bluish-grey. Mapimi, Durango, Mexico. H=?, D_m=?, D_c= 3.03, VIIC 07. Schmetzer et al., 1982. Am. Min. 68(1983), 1039 (Abst.). See also Metavivianite, Symplesite.

Metaliebigite, [D] Inadequate data. Min. Mag. 38 (1971), 103. Kiss, 1966.

Metalodevite, $Zn(UO_2)_2(AsO_4)_2 \cdot 10H_2O$, [A] Tet. Translucent pale yellow, olive. Lodève, Hérault, France. H=?, D_m=?, D_c= 4.00, VIID 20b. Agrinier et al., 1972. Am. Min. 59(1974), 210 (Abst.).

Metalomonosovite, [D] = beta-lomonosovite. Am. Min. 48(1963), 1413 (Abst.). Semenov et al., 1963.

Metamurmanite, [D] Inadequate data. Min. Mag. 36 (1967), 133. Semenov et al., 1961.

Metanováčekite, $Mg(UO_2)_2(AsO_4)_2 \cdot 4H_2O$, [G] Tet. Anton mine, Schwarzwald, Germany. H=?, D_m=?, D_c= 3.48, VIID 20b. Frondel, 1951. Strunz (1970), 352.

Metarossite, $Ca(VO_3)_2 \cdot 2H_2O$, [G] Tric. Dull/pearly light yellow. Wm. O'Neills Claim, Dolores River, Bull Pen Canyon, San Miguel Co., Colorado, USA. H=?, D_m= 2.45, D_c= 2.80, VIIC 21. Foshag & Hess, 1927. Am. Min. 37 (1952), 407. Can. Min. 6 (1960), 448.

Metaschoderite, $Al(PO_4,VO_4) \cdot 3H_2O$, [G] Mon. Translucent yellow-orange. Van Hansand claim, Fish Creek Range, Eureka (S), Nevada, USA. H= 2, D_m= 1.88, D_c= 1.61, VIIC 24. Hausen, 1962. Am. Min. 47 (1962), 637.

Metaschoepite, $UO_3 \cdot 1\text{–}2H_2O$, [G] Orth. Translucent yellow. Shinkolobwe, Shaba, Zaïre. H=?, D_m=?, D_c= 4.69, IVF 13. Christ & Clark, 1960. Am. Min. 50 (1965), 235.

Metasideronatrite, $Na_2Fe(SO_4)_2(OH) \cdot 2H_2O$, [G] Orth. Silky yellow. Sierro Gordo, Chuquicamata, Antofagasta, Chile. H= 2.5, D_m=?, D_c= 2.76, VID 13. Bandy, 1938. N.Jb.Min.Mh. (1982), 255.

Meta-Na-uranospinite, $Na_2(UO_2)_2(AsO_4)_2 \cdot 8H_2O$, [Q] Tet. H=?, D_m=?, D_c= 3.83, VIID 20b. Kopchenova et al., 1957. Strunz (1970), 353. See also Metauranospinite, Uranospinite, Sodium-uranospinite.

Metastibnite, Sb_2S_3, [Q] Amor. Translucent red. Steamboat Springs, Washoe Co., Nevada, USA. H=?, D_m=?, D_c=?, IIC 02. Becker, 1888. Min. Abstr. 81-1855. See also Stibnite.

Metastudtite, $UO_4 \cdot 2H_2O$, [A] Orth. Silky yellowish. Shinkolobwe, Shaba, Zaïre. H=?, D_m=?, D_c= 4.67, IVF 11. Deliens & Piret, 1983. Am. Min. 68 (1983), 456.

Metaswitzerite, $(Mn,Fe)_3(PO_4)_2 \cdot 4H_2O$, [A] Mon. Pearly/adamantine pale pink, light golden-brown, brown. Foote Mineral Company spodumene mine, Kings Mt. (near), Cleveland Co., North Carolina, USA. H= 2.5, D_m= 2.95, D_c= 2.96, VIIC 07. Leavens & White, 1967. Am. Min. 71 (1986), 1221. Min. Abstr. 81-1248. See also Switzerite.

Metatorbernite, $Cu(UO_2)_2(PO_4)_2 \cdot 8H_2O$, [G] Tet. Vitreous/sub-adamantine pale/dark green. Schneeberg, Sachsen (Saxony), Germany. H= 2.5, D_m= 3.7,

D_c= 3.70, VIID 20b. Werner, 1786. Dana, 7th ed.(1951), v.2, 991. Am. Min. 49 (1964), 1603.

Metatyuyamunite, $Ca(UO_2)_2(VO_4)_2 \cdot 3H_2O$, G Orth. Adamantine yellow, greenish-yellow. Jo Dandy mine, Montrose Co., Colorado, USA. H= 2, D_m= 3.8, D_c= 3.81, VIID 23. Stern et al., 1956. Am. Min. 41 (1956), 187.

Meta-uramphite, $(NH_4)_2(UO_2)_2(PO_4)_2 \cdot 6H_2O$, Q Tet. H=?, D_m=?, D_c=?, VIID 20b. Nekrasova, 1957. Strunz (1970), 352. See also Uramphite.

Meta-uranocircite I, $Ba(UO_2)_2(PO_4)_2 \cdot 8H_2O$, G Tet. Translucent, pearly yellow-green. H= 2-2.5, D_m= 4.08, D_c= 4.06, VIID 20b. Gaubert, 1904. Am. Min. 38 (1953), 476. See also Uranocircite.

Meta-uranocircite II, $Ba(UO_2)_2(PO_4)_2 \cdot 6H_2O$, Q Mon. Bergen, Falkenstein, Voigtland, Sachsen (Saxony), Germany. H=?, D_m=?, D_c= 3.994, VIID 20b. Walenta, 1964. Am. Min. 68(1983), 472 (Abst.). Struct. Repts. 49A (1982), 274. See also Uranocircite.

Meta-uranopilite, $(UO_2)_6SO_4(OH)_{10} \cdot 5H_2O$, Q Syst=? Translucent yellow, greyish, brown, green. Jáchymov (St. Joachimsthal), Západočeský kraj, Čechy (Bohemia), Czechoslovakia. H=?, D_m=?, D_c=?, VID 08. Frondel, 1952. Am. Min. 37 (1952), 950. See also Uranopilite.

Meta-uranospinite, $Ca(UO_2)_2(AsO_4)_2 \cdot 8H_2O$, G Tet. Sophia mine, Wittichen, Schwarzwald, Germany. H=?, D_m=?, D_c= 3.65, VIID 20b. Walenta, 1958. Am. Min. 45(1960), 254 (Abst.). See also Uranospinite.

Metavandendriesscheite, $PbU_7O_{22} \cdot nH_2O$, G Orth. Translucent yellowish-orange, orange. Shinkolobwe, Shaba, Zaïre. H=?, D_m= 5.45, D_c=?, IVF 13. Christ & Clark, 1960. Am. Min. 45 (1960), 1026. See also Vandendriesscheite.

Metavanmeersscheite, $U(UO_2)_3(PO_4)_2(OH)_6 \cdot 2H_2O$, A Orth. Translucent yellow. Kobokobo pegmatite, Kivu, Zaïre. H=?, D_m=?, D_c= 4.49, VIID 21. Piret & Deliens, 1982. Bull. Min. 105 (1982), 125.

Metavanuralite, $Al(UO_2)_2(VO_4)_2(OH) \cdot 8H_2O$, A Tric. Translucent yellow. Mounana mine, Franceville, Haut-Ogooué, Gabon. H=?, D_m=?, D_c= 3.66, VIID 23. Cesbron, 1970. Bull. Min. 93 (1970), 242.

Metavariscite, $AlPO_4 \cdot 2H_2O$, G Mon. Vitreous pale green. Edison-Bird mine, Utahlite Hill, Lucin (NW), Box Elder Co., Utah, USA. H= 3.5, D_m= 2.54, D_c= 2.535, VIIC 05. Larsen & Schaller, 1925. Dana, 7th ed.(1951), v.2, 769. Acta Cryst. B29 (1973), 2292. See also Variscite, Kolbeckite, Phosphosiderite.

Metavauxite, $FeAl_2(PO_4)_2(OH)_2 \cdot 8H_2O$, G Mon. Vitreous/silky colorless, white, pale green. Llallagua, Potosí, Bolivia. H= 3, D_m= 2.345, D_c= 2.36, VIID 03. Gordon, 1922. Dana, 7th ed.(1951), v.2, 971. Struct. Repts. 32A (1967), 367. See also Paravauxite.

Metavivianite, $Fe_3(PO_4)_2(OH)_x \cdot 6H_2O$, A Tric. Opaque/translucent green. Big Chief Quarry, Glendale (near), Pennington Co., South Dakota, USA. H=?, D_m=?, D_c= 2.69, VIIC 07. Ritz et al., 1974. Min. Mag. 50 (1986), 687. See also Metaköttigite, Symplesite.

Metavoltine, $K_2Na_6Fe_7O_2(SO_4)_{12} \cdot 18H_2O$, G Rhom. Resinous yellowish-brown, orange-brown, greenish-brown. Madeni Zakh (near), Iran. H= 2.5, D_m= 2.5, D_c= 2.435, VID 13. Blaas, 1883. Min. Mag. 41 (1977), 371. Min. Abstr. 77-4074.

Metazellerite, $Ca(UO_2)(CO_3)_2 \cdot 3H_2O$, A Orth. Dull yellow. Lucky MC mine, Gas Hill deposit, Fremont Co., Wyoming, USA. H=?, D_m=?, D_c= 3.41, VbD 04. Coleman et al., 1966. Am. Min. 51 (1966), 1567.

Metazeunerite, $Cu(UO_2)_2(AsO_4)_2 \cdot 8H_2O$, G Tet. Translucent green. Centennial Eureka mine, Tintic, Utah, USA and/or Weisser Hirsch mine, Schneeberg, Sachsen (Saxony), Germany. H=?, D_m=?, D_c= 3.85, VIID 20b. Frondel, 1951. Am. Min. 42 (1957), 222. Am. Min. 49 (1964), 1619.

Meyerhofferite, $CaB_3O_3(OH)_5 \cdot H_2O$, G Tric. Vitreous/silky colorless, white. Mt Blanco deposit, Furnace Creek, Death Valley, Inyo Co., California, USA. H= 2, D_m= 2.120, D_c= 2.125, Vc 05. Schaller, 1914. Dana, 7th ed.(1951), v.2, 356. Zts. Krist. 114 (1960), 321.

Meymacite, $WO_3 \cdot 2H_2O$, A Amor. Resinous/vitreous yellow-brown. Meymac, Corrèze, France. H=?, D_m= 4.0, D_c=?, IVF 10. van Tassel, 1961. Am. Min. 53(1968), 1065 (Abst.).

Mgriite, Cu_3AsSe_3, A Cub. Metallic brownish-grey. Erzgebirge (SW), Sachsen (Saxony), Germany. H= 4.5, D_m=?, D_c= 4.9, IID 01a. Dymkov et al., 1982. Am. Min. 68(1983), 280 (Abst.).

Miargyrite, $AgSbS_2$, G Mon. Metallic adamantine black/grey; white in reflected light. Braunsdorf, Freiberg (near), Sachsen (Saxony), Germany. H= 2.5, D_m= 5.25, D_c= 5.261, IID 07. Rose, 1829. Am. Min. 36 (1951), 436. Acta Cryst. 17 (1964), 847.

Mica, A group name for sheet silicates with the general formula $(K, Na, Ca, Ba, H_3O, NH_4)(Al, Mg, Fe, Li, Cr, Mn, V, Zn)_{2-3}(Si, Al, Fe)_4O_{10}(OH, F)_2$.

Michenerite, $(Pd, Pt)BiTe$, A Cub. Metallic creamy-white. Frood mine, McKim Twp., Sudbury Dist., Ontario, Canada. H= 4.3, D_m=?, D_c= 9.84, IIC 06b. Hawley & Berry, 1958. Can. Min. 11 (1973), 903. Can. Min. 12 (1973), 61. See also Maslovite, Testibiopalladite.

Microcline, $KAlSi_3O_8$, G Tric. Vitreous/pearly colorless, white, yellow, red, green. Fredriksvärn, Norway. H= 6-6.5, D_m= 2.56, D_c= 2.593, VIIIF 03. Breithaupt, 1830. DHZ (1963), v.4, 6. Bull. Min. 107 (1984), 401. See also Orthoclase.

Microlite, $(Ca, Na)_2Ta_2O_6(O, OH, F)$, G Cub. Vitreous/resinous yellow, brown. Chesterfield, Hampshire Co., Massachusetts, USA. H= 5-5.5, D_m= 6.42, D_c= 6.34, IVC 09a. Shepard, 1835. Dana, 7th ed.(1944), v.1, 748. See also Pyrochlore.

Microsommite, $(Na, Ca)_{7-8}(Si, Al)_{12}O_{24}(Cl, SO_4, CO_3)_{2-3}$, G Hex. Vitreous/silky colorless. Monte Somma, Mt. Vesuvius, Napoli, Campania, Italy. H= 6, D_m= 2.4, D_c= 2.30, VIIIF 05. Scacchi, 1872. Bull. Min. 91 (1968), 34.

Miersite, α-$(Ag, Cu)I$, G Cub. Adamantine yellow. Broken Hill, New South Wales, Australia. H= 2.5, D_m= 5.64, D_c= 5.67, IIIA 01. Spencer, 1898. Dana 7th ed.(1951), v.2, 19. See also Marshite, Nantokite.

Miharaite, $Cu_4FePbBiS_6$, A Orth. Metallic grey/greyish white in reflected light. Mihara mine, Yoshii-cho, Shitsuki-gun, Okayama, Japan. H= 3.8, D_m=?, D_c= 6.06, IID 08. Sugaki et al., 1980. Am. Min. 65 (1980), 784.

Milarite, $(K, Na)Ca_2(Be, Al)_3Si_{12}O_{30} \cdot H_2O$, G Hex. Vitreous colorless, pale green. Val Giuf, Tavetsch, Grischun (Graubünden), Switzerland. H= 5.5-6, D_m= 2.560, D_c= 2.565, VIIIC 10. Kenngott, 1870. Europ.Jour.Min. 1 (1989), 353. Can. Min. 18 (1980), 41.

Millerite, β-NiS, G Rhom. Metallic yellow. Germany. H= 3-3.5, D_m= 5.27, D_c= 5.374, IIB 10. Haidinger, 1845. Can. Min. 12 (1974), 248. Can. Min. 12 (1974), 253. See also Mäkinenite.

Millisite, $(Na, K)CaAl_6(PO_4)_4(OH)_9 \cdot 3H_2O$, G Tet. Translucent white, light grey. Clay Canyon (near), Fairfield, Utah Co., Utah, USA. H= 5.5, D_m= 2.83, D_c= 2.88, VIID 13. Larsen & Shannon, 1930. Am. Min. 45 (1960), 547.

Millosevichite, $(Al, Fe)_2(SO_4)_3$, G Rhom. Vitreous violet-blue. Alum Grotto, Vulcano Island, Lipari Islands, Italy. H=?, D_m=?, D_c= 2.86, VIA 03. Panichi, 1913. NBS Monogr. 15 (1978), 8.

Mimetite, $Pb_5(AsO_4)_3Cl$, G Mon. Resinous/sub-adamantine yellow, orange, white, etc. H= 3.5-4, D_m= 7.24, D_c= 7.305, VIIB 16. Beudant, 1832. Am. Min. 54(1969), 993 (Abst.). Zts. Krist. 81 (1931), 352.

Minamiite, $(Na, Ca, K)Al_3(SO_4)_2(OH)_6$, A Rhom. Vitreous/pearly white, colorless. Monzo Hot Springs, Kusatsu-Shrane volcano, Gumma, Japan. H= 3.5,

D_m= 2.8, D_c= 2.81, VIB 03a. Ossaka et al., 1982. Am. Min. 67 (1982), 114. Am. Min. 67 (1982), 114.

Minasgeraisite-(Y), $CaBe_2Y_2Si_2O_{10}$, [A] Mon. Earthy/subvitreous lavender, purple. Timoteo, Jacuaracu, Minas Gerais, Brazil. H= 6-7, D_m=?, D_c= 4.90, VIIIA 23. Foord et al., 1986. Am. Min. 71 (1986), 603.

Minasragrite, $VO(SO_4)\cdot 5H_2O$, [G] Mon. Vitreous blue. Minasragra, Junin, Peru. H=?, D_m=?, D_c= 2.06, VID 12. Schaller, 1915. Am. Min. 58 (1973), 531. Acta Cryst. B35 (1979), 1545.

Minehillite, $(K,Na)_{2-3}Ca_{28}Zn_4Al_4Si_{40}O_{112}(OH)_{16}$, [A] Hex. Pearly colorless, white. Franklin, Sussex Co., New Jersey, USA. H= 4, D_m= 2.93, D_c= 2.94, VIIIE 14. Dunn et al., 1984. Am. Min. 69 (1984), 1150. See also Reyerite, Truscottite.

Minguzzite, $K_3Fe(C_2O_4)_3\cdot 3H_2O$, [G] Mon. Vitreous green, yellow-green. Elba, Toscana, Italy. H=?, D_m= 2.142, D_c= 2.156, IXA 01. Garavelli, 1955. Am. Min. 41(1956), 370 (Abst.). Bull. Min. 81 (1958), 245.

Minium, Pb_3O_4, [G] Tet. Greasy red, brownish-red. H= 2.5, D_m= 9.05, D_c= 8.93, IVB 03. Dana, 7th ed.(1944), v.1, 517. Struct. Repts. 41A (1975), 212.

Minnesotaite, $(Fe,Mg)_3Si_4O_{10}(OH)_2$, [G] Tric. Greasy/waxy greenish-grey. Mesabi and Cuyuna iron ranges, Minnesota, USA. H= < 3, D_m= 3.01, D_c= 2.99, VIIIE 04. Gruner, 1944. Can. Min. 20 (1982), 579. Can. Min. 24 (1986), 479. See also Talc, Willemseite.

Minrecordite, $CaZn(CO_3)_2$, [A] Rhom. Pearly white, colorless. Tsumeb, Namibia. H=?, D_m= 3.45, D_c= 3.445, VbA 03a. Garavelly et al., 1982. Min Rec. 13 (1982), 131.

Minyulite, $KAl_2(PO_4)_2(OH,F)\cdot 4H_2O$, [G] Orth. Silky colorless, white. Dandaragan, Minyulo Well (near), Western Australia, Australia. H= 3.5, D_m= 2.46, D_c= 2.47, VIID 14b. Simpson & LeMesurier, 1933. Dana, 7th ed.(1951), v.2, 970. Am. Min. 62 (1977), 256.

Miomirite, [D] Name also given as miromirite; = plumboan davidite. Min. Mag. 43 (1980), 1055. Vujanovic, 1969.

Mirabilite, $Na_2SO_4\cdot 10H_2O$, [G] Mon. Vitreous colorless, white. H= 1.5-2, D_m= 1.49, D_c= 1.465, VIC 13. Haidinger, 1845. Min. Rec. 1 (1970), 12. Acta Cryst. 34B (1978), 3502.

Mirupolskite, [D] = bassanite. Min. Mag. 43 (1980), 1055. Yurgenson et al., 1968.

Misenite, $H_6K_8(SO_4)_7$(?), [Q] Mon. Pearly/silky colorless, greyish-white. Cape Miseno, Napoli (Naples), Italy. H=?, D_m= 2.32, D_c=?, VIA 04. Scacchi, 1849. Dana, 7th ed.(1951), v.2, 396.

Miserite, $KCa_5Si_8O_{22}(OH)F$, [G] Tric. Translucent pink. Wilson Mineral Springs, Arkansas, USA. H=?, D_m= 2.92, D_c= 2.646, VIIID 18. Schaller, 1950. Am. Min. 35 (1950), 911. Can. Min. 14 (1976), 515.

Mitridatite, $Ca_2Fe_3O_2(PO_4)_3\cdot 3H_2O$, [G] Mon. Earthy/resinous greenish-yellow, green, brownish-green, etc. Mitridot Mtn., Kerch peninsula, Crimea, USSR. H= 2.5, D_m= 3.24, D_c= 3.08, VIID 15a. Dvoichenko, 1914. Am. Min. 59 (1974), 48. Min. Mag. 41 (1977), 527. See also Arseniosiderite, Robertsite.

Mitscherlichite, $K_2CuCl_4\cdot 2H_2O$, [G] Tet. Vitreous greenish-blue. Mt. Vesuvius, Napoli, Campania, Italy. H= 2.5, D_m= 2.418, D_c= 2.41, IIIB 07. Zambonini & Carobbi, 1925. Dana, 7th ed.(1951), v.2, 101.

Mixite, $Cu_6Bi(AsO_4)_3(OH)_6\cdot 3H_2O$, [G] Hex. Dull/brilliant green, bluish-green, pale green, whitish. Jáchymov (St. Joachimsthal), Západočeský kraj, Čechy (Bohemia), Czechoslovakia. H= 3-4, D_m= 3.79, D_c= 4.045, VIID 17. Schrauf, 1879. Dana, 7th ed.(1951), v.2, 943. Struct. Repts. 54A (1987), 250.

Miyashiroite, [D] Hypothetical end-member amphibole. Min. Mag. 36 (1968), 1144. Phillips & Layton, 1964.

Moctezumite, $Pb(UO_2)(TeO_3)_2$, [A] Mon. Translucent bright orange. Moctezuma mine, Moctezuma, Sonora, Mexico. H= 3, D_m= 5.73, D_c= 5.41, VIG 03. Gaines, 1965. Am. Min. 50 (1965), 1158.

Modderite, (Co, Fe)As, [G] Orth. Metallic. Johannesburg (E of), Modderfontein, Transvaal, South Africa. H=?, D_m=?, D_c= 8.28, IIB 09d. Cooper, 1923. JCPDS 9-94.

Mohite, Cu_2SnS_3, [A] Tric. Metallic greenish-grey. Kochbulak deposit, Uzbekistan (E), USSR. H= 4.5, D_m=?, D_c= 4.9, IIB 03b. Kovalenker et al., 1982. Am. Min. 68,(1983), 281 (Abst.).

Mohrite, $(NH_4)_2Fe(SO_4)_2 \cdot 6H_2O$, [A] Mon. Vitreous pale green. Travale, Montieri, Grosseto, Toscana, Italy. H=?, D_m=?, D_c= 1.86, VIC 11. Garavelli, 1964. NBS Monogr. 25(21) (1985), 10.

Moissanite-5H, SiC, [P] Hex. Metallic green/black. Dzoraget River, Sevan-Amasii zone, USSR. H=?, D_m=?, D_c= 3.38, IC 04. Gevorkyan et al., 1974. Am. Min. 61(1976), 1054 (Abst.). Acta Cryst. B25 (1969), 477. See also Moissanite-6H, Beta-Moissanite.

Moissanite-6H, α–SiC, [G] Hex. Metallic green/black; grey in reflected light. Canyon Diablo meteorite, Meteor Crater, Coconino Co., Arizona, USA. H= 9.5, D_m= 3.1, D_c= 3.21, IC 04. Kunz, 1905. Am. Min. 48 (1963), 620. J. Appl. Cryst. 2 (1969), 45. See also Moissanite-5H, Beta-Moissanite.

Moluranite, $H_4U(UO_2)_3(MoO_4)_7 \cdot 18H_2O$, [G] Syst=? Resinous black. Unspecified locality, USSR. H= 3-4, D_m= 4, D_c=?, VIF 02. Epprecht et al., 1959. Am. Min. 45(1960), 258 (Abst.).

Molybdenite-2H, MoS_2, [G] Hex. Metallic grey. H= 1-1.5, D_m= 4.7, D_c= 5.00, IIC 10. Hielm, 1782. NBS Circ. 539, v.5 (1955), 47. See also Jordisite, Molybdenite-3R, Drysdallite, Tungstenite.

Molybdenite-3R, MoS_2, [P] Rhom. Metallic grey. Con mine, Yellowknife (S), Northwest Territories, Canada. H=?, D_m=?, D_c= 5.03, IIC 10. Traill, 1963. Can. Min. 7 (1963), 524. See also Jordisite, Molybdenite-2H, Tungstenite-3R.

Molybdite, MoO_3, [A] Orth. Adamantine light greenish-yellow. Krupka, Středočeský kraj, Čechy (Bohemia), Czechoslovakia. H=?, D_m=?, D_c= 4.72, IVE 01. Greg & Lettsom, 1858. Am. Min. 49(1964), 1497 (Abst.).

Molybdofornacite, $CuPb_2(Mo,Cr)O_4(As,P)O_4(OH)$, [A] Mon. Adamantine light green. Tsumeb mine, Tsumeb, Namibia. H= 2-3, D_m=?, D_c= 6.6, VIE 02. Medenbach et al., 1983. Am. Min. 69(1984), 567 (Abst.). See also Fornacite, Vauquelinite.

Molybdomenite, $PbSeO_3$, [G] Mon. Pearly/adamantine colorless, white. Cacheuta, Mendoza, Argentina. H=?, D_m= 7.07, D_c= 7.12, VIA 08. Bertrand, 1882. Can. Min. 8 (1965), 149. See also Scotlandite.

Molybdophyllite, $Mg_2Pb_2Si_2O_7(OH)_2$, [Q] Rhom. Translucent colorless, pale green. Långban mine, Filipstad (near), Värmland, Sweden. H= 3-4, D_m= 4.7, D_c=?, VIIIB 14. Flink, 1901. Strunz (1970), 396.

Molysite, $FeCl_3$, [G] Rhom. Translucent yellow, brownish-red, purple-red. Mt. Vesuvius, Napoli, Campania, Italy. H=?, D_m= 2.90, D_c= 2.912, IIIA 08. Dana, 1868. Dana, 7th ed.(1951), v.2, 47. J.Appl.Cryst. 22 (1989), 173.

Monazite-(Ce), $(Ce,La,Nd,Th)PO_4$, [G] Mon. Resinous/waxy/vitreous yellowish, brown, reddish-brown, etc. H= 5-5.5, D_m= 5.20, D_c= 5.26, VIIA 11. Breithaupt, 1829. Min. Mag. 43 (1980), 1031. Struct. Repts. 32A (1967), 358.

Monazite-(La), $(La,Ce,Nd)PO_4$, [A] Mon. H=?, D_m=?, D_c= 5.13, VIIA 11. Borovskii & Gerasimovskii,1945. JCPDS 35-731.

Monazite-(Nd), $(Nd,La,Ce)PO_4$, [A] Mon. Translucent/milky rose. Glogstafel, Val Formazza, Italy. H=?, D_m=?, D_c= 5.43, VIIA 11. Graeser & Schwander, 1987. Am. Min. 73(1988), 1495 (Abst.).

Moncheite, $(Pt,Pd)(Te,Bi)_2$, [G] Rhom. Metallic white. Monche Tundra, Monchegorsk, USSR. H= 3.5, D_m=?, D_c= 10.24, IIC 11. Genkin et al., 1963. Am. Min. 48(1963), 1181 (Abst.).

Monetite, $Ca(PO_3OH)$, [G] Tric. Vitreous pale yellowish-white. Moneta Island, Mayeguez(S), Puerto Rico, West Indies. H= 3.5, D_m= 2.929, D_c= 2.924, VIIA 09. Shepard, 1882. Dana, 7th ed.(1951), v.2, 660. Acta Cryst. B33 (1977), 1223. See also Weilite.

Mongolite, $Ca_4Nb_6Si_5O_{24}(OH)_{10}\cdot 6H_2O$, [A] Tet. Silky pale/greyish lilac. Dotozhny pegmatite, Khan-Bogdinskiĭ massif, Gobi, Mongolia. H= 2, D_m= 3.147, D_c= 3.55, VIIIB 08. Vladykin et al., 1985. Am. Min. 71(1986), 1279 (Abst.).

Mongshanite, $(Mg,Cr,Fe,Ca,K)_2(Ti,Zr,Cr,Fe)_5O_{12}$, [Q] Hex. Unspecified locality, China(?). H=?, D_m=?, D_c=?, IVC 11. Zhou et al., 1982. Am. Min. 73(1988), 441 (Abst.).

Monimolite, $Pb_3Sb_2O_7$, [Q] Cub. Greasy/adamantine yellow, grey-green, brown. Harstigen mine, Pajsberg, Värmland, Sweden. H= 4.5-6, D_m= 7, D_c= 11.74, IVC 08. Igelström, 1865. Dana, 7th ed.(1951), v.2, 1023. See also Bindheimite.

Monohydrocalcite, $CaCO_3\cdot H_2O$, [G] Rhom. Translucent white, grey. Lake Issyk-Kul, Kirgizia, USSR. H=?, D_m= 2.38, D_c= 2.42, VbC 01. Semenov, 1964. Am. Min. 58 (1973), 1102. Struct. Repts. 48A (1981), 266.

Monsmedite, $H_8K_2Tl_2(SO_4)_8\cdot 11H_2O$, [G] Syst=? Translucent dark green, black. Rotmundi vein, Baia Sprie (Felsöbánya), Romania. H= 2, D_m= 3.00, D_c=?, VIC 18. Götz et al., 1969. Am. Min. 54(1969), 1496 (Abst.).

Montanite, $(BiO)_2TeO_4\cdot 2H_2O$, [Q] Syst=? Dull/waxy/earthy yellowish, greenish, white. Highland, Silver Bow Co., Montana, USA. H=?, D_m=?, D_c=?, VIG 07. Genth, 1868. Dana, 7th ed.(1951), v.2, 636.

Montbrayite, $(Au,Sb)_2Te_3$, [G] Tric. Metallic white. Robb-Montbray mine, Montbray Twp., Abitibi Co., Québec, Canada. H= 2.5, D_m= 9.94, D_c= 9.81, IIC 04. Peacock & Thompson, 1945. Am. Min. 57 (1972), 146.

Montdorite, [D] An unnecessary name for a manganoan biotite. Am. Min. 64(1979), 1331 (Abst.). Robert & Maury, 1979.

Montebrasite, $LiAlPO_4(OH,F)$, [G] Tric. Vitreous/greasy white, yellowish, pink, greenish, etc. Montebras, Soumans, Creuse, France. H= 5.5-6, D_m= 2.98, D_c= 3.105, VIIB 02. Des Cloizeaux, 1871. Min. Mag. 37 (1969), 414. ZVMO 118(3) (1989), 47. See also Amblygonite.

Monteponite, CdO, [G] Cub. Metallic grey in reflected light. Monteponi mine, Iglesias, Cagliari, Sardegna, Italy. H= 3, D_m= 8.15, D_c= 8.24, IVA 04. Fairbanks, 1946. NBS Circ. 539, v.2 (1953), 27.

Monteregianite-(Y), $K_2Na_4Y_2Si_{16}O_{38}\cdot 10H_2O$, [A] Mon. Vitreous/silky colorless, white, grey, mauve, pale green. Mont Saint-Hilaire, Rouville Co., Québec, Canada. H= 3.5, D_m= 2.42, D_c= 2.41, VIIIE 15. Chao, 1978. Can. Min. 16 (1978), 561. Am. Min. 72 (1987), 365.

Montgomeryite, $Ca_4MgAl_4(PO_4)_6(OH)_4\cdot 12H_2O$, [G] Mon. Vitreous deep green, pale green, colorless. Clay Canyon (near), Fairfield, Utah Co., Utah, USA. H= 4, D_m= 2.530, D_c= 2.523, VIID 35. Larsen, 1940. Am. Min. 61 (1976), 12. Am. Min. 59 (1974), 843. See also Kingsmountite.

Monticellite, $CaMgSiO_4$, [G] Orth. Translucent colorless, grey. Mt. Vesuvius, Napoli, Campania, Italy. H= 5.5, D_m= 3.2, D_c= 3.04, VIIIA 04. Brooke, 1831. DHZ, 2nd ed.(1982), v.1A, 352. Struct. Repts. 30A(1965), 424. See also Kirschsteinite, Glaucochroite.

Montmorillonite, $(Na,Ca)_{0.3}(Al,Mg)_2Si_4O_{10}(OH)_2\cdot nH_2O$, [G] Mon. Translucent white, yellow, green. Montmorillon, Vienne, France. H= 1-2, D_m= 2-3, D_c= 2.38, VIIIE 08a. Mauduyt, 1847. DHZ (1962), v.3, 226. Struct. Repts. 16 (1952), 368.

Montroseite, $(V,Fe)O(OH)$, [G] Orth. Sub-metallic black. Bitter Creek mine, Paradox Valley, Montrose Co., Colorado, USA. H=?, D_m= 4.0, D_c= 4.15, IVF 04. Weeks et al., 1953. Am. Min. 40 (1955), 861. Am. Min. 38 (1953), 1242.

Montroyalite, $Sr_4Al_8(CO_3)_3(OH,F)_{26} \cdot 10H_2O$, [A] Tric. Porcelanous/waxy white. Francon quarry, St.-Michel Dist., Montreal Island, Québec, Canada. H= 3.5, D_m= 2.677, D_c=?, VbD 02. Roberts et al., 1986. Can. Min. 24 (1986), 455.

Montroydite, HgO, [G] Orth. Vitreous red, reddish-brown. Terlingua, Brewster Co., Texas, USA. H= 2.5, D_m= 11.3, D_c= 11.2, IVA 07. Moses, 1903. Dana, 7th ed.(1944), v.1, 511. Struct. Repts. 20 (1956), 266.

Mooihoekite, $Cu_9Fe_9S_{16}$, [A] Tet. Metallic yellow. Mooihoek Farm, Lydenberg Dist., Transvaal, South Africa. H= 4, D_m= 4.36, D_c= 4.35, IIB 02. Cabri & Hall, 1972. Am. Min. 57 (1972), 689. Can. Min. 13 (1975), 168.

Moolooite, $CuC_2O_4 \cdot nH_2O$, [A] Orth. Dull/waxy blue, green. Bunbury Well (N), Mooloo Downs, Western Australia, Australia. H=?, D_m=?, D_c= 3.43, IXA 01. Clarke & Williams, 1986. Min. Mag. 50 (1986), 295.

Mooreite, $(Mg,Zn,Mn)_{15}(SO_4)_2(OH)_{26} \cdot 8H_2O$, [G] Mon. Vitreous colorless. Sterling Hill mine, Ogdensburg, Sussex Co., New Jersey, USA. H= 3, D_m= 2.47, D_c= 2.465, IVF 03e. Bauer & Berman, 1929. Am. Min. 54 (1969), 973. Acta Cryst. B36 (1980), 1304. See also Lawsonbauerite, Shigaite, Torreyite.

Moorhouseite, $(Co,Ni,Mn)SO_4 \cdot 6H_2O$, [A] Mon. Vitreous pink. Magnet Cove Barium Corp. mine, Walton, Hants Co., Nova Scotia, Canada. H= 2.5, D_m= 1.97, D_c= 2.006, VIC 03b. Jambor & Boyle, 1965. Can. Min. 8 (1965), 166. Acta Cryst. 15 (1962), 1219.

Mopungite, $NaSb(OH)_6$, [A] Tet. Vitreous colorless, white. Mopung Hills, Churchill Co., Nevada, USA. H= 3, D_m= 3.2, D_c= 3.264, IVF 06. Williams, 1985. Min. Rec. 16 (1985), 73.

Moraesite, $Be_2PO_4(OH) \cdot 4H_2O$, [G] Mon. Translucent white. Sapucaia Pegmatite, Galileia, Minas Gerais, Brazil. H=?, D_m= 1.805, D_c= 1.806, VIID 01. Lindberg et al., 1953. Am. Min. 38 (1953), 1126. See also Bearsite.

Mordenite, $K_{2.8}Na_{1.5}Ca_2(Al_9Si_{39})O_{96} \cdot 29H_2O$, [G] Orth. Vitreous/pearly/silky white, yellowish, pinkish. Morden (E), Kings Co., Nova Scotia, Canada. H= 3-4, D_m= 2.1, D_c= 2.152, VIIIF 12. How, 1864. Natural Zeolites (1985), 223. Zts. Krist. 175 (1986), 249.

Moreauite, $Al_3(UO_2)(PO_4)_3(OH)_2 \cdot 13H_2O$, [A] Mon. Translucent greenish-yellow. Kobokobo, Kivu, Zaïre. H=?, D_m= 2.64, D_c= 2.61, VIID 21. Deliens & Piret, 1985. Am. Min. 70(1985), 1330 (Abst.).

Morelandite, $(Ba,Ca,Pb)_5(AsO_4,PO_4)_3Cl$, [A] Hex. Greasy/vitreous light yellow, grey. Jakobsberg mine, Filipstad, Nordmark (near), Värmland, Sweden. H= 4.5, D_m= 5.33, D_c= 5.30, VIIB 16. Dunn & Rouse, 1978. Can. Min. 16 (1978), 601.

Morenosite, $NiSO_4 \cdot 7H_2O$, [G] Orth. Vitreous green, greenish-white. Cap Hortegal, Galicia, Spain. H= 2-2.5, D_m= 1.953, D_c= 1.93, VIC 03d. Casares, 1851. Am. Min. 33 (1964), 1110. See also Epsomite, Goslarite.

Morinite, $NaCa_2Al_2(PO_4)_2(OH)F_4 \cdot 2H_2O$, [G] Mon. Vitreous colorless, pink. Montebras, Soumans, Creuse, France. H= 4.5, D_m= 2.962, D_c= 2.981, VIID 12. Lacroix, 1891. Am. Min. 43 (1958), 585. Can. Min. 17 (1979), 93.

Morozeviczite, $Pb_3Ge_{1-x}S_4$, [A] Cub. Metallic white. Zechstein shales, Lower Silesia, Poland. H= 3, D_m=?, D_c= 6.62, IIB 03c. C. Haranczyk, 1975. Am. Min. 66(1981), 437 (Abst.). See also Polkovicite.

Mosandrite, $(Ca,Na,Ce)_{12}(Ti,Zr)_2Si_7O_{25}(OH)_6F_4$, [G] Mon. Resinous/vitreous/greasy reddish-brown, greenish, etc. Låven island, Langesundfjord, Norway. H= 4, D_m= 3.44, D_c= 3.45, VIIIB 09. Erdmann, 1841. Am. Min. 43(1958), 795 (Abst.). Acta Cryst. B27 (1971), 1277.

Moschellandsbergite, γ-Ag_2Hg_3, G Cub. Metallic white. Landsberg (Moschellandsberg), Obermoschel (near), Rheinland-Pfalz, Germany. H= 3.5, D_m= 13.6, D_c= 13.41, IA 06. Berman & Harcourt, 1938. Am. Min. 23 (1938), 761.

Mosesite, $Hg_2N(Cl, SO_4, MoO_4, CO_3) \cdot H_2O$, G Cub. Adamantine yellow, olive-green. Terlingua, Brewster Co., Texas, USA. H= 3.5, D_m= 7.72, D_c= 7.53, VIB 10. Canfield et al., 1910. Am. Min. 38 (1953), 1225.

Mottramite, $Pb(Cu, Zn)VO_4(OH)$, G Orth. Greasy brownish-red, green. Mottram, St. Andrew's, Cheshire, England. H= 3-3.5, D_m= 5.9, D_c= 6.19, VIIB 11b. Roscoe, 1876. Min. Mag. 50 (1986), 137. See also Descloizite.

Motukoreaite, $[Mg_6Al_3(OH)_{18}][Na_{0.6}(SO_4, CO_3)_2 \cdot 12H_2O]$, A Rhom. Dull pale yellowish-green. Waitemata Harbour, Brown's Island, Auckland, New Zealand. H= 1-1.5, D_m= 1.5, D_c= 1.48, IVF 03e. Rodgers et al., 1977. Min. Mag. 41 (1977), 389. Struct. Repts. 53A(1986), 182.

Mounanaite, $PbFe_2(VO_4)_2(OH)_2$, A Tric. Translucent brownish-red. Mounana mine, Franceville, Haut-Ogooué, Gabon. H=?, D_m= 4.85, D_c= 4.89, VIIB 13. Cesbron & Fritsche, 1969. Am. Min. 54(1969), 1738 (Abst.).

Mountainite, $(Ca, Na_2, K_2)_2Si_4O_{10} \cdot 3H_2O$, G Mon. Silky white. Bultfontein mine, Kimberley, South Africa. H=?, D_m= 2.36, D_c= 2.37, VIIIF 16. Gard et al., 1957. Min. Mag. 31 (1957), 611.

Mountkeithite, $(Mg, Ni)_{11}(Fe, Cr)_3(SO_4, CO_3)_{3.5}(OH)_{24} \cdot 11H_2O$, A Hex. Pearly/translucent pale pink, white. Mount Keith deposit, Western Australia, Australia. H=?, D_m= 2.12, D_c= 1.95, IVF 03e. Hudson & Bussell, 1981. Min. Mag. 44 (1981), 345.

Mourite, $UMo_5O_{12}(OH)_{10}$, G Mon. Adamantine violet. USSR. H= 3, D_m= 4.17, D_c= 4.22, VIF 02. Kopchenova et al., 1962. Am. Min. 56 (1971), 163.

Moydite-(Y), $YB(OH)_4CO_3$, A Orth. Vitreous yellow. Evans-Lou mine, St-Pierre-de-Wakefield (near), Portland Twp., Papineau Co., Québec, Canada. H=?, D_m= 3.13, D_c= 3.09, Vc 03e. Grice et al., 1986. Can. Min. 24 (1986), 665. Can. Min. 24 (1986), 675.

Mozambikite, D Inadequate data; may be thorite or thorogummite. Min. Mag. 33 (1962), 261. Neiva & Neves, 1960.

Mpororoite, $AlWO_3(OH)_3 \cdot 2H_2O$, A Tric. Powdery greenish-yellow. Mpororo tungsten deposit, Kigezi, Uganda. H=?, D_m=?, D_c=?, VIF 03. v. Knorring et al., 1972. Min. Rec. 12 (1981), 81.

Mrazekite, D Inadequate data. Min. Mag. 43 (1980), 1055. Neacsu, 1970.

Mroseite, $CaTeO_2(CO_3)$, A Orth. Adamantine colorless, white. Moctezuma mine, Sonora, Mexico. H= 4, D_m= 4.35, D_c= 4.171, VIG 11. Mandarino et al., 1975. Can. Min. 13 (1975), 286. Can. Min. 13 (1975), 383.

Muchuanite, $MoS_2 \cdot 0.5H_2O$, Q Rhom. Metallic greyish-white. Middle Jurassic bed, Muchuan Co., Sichuan, China. H= 1.5, D_m= 5.01, D_c= 5.10, IIC 10. Zhang, 1981. Am. Min. 67(1982), 856 (Abst.).

Mückeite, $CuNiBiS_3$, A Orth. Metallic grey; creamy-grey in reflected light. Grüne Au mine, Schutzbach, Siegerland, Germany. H= 3.5, D_m= 5.88, D_c= 6.04, IID 19. Schnorrer-Kohler et al., 1989. N.Jb.Min.Mh. (1989), 193. Acta Cryst. C46 (1990), 127.

Muirite, $Ba_{10}Ca_2MnTiSi_{10}O_{30}(OH, Cl, F)_{10}$, A Tet. Sub-vitreous orange. Big Creek and Rush Creek, Fresno Co., California, USA. H= 2.5, D_m= 3.86, D_c= 3.74, VIIIC 04. Alfors et al., 1965. Am. Min. 50 (1965), 1500. Min. Abstr. 81-2387.

Mukhinite, $Ca_2Al_2VSi_3O_{12}(OH)$, A Mon. Vitreous black. Tashelginsk iron ore deposit, Gornaya Shoriya, Siberia (W), USSR. H= 8, D_m=?, D_c= 3.47, VIIIB 15. Shepel & Karpenko, 1969. Am. Min. 55(1970), 321 (Abst.).

Mullite, $Al_{4+x}Si_{2-2x}O_{10-x}$, [G] Orth. Translucent colorless, white, yellow, pink, red. Mull Island, Scotland. H= 6-7, D_m= 3.2, D_c= 3.02, VIIIA 14. Bowen et al., 1924. Am. Min. 62 (1977), 747. Am. Min. 71 (1986), 1476.

Mummeite, $(Ag,Cu)_7Pb_2Bi_{13}S_{26}$, [Q] Mon. Metallic. Alaska mine, San Juan Co., Colorado, USA. H=?, D_m=?, D_c= 6.64, IID 09c. Karup-Møller & Makovicky, 1986. IMA 1986, Abstracts p. 138. See also Pavonite.

Mundite, $Al(UO_2)_3(PO_4)_2(OH)_3 \cdot 5.5H_2O$, [A] Orth. Translucent yellow. Kobokobo, Kivu, Zaïre. H=?, D_m=?, D_c= 4.295, VIID 21. Deliens & Piret, 1981. Bull. Min. 104 (1981), 669.

Mundrabillaite, $(NH_4)_2Ca(PO_3OH)_2 \cdot H_2O$, [A] Mon. Earthy, colorless. Petrogale cave, Madura, Western Australia, Australia. H=?, D_m= 2.05, D_c= 2.09, VIIC 14. Bridge & Clarke, 1983. Min. Mag. 47 (1983), 80.

Munirite, $NaVO_3 \cdot 1.9H_2O$, [A] Mon. Pearly white. Siwalik sandstone, Bhimber, Azad Kashmir, Pakistan. H=?, D_m= 2.28, D_c= 2.271, VIIC 21. Butt & Mahmood, 1983. Min. Mag. 52 (1988), 716. Struct. Repts. 44A (1978), 201.

Munkforssite, [D] Manganiferous apatite (?). Am. Min. 49 (1964), 1778 (Abst.). Igelström, 1897.

Munkrudite, [D] = kyanite. Am. Min. 49 (1964), 1778 (Abst.). Igelström, 1898.

Murataite, $(Na,Y)_4(Zn,Fe)_3(Ti,Nb)_6O_{18}(F,OH)_4$, [A] Cub. Submetallic black; grey in reflected light. St. Peter's Dome, El Paso Co., Colorado, USA. H= 6, D_m= 4.69, D_c= 4.64, IVC 09b. Adams et al., 1974. Am. Min. 59 (1974), 172.

Murdochite, $Cu_{12}Pb_2O_{15}(Cl,Br)_2$, [G] Cub. Black. Mammoth mine, Tiger, Pinal Co., Arizona, USA. H= 4, D_m= 6.47, D_c= 6.06, IVC 01. Fahey, 1953. Am. Min. 40 (1955), 905. Acta Cryst. C39 (1983), 1143.

Murgocite, [D] Inadequate data. Min. Mag. 43 (1980), 1055. Mineev et al., 1970.

Murmanite, $Na_3(Ti,Nb)_4O_4(Si_2O_7)_2 \cdot 4H_2O$, [G] Tric. Translucent violet. Kola Peninsula, USSR. H= 2-3, D_m=?, D_c= 3.00, VIIIB 11. Fersman, 1923. Min. Zhurn. 11(5) (1989), 19. Sov.Phys.Cryst. 31 (1986), 44.

Murunskite, $K_2Cu_3FeS_4$, [A] Tet. Metallic brown. Murun massif, Olekminsk (near), Yakutiya, USSR. H= 3, D_m=?, D_c= 3.81, IIA 10. Dobrovolskaya et al., 1981. ZVMO 110 (1981), 468. See also Bukovite, Thalcusite.

Muscovite-1M, $KAl_2(Si_3Al)O_{10}(OH,F)_2$, [P] Mon. Translucent colorless, pale green/red/brown. H= 2.5-3, D_m= 2.8, D_c= 2.805, VIIIE 05a. Dana, 1850. Clays Clay Mins. 36(1988), 193. Struct. Repts. 19 (1955), 468.

Muscovite-2M$_1$, $KAl_2(Si,Al)_4O_{10}(OH,F)_2$, [G] Mon. Translucent colorless, pale green/red/brown. H= 2.5-3, D_m= 2.8, D_c= 2.818, VIIIE 05a. Dana, 1850. Min. Mag. 36 (1968), 883. Am. Min. 67 (1982), 69.

Muscovite-2M$_2$, $(K,Na)Al_2(Si,Al)_4O_{10}(OH)_2$, [P] Mon. H=?, D_m=?, D_c= 2.80, VIIIE 05a. Zhoukhlistov et al., 1973. Clays Cl. Mins. 36 (1988), 193. Clays Cl. Mins. 21 (1973), 465.

Muscovite-3T, $KAl_2(Si_3Al)O_{10}(OH,F)_2$, [P] Rhom. Translucent colorless, pale green/red/brown. H= 2.5-3, D_m= 2.8, D_c= 2.831, VIIIE 05a. Dana, 1850. Clays Clay Mins. 36(1988), 193.

Musgravite, $(Mg,Fe,Zn)_2BeAl_6O_{12}$, [A] Rhom. Translucent pale olive green. Enderby Land, Mcintyre Island (near), Zircon Point, Antarctica (E) and/or Ernabella Mission (NNE), Musgrave range, South Australia, Australia. H=?, D_m= 3.68, D_c= 3.69, IVC 05c. Hudson et al., 1967. Min. Mag. 36 (1967), 305. Min. Abstr. 85M/0172. See also Pehrmanite.

Mushistonite, $(Cu,Zn,Fe)Sn(OH)_6$, [A] Cub. Earthy brownish-green. Mushiston deposit, Tadzhikistan, USSR. H= 4-4.4, D_m=?, D_c= 4.08, IVF 06. Marshukova et al., 1984. Am. Min. 70(1985), 1331 (Abst.).

Muskoxite, $Mg_7Fe_4O_{13} \cdot 10H_2O$(?), [A] Rhom. Vitreous dark reddish brown. Muskox Intrusion, Coppermine River area, Northwest Territories, Canada. H=?, D_m= 3.2, D_c=?, IVF 03e. Jambor, 1969. Am. Min. 54 (1969), 684.

Muthmannite, $AuAgTe_2$, [G] Syst=? Metallic yellow/greyish-white. Săcărâmb (Nagyág), Transylvania, Romania. H= 3.5, D_m= 5.60, D_c=?, IIB 05. Zambonini, 1911. Dokl.Earth Sci. 280(1985), 159.

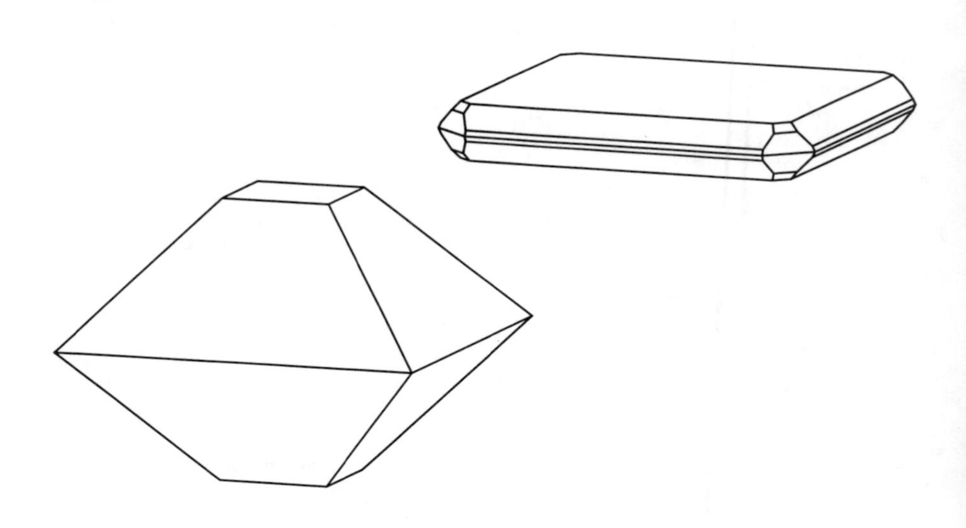

Wulfenite from the Red Cloud mine, Arizona showing (Left) {011}, {102} and (Right) {001}, {102}, {101}, {113} and {111} (both after Dana, 1892).

N

Nabaphite, $NaBaPO_4 \cdot 9H_2O$, [A] Cub. Vitreous colorless. Yukspor Mts., Khibina massif, Kola Peninsula, USSR. H= 2, D_m= 2.3, D_c= 2.26, VIIC 18. Khomyakov et al., 1982. Am. Min. 68(1983), 643 (Abst.). Am. Min. 68(1983), 643 (Abst.). See also Nastrophite.

Nabokoite, $Cu_7TeO_4(SO_4)_5 \cdot KCl$, [A] Tet. Translucent yellowish-brown. Yadovitaya fumarole, Tolbachik volcano, Kamchatka, USSR. H= 2.5, D_m= 4.18, D_c= 3.974, VIB 12. Popova et al., 1987. Min. & Petrol. 38 (1988), 291. Min. Petrol. 38 (1988), 291.

Nacaphite, $Na_2Ca(PO_4)F$, [A] Orth. Vitreous colorless. Mt. Rasvumchorr, Khibina massif, Kola Peninsula, USSR. H= 3, D_m= 2.85, D_c= 2.88, VIIB 18. Khomyakov et al., 1980. Am. Min. 66(1981), 218 (Abst.).

Nacareniobsite-(Ce), $Na_3Ca_3(Ce,La)Nb(Si_2O_7)_2OF_3$, [A] Mon. Vitreous colorless. Kvanefjeld Tunnel, Ilímaussaq Intrusion, Greenland (S). H= 5, D_m= 3.45, D_c= 3.43, VIIIB 09. Petersen et al., 1989. N.Jb.Min.Mh. (1989), 84.

Nacrite, $Al_2Si_2O_5(OH)_4$, [G] Mon. Earthy/pearly white, grey, yellowish, brownish, etc. H= 2-2.5, D_m= 2.6, D_c= 2.602, VIIIE 10a. Brongniart, 1807. Am. Min. 48 (1963), 1196. Clays Cl. Mins. 17 (1969), 185. See also Kaolinite, Dickite, Halloysite.

Nadorite, $PbSbO_2Cl$, [G] Orth. Resinous/adamantine brown, yellow. Djebel Debar, Qacentina (Constantine), Algeria. H= 3.5-4, D_m= 7.02, D_c= 7.05, IIIC 06. Flajolot, 1870. Dana, 7th ed.(1951), v.2, 1039. Struct. Repts. 11 (1947), 311. See also Perite.

Nagashimalite, $Ba_4(V,Ti)_4B_2Si_8O_{27}(O,OH)_2Cl$, [A] Orth. Sub-metallic/vitreous greenish-black. Mogurazawa mine, Gumma, Kiryu, Honshu, Japan. H= 6, D_m= 4.08, D_c= 4.14, VIIIC 04. Matsubara & Kato, 1980. Am. Min. 66(1981), 638 (Abst.). Min. Jour. 10 (1980), 131. See also Taramellite, Titantaramellite.

Nagelschmidtite, $Ca_7(SiO_4)_2(PO_4)_2$, [A] Hex. Hatrurim formation, Israel. H=?, D_m= 3.04, D_c= 3.49, VIIIC 13. Gross, 1977. Am. Min. 63 (1978), 425 (Abst.).

Nagyágite, $Au(Pb,Sb,Fe)_8(Te,S)_{11}$, [G] Tet. Metallic blackish-grey; grey-white in reflected light. Săcărâmb (Nagyág), Transylvania, Romania. H= 1-1.5, D_m= 7.41, D_c=?, IID 18. Haidinger, 1845. Am. Min. 34 (1949), 342.

Nahcolite, $NaHCO_3$, [G] Mon. Vitreous/resinous colorless, white. Napoli (Naples), Italy. H= 2.5, D_m= 2.24, D_c= 2.22, VbA 01. Bannister, 1928. Min. Mag. 39 (1974), 564.

Nahpoite, $Na_2(PO_3OH)$, [A] Mon. Earthy white. Big Fish River, Yukon Territory, Canada. H=?, D_m=?, D_c= 2.58, VIIA 14. Coleman & Robertson, 1981. Can. Min. 19 (1981), 373.

Nakaséite, [D] Possibly = andorite. N.Jb.Min.Abh. 131 (1977) 187. Tei-Ichi & Muraoka.

Nakauriite, $Cu_8(SO_4)_4(CO_3)(OH)_6 \cdot 48H_2O$, [A] Orth. Translucent blue. Nakauri mine, Aichi, Shinshiro, Honshu, Japan. H=?, D_m= 2.39, D_c= 2.35, VID 15. Suzuki et al., 1976. Am. Min. 62 (1977), 594 (Abst.).

Na-komarovite, $(Na,Ca,H)_2Nb_2Si_2O_{10}(OH,F)_2 \cdot H_2O$, [Q] Orth. Translucent white. Kola Peninsula, USSR. H=?, D_m= 3.3, D_c= 3.26, VIIIB 08. Krivokoneva et al., 1979. Dokl.Earth Sci. 248(1979), 127. See also Komarovite.

Nambulite, $(Li,Na)Mn_4Si_5O_{14}(OH)$, [A] Tric. Vitreous reddish-brown. Tunakozowa mine, Kitakami Mts., Japan. H= 6.5, D_m= 3.51, D_c= 3.55, VIIID 13. Yoshi et al., 1972. Am. Min. 58(1973), 1112 (Abst.). Acta Cryst. B31 (1975), 2422. See also Natronambulite.

Namibite, $CuVBi_2O_6$, [A] Mon. Translucent dark green. Khorixas, Namibia (NW). H= 4.5-5, D_m= 6.86, D_c= 6.76, IVC 14. v. Knorring & Sahama, 1981. Am. Min. 67(1982), 857 (Abst.).

Namuwite, $(Zn,Cu)_4SO_4(OH)_6 \cdot 4H_2O$, [A] Hex. Pearly sea-green. Aberllyn mine, Llanrwst, Bettws-y-coed, Gwynedd, Wales. H= 2, D_m= 2.77, D_c= 2.81, VID 01. Bevins et al., 1982. Min. Mag. 46 (1982), 51.

Nanlingite, $CaMg_4(AsO_3)_2F_4$, [A] Rhom. Vitreous brownish-red. Nan Ling area, China. H= 2.3, D_m= 3.927, D_c= 3.993, IIID 03. Gu et al., 1976. Am. Min. 62(1977), 1058 (Abst.).

Nanpingite, $CsAl_2(Si,Al)_4O_{10}(OH,F)_2$, [A] Mon. White to silver white, transparent in flakes. Nanping, Fujian, China. H=?, D_m= 3.11, D_c= 3.19, Yang et al, 1988. Am. Min. 75 (1990), 708 (Abst.).

Nantokite, CuCl, [G] Cub. Adamantine colorless. Nantoko, Copiapó, Chile. H= 2.5, D_m= 4.136, D_c= 4.22, IIIA 01. Breithaupt, 1868. Dana, 7th ed.(1951), 18. See also Marshite, Miersite.

Narsarsukite, $Na_2(Ti,Fe)Si_4(O,F)_{11}$, [G] Tet. Translucent yellow, brownish-grey. Narsarsuk, Greenland (S). H= 7, D_m= 2.783, D_c= 2.77, VIIID 18. Flink, 1900. Dana/Ford (1932), 691. Am. Min. 47 (1962), 539.

Nasinite, $Na_2B_5O_8(OH) \cdot 2H_2O$, [G] Orth. Lardarello, Val di Cecina, Pisa, Toscana, Italy. H=?, D_m= 2.12, D_c= 2.134, Vc 10. Cipriani & Vannuccini, 1961. Am. Min. 48(1963), 709 (Abst.). Acta Cryst. B31 (1975), 2405.

Nasledovite, $PbMn_3Al_4O_5(SO_4)(CO_3)_4 \cdot 5H_2O$, [Q] Syst=? Silky white. Altyn-Topken mine, Sardob, Central Asia, USSR. H= 2, D_m= 3.069, D_c=?, VbD 02. Enikeev, 1958. Am. Min. 44(1959), 1325 (Abst.).

Nasonite, $Ca_4Pb_6(Si_2O_7)_3Cl_2$, [G] Hex. Translucent white. Franklin Furnace, Sussex Co., New Jersey, USA. H= 4, D_m= 5.4, D_c= 5.63, VIIIA 29. Penfield & Warren, 1899. Am. Min. 56 (1971), 1174. Acta Cryst. B43 (1987), 171. See also Ganomalite.

Nastrophite, $Na(Sr,Ba)PO_4 \cdot 9H_2O$, [A] Cub. Vitreous colorless. Alluaiv, Lovozero massif, Kola Peninsula, USSR. H= 2, D_m= 2.05, D_c= 2.03, VIIC 18. Khomyakov et al., 1981. Am. Min. 67(1982), 857 (Abst.). Min. Abstr. 83M/4251. See also Nabaphite.

Natalyite, $Na(V,Cr)Si_2O_6$, [A] Mon. Vitreous green. Slyudyanka complex, Lake Baikal region (S), USSR. H=?, D_m=?, D_c= 3.55, VIIID 01a. Reznitsky et al., 1985. Am. Min. 72(1987), 223 (Abst.).

Natanite, $FeSn(OH)_6$, [A] Cub. Vitreous greenish-brown. Trudov deposit and Mushiston deposit, Tadzhikistan, USSR. H= 4.7, D_m=?, D_c= 4.035, IVF 06. Marshukova et al., 1981. Am. Min. 67(1982), 1077 (Abst.).

Natisite, Na_2TiSiO_5, [A] Tet. Vitreous/adamantine yellow-green, greenish-grey. Mt. Karnasurt, Lovozero massif, Kola Peninsula, USSR. H= 3-4, D_m=?, D_c= 3.15, VIIID 28. Men'shikov et al., 1975. Am. Min. 61(1976), 339 (Abst.).

Natramblygonite, $(Na,Li)AlPO_4(F,OH)$, [Q] Tric. Canyon City, Fremont Co., Colorado, USA. H=?, D_m=?, D_c=?, VIIB 02. Schaller, 1911. Strunz (1970), 316. See also Amblygonite, Montebrasite, Natromontebrasite.

Natrite, γ-Na_2CO_3, [A] Mon. Vitreous, rose/yellow-orange. Khibina massif and Lovozero massif, Kola Peninsula, USSR. H= 3.5, D_m= 2.54, D_c= 2.546, VbA 06. Khomyakov, 1983. Am. Min. 68(1983), 281 (Abst.).

Natroalunite, $NaAl_3(SO_4)_2(OH)_6$, [G] Rhom. Vitreous white, greyish, yellowish, reddish, etc. National Bell mine, Silverton (near), Hinsdale Co., Colorado, USA. H= 3.5-4, D_m= 2.8, D_c= 2.819, VIB 03a. Hillebrand & Penfield, 1902. Dana, 7th ed.(1951), v.2, 556. N.Jb.Min.Mh. (1982), 534.

Natroapophyllite, $NaCa_4Si_8O_{20}F \cdot 8H_2O$, [A] Orth. Vitreous/pearly brownish-yellow, yellowish-brown, colorless. Sampo mine, Takahashi (W), Okayama, Honshu, Japan. H= 4-5, D_m= 2.50, D_c= 2.30, VIIIE 02. Matsueda et al., 1981. Am. Min. 66 (1981), 410. Am. Min. 66 (1981), 416. See also Fluorapophyllite, Hydroxyapophyllite.

Natrobistantite, $(Na,Cs)Bi(Ta,Nb,Sb)_4(O,OH)_{12}$, A Cub. Translucent blue-green, yellow-green. Kyokbogor, China. H=?, D_m= 6.2, D_c= 6.32, IVC 09a. Voloshin et al., 1983. Am. Min. 69(1984), 406 (Abst.). See also Cesstibtantite.

Natrochalcite, $NaCu_2(SO_4)_2(OH)\cdot H_2O$, G Mon. Vitreous green. Chuquicamata, Antofagasta, Chile. H= 4.5, D_m= 3.49, D_c= 3.487, VID 05. Palache & Warren, 1908. Dana, 7th ed.(1951), v.2, 602. Zts. Krist. 187 (1989), 239.

Natrodufrenite, $NaFe_6(PO_4)_4(OH)_6\cdot 2H_2O$, A Mon. Translucent pale blue-green. Rochefort-en-Terre, France. H=?, D_m= 3.23, D_c= 3.20, VIID 18. Fontan et al., 1982. Am. Min. 68(1983), 1039 (Abst.). See also Dufrenite, Burangaite.

Natrofairchildite, $Na_2Ca(CO_3)_2$, Q Hex. Vitreous white. Vourijärvi massif, Kola Peninsula, USSR. H= 2.5, D_m=?, D_c=?, VbA 05. Kapustin, 1971. Am. Min. 60(1975), 487 (Abst.). See also Nyerereite.

Natrojarosite, $NaFe_3(SO_4)_2(OH)_6$, G Rhom. Vitreous yellow, brown. Soda Springs Valley, Esmeralda Co., Nevada, USA. H= 3, D_m= 3.18, D_c= 3.122, VIB 03a. Hillebrand & Penfield, 1902. Dana, 7th ed.(1951), v.2, 563. Min. Abstr. 77-1496.

Natrolite, $Na_2(Al_2Si_3)O_{10}\cdot 2H_2O$, G Orth. Vitreous/pearly white, colorless, greyish, yellowish, etc. Hegau (Högau), Germany. H= 5-5.5, D_m= 2.2, D_c= 2.25, VIIIF 10. Klaproth, 1803. DHZ (1963), v.4, 358. Acta Cryst. C40 (1984), 1658. See also Tetranatrolite.

Natromontebrasite, $(Na,Li)AlPO_4(OH,F)$, G Tric. Vitreous/greasy white, yellowish, pink, greenish, etc. Canyon City, Fremont Co., Colorado, USA. H= 5.5-6, D_m= 3.1, D_c=?, VIIB 02. Gonnard, 1913. Dana, 7th ed. (1951), v.2, 823. See also Natramblygonite.

Natron, $Na_2CO_3\cdot 10H_2O$, G Mon. Vitreous colorless, white. Egypt (Ancient). H= 1-1.5, D_m= 1.44, D_c= 1.46, VbC 02. Wallerius, 1747. JCPDS 15-800.

Natronambulite, $(Na,Li)Mn_4Si_5O_{14}(OH)$, A Tric. Vitreous pinkish-orange. Tanohata mine, Iwate, Japan. H= 5.5-6, D_m= 3.51, D_c= 3.50, VIIID 13. Matsubara et al., 1985. Am. Min. 72(1987), 224 (Abst.). See also Nambulite.

Natroniobite, $NaNbO_3$, Q Mon. Translucent yellowish, brownish, blackish. Lesnaya Baraka and Sallanlatvi massifs, Kola Peninsula, USSR. H= 5.5-6, D_m= 4.40, D_c=?, IVC 07. Bulakh et al., 1960. Am. Min. 47(1962), 1483 (Abst.). See also Lueshite.

Natrophilite, $NaMnPO_4$, G Orth. Resinous/adamantine yellow. Branchville, Fairfield Co., Connecticut, USA. H= 4.5-5, D_m= 3.41, D_c= 3.47, VIIA 02. Brush & Dana, 1890. Dana, 7th ed.(1951), v.2, 670. Am. Min. 57 (1972), 1333.

Natrophosphate, $Na_7(PO_4)_2(F,OH)\cdot 19H_2O$, A Cub. Vitreous/greasy white. Yukspor Mts., Khibina massif, Kola Peninsula, USSR. H= 2.5, D_m= 1.76, D_c= 1.77, VIIC 11. Kapustin et al., 1972. Am. Min. 66(1981), 1281. Acta Cryst. 30 (1974), 2218.

Natrosilite, $Na_2Si_2O_5$, A Mon. Translucent colorless. Mt. Karnasurt, Lovozero massif, Kola Peninsula, USSR. H=?, D_m= 2.48, D_c= 2.51, VIIIE 03. Timoshenkov et al., 1975. Am. Min. 61(1976), 339 (Abst.).

Natrotantite, $Na_2Ta_4O_{11}$, A Rhom. Adamantine colorless, yellowish. Kola Peninsula, USSR. H= 7, D_m=?, D_c= 7.706, IVD 17. Voloshin et al., 1981. Am. Min. 67(1982), 413 (Abst.). Bull. Min. 108 (1985), 541.

Naujakasite, $Na_6(Fe,Mn)Al_4Si_8O_{26}$, G Mon. Pearly grey, white, greenish when fresh. Naujakasik, Ilímaussaq, Greenland (S). H= 2-3, D_m= 2.622, D_c= 2.71, VIIIE 23. Bøggild, 1933. Am. Min. 53(1968), 1780 (Abst.). Min. Abstr. 76-1417.

Naumannite, Ag_2Se, G Orth. Metallic black; white in reflected light. Tilkerode, Harz, Germany. H= 2.5, D_m= 7.4, D_c= 7.866, IIA 03. Haidinger, 1845. Can. Min. 12 (1974), 365. Am. Min. 56 (1971), 1882.

Naurodite, D An alkali amphibole. Am. Min. 63 (1978), 1023. von Knebel, 1903.

Navajoite, $V_2O_5 \cdot 3H_2O$, G Mon. Silky dark brown. Monument #2 mine, Monument Valley, Apache Co., Arizona, USA. H= < 2, D_m= 2.56, D_c= 3.038, IVF 09. Weeks et al., 1955. Am. Min. 40 (1955), 207.

Nealite, $FePb_4(AsO_4)_2Cl_4$, A Tric. Adamantine orange. Lávrion (Laurium), Attikí, Greece. H=?, D_m=?, D_c= 5.88, IIID 03. Dunn & Rouse, 1980. Min. Record 11 (1980), 299.

Nefedovite, $Na_5Ca_4(PO_4)_4F$, A Tric. Vitreous colorless. Khibina massif, Kola Peninsula, USSR. H= 4.5, D_m= 3.01, D_c= 3.05, VIIB 18. Khomyakov et al., 1983. Am. Min. 69(1984), 812 (Abst.). Min. Abstr. 89M/4119.

Neighborite, $NaMgF_3$, G Orth. Vitreous colorless, pink, brown, cream. South Ouray, Green River Formation, Uintah Co., Utah, USA. H= 4.5, D_m= 3.03, D_c= 3.06, IIIB 05. Chao et al., 1961. Am. Min. 46 (1961), 379.

Nekoite, $Ca_3Si_6O_{15} \cdot 7H_2O$, G Tric. Wet Weather Quarry, Crestmore, Riverside Co., California, USA. H=?, D_m= 2.23, D_c= 2.21, VIIID 10. Gard & Taylor, 1956. Min. Mag. 31 (1956), 5. Am. Min. 65 (1980), 1270.

Nekrasovite, $Cu_{26}V_2Sn_6S_{32}$, A Cub. Metallic pale brown. Kairagach deposit, Kuramin Mts., Uzbekistan (E), USSR. H= 4.5, D_m=?, D_c= 4.62, IIB 03c. Kovalenker et al., 1984. Am. Min. 70(1985), 437 (Abst.). See also Colusite, Germanite.

Nelenite, $(Mn,Fe)_{16}As_3Si_{12}O_{36}(OH)_{17}$, A Mon. Vitreous/resinous light/medium brown. Trotter Shaft, Franklin, Sussex Co., New Jersey, USA. H= 5, D_m= 3.46, D_c= 3.45, VIIIE 12. Dunn & Peacor, 1984. Min. Mag. 48 (1984), 271. See also Schallerite.

Neltnerite, $CaMn_6SiO_{12}$, A Tet. Sub-metallic black; grey/brownish-grey in reflected light. Tachgagalt, Morocco. H= 6, D_m= 4.63, D_c= 4.65, IVC 03. Baudracco-Gritti et al., 1982. Am. Min. 68(1983), 282 (Abst.). See also Braunite.

Nenadkevichite, $Na(Nb,Ti)Si_2O_6(O,OH) \cdot 2H_2O$, G Orth. Dull dark brown, brown, rose. Kola Peninsula, USSR. H= 5, D_m= 2.78, D_c= 2.73, VIIIB 11. Kuz'menko & Kazakova, 1955. Min. Abstr. 87M/1267. Acta Cryst. B29 (1973), 1432.

Nenadkevite, D A mixture of uraninite, boltwoodite and uranium hydroxides. Am. Min. 62 (1977), 1261 (Abst.). Polykarpova, 1956.

Neotocite, $(Mn,Fe)SiO_3 \cdot H_2O(?)$, G Amor. Vitreous/resinous red, brown, black. Gestrikland or Pajsberg, Filipstad, Värmland, Sweden. H= 3-4, D_m= 2.7, D_c=?, VIIIE 08d. Nordenskiöld, 1849. Min. Mag. 42 (1978), 279. Min. Abstr. 83M/2626.

Nepheline, $(Na,K)AlSiO_4$, G Hex. Vitreous/greasy white, yellowish, green, grey, etc. Monte Somma, Mt. Vesuvius, Napoli, Campania, Italy. H= 5.5-6, D_m= 2.63, D_c= 2.63, VIIIF 01. Haüy, 1800. DHZ (1963), v.4, 231. Bull. Min. 107 (1984), 499.

Népouite, $Ni_3Si_2O_5(OH)_4$, G Mon. Translucent green. Népoui, New Caledonia. H= 2-2.5, D_m= 3, D_c= 3.562, VIIIE 10b. Glasser, 1907. Am. Min. 60 (1975), 863. See also Lizardite, Pecoraite.

Neptunite, $KNa_2Li(Fe,Mg,Mn)_2Ti_2Si_8O_{24}$, G Mon. Black. Narsarsuk, Greenland (S). H= 5-6, D_m= 3.23, D_c= 3.14, VIIID 22. Flink, 1893. Am. Min. 57 (1972), 85. Acta Cryst. 21 (1966), 200. See also Mangan-neptunite.

Nesquehonite, $MgCO_3 \cdot 3H_2O$, G Mon. Vitreous colorless, white. Nesquehoning, Lansford (near), Carbon Co., Pennsylvania, USA. H= 2.5, D_m= 1.852, D_c= 1.85, VbC 01. Genth & Penfield, 1890. Dana, 7th ed.(1951), v.2, 225. Acta Cryst. B28 (1972), 1031.

Nevskite, $Bi(Se,S)$, A Hex. Metallic grey. Nevskii ore deposit, USSR (NE). H= 2-3, D_m=?, D_c= 7.85, IIB 18. Nechelyustov et al., 1984. Am. Min. 70(1985), 875 (Abst.).

Newberyite, $Mg(PO_3OH) \cdot 3H_2O$, G Orth. Vitreous colorless. Skipton Caves, Ballarat, Victoria, Australia. H= 3-3.5, D_m= 2.123, D_c= 2.119, VIIC 08. vom Rath, 1879. Min. Abstr. 84M/3849. Acta Cryst. 23 (1967), 418.

Neyite, $(Cu,Ag)_2Pb_7Bi_6S_{17}$, [A] Mon. Metallic grey; white in reflected light. Lime Creek Molybdenum deposit, Patsy Creek, Alice Arm (S), British Columbia, Canada. H= 2.5, D_m= 7.02, D_c= 7.16, IID 08. Drummond et al., 1969. Can. Min. 10 (1969), 90.

Niahite, $(NH_4)(Mn,Mg,Ca)PO_4 \cdot H_2O$, [A] Orth. Translucent pale orange. Niah Great Cave, Sarawak, Malaysia. H=?, D_m= 2.39, D_c= 2.437, VIIC 14. Bridge & Robinson, 1983. Min. Mag. 47 (1983), 79.

Nichromite, $(Ni,Co,Fe)(Cr,Fe,Al)_2O_4$, [Q] Cub. Metallic dark green. Barberton, Transvaal, South Africa. H=?, D_m=?, D_c= 5.24, IVB 01. DeWaal, 1978. JCPDS 23-1271.

Nickel, Ni, [G] Cub. Metallic white. New Caledonia. H= 3.5, D_m= 8.90, D_c= 8.909, IA 05. Ramdohr, 1967. Min. Mag. 40 (1975), 247.

Nickelalumite, $(Ni,Cu)Al_4(SO_4)(NO_3)_2(OH)_{12} \cdot 3H_2O$, [A] Mon. Mbobo Mkulu Cave, Transvaal (E), South Africa. H=?, D_m= 2.24, D_c= 2.28, IVF 03e. Martini, 1980. Am. Min. 67(1982), 416 (Abst.). See also Chalcoalumite, Mbobomkulite.

Nickelaustinite, $Ca(Ni,Zn)AsO_4(OH)$, [A] Orth. Silky/sub-adamantine yellowish-green, green. Bou Azzer, Morocco. H= 4, D_m=?, D_c= 4.27, VIIB 11b. Cesbron et al., 1987. Can. Min. 25 (1987), 401. Can. Min. 25 (1987), 401.

Nickelbischofite, $NiCl_2 \cdot 6H_2O$, [A] Mon. Vitreous green. Dumont Intrusion, Amos (W), Launay Twp., Abitibi Co., Québec, Canada. H= 1.5, D_m= 1.929, D_c= 1.932, IIIC 10a. Crook & Jambor, 1979. Can. Min. 17 (1979), 107. See also Bischofite.

Nickelblödite, $Na_2(Ni,Mg)(SO_4)_2 \cdot 4H_2O$, [A] Mon. Translucent pale yellowish-green, pale green. Carr Boyd mine, Kalgoorlie, Western Australia, Australia. H= 1.5, D_m= 2.43, D_c= 2.455, VIC 10. Nickel & Bridge, 1977. Min. Mag. 41 (1977), 37. See also Blödite.

Nickelboussingaultite, $(NH_4)_2(Ni,Mg)(SO_4)_2 \cdot 6H_2O$, [A] Mon. Translucent greenish-blue. Talnakh, Noril'sk, Siberia (N), USSR. H= 2.5, D_m=?, D_c= 1.86, VIC 11. Yakhontova et al., 1976. Am. Min. 71(1986), 1545 (Abst.). Acta Cryst. 17 (1964), 1478. See also Boussingaultite.

Nickelemelane, [D] Mixture. Min. Mag. 33 (1962), 261. Ginzberg & Rukavishnikova 1951.

Nickelhexahydrite, $(Ni,Mg,Fe)SO_4 \cdot 6H_2O$, [G] Mon. Vitreous bluish-green. Severnaya mine, Noril'sk, Siberia (N), USSR. H=?, D_m= 2.036, D_c= 2.05, VIC 03b. Oleinikov et al., 1965. Am. Min. 51(1966), 529 (Abst.). Acta Cryst. C44 (1988), 1869. See also Retgersite.

Nickeline, NiAs, [G] Hex. Metallic copper-red. H= 5-5.5, D_m= 7.78, D_c= 7.77, IIB 09a. Beudant, 1832. Econ. Geol. 43 (1948), 408. Zts. Krist. 84 (1933), 408.

Nickellinnaeite, $NiCo_2S_4$, [Q] Cub. H=?, D_m=?, D_c= 4.895, IIC 01. Minceva-Stefanova & Kostov, 1976. Min. Abstr. 80-4873.

Nickel-skutterudite, $NiAs_{2-3}$, [G] Cub. Bullard's Peak dist., Grant Co., New Mexico, USA. H=?, D_m=?, D_c= 6.40, IIC 12. Breithaupt, 1845. Strunz (1970), 138. See also Skutterudite.

Nickel-zippeite, $Ni_2(UO_2)_6(SO_4)_3(OH)_{10} \cdot 16H_2O$, [A] Syst=? Translucent orange, tan. Jáchymov (St. Joachimsthal), Západočeský kraj, Čechy (Bohemia), Czechoslovakia. H= 2, D_m=?, D_c=?, VID 08. Frondel et al., 1976. Can. Min. 14 (1976), 429. See also Zippeite, Sodium-zippeite, Zinc-zippeite, Magnesium-zippeite, Cobalt-zippeite.

Nifontovite, $Ca_3[BO(OH)_2]_3 \cdot 2H_2O$, [G] Mon. Vitreous colorless. Ural Mts., USSR. H= 3.5, D_m= 2.34, D_c= 2.36, Vc 04c. Malinko & Lisitsyn. Am. Min. 47(1962), 172 (Abst.). Min. Abstr. 79-2130.

Nigerite-24R, $(Fe,Zn)_4Sn_2(Al,Fe)_{15}O_{30}(OH)_2$, [P] Rhom. Mt. Garnet, Queensland (N), Australia. H=?, D_m=?, D_c= 3.33, IVC 05b. Grey & Gatehouse, 1979. Am. Min. 52 (1967), 864. Am. Min. 64 (1979), 1255. See also Nigerite-6H.

Nigerite-6H, $(Fe,Zn)_2Sn_2(Al,Fe)_{12}O_{22}(OH)_2$, [G] Rhom. Translucent light brown. Kabba province, Nigeria. H= 8.5, D_m= 4.51, D_c= 4.47, IVC 05b. Jacobsen & Webb, 1947. Am. Min. 64 (1979), 1255. See also Nigerite-24R.

Niggliite, PtSn, [G] Hex. Metallic pinkish-cream/blue. Waterfall Gorge, Insizwa, East Griqualand, South Africa. H= 5.5, D_m=?, D_c= 13.14, IA 11. Scholtz, 1936. Min. Mag. 38 (1972), 794.

Nimite, $(Ni,Mg)_6(Si,Al)_4O_{10}(OH)_8$, [G] Mon. Yellowish-green. Scotia Talc mine, Bon Accord, Barberton Dist., Transvaal, South Africa. H= 3, D_m= 3.123, D_c= 3.2, VIIIE 09a. De Waal, 1968. Am. Min. 55 (1970), 18.

Ningyoite, $(U,Ca,Ce)_2(PO_4)_2 \cdot 1\text{–}2H_2O$, [G] Orth. Translucent brownish-green, brown. Ningyo-Toge mine, Tottori, Honshu, Japan. H=?, D_m=?, D_c= 4.75, VIIC 19. Muto et al., 1959. Am. Min. 44 (1959), 633.

Niningerite, (Mg, Fe, Mn)S, [A] Cub. Metallic grey. Chondrite meteorites; St. Sauveur, Adhi-Kot, Indarch, St. Marks and Kota-Kota. H=?, D_m=?, D_c= 3.64, IIB 11. Keil & Snetsinger, 1967. Am. Min. 52(1967), 925 (Abst.). See also Alabandite.

Niobo-aeschynite-(Ce), $(Ce,Ca)(Nb,Ti)_2(O,OH)_6$, [G] Orth. Resinous red-brown, dark brown. Vishnevye Gory, Ural Mts., USSR. H= 4-5, D_m= 5.06, D_c=?, IVD 11. Zhabin et al., 1960. Am. Min. 60 (1975), 309. See also Aeschynite-(Ce), Niobo-aeschynite-(Nd), Aeschynite-(Y), Aeschynite-(Nd).

Niobo-aeschynite-(Nd), $(Nd,Ce)(Nb,Ti)_2(O,OH)_6$, [Q] Orth. H=?, D_m=?, D_c=?, IVD 11. Fleischer, 130 (1986). See also Aeschynite-(Nd), Aeschynite-(Y), Niobo-aeschynite-(Ce), Aeschynite-(Ce).

Nioboloparite, $(Na,Ce)(Ti,Nb)O_3$, [Q] Cub. Metallic black. Kola Peninsula, USSR. H= 5.5-6, D_m= 4.657, D_c= 4.603, IVC 07. Tikhonenkov & Kazakova, 1957. Am. Min. 43(1958), 792(Abst.).

Niobophyllite, $(K,Na)_3(Fe,Mn)_6(Nb,Ti)_2Si_8(O,OH)_{31}$, [A] Tric. Translucent brown. Ten Mile Lake area, Seal Lake, Labrador, Newfoundland, Canada. H=?, D_m= 3.42, D_c= 3.406, VIIID 25. Nickel et al., 1964. Can. Min. 8 (1964), 40.

Niocalite, $Ca_7Nb(Si_2O_7)_2O_3F$, [G] Mon. Vitreous light yellow. Bond zone, Oka, Deux-Montagnes Co., Québec, Canada. H= 6, D_m= 3.32, D_c= 3.051, VIIIB 08. Nickel, 1956. Can. Min. 6 (1958), 191. Min. Abstr. 84M/3796.

Nisbite, $NiSb_2$, [A] Orth. Metallic white. Trout Bay, Red Lake area, Mulcahy Twp., Kenora District, Ontario, Canada. H= 5, D_m=?, D_c= 8.0, IIC 07. Cabri et al., 1970. Can. Min. 10 (1970), 232.

Nissonite, $CuMgPO_4(OH) \cdot 2.5H_2O$, [A] Mon. Translucent bluish-green. Panoche Valley, San Benito Co., California, USA. H= 2.5, D_m= 2.73, D_c= 2.762, VIID 02. Mrose et al., 1966. Powd. Diff. 1 (1986), 331.

Niter, KNO_3, [G] Orth. Vitreous colorless, white, grey. H= 2, D_m= 2.10, D_c= 2.11, Va 01. NBS Circ. 539, v.3 (1954), 58.

Nitrammite, NH_4NO_3, [G] Orth. Nicojack Cave, Marion Co., Tennesseee, USA. H=?, D_m= 1.72, D_c= 1.721, Va 01. Shepard, 1857. Min. Abstr. 75-176. Acta Cryst. B28 (1972), 1357.

Nitratine, $NaNO_3$, [G] Rhom. Vitreous colorless, white. Tarapacá dist., Chile (N). H= 1.5-2, D_m= 2.26, D_c= 2.26, Va 01. Glocker, 1847. NBS Circ. 539, v.6 (1956), 50.

Nitrobarite, $Ba(NO_3)_2$, [G] Cub. Transparent colorless. Unknown locality, Chile. H=?, D_m= 3.250, D_c= 3.24, Va 02. Lewis, 1882. NBS Circ. 539, v.1 (1953), 81.

Nitrocalcite, $Ca(NO_3)_2 \cdot 4H_2O$, [G] Mon. Transparent white, grey. Nicojack Cave, Marion Co., Tennessee, USA. H=?, D_m= 1.90, D_c= 1.907, Va 03. Shepard, 1835. Dana, 7th ed.(1951), v.2, 306. Acta Cryst. 33B (1977), 1861.

Nitromagnesite, $Mg(NO_3)_2 \cdot 6H_2O$, G Mon. Vitreous colorless, white. Nicojack Cave, Marion Co., Tennessee, USA. H=?, D_m= 1.58, D_c= 1.64, Va 03. Shepard, 1835. Acta Cryst. 14 (1961), 1296.

Nobleite, $CaB_6O_9(OH)_2 \cdot 3H_2O$, G Mon. Sub-vitreous/pearly colorless, white. De Bely mine (NNW), Death Valley National Monument, Inyo Co., California, USA. H= 3, D_m= 2.09, D_c= 2.09, Vc 12a. Erd et al., 1961. Am. Min. 46 (1961), 560.

Nolanite, $(V,Fe,Ti)_{10}O_{14}(OH)_2$, G Hex. Black; dark brown/deep blue in reflected light. Fish Hook Bay, Lake Athabasca, Beaverlodge region, Saskatchewan, Canada. H= 6, D_m= 4.65, D_c= 4.60, IVC 05b. Robinson et al., 1957. Am. Min. 42 (1957), 619. Am. Min. 68 (1983), 833.

Nontronite, $Na_{0.3}Fe_2(Si,Al)_4O_{10}(OH)_2 \cdot nH_2O$, G Mon. Translucent yellow, yellow-green. Saint-Pardoux, Nontron, Dordogne, France. H=?, D_m=?, D_c= 1.92, VIIIE 08a. Berthier, 1827. Am. Min. 60 (1975), 840.

Noonkanbahite, D May be shcherbakovite. Min. Mag. 36 (1968), 1144. Prider, 1965.

Norbergite, $Mg_3(SiO_4)(F,OH)_2$, G Orth. Translucent tawny. Norberg, Sweden. H= 6.5, D_m= 3.177, D_c= 3.186, VIIIA 16. Geijer, 1926. DHZ, 2nd ed.(1982), v.1B, 378. Am. Min. 54 (1969), 376.

Nordenskiöldine, $CaSn(BO_3)_2$, G Rhom. Vitreous colorless, yellow. Arö, Langesundfjord, Norway. H= 5.5-6, D_m= 4.20, D_c= 4.314, Vc 01b. Brögger, 1887. Dana, 7th ed.(1951), v.2, 332. N.Jb.Min.Mh. (1986), 111. See also Tusionite, Dolomite.

Nordite-(Ce), $Na_3(Sr,Ca)(Ce,La)(Zn,Mg)Si_6O_{17}$, A Orth. Lovozero, Kola peninsula, USSR. H=?, D_m=?, D_c=?, VIIIB 02b. Semenov & Barinskii, 1958. Am. Min. 51 (1966), 152.

Nordite-(La), $Na_3(Sr,Ca)(La,Ce)(Zn,Mg)Si_6O_{17}$, G Orth. Semi-transparent light brown. Chinglusuai River, Lovozero massif, Kola Peninsula, USSR. H= 5-6, D_m= 3.430, D_c= 3.58, VIIIB 02b. Gerasimovsky, 1941. Am. Min. 28(1943), 282 (Abst.). Am. Min. 55 (1970), 1167.

Nordstrandite, $Al(OH)_3$, G Tric. Vitreous colorless, white, pink, beige, pale green. Bau, Gunony Kapor, Sarawak, Malaysia; Mt. Alifan-Mt. Lamlan Ridge, Guam (S). H= 3, D_m= 2.42, D_c= 2.454, IVF 01. Papée et al., 1958. Can. Min. 20 (1982), 77. See also Bayerite, Doyleite, Gibbsite.

Nordströmite, $CuPb_3Bi_7(S,Se)_{14}$, A Mon. Metallic grey; metallic white in reflected light. Falun, Kopparberg, Sweden. H= 2-2.5, D_m=?, D_c= 7.12, IID 08. Mumme, 1980. Am. Min. 65 (1980), 789. Can. Min. 18 (1980), 343.

Norrishite, $KLiMn_2Si_4O_{12}$, A Mon. Lustrous black. Hoskins mine, New South Wales, Australia. H= 2.5, D_m= 3.264, D_c= 3.21, VIIIE 05b. Eggleton & Ashley, 1989. Am. Min. 74 (1989), 1360.

Norsethite, $BaMg(CO_3)_2$, G Rhom. Vitreous colorless, white. Westvaco Trona mine, Green River Formation, Sweetwater Co., Wyoming, USA. H= 3.5, D_m= 3.84, D_c= 3.83, VbA 03a. Mrose et al., 1961. Am. Min. 46 (1961), 420. Zts. Krist. 171 (1985), 275.

Northupite, $Na_3Mg(CO_3)_2Cl$, G Cub. Vitreous colorless, pale yellow, grey, brown. Searles Lake, San Bernardino Co., California, USA. H= 3.5-4, D_m= 2.380, D_c= 2.373, VbB 04. Foote, 1895. Dana, 7th ed.(1951), v.2, 278. TMPM 22 (1975), 158. See also Tychite.

Nosean, $Na_8(Si_6Al_6)O_{24}(SO_4) \cdot H_2O$, G Cub. Translucent grey, brown, blue. H= 5.5, D_m= 2.3, D_c= 2.21, VIIIF 07. Klaproth, 1815. Am. Min. 74 (1989), 394. Can. Min. 27 (1989), 165.

Nováčekite, $Mg(UO_2)_2(AsO_4)_2 \cdot 9H_2O$, G Tet. Translucent yellow. Schneeberg, Sachsen (Saxony), Germany. H= 2.5, D_m= 3.25, D_c= 3.28, VIID 20a. Frondel, 1951. Am. Min. 36 (1951), 680.

Novákite, $(Cu,Ag)_{21}As_{10}$, G Mon. Metallic grey; creamy-white in reflected light. Černý důl (Schwarzenthal), Krkonoše (Giant Mtns.), Vychodočeský kraj, Čechy (Bohemia), Czechoslovakia. H= 3-3.5, D_m= 8.1, D_c= 8.01, IIG 01. Johan & Hak, 1961. TMPM 34 (1985), 167.

Nowackiite, $Cu_6Zn_3As_4S_{12}$, G Rhom. Lengenbach quarry, Binntal, Valais (Wallis), Switzerland. H=?, D_m=?, D_c= 4.38, IID 01c. Marumo & Burri, 1965. Am. Min. 51(1966), 532 (Abst.). Zts. Krist. 124 (1967), 352. See also Aktashite, Gruzdevite.

Nsutite, $Mn(O,OH)_2$, G Hex. Earthy/metallic grey, brown. Nsuta, Ghana. H= 6.5-8.5, D_m= 4.5, D_c= 4.86, IVD 03a. Zwicker et al., 1962. Am. Min. 50 (1965), 170. Acta Cryst. 12 (1959), 341. See also Pyrolusite, Ramsdellite, Vernadite.

Nuffieldite, $CuPb_2(Pb,Bi)Bi_2S_7$, A Orth. Metallic grey; creamy-white in reflected light. Patsy Creek, Lime Creek stock, Alice Arm (S), British Columbia, Canada. H= 3.5, D_m= 7.01, D_c= 7.21, IID 05a. Kingston, 1968. Can. Min. 9 (1968), 439 343. Zts. Krist. 138 (1973), 343.

Nukundamite, $Cu_{3.4}Fe_{0.6}S_4$, A Rhom. Metallic reddish-orange/pale grey. Undu mine (dumps), Vanu Levu, Nukundamu, Fiji. H= 3, D_m= 4.49, D_c= 4.53, IIB 15. Rice et al., 1979. Min. Mag. 43 (1979), 193. Am. Min. 66 (1981), 398. See also Idaite.

Nullaginite, $Ni_2CO_3(OH)_2$, A Mon. Silky green. Otway deposit, Nullagine dist., Western Australia, Australia. H= 2, D_m= 3.56, D_c= 4.07, VbB 01. Nickel & Berry, 1981. Can. Min. 19 (1981),315.

Nuolaite, D A mixture of yttropyrochlore and other niobium minerals. Am. Min. 62 (1977), 403. Lokka, 1928.

Nyböite, $Na_3Mg_3Al_2(Si_7Al)O_{22}(OH)_2$, A Mon. Nybö pod, Norway. H=?, D_m=?, D_c= 3.031, VIIID 05d. Ungaretti et al., 1981. Bull. Min. 104 (1981), 400. Am. Min. 67(1982), 858 (Abst.).

Nyerereite, $Na_2Ca(CO_3)_2$, A Orth. Transparent colorless. Carbonatite lava (1963 flow), Ol Doinyo Lengai Volcano, Tanzania. H=?, D_m= 2.541, D_c= 2.417, VbA 05. Milton & Ingram, 1963. Am. Min. 60(1975), 487 (Abst.). Zts. Krist. 145 (1977), 73.

O

Oboyerite, $H_6Pb_6(TeO_3)_3(TeO_6)_2 \cdot 2H_2O$, [A] Tric. Translucent white. Grand Central mine, Tombstone, Cochise Co., Arizona, USA. H= 1.5, D_m= 6.4, D_c= 6.707, VIG 08. Williams, 1979. Am. Min. 66(1981), 220 (Abst.).

Obradovicite, $H_4(K,Na)CuFe_2(AsO_4)(MoO_4)_5 \cdot 12H_2O$, [A] Orth. Translucent green. Chuquicamata, Antofagasta, Chile. H= 2.5, D_m= 3.55, D_c= 3.68, VIID 15b. Finney et al., 1986. Min. Mag. 50 (1986), 283.

O'Danielite, $H_2Na(Zn,Mg)_3(AsO_4)_3$, [A] Mon. Vitreous pale violet. Tsumeb, Namibia. H= 3, D_m=?, D_c= 4.83, VIIA 07. Keller et al., 1981. Am. Min. 66(1981), 1276 (Abst.). N.Jb.Min.Mh. (1988), 395. See also Johillerite.

Odinite-1M, $(Fe,Mg,Al)_{2.4}(Si,Al)_2O_5(OH)_4$, [A] Mon. Translucent silky-green/dark green. New Caledonia (near), Pacific Ocean. H=?, D_m= 2.6, D_c= 2.77, VIIIE 10a. Bailey, 1988. Clay Minerals 23 (1988), 237.

Odinite-1T, $(Fe,Mg,Al)_{2.4}(Si,Al)_2O_5(OH)_4$, [P] Tet. Translucent silky-green/dark green. Los Islands, Guinea. H=?, D_m= 2.6, D_c= 2.78, VIIIE 10a. Bailey, 1988. Clay Minerals 23 (1988), 237.

Offrétite, $KCaMg(Al_5Si_{13})O_{36} \cdot 15H_2O$, [G] Hex. Vitreous colorless, white. Mont Semiol (Simiouse), Montbrison, Loire, France. H= 4, D_m= 62.1, D_c= 2.076, VIIIF 15. Gonnard, 1890. Natural Zeolites (1985), 209. Acta Cryst. B28 (1972), 825.

Ogdensburgite, $(Ca,Zn,Mn)_4Fe_6(AsO_4)_5(OH)_{11} \cdot 5H_2O$, [A] Orth. Resinous dark brownish-red. Sterling Hill mine, Ogdensburg, Sussex Co., New Jersey, USA. H= 2, D_m= 3.11, D_c= 3.39, VIID 12. Dunn, 1981. Am. Min. 72 (1987), 409.

Ohmilite, $Sr_3(Ti,Fe)(Si_2O_6)_2(O,OH) \cdot 2H_2O$, [A] Mon. Translucent pink, pinkish-brown. Ohmi, Niigata, Japan. H= 3.5, D_m= 3.38, D_c= 3.37, VIIID 21. Komatsu et al., 1973. Min. Journ. 7 (1973), 298. Am. Min. 68 (1983), 811.

Ojuelaite, $ZnFe_2(AsO_4)_2(OH)_2 \cdot 4H_2O$, [A] Mon. Translucent greenish-yellow. Ojuela mine, Mapimi, Durango, Mexico. H= 3, D_m= 2.39, D_c= 2.39, VIID 27. Cesbron et al., 1981. Bull. Min. 104 (1981), 582. See also Arthurite, Earlshannonite, Whitmoreite.

Okanoganite-(Y), $(Na,Ca)_3(Y,Ce)_{12}B_2Si_6O_{27}F_{14}$, [A] Rhom. Translucent tan, pale pink. Washington Pass, Okanogan Co., Washington, USA. H= 4, D_m= 4.35, D_c= 4.37, VIIIA 23. Boggs, 1980. Am. Min. 65 (1980), 1138.

Okenite, $Ca_{10}Si_{18}O_{46} \cdot 18H_2O$, [G] Tric. Sub-pearly white, yellowish, bluish. Disko Island, Greenland (W). H= 4.5-5, D_m= 2.28, D_c= 2.29, VIIID 10. Kobell, 1828. Dana, 6th ed. (1892), 565. Am. Min. 68 (1983), 614.

Okhotskite, $Ca_2(Mn,Mg)(Mn,Al,Fe)_2Si_3O_{10}(OH)_4$, [A] Mon. Vitreous orange. Kokuriki mine, Hokkaido, Japan. H= 6, D_m=?, D_c= 3.40, VIIIB 26. Togari & Akasaka, 1987. Min. Mag. 51 (1987), 611.

Oldhamite, $(Ca,Mg)S$, [G] Cub. Transparent brown. Bustee meteorite, Gorakhpur (near), Basti dist., Uttar Pradesh, India. H= 4, D_m= 2.58, D_c= 2.591, IIB 11. Maskelyne, 1862. Dana, 7th ed.(1944), v.1, 208.

Olenite, $Na_{0.5}Al_9(BO_3)_3Si_6O_{18}(O,OH)_4$, [A] Rhom. Vitreous pale pink. Olenëk River Basin, USSR. H= 7, D_m= 3.010, D_c= 3.12, VIIIC 08. Sokolov et al., 1986. Am. Min. 73 (1988), 441.

Olgite, $Na(Sr,Ba)PO_4$, [A] Rhom. Vitreous bright blue, bluish-green. Mt. Karnasurt, Lovozero massif, Kola Peninsula, USSR. H= 4.5, D_m= 3.94, D_c= 3.52, VIIA 02. Khomyakov et al., 1980. Am. Min. 66(1981), 438 (Abst.). Sov.Phys.Cryst. 29(1984), 633.

Oligoclase, $(Na,Ca)(Si,Al)_4O_8$, [G] Tric. Vitreous/pearly whitish, greyish, green, reddish, etc. Danviks-Zoll, Norway(?). H= 6-7, D_m= 2.66, D_c= 2.65, VIIIF

03. Breithaupt, 1826. DHZ (1963), v.4, 94. Zts. Krist. 133 (1971), 43. See also Plagioclase.

Olivenite, $Cu_2AsO_4(OH)$, G Mon. Adamantine/vitreous olive-green, brown, yellow, grey, etc. Cornwall, England. H= 3, D_m= 4.46, D_c= 4.45, VIIB 04. Jameson, 1820. Acta Cryst. B34 (1978), 715. Acta Cryst. B33 (1977), 2628. See also Adamite, Libethenite.

Olivine, A group name for orthosilicates with the general formula $(Mg, Fe, Mn, Ni)_2SiO_4$. See Forsterite, Fayalite, Liebenbergite, Tephroite.

Olmsteadite, $KFe_2(Nb, Ta)O_2(PO_4)_2 \cdot 2H_2O$, A Orth. Sub-adamantine brown, red-brown, black. Hesnard mine, Keystone (near), Pennington Co., South Dakota, USA. H= 4, D_m= 3.33, D_c= 3.41, VIID 34. Moore et al., 1976. Am. Min. 61 (1976), 5. Am. Min. 61 (1976), 5. See also Johnwalkite.

Olovotantalite, D Unnecessary name for stannian manganotantalite. Min. Mag. 36 (1967), 133. Matias, 1961.

Olsacherite, $Pb_2(SeO_4)(SO_4)$, A Orth. Vitreous colorless. Pacajake mine, Colquechaca (Near), Potosí, Bolivia. H= 3-3.5, D_m=?, D_c= 6.55, VIA 07. Hurlbut & Aristarain, 1969. Am. Min. 54 (1969), 1519.

Olshanskyite, $Ca_3[B(OH)_4]_4(OH)_2$, A Mon.(?) Translucent colorless. Siberia (E), USSR. H= 4, D_m= 2.23, D_c=?, Vc 03a. Bogomolov et al., 1969. Am. Min. 54(1969), 1737 (Abst.).

Olympite, Na_3PO_4, A Orth. Vitreous colorless. Mt. Rasvumchorr, Khibina massif, Kola Peninsula, USSR. H= 4, D_m= 2.8, D_c= 2.85, VIIA 14. Khomyakov et al., 1980. Am. Min. 66(1981), 438 (Abst.).

Omeiite, $(Os, Ru)As_2$, Q Orth. Metallic white. Omeishan Mt. (Near), Sichuan, China. H=?, D_m=?, D_c= 11.20, IIC 07. Ren et al., 1978. Am. Min. 64(1979), 464 (Abst.). See also Anduoite.

Omphacite, $(Ca, Na)(Mg, Fe, Al)Si_2O_6$, G Mon. Vitreous green. Hof (near), Bayern (Bavaria), Germany. H= 5-6, D_m= 3.3, D_c= 3.39, VIIID 01a. Werner, 1815. DHZ, 2nd ed.(1978), v.21, 424. Am. Min. 60 (1975), 634. See also Augite, Jadeite, Acmite.

Ondrejite, D A mixture of huntite and magnesite. Am. Min. 49 (1964), 1502. Kaspar, 1944.

Onoratoite, $Sb_8O_{11}Cl_2$, A Mon. Transparent colorless. Cetine mine, Rosia, Siena, Toscana, Italy. H=?, D_m= 5.3, D_c= 5.425, IVC 08. Belluomini et al., 1968. Min. Mag. 36 (1968), 1037. Acta Cryst. C40 (1984), 1506.

Oosterboschite, $(Pd, Cu)_7Se_5$, A Orth. Metallic white-yellow. Musonoi deposit, Shaba, Zaïre. H= 4.5, D_m=?, D_c= 8.48, IIA 14. Johan et al., 1970. Bull. Min. 93 (1970), 476.

Opal, $SiO_2 \cdot nH_2O$, G Amor. Transparent, colorless, white, yellow, brown, etc. H= 5.5-6.5, D_m= 2, D_c=?, IVD 01. Am. Min. 60 (1975), 749.

Orcelite, $Ni_{5-x}As_2$, G Hex. Metallic yellowish-white in reflected light. Tiebaghi massif, New Caledonia. H=?, D_m= 6.5, D_c= 8.44, IIG 03. Caillére et al., 1959. Bull. Min. 84 (1961), 9.

Ordoñezite, $ZnSb_2O_6$, G Tet. Adamantine light brown, dark brown, colorless, grey, etc. Santin mine, Santa Catarina, Cerro de las Fajas, Guanajuato, Mexico. H= 6.5, D_m= 6.635, D_c= 6.657, IVD 04. Switzer & Foshag, 1955. Am. Min. 40 (1955), 64. See also Byströmite.

Örebroite, $Mn_6(Fe, Sb)_2(SiO_4)_2(O, OH)_6$, A Hex. Vitreous dark brown. Sjögruvan, Grythyttan (near), Örebro, Sweden. H= 4, D_m=?, D_c= 4.77, VIIIA 17. Dunn et al., 1986. Am. Min. 71 (1986), 1522.

Oregonite, $FeNi_2As_2$, G Hex. Metallic white. Josephine Creek, Oregon, USA. H=?, D_m=?, D_c= 6.92, IIA 06. Ramdohr & Schmitt, 1959. Am. Min. 45(1960), 1130 (Abst.).

Orickite, $CuFeS_2 \cdot nH_2O$, [A] Hex. Metallic yellow. Coyote Peak, Humboldt Co., California, USA. H=?, D_m=?, D_c= 4.212, IIE 01. Erd & Czamanske, 1983. Am. Min. 68 (1983), 245.

Orientite, $Ca_2Mn_3Si_3O_{10}(OH)_4$, [A] Orth. Translucent brown, brownish-red. Oriente Province, Cuba. H=?, D_m= 3.33, D_c= 3.48, VIIIB 19. Moore et al., 1979. Min. Mag. 43 (1979), 325. Am. Min. 70 (1985), 171.

Orpheite, $H_6Pb_{10}Al_{20}(PO_4)_{12}(SO_4)_5(OH)_{40} \cdot 11H_2O$(?), [A] Rhom. Vitreous colorless, grey, pale blue, yellow-green. Madyarovo deposit, Rhodope Mts., Bulgaria. H= 3.5, D_m= 3.75, D_c= 4.02, VIB 03b. Kolkovski, 1971-72. Am. Min. 61(1976), 176 (Abst.).

Orpiment, As_2S_3, [G] Mon. Translucent resinous yellow. H= 1.5-2, D_m= 3.49, D_c= 3.48, IIB 19. Pliny, 77. Dana, 7th ed.(1944), v.1, 266. Zts. Krist. 136 (1972), 48. See also Laphamite.

Ortho-armalcolite, [D] Unnecessary name for a variety of armalcolite. Min. Mag. 43 (1980), 1055. Haggerty, 1973.

Orthobrannerite, $UTi_2O_6(OH)$, [G] Orth. Adamantine black; greyish-white in reflected light. China. H= 5.3, D_m= 5.46, D_c= 5.46, IVD 12. Am. Min. 64(1979), 656 (Abst.). See also Brannerite.

Orthochamosite, $(Fe, Mg)_5Al(Si_3Al)O_{10}(O, OH)_8$, [G] Orth. Translucent green. Kaňk, Kutná Hora (Near), Středočeský kraj, Čechy (Bohemia), Czechoslovakia. H= 2, D_m= 3.078, D_c= 3.21, VIIIE 09a. Novak et al., 1957. Am. Min. 43(1958), 792 (Abst.). See also Chamosite.

Orthochrysotile, $Mg_3Si_2O_5(OH)_4$, [G] Orth. Translucent yellow, white, grey, green. Cuddapah, Madras, India. H= 2.5, D_m= 2.5, D_c= 2.56, VIIIE 10b. Whittaker, 1951. Rev. Min. (1988), 91. Acta Cryst. 9 (1956), 862. See also Antigorite, Clinochrysotile, Lizardite, Parachrysotile.

Orthoclase, $KAlSi_3O_8$, [G] Mon. Vitreous colorless, white, pale yellow, pink, grey, etc. H= 6-6.5, D_m= 2.563, D_c= 2.543, VIIIF 03. Breithaupt, 1823. DHZ (1963), v.4, 6. Am. Min. 58 (1973), 500. See also Microcline, Celsian, Hyalophane.

Orthoericssonite, $(Ba, Sr)(Fe, Ti)(Mn, Fe)Si_2O_7(O, OH)_2$, [A] Orth. Translucent reddish-black. Långban mine, Filipstad (near), Värmland, Sweden. H= 4.5, D_m= 4.21, D_c= 4.26, VIIIB 10. Moore, 1971. Min. Jour. 7 (1975), 513. Min. Jour. 10 (1980), 107. See also Ericssonite.

Orthojoaquinite-(Ce), $NaBa_2FeCe_2Ti_2(SiO_3)_8O_2(O, OH) \cdot H_2O$, [A] Orth. New Idria (near), San Benito Co., California, USA. H=?, D_m=?, D_c= 4.14, VIIIC 05. Wise, 1982. Am. Min. 67 (1982), 809. See also Joaquinite-(Ce), Bario-orthojoaquinite.

Orthopinakiolite, $(Mg, Mn, Fe)_3O_2(BO_3)$, [G] Orth. Metallic black. Långban mine, Filipstad (near), Värmland, Sweden. H= 6, D_m= 4.03, D_c= 4.06, Vc 01c. Randmets, 1960. Am. Min. 46(1961), 768 (Abst.). Can. Min. 16 (1978), 475. See also Fredrikssonite, Pinakiolite, Takeuchiite.

Orthorhombic låvenite, [D] Insufficient evidence that the mineral is orthorhombic. Min. Mag. 36 (1968), 1144. Portnov et al., 1966.

Orthoserpierite, $Ca(Cu, Zn)_4(SO_4)_2(OH)_6 \cdot 3H_2O$, [A] Orth. Vitreous blue. Chessy mine, Rhône, France. H=?, D_m= 3.00, D_c= 3.07, VID 14. Sarp, 1985. Am. Min. 72(1987), 1026 (Abst.). See also Serpierite.

Orthozoisite, [D] = zoisite. Min. Mag. 38 (1971), 103. Nitsch & Winkler.

Osarizawaite, $CuPb(Al, Fe)_2(SO_4)_2(OH)_6$, [G] Rhom. Earthy greenish-yellow, pale green. Osarizawa mine, Akita, Japan. H=?, D_m= 3.9, D_c= 4.20, VIB 03a. Taguchi, 1961. Am. Min. 47(1962), 1216 (Abst.). Min. Abstr. 81-1233.

Osarsite, $(Os, Ru)AsS$, [A] Mon. Metallic grey. Gold Bluff, Humboldt Co., California, USA. H=?, D_m=?, D_c= 8.44, IIC 09. Snetsinger, 1972. Am. Min. 57 (1972), 1029. See also Ruarsite.

Osbornite, TiN, G Cub. Golden yellow. Bustee meteorite, Gorakhpur (near), Basti dist., Uttar Pradesh, India. H=?, D_m= 5.25, D_c= 5.386, IC 01. Maskelyne, 1870. Min. Mag. 26 (1941), 36. Struct. Repts. 18 (1954), 251.

Osmiridium, (Ir, Os), A Cub. Metallic white. H= 6-7, D_m= 20, D_c= 21.69, IA 07. Glocker, 1831. Can. Min. 12 (1973), 104. See also Iridium.

Osmium, Os, A Hex. Metallic white. H= 7, D_m= 22.48, D_c= 22.627, IA 08. Berzelius, 1819. Can. Min. 12 (1973), 104.

Osumilite, $K(Fe,Mg)_2(Al,Fe)_3(Si,Al)_{12}O_{30}$, G Hex. Black, dark blue, dark brown, grey. Osumi, Japan. H=?, D_m= 2.6, D_c= 2.34, VIIIC 10. Miyashiro, 1953. DHZ, 2nd ed.(1986), v.1B, 541. Am. Min. 73 (1988), 585.

Osumilite-(Mg), $KMg_2(Al,Fe)_3(Si,Al)_{12}O_{30}$, G Hex. Translucent black, blue, brown, grey, pink. Tieveragh, Antrim Co., Northern Ireland. H=?, D_m= 2.6, D_c= 2.67, VIIIC 10. Miyashiro, 1953. DHZ (1986), 2 ed., v.IB, 541. Am. Min. 73 (1988), 585.

Otavite, $CdCO_3$, G Rhom. Adamantine white, yellow-brown, reddish. Otavi, Namibia. H=?, D_m= 5.03, D_c= 5.02, VbA 02. Schneider, 1906. NBS Circ. 539, v.7 (1957), 11.

Otjisumeite, $PbGe_4O_9$, A Tric. Greasy white, colorless. Tsumeb, Namibia. H= 3, D_m=?, D_c= 5.77, VIIIH 01. Keller et al., 1981. Am. Min. 72(1987), 1026 (Abst.).

Ottemannite, Sn_2S_3, G Orth. Metallic grey. Maria Teresa mine, Huari, Oruro, Bolivia. H= 2, D_m= 4.87, D_c= 4.75, IIC 17. Moh, 1966. Am. Min. 51(1966), 1551 (Abst.). Acta Cryst. B38 (1982), 2022.

Ottrelite, $(Mn,Fe,Mg)Al_2SiO_5(OH)_2$, G Mon. Adamantine yellow-green, blackish-grey, greenish-grey, black. Ottrez, Ardennes, Belgium. H= 6-7, D_m= 3.52, D_c= 3.49, VIIIA 22. Dethier, 1819. Bull. Mineral. 101 (1978), 548. See also Chloritoid, Magnesiochloritoid.

Otwayite, $Ni_2CO_3(OH)_2 \cdot H_2O$, A Orth. Silky green. Otway nickel deposit, Nullagine dist., Western Australia, Australia. H= 4, D_m= 3.41, D_c= 3.346, VbD 01. Nickel et al., 1977. Am. Min. 62 (1977), 999.

Ourayite, $Ag_3Pb_4Bi_5S_{13}$, A Orth. Metallic grey; white in reflected light. Old Lout mine, Ouray, San Juan Co., Colorado, USA. H=?, D_m=?, D_c= 7.24, IID 09b. Makovicky & Karup-Møller, 1977. Can. Min. 22 (1984), 565.

Oursinite, $(Co,Mg)(UO_2)_2Si_2O_7 \cdot 6H_2O$, A Orth. Translucent pale yellow. Shinkolobwe, Shaba, Zaïre. H=?, D_m=?, D_c= 3.674, VIIIA 25. Deliens & Piret, 1983. Am. Min. 69(1984), 567 (Abst.).

Overite, $CaMgAl(PO_4)_2(OH) \cdot 4H_2O$, G Orth. Vitreous light green, colorless. Clay Canyon (near), Fairfield, Utah Co., Utah, USA. H= 3.5-4, D_m= 2.53, D_c= 2.51, VIID 25. Larsen, 1940. Am. Min. 59 (1974), 48. Am. Min. 62 (1977), 692.

Owyheeite, $Ag_{3+x}Pb_{10-2x}Sb_{11+x}S_{28}$, G Orth. Metallic grey; yellowish-white/grey in reflected light. Poorman mine, Silver City Dist., Owyhee Co., Idaho, USA. H= 2.5, D_m= 6.25, D_c= 6.43, IID 07. Shannon, 1921. Am. Min. 70(1985), 440 (Abst.).

Oxammite, $(NH_4)_2C_2O_4 \cdot H_2O$, G Orth. Pulverulent colorless, yellowish-white. Guañape Islands, Peru. H= 2.5, D_m= 1.5, D_c= 1.501, IXA 01. Shepard, 1870. Am. Min. 36 (1951), 590. Acta Cryst. B28 (1972), 3340.

Oxy-apatite, $Ca_{10}(PO_4)_6O$, Q Hex. H=?, D_m=?, D_c= 3.084, VIIB 16. Rogers, 1912. Strunz (1970), 327. Zts. Krist. 81 (1931), 352. See also Apatite.

Oyelite, $Ca_{10}B_2Si_8O_{29} \cdot 12H_2O$, A Orth. Vitreous white. Jiro Oye, Fuka, Okayama, Honshu, Japan. H= 5, D_m= 2.62, D_c= 2.64, VIIID 10. Kusachi et al., 1984. Am. Min. 71(1986), 230 (Abst.).

P

Pääkkönenite, Sb_2AsS_2, [A] Mon. Metallic grey. Seinäjoki, Finland. H= 2.5, D_m=?, D_c= 5.21, IIB 19. Borodaev et al., 1981. Am. Min. 67(1982), 858 (Abst.).

Pabstite, $Ba(Sn,Ti)Si_3O_9$, [A] Hex. Vitreous colorless, white. Pacific Limestone Products Quarry, Santa Cruz, Santa Cruz Co., California, USA. H= 6, D_m= 4.03, D_c= 4.07, VIIIC 01. Gross et al., 1965. Am. Min. 50 (1965), 1164. N.Jb.Min.Mh. (1987), 16. See also Bazirite, Benitoite.

Pachnolite, $NaCaAlF_6 \cdot H_2O$, [G] Mon. Vitreous colorless, white. Ivigtut, Greenland (SW). H= 3, D_m= 2.983, D_c= 2.983, IIIC 09. Knop, 1863. Dana, 7th ed.(1951), v.2, 114. Can. Min. 21 (1983), 561. See also Thomsenolite.

Paděraite, $Ag_{1.3}Cu_{5.9}Pb_{1.6}Bi_{11.2}S_{22}$, [A] Mon. Metallic grey; creamy white in reflected light. Baita Bihorului, Romania. H=?, D_m=?, D_c= 6.91, IID 06a. Mumme & Zak, 1985. Am. Min. 72(1987), 224 (Abst.). Can. Min. 24 (1986), 513. See also Cuprobismutite, Hodrushite.

Pahasapaite, $Li_8(Ca,Li,K)_{10.5}Be_{24}(PO_4)_{24} \cdot 38H_2O$, [A] Cub. Vitreous colorless, light pink. Tip Top Pegmatite, Custer (near), Custer Co., South Dakota, USA. H= 4.5, D_m= 2.28, D_c= 2.241, VIIC 13. Rouse et al., 1987. Am. Min. 73(1988), 496 (Abst.). Am. Min. 74 (1989), 1195.

Painite, $CaZrBAl_9O_{18}$, [G] Hex. Translucent red, brownish, orange-red. Mogok, Ohngaing, Sagaing, Burma. H= 8, D_m= 4.01, D_c= 3.996, IVC 17. Claringbull et al., 1957. Min. Mag. 50 (1986), 267. Am. Min. 61 (1976), 88.

Palarstanide, Pd_5SnAs, [A] Rhom. Metallic grey; greyish-white in reflected light. Talnakh, Noril'sk (near), Siberia (N), USSR. H= 5, D_m=?, D_c= 9.57, IIG 04. Begizov et al., 1981. Am. Min. 74(1989), 1219 (Abst.).

Palenzonaite, $NaCa_2Mn_2(VO_4)_3$, [A] Cub. Adamantine red. Molinello mine, Val Graveglia, Liguria, Italy. H= 5-5.5, D_m= 3.63, D_c= 3.78, VIIA 07. Basso, 1987. Am. Min. 73(1988), 930 (Abst.). N.Jb.Min.Mh. (1987), 136.

Palermoite, $(Li,Na)_2(Sr,Ca)Al_4(PO_4)_4(OH)_4$, [G] Orth. Vitreous/sub-adamantine colorless, white. Palermo mine, North Groton, New Hampshire, USA. H= 5.5, D_m= 3.22, D_c= 3.26, VIIB 13. Mrose, 1952. Am. Min. 50 (1965), 777. Am. Min. 60 (1975), 460. See also Bertossaite.

Palladium, Pd, [G] Cub. Metallic white/grey; white in reflected light. Brazil. H= 4.5-5, D_m= 11.9, D_c= 12.036, IA 07. Wollaston, 1803. Can. Min. 19 (1981), 599.

Palladium arsenostannide, [D] Inadequate characterization. Am. Min. 64(1979), 1333 (Abst.). Razin & Dubakina, 1974.

Palladoarsenide, Pd_2As, [A] Mon. Metallic grey; greyish-white in reflected light. Oktyabr deposit, Talnakh, Noril'sk (near), Siberia (N), USSR. H= 4.5, D_m=?, D_c= 10.42, IIG 04. Begizov et al., 1974. Am. Min. 60(1975), 162 (Abst.).

Palladobismutharsenide, $Pd_2(As,Bi)$, [A] Orth. Metallic cream. Stillwater Complex, Montana, USA. H= 5, D_m=?, D_c= 10.8, IIG 04. Cabri et al., 1976. Can. Min. 14 (1976), 410.

Palladseite, $Pd_{17}Se_{15}$, [A] Cub. Metallic white. Itabira, Minas Gerais, Brazil. H= 4.5-5, D_m= 8.30, D_c= 8.33, IIA 14. Davis et al., 1977. Min. Mag 41 (1977), 123. See also Prassoite.

Palmierite, $(K,Na)_2Pb(SO_4)_2$, [G] Rhom. Vitreous/pearly colorless, white. Mt. Vesuvius, Napoli, Campania, Italy. H=?, D_m= 4.195, D_c= 4.26, VIA 06. Lacroix, 1907. Per. Mineral. 54 (1985), 34. See also Kalistrontite.

Palygorskite, $(Mg,Al)_2Si_4O_{10}(OH) \cdot 4H_2O$, [G] Mon. Translucent white. Molotov mining dist., Ural Mts., USSR. H=?, D_m= 2.217, D_c= 2.31, VIIIE 13. Savchenkov, 1862. Rev. Min. 19 (1988), 631. Am. Min. 62 (1977), 784. See also Tuperssautsiaite, Yofortierite.

Panasqueiraite, $Ca(Mg,Fe)PO_4(OH,F)$, A Mon. Vitreous pink. Panasqueira mine, Panasqueira, Beira Baixa, Portugal. H= 5, D_m= 3.27, D_c= 3.22, VIIB 11a. Isaacs & Peacor, 1981. Can. Min. 23 (1985), 131. See also Isokite.

Panethite, $(Na,Ca,K)(Mg,Fe,Mn)PO_4$, A Mon. Transparent pale amber. Dayton meteorite, Montgomery Co., Ohio, USA. H=?, D_m= 2.9, D_c= 2.99, VIIA 04. Fuchs et al., 1967. Am. Min. 53(1968), 509 (Abst.).

Panunzite, $(K,Na)AlSiO_4$, A Hex. Vitreous colorless. Monte Somma, Mt. Vesuvius, Napoli, Campania, Italy. H= 5.5, D_m= 2.59, D_c= 2.62, VIIIF 01. Franco & Gennaro, 1988. Am. Min. 73 (1988), 420. N.Jb.Min.Mh. (1985), 322. See also Kaliophilite, Kalsilite, Trikalsilite.

Paolovite, Pd_2Sn, A Orth. Metallic lilac-rose. Oktyabr deposit, Talnakh, Noril'sk (near), Siberia (N), USSR. H= 4.5, D_m=?, D_c= 11.09, IA 10. Genkin et al., 1974. Am. Min. 59(1974), 1331 (Abst.).

Papagoite, $CaCuAlSi_2O_6(OH)_3$, G Mon. Translucent blue. New Cornelia mine, Ajo, Pima Co., Arizona, USA. H= 5-5.5, D_m= 3.25, D_c= 3.234, VIIIC 04. Hutton & Vlisidis, 1960. Bull. Min. 88 (1965), 119. Min. Abstr. 88M/3464.

Para-armalcolite, D Unnecessary name for variety of armalcolite. Min. Mag. 43 (1980), 1055. Haggerty, 1973.

Para-alumohydrocalcite, $CaAl_2(CO_3)_2(OH)_4 \cdot 6H_2O$, Q Syst=? Translucent white. Vodinsk and Gaurdok deposits, USSR. H= 1.8, D_m= 2.0, D_c=?, VbD 02. Srebrodol'skii, 1977. Am. Min. 63(1978), 794 (Abst.). See also Alumohydrocalcite.

Parabariomicrolite, $BaTa_4O_{10}(OH)_2 \cdot 2H_2O$, A Rhom. Translucent white, pale pink. Onca mine, Parelhas, Rio Grande do Norte, Brazil. H= 4, D_m=?, D_c= 5.97, IVD 17. Ercit et al., 1986. Can. Min. 24 (1986), 655. Can. Min. 24 (1986), 655.

Paraboléite, D Unnecessary name for member of boléite group. Min. Mag. 43 (1980), 1055. Mücke, 1972.

Parabrandtite, $Ca_2Mn(AsO_4)_2 \cdot 2H_2O$, A Tric. Vitreous, colorless; SG = 3.55; H = 3-4. Sterling Hill mine, Ogdensburg, Sussex Co., New Jersey, USA. H= 3-4, D_m= 3.55, D_c= 3.60, VIIC 12. Dunn et al., 1987. Am. Min. 73(1988), 1496 (Abst.).

Parabutlerite, $FeSO_4(OH) \cdot 2H_2O$, G Orth. Vitreous light orange, orange-brown. Chuquicamata, Antofagasta, Chile. H= 2.5, D_m= 2.55, D_c= 2.47, VID 02. Bandy, 1938. Dana, 7th ed.(1951), v.2, 610. Bull. Min. 93 (1970), 185. See also Butlerite.

Paracelsian, $BaAl_2Si_2O_8$, G Mon. Vitreous colorless, white. Candoglia, Piemonte, Italy. H= 5.5-6, D_m= 3.29, D_c= 3.351, VIIIF 04. Tacconi, 1905. DHZ (1963), v.4, 172. Am. Min. 70 (1985), 969. See also Celsian.

Parachrysotile, $Mg_3Si_2O_5(OH)_4$, G Orth. Translucent yellow, white, grey, green. H= 2.5, D_m= 2.5, D_c= 2.56, VIIIE 10b. Whittaker & Zussman, 1955. Min. Rev. (1988), 91. Acta Cryst. 9 (1956), 865. See also Antigorite, Clinochrysotile, Lizardite, Orthochrysotile.

Paracoquimbite, $Fe_2(SO_4)_3 \cdot 9H_2O$, G Rhom. Vitreous pale violet. Troja, Praha (Prague), Czechoslovakia. H= 2.5, D_m= 2.11, D_c= 2.110, VIC 17. Ungemach, 1935. Dana, 7th ed.(1951), v.2, 534. Am. Min. 56 (1971), 1567. See also Coquimbite.

Paracostibite, $CoSbS$, A Orth. Metallic white. Trout Bay, Red Lake area, Mulcahy Twp., Kenora Dist., Ontario, Canada. H= 6.5, D_m= 6.9, D_c= 6.97, IIC 08. Cabri et al., 1970. Can. Min. 10 (1970), 232. Can. Min. 13 (1975), 188. See also Costibite.

Paradamite, $Zn_2AsO_4(OH)$, G Tric. Vitreous pale yellow. Ojuela mine, Mapimi, Durango, Mexico. H=?, D_m= 4.55, D_c= 4.595, VIIB 04. Switzer, 1956. Am. Min. 65 (1980), 353. Acta Cryst. B35 (1979), 720. See also Adamite.

Paradocrasite, Sb_3As, A Mon. Metallic white. Broken Hill, New South Wales, Australia. H= 3, D_m= 6.52, D_c= 6.44, IB 01. Leonard et al., 1971. Am. Min. 56 (1971), 1127.

Paragearksutite, $Ca_4Al_4(F,OH)_{12}F_8 \cdot 3H_2O$, Q Syst=? Earthy white. Transbaikal, USSR. H=?, D_m=?, D_c=?, IIIC 09. Smolyaninov & Isakov, 1946. Dana, 7th ed.(1951), v.2, 119.

Paragonite-2M, $NaAl_2(Si_3Al)O_{10}(OH)_2$, G Mon. Translucent colorless, pale yellow. Monte Campione, Faido (Near), Tessin, Switzerland. H= 2.5, D_m= 2.85, D_c= 2.893, VIIIE 05a. Schafhautl, 1843. Sov.Phys.Cryst. 22 (1977), 557. Am. Min. 69 (1984), 122.

Paraguanajuatite, $Bi_2(Se,S)_3$, G Rhom. Metallic grey. Sierra de Sta. Rosa, Santa Catarina mine, Guanajuato (near), Mexico. H= 2, D_m=?, D_c= 7.68, IIC 03. Ramdohr, 1948. Min. Jour. 14 (1988), 92. See also Guanajuatite.

Parahopeite, $Zn_3(PO_4)_2 \cdot 4H_2O$, G Tric. Vitreous colorless. Kabwe (Broken Hill), Central province, Zambia. H= 3.7, D_m= 3.31, D_c= 3.126, VIIC 06. Spencer, 1907. Dana, 7th ed.(1951), v.2, 733. Min. Mag. 36 (1968), 621. See also Hopeite.

Parajamesonite, $Pb_4FeSb_6S_{14}$, Q Syst=? Metallic grey. Herja (Kisbanya), Judetul Maramures, Romania. H=?, D_m= 5.48, D_c=?, IID 03. Zsivny & Naray-Szabo, 1947. Am. Min. 34(1949), 133 (Abst.). See also Jamesonite.

Parakeldyshite, $Na_2ZrSi_2O_7$, A Tric. Vitreous white. Bratthagen, Lågendalen, Larvik (near), Norway; Lovozero massif, Kola peninsula, USSR. H= 5.5-6, D_m= 3.38, D_c= 3.40, VIIIB 22. Khomyakov, 1977. Powd. Diffr. 2 (1987), 176. Min. Abstr. 77-292. See also Keldyshite.

Parakhinite, $Cu_3PbTeO_4(OH)_6$, A Hex. Translucent dark green. Tombstone, Cochise Co., Arizona, USA. H= 3.5, D_m=?, D_c= 6.69, VIG 09. Williams, 1978. Am. Min. 63 (1978), 1016. See also Khinite.

Paralaurionite, $PbCl(OH)$, G Mon. Translucent white, pale greenish, violet. Lávrion (Laurium), Attikí, Greece. H=?, D_m= 6.15, D_c= 6.28, IIIC 05. Smith, 1899. Dana, 7th ed.(1951), v.2, 64. See also Laurionite.

Paralstonite, $BaCa(CO_3)_2$, A Hex. Vitreous colorless, smoky white, grey-white. Minerva #1 mine, Cave-in-rock, Hardin Co., Illinois, USA. H= 4-4.5, D_m= 3.75, D_c= 3.62, VbA 04. Roberts, 1979. Am. Min. 66(1981), 219 (Abst.). N.Jb.Min.Mh. (1980), 353. See also Alstonite, Barytocalcite.

Paramelaconite, Cu_4O_3, G Tet. Adamantine black; white in reflected light. Copper Queen mine, Bisbee, Cochise Co., Arizona, USA. H= 4.5, D_m= 5.9, D_c= 5.93, IVA 05. Koenig, 1891. Dana, 7th ed.(1944), v.1, 510. Am. Min. 63 (1978), 180.

Paramendozavilite, $NaAl_4Fe_7(PO_4)_5(PMo_{12}O_{40})(OH)_{16} \cdot 56H_2O$, A Syst=? Vitreous pale yellow. Cumobabi, Sonora, Mexico. H= 1, D_m= 3.35, D_c=?, VIID 15b. Williams, 1986. Am. Min. 73(1988), 194 (Abst.).

Paramontroseite, VO_2, G Orth. Sub-metallic greyish-black. Paradox Valley, Montrose Co., Colorado, USA. H=?, D_m=?, D_c= 4.18, IVF 04. Evans & Mrose, 1955. Am. Min. 40 (1955), 861. Am. Min. 40 (1955), 861.

Paranatrolite, $Na_2(Al_2Si_3)O_{10} \cdot 3H_2O$, A Orth. Translucent white. Mont Saint-Hilaire, Rouville Co., Québec, Canada. H= 5-5.5, D_m= 2.21, D_c= 2.204, VIIIF 10. Chao, 1980. Can. Min. 18 (1980), 85.

Paraotwayite, $Ni(OH)_{2-x}(SO_4,CO_3)_{0.5x}$, A Mon. Translucent silky green. Otway nickel deposit, Pilbara Region, Western Australia, Australia. H= 4, D_m= 3.30, D_c= 3.520, IVF 03a. Nickel & Graham, 1987. Can. Min. 25 (1987), 409.

Parapectolite, D A polytype of pectolite. Min. Mag. 43 (1980), 1055. Muller, 1976.

Paraphane, D Inadequately described. Min. Mag. 36 (1968), 1144. Sidorenko, 1960.

Parapierrotite, $Tl(Sb,As)_5S_8$, A Mon. Semi-metallic black; white in reflected light. Alšar (Allchar), Rožden (near), Makedonija (Macedonia), Yugoslavia. H= 2.5-3, D_m= 5.07, D_c= 5.00, IID 14. Johan et al., 1975. Am. Min. 61(1976), 504 (Abst.). Zts. Krist. 151 (1980), 203.

Pararammelsbergite, $NiAs_2$, G Orth. Metallic white. Moose Horn mine, Hudson Bay mine and Keeley mine, Cobalt, Ontario, Canada. H= 5, D_m= 7.12,

D_c= 7.24, IIC 08. Peacock, 1939. Dana, 7th ed.(1944), v.1, 310. Am. Min. 57 (1972), 1. See also Rammelsbergite, Krutovite.

Pararealgar, AsS, [A] Mon. Vitreous/resinous yellow/orange-yellow. Mount Washington copper deposit, Vancouver Island, British Columbia, Canada. H= 1-1.5, D_m= 3.52, D_c= 3.499, IIB 19. Roberts et al., 1980. Can. Min. 18 (1980), 525. See also Realgar.

Pararobertsite, $Ca_2Mn_3(PO_4)_3O_2 \cdot 3H_2O$, [A] Mon. Vitreous red. Tip Top pegmatite, Custer, Custer Co., South Dakota, USA. H=?, D_m= 3.22, D_c= 3.21, VIID 15a. Roberts et al., 1989. Can. Min. 27 (1989), 451.

Paraschachnerite, $Ag_{1.2}Hg_{0.8}$, [A] Orth. Metallic grey. Vertrauen auf Gott mine and Moschellandsberg, Obermoschel, Rheinland-Pfalz, Germany. H= 4, D_m=?, D_c= 12.98, IA 06. Seeliger & Mücke, 1972. Am. Min.58(1973), 347 (Abst.).

Paraschoepite, $UO_3 \cdot 2H_2O$(?), [Q] Orth. Translucent yellow. Shinkolobwe, Shaba, Zaïre. H=?, D_m=?, D_c=?, IVF 13. Schoep & Stradiot, 1947. Am. Min. 45 (1960), 1026.

Parascholzite, $CaZn_2(PO_4)_2 \cdot 2H_2O$, [A] Mon. Translucent white, colorless. Hagendorf pegmatite, Waidhaus, Oberpfalz, Bayern (Bavaria), Germany. H= 4, D_m= 3.12, D_c= 3.10, VIIC 03. Sturman et al., 1981. Am. Min. 66 (1981), 843. See also Scholzite.

Paraspurrite, $Ca_5(SiO_4)_2(CO_3)$, [A] Mon. Translucent colorless. Darwin (near), Inyo Co., California, USA. H=?, D_m= 3.00, D_c= 3.01, VIIIA 10. Colville & Colville, 1977. Am. Min. 62 (1977), 1003. See also Spurrite.

Parastrengite, [D] Inadequate data. Min. Mag. 43 (1980), 1055. Gevorkyan, 1974.

Parasymplesite, $Fe_3(AsO_4)_2 \cdot 8H_2O$, [G] Mon. Japan. H= 2, D_m= 3.07, D_c= 3.043, VIIC 10. Ito et al., 1954. Bull. Min. 100 (1977), 310. See also Symplesite, Köttigite.

Paratacamite, $Cu_2Cl(OH)_3$, [G] Rhom. Vitreous green. Sierra Gorda, Copiapó (near), Chile. H= 3, D_m= 3.74, D_c= 3.754, IIIC 01. Smith, 1905. Dana, 7th ed.(1951), v.2, 74. Acta Cryst. B31 (1975), 183. See also Atacamite, Botallackite.

Paratellurite, TeO_2, [G] Tet. Resinous/waxy greyish-white. Moctezuma, Sonora, Mexico. H= 1, D_m= 5.60, D_c= 6.018, IVD 06. Switzer & Swanson, 1960. Am. Min. 45 (1960), 1272. Sov.Phys.Cryst. 32 (1987), 354. See also Tellurite.

Paraumbite, $K_3Zr_2H(Si_3O_9)_2 \cdot 3H_2O$, [A] Orth. Vitreous/pearly colorless, white. Mt. Eveslogchorr, Khibina massif (E), Kola Peninsula, USSR. H= 4.5, D_m= 2.59, D_c= 2.69, VIIID 09. Khomyakov, 1984. Am. Min. 69(1984), 813 (Abst.).

Paravariscite, [D] Inadequate data. Min. Mag. 43 (1980), 1055. Gevorkyan et al., 1974.

Paravauxite, $FeAl_2(PO_4)_2(OH)_2 \cdot 8H_2O$, [G] Tric. Vitreous/pearly colorless, pale greenish-white. Llallagua, Potosí, Bolivia. H= 3, D_m= 2.36, D_c= 2.479, VIID 03. Gordon, 1922. Am. Min. 47 (1962), 1. Struct. Repts. 34A (1969), 332. See also Metavauxite.

Pargasite, $NaCa_2(Mg,Fe)_4Al(Si_6Al_2)O_{22}(OH)_2$, [A] Mon. Vitreous light brown, brown. Pargas Valley, Finland. H= 5-6, D_m= 3.209, D_c= 3.182, VIIID 05b. Steinheil, 1814. Am. Min. 63 (1978), 1023. Am. Min. 74 (1989), 1097. See also Ferropargasite.

Parisite-(Ce), $Ca(Ce,La)_2(CO_3)_3F_2$, [G] Rhom. Vitreous/resinous brownish-yellow, brown, yellow. Muzo dist., Bogota, Columbia. H= 4.5, D_m= 4.2, D_c= 4.29, VbB 04. Medici-Spada, 1845. Am. Min. 38 (1953), 932. Am. Min. 38 (1953), 932.

Parisite-(Nd), $Ca(Nd,Ce,La)_2(CO_3)_3F_2$, [Q] Syst=? Vitreous yellowish-brown. Bayan Obo, Inner Mongolia, China (N). H= 4-5, D_m= 4.3, D_c=?, VbB 04. Zhang & Tao, 1986. Am. Min. 73(1988), 1496 (Abst.).

Parkerite, $Ni_3(Bi,Pb)_2S_2$, [G] Orth. Metallic creamy-white. Waterfall Gorge, Insizwa, East Griqualand, South Africa. H= > 3, D_m= 8.4, D_c= 8.50, IIA 05. Scholtz, 1936. Am. Min. 59 (1974), 296. Am. Min. 58 (1973), 435. See also Rhodplumsite, Shandite.

Parnauite, $Cu_9(AsO_4)_2(SO_4)(OH)_{10}\cdot 7H_2O$, [A] Orth. Translucent pale blue, green, blue-green. Majuba Hill mine, Cross Cut, Pershing Co., Nevada, USA. H= 2, D_m= 3.09, D_c= 3.216, VIID 09. Wise, 1978. Am. Min. 63 (1978), 704.

Parsettensite, $Mn_5Si_6O_{13}(OH)_8$, [G] Orth. Translucent copper-red. Alp Parsettens, Sursass (Oberhalbstein), Tinizong (Tinzen), Grischun (Graubünden), Switzerland. H=?, D_m= 2.59, D_c= 2.58, VIIIE 07b. Jacob, 1923. IMA (1986), 194.

Parsonsite, $Pb_2(UO_2)(PO_4)_2\cdot 0\text{–}2H_2O$, [G] Tric. Sub-adamantine pale yellow. Kasolo, Shaba, Zaïre. H= 2.5-3, D_m= 5.37, D_c= 6.3, VIID 19. Schoep, 1923. Am. Min. 35 (1950), 245. Struct. Repts. 22 (1958), 422.

Parthéite, $Ca_2Al_4Si_4O_{15}(OH)_2\cdot 4H_2O$, [A] Mon. Vitreous white. Belenköysirti Hill, Doğanbaba, Burdur, Taurus Mts., Turkey (SW). H=?, D_m= 2.39, D_c= 2.41, VIIIF 15. Sarp et al. 1979. Am. Min. 65(1980), 1068 (Abst.). Zts. Krist. 169 (1984), 165. See also Lawsonite, Zeolite.

Partzite, $Cu_2Sb_2(O,OH)_7(?)$, [Q] Cub. Olive-green, yellowish-green, blackish-green, black. Blind Spring Hill dist., Benton (near), Mono Co., California, USA. H= 3-4, D_m= 3.8, D_c= 5.95, IVC 08. Arents, 1867. Min. Mag. 30 (1953), 100.

Parwelite, $Mn_{10}Sb_2As_2Si_2O_{24}$, [A] Mon. Transparent tan. Långban mine, Filipstad (near), Värmland, Sweden. H= 5.5, D_m= 4.62, D_c= 4.860, VIIB 10d. Moore, 1968. Am. Min. 55(1970), 323 (Abst.). Struct. Repts. 43A (1977), 319.

Pascoite, $Ca_3V_{10}O_{28}\cdot 17H_2O$, [G] Mon. Vitreous/sub-adamantine red/yellow orange. Minasragra, Junin, Peru. H= 2.5, D_m= 2.45, D_c= 2.465, VIIC 23. Hillebrand et al., 1914. Dana, 7th ed.(1951), v.2, 1055. Acta Cryst. 21 (1966), 397.

Patrónite, VS_4, [G] Mon. Grey-black; grey in reflected light. Minasragra, Cerro de Pasco (near), Junin, Peru. H= 2, D_m= 2.81, D_c= 2.834, IIC 16. Hewett, 1906. Ramdohr (1980), 887. Struct. Repts. 29 (1964), 89.

Paulingite, $(K,Ca,Na,Ba)_{12}(Si,Al)_{24}O_{48}\cdot 25H_2O$, [G] Cub. Vitreous colorless, yellowish, orange. Rock Island Dam, Columbia River, Wenatchee, Douglas Co., Washington, USA. H= 5, D_m=?, D_c= 2.20, VIIIF 14. Kamb & Oke, 1960. Natural Zeolites (1985), 164. Struct. Repts. 50A (1983), 333.

Paulite, [D] Inadequate data. Min. Mag. 33 (1962), 261. Bultemann, 1960.

Paulkellerite, $FeBi_2O_2PO_4(OH)_2$, [A] Mon. Vitreous/adamantine greenish-yellow. Schneeberg, Sachsen (Saxony), Germany. H= 4, D_m= 6.17, D_c= 6.306, VIIB 17. Dunn et al., 1988. Am. Min. 73 (1988), 870. Am. Min. 73 (1988), 873.

Paulkerrite, $K(Mg,Mn)_2Ti(Fe,Al)_2(PO_4)_4(OH)_3\cdot 15H_2O$, [A] Orth. Vitreous yellow-brown. Bagdad Copper mine (near), 7-U-7 Ranch, Hillside, Yavapai Co., Arizona, USA. H= 3, D_m= 2.36, D_c= 2.36, VIID 29. Peacor et al., 1984. Am. Min. 70(1985), 875 (Abst.). See also Mantienneite.

Paulmooreite, $Pb_2As_2O_5$, [A] Mon. Adamantine colorless, light orange. Långban mine, Filipstad (near), Värmland, Sweden. H= 3, D_m= 6.9, D_c= 6.886, VIIA 15. Dunn & Peacor, 1979. Am. Min. 64 (1979), 352. Am. Min. 65 (1980), 340.

Pavonite, $(Ag,Cu)(Bi,Pb)_3S_5$, [G] Mon. Metallic grey; white in reflected light. Bolivar mine, Cerro Bonete, Sur-Lipez, Bolivia. H= 3.5, D_m= 6.74, D_c= 6.79, IID 09c. Nuffield, 1954. Can. Min. 13 (1975), 408. Can. Min. 15 (1977), 339. See also Mummeite.

Paxite, $CuAs_2$, [G] Mon. Metallic grey. Černý důl (Schwarzenthal), Krkonoše (Giant Mtns.), Vychodočeský kraj, Čechy (Bohemia), Czechoslovakia. H= 3.2, D_m= 5.4, D_c= 5.97, IIC 08. Johan, 1961. TMPM 34 (1985), 167.

Pearceite, $(Ag,Cu)_{16}As_2S_{11}$, G Mon. Metallic black; white in reflected light. Veta Rica mine, Sierra Mojada, Coahuila, Mexico. H= 3, D_m= 6.15, D_c= 6.20, IID 11. Penfield, 1896. Am. Min. 48 (1963), 565. See also Polybasite.

Pecoraite, $Ni_3Si_2O_5(OH)_4$, A Mon. Translucent green. Wolf Creek Meteorite Crater, Western Australia, Australia. H=?, D_m=?, D_c= 3.08, VIIIE 10b. Faust et al., 1969. Min. Mag. 44 (1981), 153. See also Nepouite, Clinochrysotile.

Pectolite, $NaCa_2Si_3O_8(OH)$, G Tric. Silky/sub-vitreous whitish, greyish. Mt. Baldo and Mt. Monzoni, Italy. H= 5, D_m= 2.7, D_c= 2.906, VIIID 08. Kobell, 1828. Am. Min. 63 (1978), 274. Zts. Krist. 146 (1977), 281. See also Serandite.

Pectolite-M2abc, $NaCa_2Si_3O_8OH$, P Mon. H=?, D_m=?, D_c=?, VIIID 08a. Müller, 1976. Am. Min. 63(1978), 427 (Abst.).

Pehrmanite-9R, $Be(Fe,Zn,Mg)_2Al_6O_{12}$, A Rhom. Vitreous green. Kemiö Island, Finland (SW). H= 8-8.5, D_m=?, D_c= 4.07, IVC 05c. Burke & Lustenhouwer, 1981. Can. Min. 19 (1981), 311. See also Musgravite.

Peisleyite, $Na_3Al_{16}(PO_4)_{10}(SO_4)_2(OH)_{17}\cdot 20H_2O$, A Mon. Earthy white. Tom's Phosphate Quarry, Kapunda (near), South Australia, Australia. H= 3, D_m= 2.12, D_c= 2.11, VIID 15c. Pilkington et al., 1982. Min. Mag. 46 (1982), 449.

Pekoite, $CuPbBi_{11}(S,Se)_{18}$, A Orth. Metallic grey; white/cream in reflected light. Peko or Juno mine, Tennant Creek, Northern Territory, Australia. H=?, D_m=?, D_c= 6.8, IID 05a. Mumme & Watts, 1976. N.Jb.Min.Mh. (1990), 35. Can. Min. 14 (1976), 322.

Pellyite, $Ba_2Ca(Fe,Mg)_2Si_6O_{17}$, A Orth. Vitreous colorless, pale yellow. Ross River and Pelly River, Gillespite Lake (S), Yukon Territory, Canada. H= 6, D_m= 3.51, D_c= 3.48, VIIIF 23. Montgomery et al., 1972. Am. Min. 61 (1976), 67.

Penfieldite, $Pb_2Cl_3(OH)$, G Hex. Adamantine/greasy colorless, white, yellowish, bluish. Lávrion (Laurium), Attikí, Greece. H=?, D_m= 5.82, D_c= 6.00, IIIC 05. Genth, 1892. Bull. Min. 91 (1968), 407.

Penikisite, $Ba(Mg,Fe)_2Al_2(PO_4)_3(OH)_3$, A Tric. Vitreous blue, green. Rapid Creek, Big Fish River–Blow River area, Yukon Territory, Canada. H= 4, D_m= 3.79, D_c= 3.82, VIIB 19. Mandarino et al., 1977. Can. Min. 15 (1977), 393. See also Kulanite.

Penkvilksite, $Na_4Ti_2Si_8O_{22}\cdot 5H_2O$, A Syst=? Dull/pearly/silky white. Lovozero massif, Kola Peninsula, USSR. H= 5, D_m= 2.58, D_c=?, VIIID 18. Bussen et al., 1974. Am. Min. 60(1975), 340 (Abst.).

Pennantite, $(Mn,Al)_6(Si,Al)_4O_{10}(OH)_8$, G Mon. Orange, orange-brown. Benallt mine, Rhiw, Caernavonshire, Lleyn Peninsula, Gwynedd, Wales. H=?, D_m= 3.06, D_c= 3.18, VIIIE 09a. Campbell et al., 1946. Can. Min. 21 (1983), 545.

Penroseite, $NiSe_2$, G Cub. Metallic grey; white in reflected light. Pacajake mine, Colquechaca (Near), Potosí, Bolivia. H= 2.5-3, D_m= 6.9, D_c= 6.785, IIC 05. Gordon, 1926. Am. Min. 74 (1989), 1168.

Pentagonite, $Ca(VO)Si_4O_{10}\cdot 4H_2O$, A Orth. Vitreous greenish-blue. Lake Owyhee State Park, Malheur Co., Oregon, USA. H= 3-4, D_m=?, D_c= 2.33, VIIIE 26. Staples et al., 1973. Am. Min. 58 (1973), 405. Am. Min. 58 (1973), 412. See also Cavansite.

Pentahydrite, $MgSO_4\cdot 5H_2O$, G Tric. Translucent colorless, pale blue. Cripple Creek, Teller Co., Colorado, USA. H=?, D_m= 1.90, D_c= 1.93, VIC 03a. Frondel, 1951. Dana, 7th ed.(1951), v.2, 492. Acta Cryst. B28 (1972), 1448.

Pentahydroborite, $CaB_2O(OH)_6\cdot 2H_2O$, G Tric. Translucent colorless. Ural Mts., USSR. H= 2.5, D_m= 2.00, D_c= 2.140, Vc 04b. Malinko, 1961. Am. Min. 47(1962), 1482 (Abst.). Min. Abstr. 74-959.

Pentlandite, $(Ni,Fe)_9S_8$, G Cub. Metallic yellow. Lillehammer, Norway (S). H= 3.5-4, D_m= 4.8, D_c= 4.956, IIA 07. Dufrenoy, 1856. Min. Mag. 41 (1977), 345. Can. Min. 12 (1973), 169. See also Cobalt, pentlandite.

Penzhinite, $(Ag,Cu)_4Au(S,Se)_4$, [A] Hex. Metallic greyish-white. Pengina River, Kamchatka (N), USSR. H=?, D_m=?, D_c= 8.35, IIA 12. Bochek et al., 1984. ZVMO 113 (1984), 356.

Percylite, $CuPbCl_2(OH)_2$, [Q] Cub. Vitreous blue. Sonora, Mexico. H= 2.5, D_m=?, D_c=?, IIIC 04. Brooke, 1850. Min. Rec. 5 (1974), 280.

Peretaite, $CaSb_4O_4(SO_4)_2(OH)_2 \cdot 2H_2O$, [A] Mon. Transparent colorless. Pereta mine, Grosseto, Toscana, Italy. H= 3.5, D_m= 4.0, D_c= 4.06, VID 11. Cipriani et al., 1980. Am. Min. 65 (1980), 936. Am. Min. 65 (1980), 940.

Perhamite, $Ca_3Al_7(SiO_4)_3(PO_4)_4(OH)_3 \cdot 16.5H_2O$, [A] Hex. Vitreous light brown, white. Bell Pit, Newry, Oxford Co., Maine, USA. H= 5, D_m= 2.64, D_c= 2.53, VIID 13. Dunn & Appleman, 1977. Min. Mag. 41 (1977), 437.

Periclase, MgO, [G] Cub. Vitreous colorless, greyish-white, yellow, green, black. Monte Somma, Mt. Vesuvius, Napoli, Campania, Italy. H= 5.5, D_m= 3.56, D_c= 3.58, IVA 04. Scacchi, 1840. Am. Min. 61 (1976), 266.

Perite, $PbBiO_2Cl$, [G] Orth. Adamantine yellow. Långban mine, Filipstad (near), Värmland, Sweden. H= 3, D_m= 8.16, D_c= 8.24, IIIC 06. Gillberg, 1960. Am. Min. 46(1961), 765 (Abst.). Struct. Repts. 24 (1960), 369. See also Nadorite.

Perlialite, $K_9Na(Ca,Sr)(Al_{12}Si_{24})O_{72} \cdot 15H_2O$, [A] Hex. Pearly white. Mt. Eveslogchorr and Yukspor Mts., Khibina massif, Kola Peninsula, USSR. H= 4-5, D_m= 2.14, D_c= 2.15, VIIIF 14. Men'shikov, 1984. Am. Min. 70(1985), 1331 (Abst.).

Perloffite, $Ba(Mn,Fe)_2Fe_2(PO_4)_3(OH)_3$, [A] Mon. Vitreous/sub-adamantine dark brown, greenish-brown. Big Chief mine, Glendale (near), Pennington Co., South Dakota, USA. H= 5, D_m=?, D_c= 3.996, VIIB 19. Kampf, 1977. Am. Min. 62(1977), 1059 (Abst.).

Permingeatite, $Cu_3(Sb,As)Se_4$, [A] Tet. Metallic brownish-pink. Předbořice, Středočeský kraj, Čechy (Bohemia), Czechoslovakia. H= 4-4.5, D_m=?, D_c= 5.82, IIB 03a. Johan et al., 1971. Am. Min. 57(1972), 1554 (Abst.).

Perovskite, $CaTiO_3$, [G] Orth. Adamantine/metallic black, brown, yellow. Slatoust dist., Ural Mts., USSR. H= 5.5, D_m= 4.01, D_c= 4.031, IVC 07. Rose, 1839. Dana, 7th ed.(1944), v.1, 730. Acta Cryst. C39 (1983), 1323.

Perrierite, $(Ca,Ce,Th)_4(Mg,Fe)_2(Ti,Fe)_3O_8(Si_2O_7)_2$, [G] Mon. Resinous black, brownish. Nettuno, Roma, Italy. H= 5.5, D_m= 4.3, D_c=?, VIIIB 16. Bonatti & Gottardi, 1950. Am. Min. 63 (1978), 499. Am. Min. 59 (1974), 1277. See also Chevkinite.

Perroudite, $Ag_4Hg_5S_4(Cl,I,Br)_4$, [A] Orth. Vitreous/adamantine red. Cap Garonne, Le Pradet, Var, France; Broken Hill, New South Wales and Coppin Pool, Western Australia, Australia. H=?, D_m=?, D_c= 6.60, IIF 01. Sarp et al., 1987. Am. Min. 72 (1987), 1251. Am. Min. 72 (1987), 1257.

Perryite, $(Ni,Fe)_5(Si,P)_2$, [G] Hex. Thin lamellae in kamacite. Horse Creek iron meteorite, Baca Co., Colorado and St. Mark's enstatite chondrite. H=?, D_m=?, D_c= 7.37, IC 03. Fredriksson & Henderson, 1965. Min. Mag. 36 (1968), 850.

Petalite, $LiAlSi_4O_{10}$, [G] Mon. Vitreous/pearly colorless, white, grey. Utö, Sweden. H= 6-6.5, D_m= 2.4, D_c= 2.378, VIIIE 29. d'Andrada, 1800. Dana, 6th ed. (1892), 311. Zts. Krist. 160 (1982), 159.

Petarasite, $Na_5Zr_2Si_6O_{18}(Cl,OH) \cdot 2H_2O$, [A] Mon. Vitreous greenish-yellow. Mont Saint-Hilaire, Rouville Co., Québec, Canada. H= 5-5.5, D_m= 2.9, D_c= 2.88, VIIIC 07. Chao et al., 1980. Can. Min. 18 (1980), 497. Min. Abstr. 83M/5067.

Petedunnite, $CaZnSi_2O_6$, [A] Mon. Translucent dark green. Franklin, Sussex Co., New Jersey, USA. H=?, D_m=?, D_c= 3.68, VIIID 01a. Essene & Peacor, 1987. Am. Min. 72 (1987), 157.

Petersite-(Y), $Cu_6(Y,Ca)(PO_4)_3(OH)_6 \cdot 3H_2O$, [A] Hex. Vitreous yellowish-green. Secaucus, Laurel Hill, Hudson Co., New Jersey, USA. H=?, D_m= 3.41, D_c= 3.40, VIID 17. Peacor & Dunn, 1982. Am. Min. 67 (1982) ,1039.

Petrovicite, $Cu_3HgPbBiSe_5$, A Orth. Metallic cream in reflected light. Petrovice Žd'ar, Jihomoravský jraj, Morava (Moravia), Czechoslovakia. H= 3, D_m=?, D_c= 7.707, IID 08. Johan et al., 1976. Am. Min. 62(1977), 594 (Abst.).

Petrovskaite, AuAg(S, Se), A Mon. Metallic grey. Maikan deposit, Kazakhstan (central), USSR. H= 2-2.5, D_m=?, D_c= 9.5, IIA 12. Nesterenko et al., 1984. ZVMO 113 (1984), 602.

Petrukite, $(Cu, Fe, Sn)_3SnS_4$, A Orth. Metallic; greenish-brown/grey in reflected light. Herb claim, Cassiar District, British Columbia, Canada. H= 4.5, D_m=?, D_c= 4.61, IIB 03a. Kissin & Owens, 1989. Can. Min. 27 (1989), 673.

Petscheckite, $UFe(Nb, Ta)_2O_8$, A Rhom. Semi-metallic dark brown. Berere, Tananarive (NNW), Madagascar. H= 5, D_m= 7, D_c= 6.99, IVD 18. Mücke & Strunz, 1978. Am. Min. 63 (1978), 941.

Petzite, Ag_3AuTe_2, G Cub. Metallic grey; light pinkish-grey in reflected light. Săcărâmb (Nagyág), Transylvania, Romania. H= 2.5-3, D_m= 9.0, D_c= 9.213, IIA 16. Haidinger, 1845. Dana, 7th ed.(1944), 186. Min. Abstr. 79-3411. See also Fischesserite.

Pharmacolite, $Ca(AsO_3OH){\cdot}2H_2O$, G Mon. Vitreous/pearly colorless, white, greyish. Wittichen, Schwarzwald, Germany. H= 2-2.5, D_m= 2.725, D_c= 2.731, VIIC 15. Karsten, 1800. Am. Min. 64 (1979), 1248. Acta Cryst. B25 (1969), 1544. See also Ardealite, Brushite, Gypsum.

Pharmacosiderite, $KFe_4(AsO_4)_3(OH)_4{\cdot}6\text{–}7H_2O$, G Cub. Adamantine/greasy olive-green, yellow, brown, red, etc. Cornwall, England; St. Leonard, Haute Vienne, France. H= 2.5, D_m= 2.797, D_c= 2.60, VIID 14b. Hausmann, 1813. N.Jb.Min.Mh. (1984), 183. Zts. Krist. 125 (1967), 92. See also Alumopharmacosiderite, Sodiumpharmacosiderite.

Phaunouxite, $Ca_3(AsO_4)_2{\cdot}11H_2O$, A Tric. Vitreous colorless. Gabe-Gottes vein, Ste.-Marie-aux-Mines, Vosges Mtns., Haut-Rhin, Alsace, France. H=?, D_m= 2.28, D_c= 2.274, VIIC 16. Bari et al., 1982. Bull. Min. 105 (1982), 327. Acta Cryst. B39 (1983), 4.

Phenakite, Be_2SiO_4, G Rhom. Vitreous colorless, yellow, red, brown. Takovaya, Sverdlovsk (Ekaterinburg), Ural Mts., USSR. H= 7.5-8, D_m= 3.0, D_c= 2.960, VIIIA 01. Nordenskiöld, 1833. Dana/Ford (1932), 602. Sov.Phys.Cryst. 15(1970), 1021.

Philipsbornite, $PbAl_3(AsO_4)(AsO_3OH)(OH)_6$, A Rhom. Earthy greyish-green. Dundas, Tasmania, Australia. H= 4.5, D_m=?, D_c= 4.33, VIIB 15a. Walenta et al., 1982. Am. Min. 67(1982), 859 (Abst.).

Philipsburgite, $(Cu, Zn)_6(AsO_4, PO_4)_2(OH)_6{\cdot}H_2O$, A Mon. Vitreous green. Black Pine mine, Philipsburg (near), Granite Co., Montana, USA. H= 3-4, D_m= 4.07, D_c= 4.04, VIID 09. Peacor et al., 1985. Can. Min. 23 (1985), 255. See also Kipushite.

Phillipsite, $K(Ca_{0.5}, Na)_2(Si_5Al_3)O_{16}{\cdot}6H_2O$, G Mon. Vitreous white, reddish, yellowish. Aci Castello, Sicilia (Sicily), Italy. H= 4-4.5, D_m= 2.20, D_c= 2.242, VIIIF 14. Lévy, 1825. Natural Zeolites (1985), 134. Acta Cryst. B30 (1974), 2426. See also Harmotome, Wellsite.

Phlogopite, $KMg_3(Si_3Al)O_{10}(F, OH)_2$, G Mon. Colorless, brown, green, yellowish/reddish-brown. H= 2-2.5, D_m= 2.8, D_c= 2.868, VIIIE 05b. Breithaupt, 1841. DHZ (1962), v.3, 42. Am. Min. 58 (1973), 889. See also Biotite.

Phoenicochroite, $Pb_2O(CrO_4)$, G Mon. Translucent dark red. Berezov, Sverdlovsk, Ural Mts., USSR. H= 2.5, D_m= 7.01, D_c= 7.073, VIE 02. Glocker, 1839. Am. Min. 55 (1970), 784. Am. Min. 55 (1970), 784. See also Lanarkite.

Phosgenite, $Pb_2CO_3Cl_2$, G Tet. Adamantine colorless, yellowish-white, brown, grey, etc. H= 2-3, D_m= 6.133, D_c= 6.123, IIID 02. Breithaupt, 1841. Dana, 7th ed.(1951), v.2, 256. Struct. Repts. 41A (1975), 293.

Phosinaite, $Na_{11}(Na,Ca)_2Ca_2Ce(SiO_3)_4(PO_4)_4$, A Orth. Vitreous pale rose, brownish-rose. Mt. Karnasurt, Lovozero massif and Mt. Koashva, Khibina, Kola Peninsula, USSR. H= 3.5, D_m= 2.8, D_c= 3.07, VIIIC 13. Kapustin et al., 1974. Am. Min. 60(1975), 488 (Abst.). Sov.Phys.Cryst. 26 (1981), 679. See also Clinophosinaite.

Phosphammite, $(NH_4)_2(PO_3OH)$, G Mon. Translucent colorless. Guañape Islands, Peru. H=?, D_m= 1.62, D_c= 1.61, VIIA 14. Shepard, 1870. Min. Mag. 39 (1973), 346. Acta Cryst. 10 (1957), 709.

Phosphate-walpurgite, $Bi_4O_4(UO_2)(PO_4)_2 \cdot 2H_2O$, Q Syst=? H=?, D_m=?, D_c=?, VIID 19. Melkov, 1946. Strunz (1970), 350. See also Walpurgite.

Phosphoferrite, $(Fe,Mn)_3(PO_4)_2 \cdot 3H_2O$, G Orth. Vitreous/sub-resinous pale green, dark reddish-brown. Hagendorf pegmatite, Oberpfalz, Bayern (Bavaria), Germany. H= 3-3.5, D_m= 3.1, D_c= 3.32, VIIC 04. Laubmann & Steinmetz, 1920. Min. Mag. 43 (1980),789. Struct. Repts. 42A (1976), 346. See also Reddingite.

Phosphofibrite, $KCuFe_{15}(PO_4)_{12}(OH)_{12} \cdot 12H_2O$, A Orth. Vitreous yellow, yellowish-green. Clara mine (Grube Clara), Wolfach, Schwarzwald, Baden-Württemberg, Germany. H= 4, D_m= 2.90, D_c= 2.94, VIID 15c. Walenta & Dunn, 1984. Am. Min. 69(1984), 1192 (Abst.).

Phosphophyllite, $Zn_2(Fe,Mn)(PO_4)_2 \cdot 4H_2O$, G Mon. Vitreous colorless, pale bluish-green. Hagendorf pegmatite, Oberpfalz, Bayern (Bavaria), Germany. H= 3-3.5, D_m= 3.109, D_c= 3.058, VIIC 06. Laubmann & Steinmetz, 1920. J. Appl. Cryst. 9 (1975), 503. Am. Min. 62 (1977), 812.

Phosphorrösslerite, $Mg(PO_3OH) \cdot 7H_2O$, G Mon. Vitreous/dull colorless, yellowish. Schellgaden (near), Salzburg province, Austria. H= 2.5, D_m= 1.736, D_c= 1.741, VIIC 11. Friedrich & Robitsch, 1939. Dana, 7th ed.(1951), v.2, 713. Zts. Krist. 137 (1973), 246. See also Rösslerite.

Phosphosiderite, $FePO_4 \cdot 2H_2O$, G Mon. Vitreous/sub-resinous red, reddish-violet, green. Kalterborn mine, Eiserfeld (near), Siegen, Germany. H= 3.5-4, D_m= 2.76, D_c= 2.76, VIIC 05. Bruhns & Busz, 1890. Am. Min. 37(1952), 362 (Abst.). Am. Min. 51 (1966), 168. See also Strengite, Metavariscite, Kolbeckite.

Phosphothorogummite, D = phosphatian thorogummite. Min. Mag. 38 (1971), 103. Karpenko et al., 1958.

Phosphuranylite, $Ca(UO_2)_3(PO_4)_2(OH)_2 \cdot 6H_2O$, G Orth. Translucent yellow. Flat Rock pegmatite and Buchanan pegmatite, Mitchell Co., North Carolina, USA. H= 2.5, D_m= 4.10, D_c= 4.14, VIID 21. Genth, 1879. Am. Min. 39 (1954), 444. Dokl.Earth Sci. 220(1975), 123. See also Kivuite.

Phuralumite, $Al_2(UO_2)_3(PO_4)_2(OH)_6 \cdot 10H_2O$, A Mon. Translucent yellow. Kobokobo, Kivu, Zaïre. H= 3, D_m= 3.5, D_c= 3.52, VIID 21. Deliens & Piret, 1979. Am. Min. 65 (1980), 208 (Abst.). Acta Cryst. B35 (1979), 1880.

Phurcalite, $Ca_2(UO_2)_3(PO_4)_2(OH)_4 \cdot 4H_2O$, A Orth. Vitreous/adamantine yellow. Bergen an den Trieb, Sachsen (Saxony), Germany. H= 3, D_m=?, D_c= 4.14, VIID 21. Deliens & Piret, 1978. Am. Min. 64(1979), 243 (Abst.). Acta Cryst. B34 (1978), 1677.

Phylloretine, $C_{18}H_{18}$, G Orth. Holtegaard (near), Denmark. H=?, D_m= 1.13, D_c= 1.244, IXB 01. Forchhammer, 1839. Strunz (1970), 496.

Phyllotungstite, $HCaFe_3(WO_4)_6 \cdot 10H_2O$, A Orth. Pearly yellow. Clara mine (Grube Clara), Wolfach, Schwarzwald, Baden-Württemberg, Germany. H= 2, D_m=?, D_c= 5.26, VIF 03. Walenta, 1984. Am. Min. 71(1986), 846 (Abst.).

Pianlinite, D Inadequate data. Am. Min. 65(1980), 1068 (Abst.). Liu et al., 1963.

Pickeringite, $MgAl_2(SO_4)_4 \cdot 22H_2O$, G Mon. Vitreous colorless, white, yellowish, reddish. Cerros Pintados, Chile (N). H= 1.5, D_m= 1.76, D_c= 1.84, VIC 06. Hayes, 1844. Dana, 7th ed.(1951), v.2, 523. See also Halotrichite.

Picotpaulite, $TlFe_2S_3$, [A] Orth. Metallic creamy-white. Alšar (Allchar), Roždeп (near), Makedonija (Macedonia), Yugoslavia. H= 2, D_m=?, D_c= 5.20, IIB 07. Johan et al., 1970. Am. Min. 57(1972), 1909 (Abst.).

Picromerite, $K_2Mg(SO_4)_2 \cdot 6H_2O$, [G] Mon. Vitreous colorless, white, reddish, yellowish, grey. Mt. Vesuvius, Napoli, Campania, Italy. H= 2.5, D_m= 2.028, D_c= 2.043, VIC 11. Scacchi, 1855. Dana, 7th ed.(1951), v.2, 453. Zts. Krist. 122 (1965), 161.

Picropharmacolite, $Ca_4Mg(AsO_3OH)_2(AsO_4)_2 \cdot 11H_2O$, [G] Tric. Pearly/silky white. Richelsdorf, Hessen, Germany. H=?, D_m= 2.62, D_c= 2.60, VIIC 02. Stromeyer, 1819. Am. Min. 61 (1976), 326. Am. Min. 66 (1981), 385.

Piemontite, $Ca_2(Mn,Fe)Al_2(Si_2O_7)(SiO_4)(O,OH)_2$, [G] Mon. Vitreous reddish-brown/black. Pravorna mine, St. Marcel, Aosta Valley, Piemonte, Italy. H= 6, D_m= 3.5, D_c= 3.45, VIIIB 15. Kenngott, 1853. DHZ, 2nd ed.(1986), v.1B, 135. Am. Min. 54 (1969), 710.

Pierrotite, $Tl_2(Sb,As)_{10}S_{16}$, [A] Orth. Metallic greyish-black; white in reflected light. Jas Roux, Valgaudemar, Hautes-Alpes, France. H= 3.5, D_m= 4.97, D_c= 4.75, IID 14. Guillemin et al., 1972. Am. Min. 57(1972), 1909 (Abst.). Zts. Krist. 165 (1983), 209.

Pigeonite, $(Mg,Fe,Ca)SiO_3$, [G] Mon. Vitreous brown, greenish-brown, black. Pigeon Point, Minnesota, USA. H= 6, D_m= 3.30, D_c= 3.44, VIIID 01a. Winchell, 1900. DHZ, 2nd ed.(1978), v.2A, 162. Am. Min. 55 (1970), 1195.

Pigotite, $Al_4C_6H_5O_{10} \cdot 13H_2O$(?), [Q] Syst=? Cornwall, England. H=?, D_m=?, D_c=?, IXA 02. Johnston, 1840. Strunz (1970), 495.

Pilite, [D] = Actinolite pseudomorph. Am. Min. 63 (1978), 1023. Becke, 19th century.

Pilsenite, Bi_4Te_3, [A] Rhom. Metallic white. Plseň (Pilsen), Czechoslovakia. H= 2, D_m= 8.6, D_c= 8.45, IIA 15. Kenngott, 1853. Am. Min. 69(1984), 215 (Abst.).

Pimelite, $Ni_3Si_4O_{10}(OH)_2 \cdot H_2O$, [G] Syst=? Translucent/greasy green. Poland(?). H= 2.5, D_m= 2-3, D_c=?, VIIIE 04. Karsten, 1800. Am. Min. 64 (1979), 615.

Pinakiolite, $Mg_2MnO_2(BO_3)$, [G] Mon. Metallic black. Långban mine, Filipstad (near), Värmland, Sweden. H= 6, D_m= 3.88, D_c= 4.09, Vc 01c. Flink, 1890. Dana, 7th ed.(1951), v.2, 324. Am. Min. 59 (1974), 985. See also Fredrikssonite, Orthopinakiolite, Takeuchiite.

Pinalite, $Pb_3(WO_4)OCl_2$, [A] Orth. Adamantine yellow. Mammoth-St. Anthony mine, Tiger, Pinal Co., Arizona, USA. H=?, D_m=?, D_c= 7.78, IIID 03. Dunn et al., 1989. Am. Min. 74 (1989), 934.

Pinchite, $Hg_5O_4Cl_2$, [A] Orth. Translucent black, dark brown. Terlingua, Brewster Co., Texas, USA. H=?, D_m= 9.5, D_c= 9.25, IIIC 03. Sturman & Mandarino, 1974. Can. Min. 12 (1974), 417.

Pinnoite, $MgB_2O(OH)_6$, [G] Tet. Vitreous yellow, greenish-yellow, pistachio-green. Stassfurt, Sachsen (Saxony), Germany. H= 3.5, D_m= 2.27, D_c= 2.293, Vc 04b. Staute, 1884. Dana, 7th ed.(1951), v.2, 334. Sov.Phys.Cryst. 28 (1983), 475.

Pintadoite, $Ca_2V_2O_7 \cdot 9H_2O$, [Q] Syst=? Translucent light/dark green. Ed Whitney Claims, Monticello (N) and Frisco #2 Claim, Canyon Pintado, San Juan Co., Utah, USA. H=?, D_m=?, D_c=?, VIIC 15. Hess & Schaller, 1914. Dana, 7th ed.(1951), v.2, 1053.

Pirquitasite, Ag_2ZnSnS_4, [A] Tet. Metallic brownish-grey. Pirquitas deposit, Jujuy province, Argentina. H= 4, D_m=?, D_c= 4.82, IIB 03a. Johan & Picot, 1982. Am. Min. 68(1983), 1249 (Abst.). See also Hocartite.

Pirssonite, $Na_2Ca(CO_3)_2 \cdot 2H_2O$, [G] Orth. Vitreous colorless, white. Searles Lake, San Bernardino Co., California, USA. H= 3-3.5, D_m= 2.352, D_c= 2.367, VbC 04. Pratt, 1896. Dana, 7th ed.(1951), v.2, 232. Acta Cryst. 23 (1967), 763.

Pitticite, $[Fe,AsO_4,SO_4,H_2O]$(?), [Q] Amor. Dull/vitreous/greasy yellowish/reddish-brown, grey, etc. Christbescherung mine, Freiberg, Sachsen (Saxony),

Germany. H= 2-3, D_m= 2.3, D_c=?, VIID 04. Hausmann, 1813. Min. Mag. 46 (1982), 261.

Piypite, $K_4Cu_4O_2(SO_4)_4 \cdot (Na, Cu)Cl$, [A] Tet. Vitreous green/black. Tolbachik volcano, Kamchatka, USSR. H= 2.5, D_m= 3.10, D_c= 3.0, VIB 07. Vergasova et al., 1984. ZVMO 118(3), (1989), 88.

Plagioclase, $(Na, Ca)(Si, Al)_3O_8$, [G] Tric. Vitreous/pearly white, grey, colorless, yellow, pink, etc. H= 6-6.5, D_m= 2.6, D_c=?, VIIIF 03. Breithaupt, 1847. DHZ (1963), v.4, 94. Acta Cryst. B32 (1976), 521. See also Albite, Anorthite, Oligoclase, Andesine, Labradorite, Bytownite.

Plagionite, $Pb_5Sb_8S_{17}$, [G] Mon. Metallic blackish-grey; creamy-white in reflected light. Wolfsberg, Harz, Germany. H= 2.5, D_m= 5.54, D_c= 5.58, IID 09a. Rose, 1833. Min. Abstr. 71-1809. Nature 225 (1970), 444.

Planchéite, $Cu_8(Si_4O_{11})_2(OH)_4 \cdot H_2O$, [G] Orth. Translucent blue. Mindouli, Congo. H= 5.5, D_m= 3.3, D_c= 3.815, VIIID 04. Lacroix, 1908. Dana/Ford (1932), 687. Am. Min. 62 (1977), 491.

Planerite, $Al_6(PO_4)_2(PO_3OH)_2(OH)_8 \cdot 4H_2O$, [A] Tric. Dull verdigris-green, olive-green. Gumeshevsk, Ural Mts., USSR. H= 5, D_m= 2.68, D_c= 2.71, VIID 08. Hermann, 1862. IMA 1986, Abstr. p. 102.

Platarsite, (Pt, Rh, Ru)AsS, [A] Cub. Metallic grey. Onverwacht (#330) mine, Lydenberg Dist., Transvaal, South Africa. H= 7-7.5, D_m= 8.0, D_c= 8.437, IIC 06a. Cabri et al., 1977. Can. Min. 15 (1977), 385. Can. Min. 17 (1979), 117.

Platiniridium, (Ir, Pt), [G] Cub. Metallic white. H= 6-7, D_m= 22.7, D_c= 22.81, IA 07. Glocker, 1839. Dana, 7th ed., I (1944), 110.

Platinum, Pt, [G] Cub. Metallic white/grey; white in reflected light. Pinto River, Papayan (near), Cauca, Colombia. H= 4-4.5, D_m=?, D_c= 21.45, IA 07. Ulloa, 1748. Can. Min. 13 (1975), 117. See also Yixunite.

Plattnerite, PbO_2, [G] Tet. Metallic/adamantine black; greyish-white in reflected light. Leadhills, Lanarkshire, Scotland. H= 5.5, D_m= 9.564, D_c= 9.563, IVD 02. Haidinger, 1845. Min. Rec. 1 (1970), 75.

Platynite, $PbBi_2(Se, S)_4$, [G] Hex. Metallic grey/black; white in reflected light. Falun, Kopparberg, Sweden. H= 3.2-3.5, D_m= 7.98, D_c= 7.01, IID 09e. Flink, 1910. ZVMO 118(3) (1989), 22.

Playfairite, $Pb_{16}(Sb, As)_{18}S_{43}$, [A] Mon. Metallic grey/black; white/grey in reflected light. Taylor pit, Huntingdon Twp, Hastings Co., Ontario, Canada. H= 3.5, D_m=?, D_c= 5.72, IID 10. Jambor, 1967. Can. Min. 9 (1967), 191.

Pleurasite, [D] A mixture of sarkinite and other minerals. Am. Min. 58 (1973), 562 (Abst.). Igelström, 1889.

Plinthite, [D] A mixture of silicates with hematite. Min. Mag. 33 (1962), 262. Thomson, 1836.

Plombièrite, $Ca_5Si_6O_{16}(OH)_2 \cdot 6H_2O(?)$, [Q] Syst=? Translucent pink, red-brown. Plombières river region, Vosges, France. H=?, D_m= 2.02, D_c=?, VIIID 10. Daubrée, 1858. Min. Mag. 30 (1954), 293.

Plumalsite, [D] Inadequate data. Am. Min. 53(1968), 349 (Abst.). Gornyi et al., 1967.

Plumangite, [D] Inadequately described. Min. Mag. 43 (1980), 1055. Adib & Ottemann, 1970.

Plumboallophane, [D] An unnecessary name for allophane with some PbO. Min. Mag. 43 (1980), 1055. Bombicci, 1868.

Plumbobetafite, $(Pb, U, Ca)(Ti, Nb)_2O_6(OH, F)$, [A] Cub. Adamantine yellowish. Burpala massif, Baikal (N), USSR. H=?, D_m= 4.64, D_c=?, IVC 09a. Ganzeev et al., 1969. Am. Min. 55(1970), 1068 (Abst.).

Plumboferrite, $PbFe_4O_7$, [G] Rhom. Black. Jakobsberg mine, Nordmark, Värmland, Sweden. H= 5, D_m= 6.07, D_c= 6.55, IVC 05a. Igelström, 1881. JCPDS 22-656.

Plumbogummite, $PbAl_3(PO_4)(PO_3OH)(OH)_6$, G Rhom. Dull/resinous greyish-white, yellow, brown, greenish, etc. Huelgoat, Finistère, Brittany, France. H= 4.5-5, D_m= 4.01, D_c= 4.07, VIIB 15a. de Laumont, 1819. JCPDS 35-623.

Plumbojarosite, $PbFe_6(SO_4)_4(OH)_{12}$, G Rhom. Dull/silky golden brown, dark brown. Cooks Peak, Luna Co., New Mexico, USA. H=?, D_m= 3.66, D_c= 3.618, VIB 03a. Hillebrand & Penfield, 1902. Am. Min. 51 (1966), 443. Can. Min. 23 (1985), 659.

Plumbomicrolite, $(Pb, Na, Ca)(Ta, Nb)_2(O, OH)_7$, A Cub. Translucent greasy greenish-yellow, orange. Kivu, Zaïre. H=?, D_m= 6.5, D_c=?, IVC 09a. Safiannikoff & v.Wambeke, 1961. Am. Min. 47(1962), 1220 (Abst.).

Plumbonacrite, $Pb_{10}(CO_3)_6O(OH)_6$(?), Q Hex. Wanlockhead, Scotland. H=?, D_m=?, D_c=?, VbB 09. Heddle, 1889. Am. Min. 52(1967), 563 (Abst.). See also Hydrocerussite.

Plumbopalladinite, Pb_2Pd_3, A Hex. Metallic white. Mayak mine, Talnakh, Noril'sk (near), Siberia (N), USSR. H= 5, D_m=?, D_c= 12.31, IA 09. Genkin et al., 1970. Am. Min. 56(1971), 1121 (Abst.).

Plumbopyrochlore, $(Pb, Y, U, Ca)_{2-x}Nb_2O_6(OH)$, A Cub. Translucent brown, greenish yellow, red. Ural Mts., USSR. H=?, D_m=?, D_c=?, IVC 09a. Skorobogatova et al., 1966. Am. Min. 55(1970), 1068 (Abst.).

Plumbotellurite, α–$PbTeO_3$, A Orth. Translucent grey, yellow-grey, brown. Zhana-Tyube, Kazakhstan (N), USSR. H= 2, D_m= 7.2, D_c= 7.16, VIG 11. Spiridonov & Tananaeva, 1982. Am. Min. 67(1982), 1074 (Abst.). See also Fairbankite.

Plumbotsumite, $Pb_5Si_4O_8(OH)_{10}$, A Orth. Transparent colorless. Tsumeb, Namibia. H= 2, D_m= 5.6, D_c= 5.56, VIIIA 27. Keller & Dunn, 1982. Am. Min. 67(1982), 1075 (Abst.).

Plumbozincocalcite, D Unnecessary name for a plumboan zincian calcite. Min. Mag. 38 (1971), 103. Kantor, 1964.

Poitevinite, $(Cu, Fe, Zn)SO_4 \cdot H_2O$, G Mon. Translucent salmon. Avoca claim, Hat Creek, Bonaparte River, Lillooet Dist., British Columbia, Canada. H= 3-3.5, D_m= 3.30, D_c= 3.30, VIC 01. Jambor et al., 1964. Can. Min. 8 (1964), 109.

Pokrovskite, $Mg_2CO_3(OH)_2 \cdot 0.5H_2O$, A Mon. Dull white (fleshy tint). Zlatorgorsk intrusive, Kazakhstan, USSR. H= 3, D_m= 2.51, D_c= 2.58, VbB 01. Ivanov et al., 1984. Am. Min. 70(1985), 217 (Abst.). See also Rosasite.

Polarite, Pd(Pb, Bi), A Orth. Metallic white. Talnakh, Noril'sk (near), Polar Urals, Siberia (N), USSR. H= 4, D_m=?, D_c= 12.51, IA 09. Genkin et al., 1969. ZVMO 98 (1969), 708. See also Borishanskiite, Sobolevskite.

Polhemusite, (Zn, Hg)S, A Tet. Adamantine black. B and B deposit, Big Creek dist., Valley Co., Idaho, USA. H= 4, D_m= 4.5, D_c= 4.23, IIB 01b. Leonard & Desborough, 1978. Am. Min. 63 (1978), 1153.

Polkovicite, $(Fe, Pb)_3(Ge, Fe)_{1-x}S_4$, A Cub. Metallic white. Zechstein shales, Lower Silesia, Poland. H= 3, D_m=?, D_c= 6.62, IIB 03c. Haranczyk, 1975. Am. Min. 66(1981), 437 (Abst.). See also Morozeviczite.

Pollucite, $(Cs, Na)(AlSi_2)O_6 \cdot nH_2O$, G Cub. Vitreous colorless. Elba, Toscana, Italy. H= 6.5, D_m= 2.94, D_c= 2.97, VIIIF 11. Breithaupt, 1846. Can. Min. 12 (1974), 334. Zts. Krist. 129 (1969), 280. See also Analcime, Analcite.

Polybasite, $(Ag, Cu)_{16}Sb_2S_{11}$, G Mon. Metallic black; grey-white in reflected light. Neuer Morgenstern mine, Freiberg, Sachsen (Saxony), Germany. H= 2-3, D_m= 6.1, D_c= 6.36, IID 11. Rose, 1829. Am. Min. 48 (1963), 565. See also Pearceite.

Polycrase-(Y), $Y(Ti, Nb)_2(O, OH)_6$, G Orth. Sub-metallic black. Hitterö, Norway. H= 5.5-6, D_m= 5.0, D_c=?, IVD 10b. Scheerer, 1870. Dana, 7th ed.(1944), v.1, 787. See also Kobeite-(Y).

Polydymite, Ni_3S_4, G Cub. Metallic grey; white in reflected light. Grünau, Sayn-Altenkirchen, Westphalia, Germany. H= 4.5-5.5, D_m= 4.6, D_c= 4.83, IIC 01. Laspeyres, 1876. Min. Mag. 43 (1980), 733. See also Linnaeite.

Polyhalite, $K_2Ca_2Mg(SO_4)_4 \cdot 2H_2O$, G Tric. Vitreous/resinous colorless, white, grey, pink, red. Ischl salt mine, Oberösterreich (Upper Austria), Austria. H= 3.5, D_m= 2.78, D_c= 2.76, VIC 12. Stromeyer, 1818. Per. Mineral. 54 (1985), 35. See also Leightonite.

Polylithionite, $KLi_2AlSi_4O_{10}(F,OH)_2$, G Mon. Translucent colorless, pink, purple, pale yellow-green. Kangerdluarsuk, Ilímaussaq, Greenland (S). H= 2.5-4, D_m= 2.85, D_c= 2.823, VIIIE 05b. Lorenzen, 1884. Am. Min. 53 (1968), 1490.

Polyxene, D Unnecessary name for alloy of Pt and other elements. Can. Min. 13 (1975), 117.

Ponomarevite, $K_4Cu_4OCl_{10}$, A Mon. Clear transparent. Tolbachik volcano, Kamchatka, USSR. H=?, D_m= 2.78, D_c= 2.73, IIIB 07. Vergasova et al., 1988. Min. Mag. 52(1988), 729 (Abst.).

Portlandite, $Ca(OH)_2$, G Rhom. Pearly colorless. Scawt Hill, Antrim Co., Northern Ireland. H= 2, D_m= 2.23, D_c= 2.24, IVF 03a. Tilley, 1933. NBS Circ. 539, v.1 (1953), 58.

Posnjakite, $Cu_4SO_4(OH)_6 \cdot H_2O$, A Mon. Vitreous blue, dark blue. Nura-Talinsk tungsten deposits, Kazakhstan, USSR. H= 2-3, D_m= 3.32, D_c= 3.351, VID 01. Komkov & Nefedov, 1967. Am. Min. 52(1967), 1582 (Abst.). Zts. Krist. 149 (1979), 249.

Potarite, PdHg, G Tet. Metallic white. Amu Creek (near), Potaro River, Guyana. H= 3.5, D_m= 14.88, D_c= 15.09, IA 06. Harrison, 1926. Am. Min. 45 (1960), 1093.

Potassium pargasite, $KCa_2Mg_5(Si,Al)_8O_{23}$, Q Mon. Dark greyish-green. Einstödingen, Lützow-Holm Bay, Antarctica. H= 5.5-6, D_m= 3.12, D_c= 3.13, VIIID 05b. Matsubara & Motoyoshi, 1985. Min. Mag. 49 (1985), 703. See also Pargasite.

Potassium alum, $KAl(SO_4)_2 \cdot 12H_2O$, G Cub. Vitreous colorless/white. H= 2-2.5, D_m= 1.757, D_c= 1.753, VIC 08. Dana, 7th ed.(1951), v.2, 472. Acta Cryst. 22 (1967), 793. See also Lonecreekite.

Potosíite, $Fe_7Pb_{48}Sn_{18}Sb_{16}S_{115}$, A Tric. Metallic bluish/yellowish grey in reflected light. Andacabe deposit, Potosí, Bolivia. H= 2.5, D_m=?, D_c= 6.2, IID 13. Wolf et al., 1981. Can. Min. 24 (1986), 45. See also Cylindrite, Incaite.

Pottsite, $PbBi(VO_4)(VO_3OH) \cdot 2H_2O$, A Tet. Translucent yellow. Potts (near), Lander Co., Nevada, USA. H= 3.5, D_m= 7.0, D_c= 7.31, VIIC 20. Williams, 1988. Min. Mag. 52 (1988), 389.

Poubaite, $PbBi_2Se_2(Te,S)_2$, A Rhom. Metallic creamy-white in reflected light. Oldřichov, Tachov (near), Čechy (Bohemia), Czechoslovakia. H= 2.5-3, D_m=?, D_c= 7.88, IID 09e. Čech & Vavrin, 1978. Am. Min. 63(1978), 1283 (Abst.).

Poudretteite, $KNa_2B_3Si_{12}O_{30}$, A Hex. Vitreous colorless, pale pink. Mont Saint-Hilaire, Rouville Co., Québec, Canada. H= 5, D_m= 2.51, D_c= 2.53, VIIIC 10. Grice et al., 1987. Can. Min. 25 (1987), 763. Can. Min. 25 (1987), 763.

Poughite, $Fe_2(TeO_3)_2(SO_4) \cdot 3H_2O$, A Orth. Translucent dark yellow, brownish-yellow, greenish-yellow. Moctezuma mine, Moctezuma, Sonora, Mexico; El Plomo mine, Ojonjona Dist., Honduras. H= 2.5, D_m= 3.75, D_c= 3.76, VIG 02. Gaines, 1968. Am. Min. 53 (1968), 1075. Struct. Repts. 37A (1971), 318.

P-Ourayite, $Ag_{3.6}Pb_{2.8}Bi_{5.6}S_{13}$, Q Orth. Metallic grey; white in reflected light. Ivigtut, Greenland (SW). H=?, D_m=?, D_c= 7.215, IID 09b. Makovicky & Karup-Møller, 1984. Can. Min. 22 (1984), 565. See also Ourayite.

Powellite, $CaMoO_4$, G Tet. Sub-adamantine/greasy yellow, brown, grey, blue, etc. Peacock Lode, Seven Devils dist., Idaho, USA. H= 3.5-4, D_m= 4.23, D_c= 4.25, VIF 01. Melville, 1891. NBS Circ. 539, v.6 (1956), 22. See also Scheelite.

Poyarkovite, Hg_3OCl, [A] Mon. Vitreous/adamantine red, black. Khaidarkan deposit, Kirgizia, USSR. H= 2-2.5, D_m= 9.50, D_c= 9.88, IIIC 03. Vasil'ev et al., 1981. Am. Min. 67(1982), 860 (Abst.).

Prassoite, $Rh_{17}S_{15}$(?), [Q] Cub. Metallic grey in reflected light. Tiébaghi massif, New Caledonia. H= 5.8, D_m=?, D_c= 7.607, IIC 15. Cabri, 1981. CIM SV (1981), 132. Acta Cryst. 15 (1962), 1198. See also Palladseite.

Pravdite, [D] = altered britholite. Am. Min. 49 (1964), 1501 (Abst.). Nurl'baev, 1962.

Prehnite, $Ca_2Al(Si,Al)_4O_{10}(OH)_2$, [G] Orth. Vitreous/pearly light green, white, grey. Cape of Good Hope, South Africa. H= 6-6.5, D_m= 2.9, D_c= 2.91, VIIIE 30. Werner, 1789. Can. Min. 25 (1987), 707. Am. Min. 52 (1967), 974.

Preisingerite, $Bi_3O(AsO_4)_2(OH)$, [A] Tric. Translucent white, grey. San Francisco de los Andes and Cerro Negro de la Aguadita, Calingasta Dept., San Juan, Argentina. H=?, D_m=?, D_c= 7.24, VIIB 17. Bedivy & Mereiter, 1981. Am. Min. 67(1982), 416 (Abst.). Am. Min. 67 (1982), 833. See also Schumacherite.

Preiswerkite, $Na(Mg,Al)_3(Si_2Al_2)O_{10}(OH)_2$, [A] Mon. Translucent pale greenish. Geisspfad, Binntal, Valais (Wallis), Switzerland. H= 2.5, D_m= 2.96, D_c= 2.94, VIIIE 05b. Keusen & Peters, 1980. Am. Min. 65 (1980), 1134.

Preobrazhenskite, $Mg_3B_{11}O_{15}(OH)_9$, [G] Orth. Translucent colorless, yellow, dark grey. Inder, Kazakhstan, USSR. H= 4.5-5, D_m=?, D_c= 2.45, Vc 06. Yarzhemskii, 1956. JCPDS 31-788.

Priceite, $Ca_4B_{10}O_{19}\cdot 7H_2O$(?), [Q] Syst=? Earthy white. Chetko (near), Curry Co., Oregon, USA. H= 3-3.5, D_m= 2.42, D_c=?, Vc 02. Silliman, 1873. Am. Min. 42 (1956), 689.

Priderite, $(K,Ba)(Ti,Fe)_8O_{16}$, [G] Tet. Adamantine black, reddish. Walgidee Hills, Fitzroy Basin, Kimberley, Western Australia, Australia. H=?, D_m= 3.86, D_c= 3.92, IVD 03b. Norrish, 1951. Austral. Min. 1 (1985), 298. Acta Cryst. B38 (1982), 1056.

Proarizonite, [D] An intermediate stage in the alteration of ilmenite to arizonite. Min. Mag. 36 (1967), 133. Bykov, 1964.

Probertite, $NaCaB_5O_7(OH)_4\cdot 3H_2O$, [G] Mon. Vitreous colorless. Baker mine, Kramer dist., Kern Co., California, USA. H= 3.5, D_m= 2.14, D_c= 2.132, Vc 09. Eakle, 1929. Dana, 7th ed.(1951), v.2, 343. Acta Cryst. B38 (1982), 3072.

Prosopite, $CaAl_2(F,OH)_8$, [G] Mon. Vitreous greyish, white. Altenberg, Sachsen (Saxony), Germany. H= 4.5, D_m= 2.89, D_c= 2.898, IIIB 05. Scheerer, 1853. N.Jb.Min.Mh. (1986), 329.

Prosperite, $CaZn_2(AsO_4)_2\cdot H_2O$, [A] Mon. Vitreous/silky white, colorless. Tsumeb, Namibia. H= 4.5, D_m= 4.31, D_c= 4.408, VIIC 03. Gait et al., 1979. Can. Min. 17 (1979), 87. Zts. Krist. 158 (1982), 33.

Protasite, $Ba(UO_2)_3O_3(OH)_2\cdot 3H_2O$, [A] Mon. Translucent bright orange. Shinkolobwe, Shaba, Zaïre. H=?, D_m=?, D_c= 5.827, IVF 12. Pagoaga et al., 1986. Min. Mag. 50 (1986), 125. Am. Min. 72 (1987), 1230.

Protoferro-anthophyllite, $Fe_7Si_8O_{22}(OH)_2$, [Q] Orth. Translucent brownish-yellow. H=?, D_m=?, D_c= 3.608, VIIID 06. Sueno & Matsuura, 1986. IMA 1986, Abst. p. 241. IMA 1986, Abst. p. 241. See also Protomanganese-anthophyllite, Anthophyllite.

Protojoséite, Bi_3TeS, [Q] Hex. Metallic. Băiţa Bihorului (Rézbánya), Romania and Malishevski Izumrydrye pit, Ural Mts., USSR. H=?, D_m=?, D_c= 8.50, IIA 15. Zav'lyalov & Begizov, 1983. Am. Min. 69(1984), 1192 (Abst.). See also Joséite.

Protomanganese-anthophyllite, $(Fe,Mn)_7Si_8O_{22}(OH)_2$, [Q] Orth. Translucent brownish-yellow. H=?, D_m=?, D_c= 3.60, VIIID 06. Sueno & Matsuura, 1986. IMA 1986, Abst. p. 241. IMA 1986, Abst. p. 241. See also Protoferro-anthophyllite, Anthophyllite.

Protopartzite, [D] Inadequate data; may be partzite. Min. Mag. 38 (1971), 103. Koritnig, 1967.

Proudite, $CuPb_{7.5}Bi_{9.3}S_{15}Se_8$, [A] Mon. Metallic silver-grey; creamy-white in reflected light. Juno mine, Tennant Creek, Northern Territory, Australia. H= 2-4, D_m= 7.12, D_c= 7.08, IID 08. Mumme, 1976. Econ. Geol. 70 (1975), 369. Am. Min. 61 (1976), 839.

Proustite, Ag_3AsS_3, [G] Rhom. Adamantine translucent red. H= 2-2.5, D_m= 5.65, D_c= 5.58, IID 01b. Beudant, 1832. Am. Min. 48 (1963), 725. Struct. Repts. 31A (1966), 19. See also Xanthoconite, Pyrargyrite.

Przhevalskite, $Pb(UO_2)_2(PO_4)_2 \cdot 4H_2O$, [Q] Syst=? Adamantine/pearly yellow. Unspecified locality, USSR(?). H=?, D_m=?, D_c=?, VIID 20b. Kruglov, 1946. Am. Min. 43(1958), 381 (Abst.).

Pseudo-aenigmatite, [D] Inadequate data. Min. Mag. 36 (1968), 1144. Kukharenko et al., 1965.

Pseudo-autunite, [D] Inadequate data. Min. Mag. 36 (1968), 1144. Sergeev, 1964.

Pseudoboléite, $28PbCl_2 \cdot 2AgCl \cdot 24Cu(OH)_2 \cdot 14H_2O$(?), [G] Tet. Translucent blue. Boléo, Santa Rosalia (near), Baja California, Mexico. H= 2.5, D_m= 4.85, D_c= 4.89, IIIC 04. Lacroix, 1895. Min. Rec. 5 (1974), 280.

Pseudobrookite, Fe_2TiO_5, [A] Mon. Metallic adamantine, reddish-brown, black. Aranyer Berg, Piski (near), Transylvania, Romania. H= 6, D_m= 4.39, D_c= 4.406, IVC 11. Koch, 1878. Am. Min. 73 (1988), 1377. Struct. Repts. 51A (1984), 208. See also Armalcolite.

Pseudocotunnite, K_2PbCl_4(?), [Q] Orth. Translucent colorless, white, yellow, greenish-yellow. Mt. Vesuvius, Napoli, Campania, Italy. H=?, D_m=?, D_c= 4.25, IIIC 05. Scacchi, 1873. Strunz (1970), 168.

Pseudograndreefite, $Pb_6SO_4F_{10}$, [A] Orth. Subadamantine colorless. Grand Reef mine, Laurel Canyon, Graham Co., Arizona, USA. H= 2.5, D_m= 7.0, D_c= 7.08, IIID 04. Kampf et al., 1989. Am. Min. 74 (1989), 927. See also Grandreefite.

Pseudolaueite, $MnFe_2(PO_4)_2(OH)_2 \cdot 7\text{–}8H_2O$, [G] Mon. Translucent orange-yellow. Hagendorf pegmatite, Oberpfalz, Bayern (Bavaria), Germany. H= 3, D_m= 2.463, D_c= 2.51, VIID 03. Strunz, 1956. Am. Min. 41(1956), 815 (Abst.). Am. Min. 54 (1969), 1312.

Pseudomalachite, $Cu_5(PO_4)_2(OH)_4$, [G] Mon. Vitreous green. H= 4.5-5, D_m= 4.35, D_c= 4.367, VIIB 08. Hausmann, 1813. Am. Min. 66 (1981), 176. Am. Min. 62 (1977), 10424.

Pseudorutile, $Fe_2Ti_3O_9$, [G] Hex. Florida, New Jersey, USA; India; Brazil. H=?, D_m=?, D_c= 4.817, IVD 05. Teufer & Temple, 1966. Nature 211 (1966), 179. Am. Min. 60 (1975), 898.

Pseudowollastonite, [D] An artificial product; not known naturally. Am. Min. 58 (1973), 560 (Abst.). Lacroix, 1895.

Pucherite, $BiVO_4$, [G] Orth. Vitreous/adamantine dark brownish-red, yellowish-brown. Pucher shaft, Wolfgang mine, Schneeberg, Sachsen (Saxony), Germany. H= 4, D_m= 6.57, D_c= 6.69, VIIA 13. Frenzel, 1871. Dana, 7th ed.(1951), v.2, 1050. Zts. Krist. 169 (1984), 289. See also Clinobisvanite, Dreyerite.

Pumpellyite-(Fe^{3+}), $Ca_2Fe^{3+}Al_2(SiO_4)(Si_2O_7)(OH,O)_2 \cdot H_2O$, [A] Mon. Långban mine, Filipstad (near), Värmland, Sweden. H=?, D_m=?, D_c= 3.35, VIIIB 26. Passaglia & Gottardi, 1973. Can. Min. 12 (1973), 219. See also Pumpellyite-(Mg), Julgoldite-(Fe), Pumpellyite-(Fe^{2+}), Pumpellyite-(Mn).

Pumpellyite-(Fe^{2+}), $Ca_2Fe^{2+}Al_2(SiO_4)(Si_2O_7)(OH)_2 \cdot H_2O$, [A] Mon. Långban mine, Filipstad (near), Värmland, Sweden. H=?, D_m=?, D_c= 3.27, VIIIB 26. Passaglia & Gottardi, 1973. Can. Min. 12 (1973), 219. See also Pumpellyite-(Mg), Julgoldite-(Fe), Pumpellyite-(Fe^{3+}) , Pumpellyite-(Mn).

Pumpellyite-(Mg), $Ca_2MgAl_2(SiO_4)(Si_2O_7)(OH)_2 \cdot H_2O$, [A] Mon. Translucent colorless, bluish-green, brown. Keweenawan copper deposits, Lake Superior region, Michigan, USA. H= 5.5-6, D_m= 3.2, D_c= 3.25, VIIIB 26. Palache & Vassar, 1925. Can. Min. 12 (1973), 219. Acta Cryst. B25 (1969), 2276. See also Pumpellyite-(Fe^{2+}), Julgoldite-(Fe), Pumpellyite-(Fe^{3+}).

Pumpellyite-(Mn), $Ca_2MnAl_2(SiO_4)(Si_2O_7)(OH)_2 \cdot H_2O$, [A] Mon. Vitreous light greyish/brownish pink. Ochiai mine, Yamanashi, Japan. H= 5, D_m=?, D_c= 3.34, VIIIB 26. Kato et al., 1981. Am. Min. 68(1983), 1250 (Abst.).

Purpurite, $(Mn, Fe)PO_4$, [G] Orth. Satiny deep rose, reddish-purple, dark brown. Faires Tin mine, Kings Mt., Gaston Co., North Carolina, USA. H= 4-4.5, D_m= 3.3, D_c= 3.69, VIIA 02. Graton & Schaller, 1905. JCPDS 17-202. See also Heterosite.

Putoranite, $Cu_{1.1}Fe_{1.2}S_2$, [A] Cub. Metallic yellow. Oktyabr deposit, Talnakh, Noril'sk (near), Siberia (N), USSR. H= 4, D_m=?, D_c= 4.48, IIB 02. Filimonova et al., 1980. Am. Min. 66(1981), 638 (Abst.).

p-Veatchite, $(Sr, Ca)_2[B_5O_8(OH)]_2B(OH)_3 \cdot H_2O$, [P] Mon. Königshall-Hindenburg mine, Reyershausen, Göttingen (near), Hanover, Niedersachsen, Germany. H=?, D_m= 2.64, D_c= 2.70, Vc 10. Braitsch, 1959. Am. Min. 45 (1960), 1221. Sov.Phys.Cryst. 16(1971), 75. See also Veatchite, Veatchite-A.

Pyrargyrite, Ag_3SbS_3, [G] Rhom. Adamantine red. H= 2.5, D_m= 5.82, D_c= 5.832, IID 01b. Glocker, 1831. Am. Min. 48 (1963), 725. Struct. Repts. 31A (1966), 19. See also Pyrostilpnite, Proustite.

Pyrite, FeS_2, [G] Cub. Metallic yellow. H= 6-6.5, D_m= 5.018, D_c= 5.013, IIC 05. Dioscorides, 50. Am. Min. 74 (1989), 1168. Am. Min. 62 (1977), 1168. See also Marcasite, Cattierite.

Pyroaurite, $Mg_6Fe_2CO_3(OH)_{16} \cdot 4H_2O$, [G] Rhom. Waxy/vitreous colorless, yellowish/brownish white, green. Långban mine, Filipstad (near), Värmland, Sweden. H= 2.5, D_m= 2.11, D_c= 2.10, IVF 03b. Igelström, 1865. Am. Min. 26 (1941), 295. Acta Cryst. B24 (1968), 972. See also Sjögrenite.

Pyrobelonite, $PbMnVO_4(OH)$, [G] Orth. Adamantine red. Långban mine, Filipstad (near), Värmland, Sweden. H= 3.5, D_m= 5.58, D_c= 5.79, VIIB 11b. Flink, 1919. Min. Mag. 41 (1977), 85. Am. Min. 40 (1955), 580.

Pyrochlore, $(Na, Ca)_2Nb_2O_6(OH, F)$, [G] Cub. Vitreous/resinous brown, black. Fredriksvärn, Norway. H= 5-5.5, D_m= 4.45, D_c= 4.26, IVC 09a. Wöhler, 1826. Can. Min. 6 (1961), 610. See also Microlite.

Pyrochlore-wiikite, [D] A mixture of yttropyrochlore and other minerals. Am. Min. 62 (1977), 403. Strunz, 1957.

Pyrochroite, $Mn(OH)_2$, [G] Rhom. Pearly colorless, pale greenish/bluish. Persberg, Filipstad (near), Värmland, Sweden. H= 2.5, D_m= 3.25, D_c= 3.260, IVF 03a. Igelström, 1864. Am. Min. 50 (1965), 1296.

Pyrolusite, β-MnO_2, [G] Tet. Metallic grey, black; cream-white in reflected light. H= 2-6.5, D_m= 5, D_c= 5.193, IVD 03a. Haidinger, 1827. Dana, 7th ed.(1944), v.1, 562. Acta Cryst. B32 (1976), 2200. See also Ramsdellite, Nsutite, Vernadite.

Pyromorphite, $Pb_5(PO_4)_3Cl$, [G] Hex. Resinous sub-adamantine green, yellow, brown, etc. H= 3.5-4, D_m= 7.04, D_c= 7.109, VIIB 16. Hausmann, 1813. Am. Min. 51 (1966), 1712. Can. Min. 27 (1989), 189.

Pyrope, $Mg_3Al_2(SiO_4)_3$, [G] Cub. Vitreous/resinous deep red. Czechoslovakia. H= 6.5-7.5, D_m= 3.51, D_c= 3.53, VIIIA 06a. Werner, 1803. DHZ, 2nd ed.(1982), v.1A, 468. Am. Min. 56 (1971), 791. See also Almandine, Knorringite.

Pyrophanite, $MnTiO_3$, [G] Rhom. Metallic/sub-metallic deep blood-red. Harstigen mine, Pajsberg, Värmland, Sweden. H= 5-6, D_m= 4.54, D_c= 4.60, IVC 04b. Hamberg, 1890. NBS Monogr. 25 (1978), 15. See also Geikielite, Ilmenite.

Pyrophyllite-2M, $AlSi_2O_5(OH)$, [G] Mon. Pearly white, yellow, pale blue, greyish/brownish-green. H= 1-2, D_m= 2.66, D_c= 2.85, VIIIE 04. Hermann, 1829. JCPDS 12-203. See also Ferripyrophyllite, Pyrophyllite-1Tc.

Pyrophyllite-1Tc, $AlSi_2O_5(OH)$, [P] Tric. Pearly white, yellow, pale blue, greyish/brownish green. H= 1-2, D_m= 2.7, D_c= 2.814, VIIIE 04. Hermann, 1829. DHZ (1962), v.3, 115. Am. Min. 66 (1981), 350. See also Pyrophyllite-2M, Ferripyrophyllite.

Pyrostilpnite, Ag_3SbS_3, [G] Mon. Adamantine translucent red. St. Andreasberg, Harz, Germany. H= 2, D_m= 5.94, D_c= 5.97, IID 01d. Dana, 1868. Dana, 7th ed.(1944), v.1, 369. See also Pyrargyrite.

Pyroxene, A group name for single-chain silicates with the general formula $(Ca, Mg, Fe, Mn, Na, Li)(Al, Mg, Fe, Mn, Cr, Sc, Ti)(Si, Al)_2O_6$.

Pyroxferroite, $(Fe, Mn, Ca)SiO_3$, [A] Tric. Translucent yellow. Tranquillity Base, Moon. H=?, D_m= 3.7, D_c= 3.85, VIIID 13. Chao et al., 1970. Am. Min. 55(1970), 2137 (Abst.). Science 168 (1970), 364. See also Pyroxmangite.

Pyroxmangite, $MnSiO_3$, [G] Tric. Amber, dark brown. Iva, Anderson Co., South Carolina, USA. H= 5.5-6, D_m= 3.8, D_c= 3.766, VIIID 13. Ford & Bradley, 1913. Dana/Ford (1932), 565. Min. Journ. 8 (1977), 329. See also Pyroxferroite.

Pyrrhite, [D] Variety of pyrochlore. Am. Min. 62 (1977), 403. Rose, 1839.

Pyrrhotite-4C, Fe_7S_8, [P] Mon. Metallic bronze. Chichibu mine, Akaiwa, Japan. H= 3.5-4.5, D_m= 4.6, D_c= 4.60, IIB 09b. Breithaupt, 1835. Econ. Geol. 70 (1975), 824. Acta Cryst. 25B (1969), 673.

Pyrrhotite-5C, Fe_9S_{10}, [P] Hex. Metallic bronze. Uri, Switzerland and/or Yanahara mine, Honko, Japan. H=?, D_m=?, D_c= 4.64, IIB 09b. Engel et al., 1978. Econ. Geol. 70 (1975), 824.

Pyrrhotite-6C, $Fe_{11}S_{12}$, [P] Mon. Metallic bronze. Makimine mine, Chigusa, Japan. H=?, D_m=?, D_c= 4.67, IIB 09b. Morimoto et al., 1975. Econ. Geol. 70 (1975), 824. Acta Cryst. B31 (1975), 2759.

Pyrrhotite-7C, β-$Fe_{1-x}S$, [P] Hex. Metallic bronze. H=?, D_m=?, D_c= 4.65, IIB 09b. Desborough & Carpenter, 1965. Econ. Geol. 60 (1965), 143.

Pyrrhotite-11C, $Fe_{10}S_{11}$, [P] Hex. Metallic bronze. Kohmori mine, Suetake, Japan. H=?, D_m=?, D_c= 4.647, IIB 09b. Morimoto et al., 1975. Econ. Geol. 70 (1975), 824.

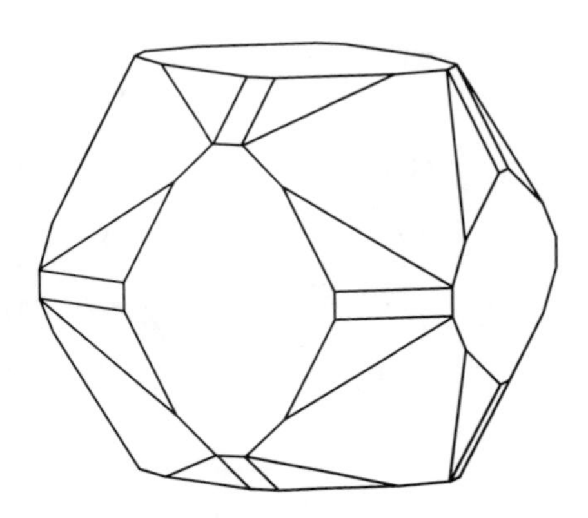

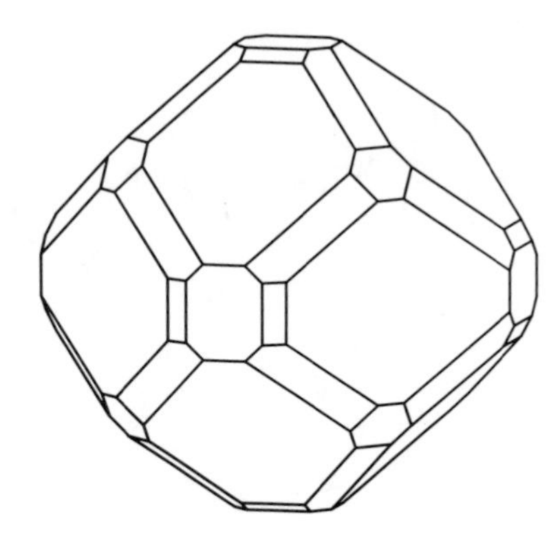

(Left) Pyrite (after Palache, Berman & Frondel, 1944) from Rossie, New York showing forms {001}, {102}, {214} and {111}.

(Right) Pyrite from Rossie showing the forms {001}, {102}, {112}, {111} and {011}.

Q

Qandilite, $(Mg,Fe)_2(Ti,Fe,Al)O_4$, [A] Cub. Metallic black; pinkish-grey in reflected light. Qala-Dizeh, Iraq (NE). H= 7, D_m= 4.03, D_c= 4.04, IVB 01. Al-Hermezi, 1985. Min. Mag. 49 (1985), 739.

Qingheiite, $Na_2NaMn_2Mg_2(Al,Fe)_2(PO_4)_6$, [A] Mon. Vitreous green. Qinghe, Altai, Xinjiang, China. H= 5.4, D_m= 3.718, D_c= 3.61, VIIA 05a. Ma et al., 1983. Am. Min. 69 (1984), 567 (Abst.). See also Ferrowyllieite, Rosemaryite, Wyllieite.

Qitianlingite, $(Fe,Mn)_2(Nb,Ta)_2WO_{10}$, [A] Orth. Metallic black. Qitianling, Hunan Province, China. H= 5-5.5, D_m=?, D_c= 6.42, IVD 07. Yang et al., 1985. Am. Min. 73(1988), 1497 (Abst.). IMA 1986 Abstracts, p. 261.

Quartz, α–SiO_2, [G] Rhom. Vitreous colorless, pink, violet, yellow, etc. H= 7, D_m= 2.655, D_c= 2.651, IVD 01. Agricola, 1529. Dana, 7th ed.(1962), v.3, 9. Acta Cryst. B32 (1976), 2456. See also Berlinite, Tridymite, Cristobalite, Coesite, Stishovite.

Quatrandorite, $Ag_{15}Pb_{18}Sb_{47}S_{96}$, [Q] Mon. Greyish-white in reflected light. Les Cougnasses, Hautes–Alpes, France. H=?, D_m=?, D_c=?, IID 09b. Moëlo et al., 1984. Am. Min. 70(1985), 219 (Abst.).

Queitite, $Zn_2Pb_4(SiO_4)(Si_2O_7)(SO_4)$, [A] Mon. Greasy pale yellow, colorless. Tsumeb, Namibia. H=?, D_m=?, D_c= 6.07, VIIIB 13. Keller et al., 1979. Am. Min. 65(1979), 407 (Abst.). Zts. Krist. 151 (1980), 287.

Quenselite, $PbMnO_2(OH)$, [G] Mon. Metallic/adamantine black. Långban mine, Filipstad (near), Värmland, Sweden. H= 2.5, D_m= 6.84, D_c= 7.07, IVF 05b. Flink, 1926. Dana, 7th ed.(1944), v.1, 729. Zts. Krist. 134 (1971), 321.

Quenstedtite, $Fe_2(SO_4)_3 \cdot 11H_2O$, [G] Tric. Transparent pale violet, reddish-violet. Tierra Amarilla, Copiapó (near), Chile. H= 2.5, D_m= 2.147, D_c= 2.145, VIC 04. Linck, 1888. Dana, 7th ed.(1951), v.2, 535. Am. Min. 59 (1974), 582.

Quetzalcoatlite, $Cu_4Zn_8(TeO_3)_3(OH)_{18}$, [A] Hex. Translucent blue. Bambollita mine, Moctezuma, Sonora, Mexico. H= 3, D_m= 6.05, D_c= 6.12, VIG 04. Williams, 1973. Min. Mag. 39 (1973), 261.

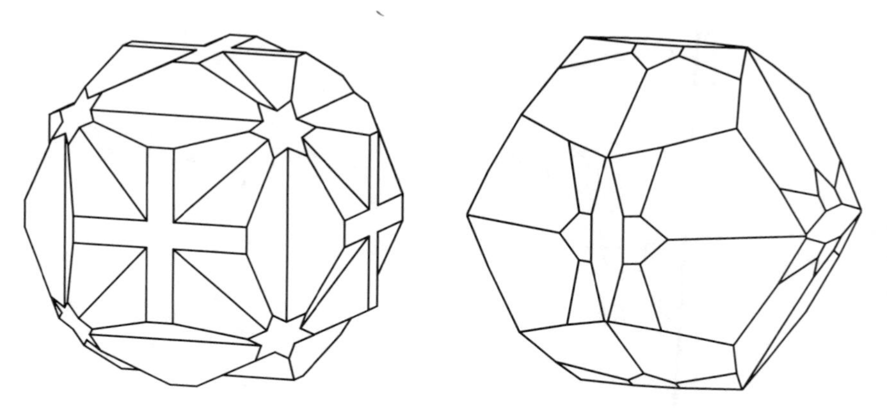

(Left) Pyrite twin from Brosso, Italy showing the forms {012}, {100} and {111} (after Strüver, 1869).

(Right) Pyrite from Arizona showing the forms {012}, {001}, {61 · 10} and {814} (after Palache, Berman & Frondel, 1944).

R

Rabbittite, $Ca_3Mg_3(UO_2)_2(CO_3)_6(OH)_4 \cdot 18H_2O$, G Mon. Silky pale green. Lucky Strike #2 mine, Emery Co., Utah, USA. H= 2, D_m= 2.57, D_c= 2.69, VbD 04. Thompson et al., 1955. Am. Min. 40 (1955), 201.

Radhakrishnaite, $PbTe_3(Cl,S)_2$, A Tet. Grey in reflected light. Champion reef lode, Kolar deposits, India. H=?, D_m=?, D_c= 8.89, IIIB 09. Genkin et al., 1985. Can. Min. 23 (1985), 501.

Raguinite, $TlFeS_2$, A Orth. Metallic bronze. Alšar (Allchar), Rožden (near), Makedonija (Macedonia), Yugoslavia. H=?, D_m= 6.40, D_c= 6.33, IIB 07. Laurent et al., 1969. Am. Min. 54(1969), 1495 (Abst.).

Raite, $(Na,Ca)_4(Mn,Ti,Fe)_3Si_8(O,OH)_{24} \cdot 9H_2O(?)$, A Orth. Translucent gold, brown. Mt. Karnasurt, Ilmajok Valley, Lovozero Tundra, Kola Peninsula, USSR. H= 3, D_m= 2.39, D_c= 2.32, VIIIE 19. Mer'kov et al., 1973. Am. Min. 58(1973), 1113 (Abst.).

Rajite, $CuTe_2O_5$, A Mon. Resinous bright green. Lone Pine mine, Silver City, Grant Co., New Mexico, USA. H= 4, D_m= 5.75, D_c= 5.77, VIG 05. Williams, 1979. Min. Mag. 43 (1979), 91.

Ralstonite, $Na_{0.4}(Al,Mg)_2(F,OH)_8 \cdot H_2O$, G Cub. Vitreous colorless, white. Ivigtut, Greenland (SW). H= 4.5, D_m= 2.64, D_c= 2.78, IIIB 05. Brush, 1871. Am. Min. 50 (1965), 1851. Min. Abstr. 85M/0180. See also Pyrochlore.

Ramdohrite, $Ag_3Pb_6Sb_{11}S_{24}$, G Mon. Metallic grey-black; white in reflected light. Guadeloupe mine, Nor-Chichas, Chocaya la Viega, Potosí, Bolivia. H= 2, D_m= 5.43, D_c= 5.55, IID 09b. Ahlfeld, 1930. Am. Min. 70(1985), 219 (Abst.). Am. Min. 69(1984), 412 (Abst.). See also Andorite.

Rameauite, $K_2CaO_8(UO_2)_6 \cdot 9H_2O$, A Mon. Translucent orange. Margnac mine, Compreignac, Haute Vienne, France. H=?, D_m= 5.60, D_c= 5.55, IVF 12. Cesbron et al., 1972. Min. Mag. 38 (1972), 781.

Rammelsbergite, $NiAs_2$, G Orth. Metallic white. Schneeberg, Sachsen (Saxony), Germany. H= 5.5-6, D_m= 7.1, D_c= 7.09, IIC 07. Dana, 1854. NBS Circ. 539, v.10 (1960), 42. Min. Abstr. 77-1487. See also Pararammelsbergite, Krutovite.

Ramsbeckite, $(Cu,Zn)_{15}(SO_4)_4(OH)_{22} \cdot 6H_2O$, A Mon. Vitreous green. Bastenberg mine, Ramsbeck, Germany. H= 3.5, D_m= 3.39, D_c= 3.41, VID 01. v. Hodenberg et al., 1985. Am. Min. 72(1987), 225 (Abst.). N.Jb.Min.Mh. (1988), 38.

Ramsdellite, γ–MnO_2, G Orth. Grey, black. Lake Valley, Sierra Co., New Mexico, USA. H= 2-4, D_m= 4.83, D_c= 4.87, IVD 03a. Fleischer & Richmond, 1943. Am. Min. 64 (1979), 1199. Struct. Repts. 15 (1951), 184. See also Pyrolusite, Nsutite, Vernadite.

Ranciéite, $(Ca,Mn)Mn_4O_9 \cdot 3H_2O$, G Hex. Metallic black, brownish, violet. Rancié, Sem, Vicdessos, Ariège, France. H=?, D_m= 3.2, D_c=?, IVF 05b. Leymerie, 1857. Min. Abstr. 80-4854. See also Takanelite.

Rankachite, $CaFeV_4O_4(WO_4)_8 \cdot 12H_2O$, A Orth. Resinous/sub-adamantine brown, yellow. Clara mine (Grube Clara), Wolfach, Schwarzwald, Baden-Württemberg, Germany. H= 2.5, D_m=?, D_c= 4.5, VIF 03. Walenta & Dunn, 1984. Am. Min. 70(1985), 876 (Abst.).

Rankamaite, $(Na,K,Pb,Li)_3(Ta,Nb,Al)_{11}(O,OH)_{30}$, A Orth. Translucent white, creamy-white. Mumba area, Kivu, Zaïre. H= 3-4, D_m= 5.5, D_c= 5.84, IVE 02. von Knorring et al., 1969. Am. Min. 55(1970),1814 (Abst.).

Rankinite, $Ca_3Si_2O_7$, G Mon. Translucent colorless. Scawt Hill, Antrim Co., Northern Ireland. H=?, D_m= 2.94, D_c= 2.99, VIIIB 04. Tilley, 1942. Min. Mag. 26 (1942), 190. Min. Jour. 8 (1976), 240. See also Kilchoanite.

Ransomite, $CuFe_2(SO_4)_4 \cdot 6H_2O$, [G] Mon. Vitreous/pearly blue. United Verde mine, Jerome, Yavapai Co., Arizona, USA. H= 2.5, D_m= 2.632, D_c= 2.735, VIC 05. Lausen, 1928. Dana, 7th ed.(1951), v.2, 519. Am. Min. 55 (1970), 729.

Ranunculite, $Al(UO_2)(PO_3OH)(OH)_3 \cdot 4H_2O$, [A] Mon. Translucent yellow. Kobokobo, Kivu, Zaïre. H= 3, D_m= 3.4, D_c= 3.39, VIID 21. Deliens & Piret, 1979. Min. Mag. 43 (1979), 321.

Rapidcreekite, $Ca_2(SO_4)(CO_3) \cdot 4H_2O$, [A] Orth. Vitreous white, colorless. Rapid Creek area, Yukon Territory, Canada. H= 2, D_m= 2.21, D_c= 2.239, VID 15. Roberts et al., 1986. Can. Min. 24 (1986), 51.

Raspite, $PbWO_4$, [G] Mon. Adamantine yellowish-brown, light yellow, grey. Broken Hill, New South Wales, Australia. H= 2.5-3, D_m= 8.46, D_c= 8.45, VIF 01. Hlawatsch, 1897. Dana, 7th ed.(1951), v.2, 1089. Acta Cryst. 33 (1977), 162. See also Stolzite.

Rasvumite, KFe_2S_3, [A] Orth. Metallic grey. Rasvumchorr pegmatite and Kukisvumchorr pegmatite, Khibina massif, Kola Peninsula, USSR. H= 4.5, D_m= 3.1, D_c= 3.029, IIB 17. Sokolova et al., 1970. Am. Min. 64 (1979), 776. Am. Min. 65 (1980), 477.

Rathite I, $(Pb,Tl)_3As_5S_{10}$, [G] Mon. Metallic grey; white in reflected light. Lengenbach quarry, Binntal, Valais (Wallis), Switzerland. H= 3, D_m= 5.37, D_c= 5.27, IID 02. Baumhauer, 1896. Dana, 7th ed.(1944), v.1, 455. Zts. Krist. 122 (1966), 433. See also Rathite IV, Rathite III.

Rathite III, $Pb_3As_5S_{10}$, [Q] Mon. Lengenbach quarry, Binntal, Valais (Wallis), Switzerland. H=?, D_m= 5.38, D_c= 5.35, IID 02. Le Bihan, 1962. Am. Min. 54(1969), 1498 (Abst.). Bull. Min. 85 (1962), 15. See also Rathite I, Rathite IV.

Rathite IV, $Pb_3As_5S_{10}$(?), [Q] Mon. Lengenbach quarry, Binntal, Valais (Wallis), Switzerland. H=?, D_m=?, D_c=?, IID 02. Ozawa & Nowacki, 1974. Min. Abstr. 75-2500. See also Rathite I, Rathite III.

Rauenthalite, $Ca_3(AsO_4)_2 \cdot 10H_2O$, [A] Tric. Translucent colorless, white. Gabe Gottes vein, Rauenthal, Ste.-Marie-aux-Mines, Vosges Mtns., Haut-Rhin, Alsace, France. H=?, D_m= 2.36, D_c= 2.362, VIIC 16. Pierrot, 1964. Am. Min. 50(1965), 805 (Abst.). Acta Cryst. B39 (1983), 4.

Rauvite, $Ca(UO_2)_2V_{10}O_{28} \cdot 16H_2O$, [Q] Syst=? Purplish/bluish black. San Rafael Swell, Temple Rock, Emery Co., Utah, USA. H=?, D_m=?, D_c=?, VIID 23. Hess, 1922. Dana, 7th ed.(1951), v.2, 1058.

Rayite, $(Ag,Tl)_2Pb_8Sb_8S_{21}$, [A] Mon. Metallic grey; white in reflected light. Rajpura-Dariba, Rajasthan, India. H=?, D_m=?, D_c= 6.13, IID 09a. Basu et al., 1983. Am. Min. 69(1984), 211 (Abst.). See also Semseyite.

Realgar, β–AsS, [G] Mon. Resinous red/orange-yellow. H= 1.5-2, D_m= 3.5, D_c= 3.59, IIB 19. Wallerius, 1747. Dana, 7th ed.(1944), v.1, 255. Zts. Krist. 136 (1972), 48. See also Pararealgar, Realgar (high).

Realgar (high), α–AsS, [P] Mon. Resinous yellow. Alacrán mine, Pampa Larga District, Copiapó (Near), Atacama Province, Chile. H= 2, D_m= 3.46, D_c= 3.46, IIB 19. Clark, 1970. Am. Min. 55 (1970), 1338. See also Realgar, Pararealgar.

Rebulite, $Tl_5Sb_5As_8S_{22}$, [Q] Mon. Sub-metallic dark grey. Alšar (Allchar), Roždeni (near), Makedonija (Macedonia), Yugoslavia. H=?, D_m= 4.81, D_c= 4.90, IID 12. Balić-Žunić & Šćavničar, 1982. Am. Min. 68(1983), 644 (Abst.). Zts. Krist. 160 (1982), 109.

Rectorite, $(Na,Ca)Al_4(Si,Al)_8O_{20}(OH)_4 \cdot 2H_2O$, [G] Mon. Hot Springs (near), Marble township, Garland Co., Arkansas, USA. H=?, D_m=?, D_c=?, VIIIE 08c. Brackett & Williams, 1891. Sov.Phys.Cryst. 16 (1971), 250. See also Mica, Smectite.

Reddingite, $Mn_3(PO_4)_2 \cdot 3H_2O$, [G] Orth. Vitreous/sub-resinous pink, yellowish-white, colorless. Reading, Branchville, Fairfield Co., Connecticut, USA. H= 3-3.5, D_m= 3.1, D_c= 3.262, VIIC 04. Brush & Dana, 1878. Min. Mag. 43 (1980), 789. Zts. Krist. 118 (1963), 327. See also Phosphoferrite.

Redingtonite, $(Fe,Mg,Ni)(Cr,Al)_2(SO_4)_4 \cdot 22H_2O$, G Syst=? Silky white, purple. Redington mine, Knoxville, Napa Co., California, USA. H=?, D_m= 1.761, D_c=?, VIC 06. Becker, 1888. Dana, 7th ed.(1951), v.2, 529.

Redledgeite, $BaTi_6Cr_2O_{16}$, G Mon. Black. Red Ledge mine, Washington Dist., Nevada Co., California, USA. H=?, D_m= 3.72, D_c= 4.413, IVD 03b. Strunz, 1961. Can. Min. 24 (1986), 55. Min. Mag. 50 (1986), 709. See also Mannardite.

Reedmergnerite, $NaBSi_3O_8$, G Tric. Vitreous colorless. Joseph Smith #1 Well, Sun Oil Co., Duchesne Co., Utah, USA. H= 6-6.5, D_m= 2.776, D_c= 2.779, VIIIF 03. Milton et al., 1960. Am. Min. 45 (1960), 188. Am. Min. 50 (1965), 1827.

Reevesite, $Ni_6Fe_2CO_3(OH)_{16} \cdot 4H_2O$, A Rhom. Translucent yellow, greenish-yellow. Wolf Creek meteorite crater, Western Australia, Australia. H=?, D_m= 2.88, D_c= 2.87, IVF 03b. White et al., 1967. Am. Min. 56 (1971), 1077.

Refikite, $C_{20}H_{32}O_2$, G Orth. Montorio, Feramo (near), Abruzze Mts., Italy. H=?, D_m=?, D_c= 1.09, IXB 01. La Cava, 1852. Am. Min. 50(1965), 2110 (Abst.).

Reichenbachite, $Cu_5(PO_4)_2(OH)_4$, A Mon. Vitreous dark green. Borstein, Reichenbach, Bensheim, Odenwald, Hessen, Germany. H= 3.5, D_m=?, D_c= 4.35, VIIB 08. Sieber et al., 1987. Am. Min. 72 (1987), 404.

Reinerite, $Zn_3(AsO_3)_2$, G Orth. Vitreous/adamantine blue, light yellow-green. Tsumeb, Namibia. H= 5-5.5, D_m= 4.270, D_c= 4.283, VIIA 15. Geier & Weber, 1958. Am. Min. 44(1959), 207 (Abst.). Am. Min. 62 (1977), 1129.

Reinhardbraunsite, $Ca_5(SiO_4)_2(OH,F)_2$, A Mon. Vitreous light pink. Bellerberg, Mayen (near), Laacher See, Eifel, Germany. H= 5-6, D_m= 2.84, D_c= 2.885, VIIIA 16. Hamm & Hentschel, 1983. Am. Min. 68(1983), 1039 (Abst.). Min. Abstr. 84M/2504.

Remondite-(Ce), $Na_3(Ca,Ce,La,Na,Sr)_3(CO_3)_5$, A Mon. Translucent red-orange. Eboundja, Kribi (near), Cameroon. H= 3-3.5, D_m= 3.43, D_c= 3.46, VbA 05. Cesbron et al., 1988. Am. Min. 75 (1990), 433 (Abst.). Acta Cryst. C45 (1989), 185.

Renardite, $Pb(UO_2)_4(PO_4)_2(OH)_4 \cdot 7H_2O$, G Orth. Transparent yellow. Shinkolobwe, Shaba, Zaïre. H=?, D_m=?, D_c= 4.34, VIID 21. Schoep, 1928. Am. Min. 39 (1954), 448.

Renierite, $(Cu,Zn)_{11}Fe_4(Ge,As)_2S_{16}$, G Tet. Metallic orange. Prince Léopold mine, Kipushi, Shaba, Zaïre. H= 4.5, D_m= 4.38, D_c= 4.40, IIB 03b. Vaes, 1948. Am. Min. 71 (1986), 210. Am. Min. 74 (1989), 1177.

Retgersite, α–$NiSO_4 \cdot 6H_2O$, G Tet. Vitreous green. Minasragra, Peru. H= 2.5, D_m= 2.04, D_c= 2.07, VIC 03b. Frondel & Palache, 1949. Am. Min. 34 (1949), 188. Acta Cryst. C44 (1988), 1869. See also Nickel-hexahydrite.

Retinostibian, D = welinite. Min. Mag. 43 (1980), 1053. Igelström.

Retzian-(Ce), $Mn_2CeAsO_4(OH)_4$, G Orth. Vitreous/greasy dark brown. Moss mine, Filipstad, Nordmark, Värmland, Sweden. H= 4, D_m= 4.15, D_c= 4.57, VIIB 10b. Sjögren, 1884. Am. Min. 67 (1982), 841.

Retzian-(La), $(Mn,Mg)_2(La,Ce,Nd)AsO_4(OH)_4$, A Orth. Vitreous reddish-brown. Sterling Hill mine, Ogdensburg, Sussex Co., New Jersey, USA. H= 3-4, D_m=?, D_c= 4.49, VIIB 10b. Dunn et al., 1984. Min. Mag. 48(1984), 533.

Retzian-(Nd), $Mn_2(Nd,Ce,La)AsO_4(OH)_4$, A Orth. Vitreous/dull pinkish-brown, reddish-brown. Sterling Hill mine, Ogdensburg, Sussex Co., New Jersey, USA. H= 3-4, D_m=?, D_c= 4.45, VIIB 10b. Dunn & Sturman, 1982. Am. Min. 67 (1982), 841.

Revdite, $Na_2Si_2O_5 \cdot 5H_2O$, A Tric. Vitreous/pearly colorless. Mt. Karnasurt, Lovozero massif, Kola Peninsula, USSR. H= 2, D_m= 1.94, D_c= 1.93, VIIIE 03. Khomyakov et al., 1980. Am. Min. 67(1982), 1076 (Abst.).

Revoredite, D Inadequately characterized. Am. Min. 44 (1959), 1070. Amstutz et al., 1957.

Reyerite, $(Na,K)_2Ca_{14}Al_2Si_{22}O_{58}(OH)_8 \cdot 6H_2O$, G Rhom. Translucent pale green. Niaqornat, Nuussuaq, Greenland. H=?, D_m= 2.578, D_c= 2.590, VIIIE 14. Giesecke, 1811. Am. Min. 58 (1973), 517. Min. Mag. 52 (1988), 247. See also Minehillite, Truscottite.

Rézbányite, $Cu_2Pb_3Bi_{10}S_{19}$, Q Orth. Metallic grey. Băiţa Bihorului (Rézbánya), Romania. H= 2.5, D_m= 6.24, D_c=?, IID 05a. Frenzel, 1882. Can. Min. 14 (1976), 194.

Rhabdophane-(Ce), $(Ce,La)PO_4 \cdot H_2O$, G Hex. Greasy brown, pinkish, yellowish-white. Fowey Consols mine, Cornwall, England. H= 3.5, D_m= 4.0, D_c= 4.72, VIIC 19. Lettsom, 1878. Min. Mag. 48 (1984), 146.

Rhabdophane-(La), $(La,Ce)PO_4 \cdot H_2O$, G Hex. Greasy brown, pinkish, yellowish-white. Cornwall, England. H= 3.5, D_m= 4.0, D_c= 4.69, VIIC 19. Bowles & Morgan, 1984. Min. Mag. 48 (1984), 146.

Rhabdophane-(Nd), $(Nd,Ce,La)PO_4 \cdot H_2O$, A Hex. Greasy brown, pinkish, yellowish-white. Salisbury, Connecticut, USA. H= 3.5, D_m= 4.0, D_c=?, VIIC 19. Levinson, 1966. Am. Min. 51 (1966), 152.

Rhenium, D Natural occurrence doubtful. Am. Min. 63(1978), 1283 (Abst.). Rafal'son & Sorokin, 1976.

Rhodesite, $(K,Na)_2Ca_4Si_{16}O_{36}(OH)_2 \cdot 10H_2O$, G Orth. Silky white. Bultfontein mine, Kimberley, South Africa. H=?, D_m= 2.36, D_c= 2.27, VIIIE 15. Gard et al., 1957. Am. Min. 54 (1969), 251. Zts. Krist. 149 (1979), 155.

Rhodium, Rh, A Cub. Metallic white. Stillwater Complex, Montana, USA. H= 3.5, D_m=?, D_c= 12.347, IA 07. Cabri & Laflamme, 1974. Can. Min. 12 (1974), 399.

Rhodizite, $(K,Cs)Be_4Al_4(B,Be)_{12}O_{28}$, G Cub. Vitreous/adamantine colorless, white, greyish, yellowish. Sarapulsk and Schaitansk, Mursinsk (near), Sverdlovsk, Ural Mts., USSR. H= 8, D_m= 3.36, D_c= 3.345, Vc 15. Rose, 1834. Am. Min. 51(1966), 533 (Abst.). Min. Mag. 50 (1986), 163.

Rhodochrosite, $MnCO_3$, G Rhom. Vitreous pink, rose, red, yellowish-grey, brown. H= 3.5-4, D_m= 3.70, D_c= 3.720, VbA 02. Hausmann, 1813. Dana, 7th ed.(1951), v.2, 171. Zts. Krist. 156 (1981), 233. See also Calcite, Siderite.

Rhodonite, $(Mn,Fe,Mg,Ca)SiO_3$, G Tric. Vitreous rose-pink, brownish-red. H= 5.5-6.5, D_m= 3.6, D_c= 3.55, VIIID 13. Jasche, 1819. Am. Min. 63 (1978), 1137. Min. Journ. 9 (1979), 286.

Rhodostannite, $Cu_2FeSn_3S_8$, A Tet. Metallic grey-brown. Vila Apacheta, Bolivia. H= 4, D_m=?, D_c= 4.79, IIB 03b. Springer, 1968. Min. Mag. 36 (1968), 1045. Acta Cryst. B35 (1979), 2195.

Rhodplumsite, $Rh_3Pb_2S_2$, A Rhom. Metallic cream-pink/grey-bluish. Umutnaya River, Ural Mts., USSR. H=?, D_m=?, D_c= 9.74, IIA 05. Genkin et al., 1983. Min. Zhurn. 5 (2) (1983), 87. See also Parkerite, Shandite.

Rhomboclase, $HFe(SO_4)_2 \cdot 4H_2O$, G Orth. Sub-vitreous/pearly colorless, white, grey, pale yellow. Smolnik (Szomolnok), Slovensko (Slovakia), Czechoslovakia. H= 2, D_m= 2.23, D_c= 2.206, VIC 04. Krenner, 1888. Min. Mag. 39 (1974), 610. Struct. Repts. 41A (1975), 350.

Rhombomagnojacobsite, D Inadequately characterized. Am. Min. 50(1965), 2101 (Abst.). Fan, 1964.

Rhönite, $Ca_2(Fe,Mg,Ti)_6(Si,Al)_6O_{20}$, G Tric. Vitreous black. Rhön Mts., Hessen, Germany. H=?, D_m= 3.64, D_c= 3.79, VIIID 07. Soelner, 1907. Am. Min. 70 (1985), 1211.

Ribbeite, $(Mn,Mg)_5(SiO_4)_2(OH)_2$, A Orth. Vitreous pink. Kombat mine, Otavi (E), Namibia. H= 5, D_m= 3.90, D_c= 3.84, VIIIA 16. Peacor et al., 1987. Am. Min. 72 (1987), 213.

Richellite, $(Ca,Fe)(Fe,Al)_2(PO_4)_2(OH,F)_2$, G Tet. Greasy reddish/yellowish brown. Richelle, Visé, Liège, Belgium. H= 2-3, D_m= 2, D_c= 1.87, VIID 13. Cesaro & Desprets, 1883. Am. Min. 48 (1963), 300.

Richelsdorfite, $Ca_2Cu_5Sb(AsO_4)_4(OH)_6Cl \cdot 6H_2O$, A Mon. Vitreous turquoise/sky-blue. Bauhaus mine, Richelsdorf (near), Hessen, Germany. H= 2, D_m= 3.2, D_c= 3.30, VIID 16. Süsse & Schnorrer-Köhler, 1983. Am. Min. 69(1984), 211 (Abst.). Zts. Krist. 179 (1987), 323.

Richetite, $PbU_4O_{13} \cdot 4H_2O$, G Tric. Black. Shaba, Zaïre. H= 3, D_m=?, D_c= 6.02, IVF 14. Vaes, 1947. Bull. Min. 107 (1984), 581.

Richterite, $Na_2Ca(Mg,Fe)_5Si_8O_{22}(OH)_2$, G Mon. Translucent brown, yellow, brownish-red, green. Långban mine and Pajsberg, Värmland, Sweden. H= 5-6, D_m= 3, D_c= 2.87, VIIID 05c. Breithaupt, 1865. Am. Min. 59 (1974), 518. See also Ferrorichterite.

Rickardite, Cu_7Te_5, G Orth. Metallic purple-red. Good Hope mine, Vulcan, Gunnison Co., Colorado, USA. H= 3.5, D_m= 7.54, D_c= 7.417, IIA 13. Ford, 1903. Min. Jour. 7 (1973), 252. Am. Min. 34 (1949), 441.

Riebeckite, $Na_2(Fe^{2+},Mg)_3Fe_2^{3+}Si_8O_{22}(OH)_2$, G Mon. Vitreous dark blue, black. Socotra island, Indian Ocean, South Yemen. H= 5, D_m= 3.2, D_c= 3.37, VIIID 05d. Sauer, 1888. Min. Mag. 33 (1963), 625. Can. Min. 16 (1978), 187. See also Magnesioriebeckite.

Rilandite, $(Cr,Al)_6SiO_{11} \cdot 5H_2O(?)$, Q Syst=? Brownish-black. J. I. Riland Claims, Meeker (NE), Coal Creek, Colorado, USA. H=?, D_m=?, D_c=?, VIIIG 01. Henderson & Hess, 1933. Min. Abstr. 5 (1933), 293.

Ringwoodite, $(Mg,Fe)_2SiO_4$, A Cub. Translucent colorless, pale purple, smoky-grey. Tenham meteorite. H=?, D_m=?, D_c= 3.90, VIIIA 03. Binns et al., 1969. Can. Min. 15 (1977), 96. See also Forsterite, Wadsleyite.

Rinkite, $(Na,Ca)_3(Ca,Ce)_4Ti_4(Si_2O_7)_2(O,F)_4$, Q Mon. Vitreous/greasy yellowish-brown, yellow. Kangerdluarsuk, Ilímaussaq, Greenland (S). H= 5, D_m= 3.44, D_c= 3.45, VIIIB 09. Lorenzen, 1884. Am. Min. 43(1958), 795 (Abst.). Acta Cryst. B27 (1971), 1277. See also Mosandrite.

Rinneite, $K_3NaFeCl_6$, G Rhom. Translucent colorless, rose, violet, yellow. Nordhausen and Hildesheim, Sachsen (Saxony), Germany. H= 3, D_m= 2.347, D_c= 2.348, IIIB 06. Boeke, 1908. Dana, 7th ed.(1951), v.2, 107. Struct. Repts. 49A (1982), 115.

Rittmannite, $(Mn,Ca)Mn(Fe,Mn,Mg)_2(Al,Fe)_2(PO_4)_4(OH)_2 \cdot 8H_2O$, A Mon. Vitreous pale yellow. Mangualde pegmatite, Viseu district, Portugal. H= 3.5, D_m= 2.81, D_c= 2.83, VIID 24. Cossato et al., 1989. Can. Min. 27 (1989), 447.

Rivadavite, $Na_6Mg[B_6O_7(OH)_6]_4 \cdot 10H_2O$, A Mon. Vitreous colorless. Tincalayu Borax deposit, Salar del Hombre Muerto, Salta, Argentina. H= 3.5, D_m= 1.905, D_c= 1.910, Vc 11. Hurlbut & Aristarian, 1967. Am. Min. 52 (1967), 326. Naturwissenschaften 60 (1973), 350.

Riversideite, $Ca_5Si_6O_{16}(OH)_2 \cdot 2H_2O$, G Orth. Silky white. Crestmore, Riverside Co., California, USA. H= 3, D_m= 2.64, D_c= 2.87, VIIID 10. Eakle, 1917. JCPDS 29-329.

Roaldite, $(Fe,Ni)_4N$, A Cub. Metallic white. Jerslev and Youndegin meteorites. H= 5.5-6.5, D_m=?, D_c= 7.168, IC 05. Buchwald & Nielsen, 1981. Am. Min. 66(1981), 1100 (Abst.). Struct. Repts. 22 (1958), 155.

Robertsite, $Ca_2Mn_3O_2(PO_4)_3 \cdot 3H_2O$, A Mon. Lustrous reddish-brown, blood-red. Tip Top Pegmatite, Custer, Custer Co., South Dakota, USA. H= 3.5, D_m= 3.15, D_c= 3.05, VIID 15a. Moore, 1974. Am. Min. 59 (1974), 48. Inorg. Chem. 16 (1977), 1096. See also Arseniosiderite, Mitridatite.

Robinsonite, $Pb_4Sb_6S_{13}$, G Tric. Metallic grey. Red Bird mine, Pershing Co., Nevada, USA. H= 2.5-3, D_m= 5.6, D_c= 5.7, IID 05b. Berry et al., 1952. Can. Min. 20 (1982), 97.

Rockbridgeite, $(Fe,Mn)_5(PO_4)_3(OH)_5$, G Orth. Translucent dark green, olive-green, brown, yellow-brown. South Mountain (near), Midvale, Rockbridge Co., Virginia, USA. H= 3.5-4.5, D_m= 3.4, D_c= 3.60, VIIB 07. Frondel, 1949. Am. Min. 34 (1949), 513. Am. Min. 55 (1970), 135. See also Frondelite.

Rodalquilarite, $H_3Fe_2(TeO_3)_4Cl$, A Tric. Greasy green. Rodalquilar deposit, Almeria, Spain. H= 2-3, D_m= 5.10, D_c= 5.14, VIG 02. Sierra Lopez et al., 1968. Am. Min. 53(1968), 2104 (Abst.). Acta Cryst. 25B (1969), 1551.

Roeblingite, $Ca_6MnPb_2(Si_3O_9)_2(SO_4)_2(OH)_2 \cdot 4H_2O$, G Mon. Translucent white. Franklin Furnace, Sussex Co., New Jersey, USA. H= 3, D_m= 3.433, D_c= 3.441, VIIIC 02. Penfield & Foote, 1897. Am. Min. 51 (1966), 504. Am. Min. 69 (1984), 1173.

Roedderite, $(Na,K)_2(Mg,Fe)_5Si_{12}O_{30}$, A Hex. Transparent colorless. Indarch meteorite, Transcaucasia, USSR. H=?, D_m= 2.6, D_c= 2.64, VIIIC 10. Fuchs et al., 1966. Am. Min. 51 (1966), 949. Europ.Jour.Min. 1 (1989), 715. See also Eifelite.

Roggianite, $Ca_{15}(Si,Al,Be)_{48}O_{90}(OH)_{16} \cdot 34H_2O$, A Tet. Translucent white, yellowish-white. Alpe Rosso, Val Vigezzo, Druogno (near), Orcesco, Novara, Piemonte, Italy. H=?, D_m= 2.02, D_c= 2.22, VIIIF 20. Passaglia, 1969. Min. Mag. 52 (1988), 201. See also Zeolite, Ginzburgite.

Rohaite, $Cu_{8.7}(Tl,Pb,K)_2Sb_2S_4$, A Orth. Yellowish-cream/bluish-grey. Ilímaussaq Intrusion, Greenland (S). H= 2.7, D_m=?, D_c= 7.78, IID 06a. Karup-Møller, 1978. N.Jb.Min.Abh. 138 (1980), 122.

Rokühnite, $FeCl_2 \cdot 2H_2O$, A Mon. Transparent colorless. Salzdetfurth mine and Sigfried mine, Zechstein basin, Germany. H=?, D_m= 2.35, D_c= 2.3, IIIC 10b. v.Hodenberg & v.Struensee,1980. Am. Min. 66(1981), 219 (Abst.). Min. Abstr. 82M/4662.

Romanèchite, $(Ba,H_2O)_2Mn_5O_{10}$, A Mon. Sub-metallic black, grey. Romanèche-Thorins, Mâcon (near), Saône et Loire, France. H= 5-6, D_m= 4.71, D_c= 4.74, IVD 03b. Lacroix, 1910. Min. Abstr. 87M/3126. Am. Min. 73 (1988), 1155.

Romarchite, SnO, A Tet. Black. Boundary Falls, Winnipeg River, Kenora Dist., Ontario, Canada. H=?, D_m=?, D_c= 6.398, IVA 07. Organ & Mandarino, 1971. Can. Min. 10(1971), 916 (Abst.).

Roméite, $(Ca,Fe,Mn,Na)_2(Sb,Ti)_2O_6(O,OH,F)$, G Cub. Vitreous/greasy yellow, brown. Paraborna mine, St. Marcel, Val d'Aoste, Piemonte, Italy. H= 5.5-6.5, D_m= 5.0, D_c= 5.326, IVC 08. Damour, 1841. Dana, 7th ed.(1951), v.2, 1020. See also Lewisite.

Römerite, $Fe_3(SO_4)_4 \cdot 14H_2O$, G Tric. Vitreous/resinous rust-brown, yellow, violet-brown. Rammelsberg mine, Harz, Germany. H= 3-3.5, D_m= 2.174, D_c= 2.173, VIC 05. Grailich, 1858. Can. Min. 6 (1959), 348. Am. Min. 55 (1970), 78.

Röntgenite-(Ce), $Ca_2(Ce,La)_3(CO_3)_5F_3$, G Rhom. Translucent yellow/brown. Narsarsuk, Greenland (S). H=?, D_m=?, D_c= 4.19, VbB 04. Donnay, 1953. Am. Min. 38 (1953), 868. Am. Min. 38 (1953), 932.

Rooseveltite, α–$BiAsO_4$, G Mon. Adamantine grey. Maragua (near), Santiaguilla, Potosí, Bolivia. H= 4-4.5, D_m= 6.86, D_c= 7.21, VIIA 11. Herzenberg, 1946. Dana, 7th ed.(1951), v.2, 697. Acta Cryst. B38 (1982), 1559.

Roquesite, $CuInS_2$, A Tet. Metallic grey. Charrier, Allier, France. H= 4, D_m=?, D_c= 4.776, IIB 02. Picot & Pierrot, 1963. Am. Min. 48(1963), 1178 (Abst.).

Rosasite, $(Cu,Zn)_2CO_3(OH)_2$, G Mon. Green, bluish-green, blue. Rosas mine, Narcao, Clagliari, Sardegna, Italy. H= 4.5, D_m= 4.1, D_c= 4.15, VbB 01. Lovisato, 1908. Powd. Diff. 1 (1986), 56.

Roscherite (monoclinic), $Ca(Mg,Fe)_2Be_2Al_x(PO_4)_3(OH)_3 \cdot 2H_2O$, G Mon. Translucent dark brown, olive-green. Greifenstein, Ehrenfriedersdorf (near), Sachsen (Saxony), Germany. H= 4.5, D_m= 2.92, D_c= 2.78, VIIC 09. Slavik, 1914. Am. Min. 43 (1958), 824. Min. Abstr. 76-3306. See also Roscherite (triclinic).

Roscherite (triclinic), $CaBe_2Mn_2Fe_x(PO_4)_3(OH)_2 \cdot 3H_2O$, P Tric. Dark brown, olive-green. Greifenstein, Ehrenfriedersdorf (near), Sachsen (Saxony), Germany. H= 4.5, D_m= 2.92, D_c= 2.89, VIIC 09. Slavik, 1914. Dana, 7th ed.(1951), v.2, 969. TMPM 24 (1977), 169. See also Roscherite (monoclinic).

Roscoelite, $K(V,Al,Mg)_2(Si,Al)_4O_{10}(OH)_2$, G Mon. Pearly dark brown, brownish-green, yellowish-green. Stuckslacker mine, Coloma (near), El Dorado Co., California, USA. H=?, D_m= 2.93, D_c= 3.08, VIIIE 05a. Blake, 1876. JCPDS 19-933.

Roseite, D Inadequate data. Min. Mag. 38 (1971), 103. Ottemann & Augustithis, 1967.

Roselite, $Ca_2(Co,Mg)(AsO_4)_2 \cdot 2H_2O$, G Mon. Vitreous pink. Rappold mine, Schneeberg, Sachsen (Saxony), Germany. H= 3.5, D_m= 3.6, D_c= 3.617, VIIC 12. Lévy, 1824. Dana, 7th ed.(1951), v.2, 723. Can. Min. 15 (1977), 36. See also Beta-Roselite, Brandtite.

Rosemaryite, $(Na,Ca,Mn)(Mn,Fe)(Fe,Mg)Al(PO_4)_3$, A Mon. Rock Ridge Pegmatite, Custer (near), Custer Co., South Dakota, USA. H=?, D_m=?, D_c=?, VIIA 05a. Moore & Ito, 1979. Min. Mag. 43 (1979), 227. See also Qingheiite, Wyllieite, Ferrowyllieite.

Rosenbuschite, $(Ca,Na)_6TiZr(Si_2O_7)_2(F,OH)_4$, G Tric. Vitreous light orange-grey. Barkevik (?), Langesundfjord, Norway. H= 5-6, D_m= 3.315, D_c= 3.25, VIIIB 09. Brögger, 1887. Dana, 6th ed. (1892), 374. Struct. Repts. 28 (1963), 254.

Rosenhahnite, $Ca_3Si_3O_8(OH)_2$, A Tric. Translucent colorless, buff, white. Russian River, Mendocino Co., California, USA. H= 4.5-5, D_m= 2.89, D_c= 2.895, VIIIB 28. Pabst et al., 1967. Am. Min. 52 (1967), 336. Am. Min. 62 (1977), 503.

Rosickýite, γ-S, G Mon. Adamantine yellow. Havírna, Letovice (near), Moravia, Czechoslovakia. H=?, D_m= 2.19, D_c= 2.08, IB 04. Sekanina, 1931. Dana, 7th ed.(1944), v.1, 145. Acta Cryst. B30 (1974), 1396. See also Sulfur.

Rosièresite, $[Pb,Cu,Al,PO_4,H_2O](?)$, Q Amor. Translucent greenish-yellow, yellow, light brown. Rosières copper mine, Carmaux, Tarn, France. H=?, D_m= 2.2, D_c=?, VIID 36. Lacroix, 1910. Dana, 7th ed.(1951), v.2, 924.

Rossite, $Ca(VO_3)_2 \cdot 4H_2O$, G Tric. Vitreous/pearly yellow. Wm. O'Neill's Claim, Bull Pen Canyon, San Miguel Co., Colorado, USA. H= 2-3, D_m= 2.45, D_c= 2.42, VIIC 21. Hess & Foshag, 1926. Min. Mag. 49 (1985), 140. Can. Min. 7 (1963), 713.

Rösslerite, $Mg(AsO_3OH) \cdot 7H_2O$, G Mon. Vitreous/dull colorless, white. Bieber, Hanau (near), Hessen, Germany. H= 2-3, D_m= 1.943, D_c= 1.948, VIIC 11. Blum, 1861. Zts. Krist. 137 (1973), 194. Acta Cryst. B29 (1973), 286. See also Phosphorrösslerite.

Rostite, $AlSO_4(F,OH) \cdot 5H_2O$, A Orth. Chalky white. Schoeller mine, Kladno, Praha (Prague), Stredočský kraj, Čechy (Bohemia), Czechoslovakia. H=?, D_m= 1.892, D_c= 1.961, VID 02. Rost, 1937. N. Jb. Min. Mh. (1988), 476. Am. Min. 66(1981), 1102 (Abst.). See also Jurbanite, Khademite.

Roubaultite, $Cu_2O_2(UO_2)_3(CO_3)_2(OH)_2 \cdot 4H_2O$, A Tric. Brilliant/greasy green. Shinkolobwe, Shaba, Zaïre. H= 3, D_m=?, D_c= 4.71, VbD 04. Cesbron et al., 1970. Am. Min. 57(1972), 1912 (Abst.). Acta Cryst. C41 (1985), 654.

Rouseite, $Pb_2Mn(AsO_3)_2 \cdot 2H_2O$, A Tric. Vitreous/dull orange-yellow. Långban mine, Filipstad (near), Värmland, Sweden. H= 3, D_m=?, D_c= 5.73, VIID 31. Dunn et al., 1986. Am. Min. 71 (1986), 1034.

Routhierite, $TlHgAsS_3$, A Tet. Translucent violet-red. Jas Roux, Valgaudemar, Hautes-Alpes, France. H= 3.5, D_m=?, D_c= 5.83, IID 05c. Johan et al., 1974. Am. Min. 60(1975), 947 (Abst.). See also Christite.

Roweite, $Ca_2Mn_2B_4O_7(OH)_6$, [G] Orth. Transparent light brown. Franklin, Sussex Co., New Jersey, USA. H= 5, D_m= 2.935, D_c= 2.962, Vc 07b. Berman & Gonyer, 1937. Dana, 7th ed.(1951), v.2, 377. Am. Min. 59 (1974), 60. See also Fedorovskite.

Rowlandite-(Y), $FeY_4(Si_2O_7)_2F_2$, [G] Tric. Translucent greyish-white, drab green. Barringer Hill, Blufton (near), Llano Co., Texas, USA. H= 6-6.5, D_m= 4.85, D_c= 4.33, VIIIB 01. Hidden, 1891. Min. Abstr. 76-734.

Roxbyite, $Cu_{1.78}S$, [A] Mon. Metallic blue-grey. Olympic Dam deposit, Roxby Downs, South Australia, Australia. H= 2.5, D_m=?, D_c= 5.557, IIA 01a. Mumme et al., 1988. Min. Mag. 52 (1988), 323.

Rozenite, $FeSO_4 \cdot 4H_2O$, [A] Mon. Vitreous white, greenish-white. Ornak, High Tatra (W), Poland; Staszic mine, Rudki, Poland. H=?, D_m= 2.27, D_c= 2.28, VIC 02. Kubisz, 1960. Can. Min. 7 (1963), 751. Acta Cryst. 15 (1962), 815.

Rozhkovite, $(Cu,Pd)_3Au_2$, [Q] Orth. Metallic pale rose. Talnakh, Noril'sk (near), Siberia (N), USSR. H= 4, D_m=?, D_c=?, IA 01c. Razin et al., 1975. Am. Min. 62(1977), 595 (Abst.).

Ruarsite, RuAsS, [Q] Mon. Metallic greyish-white. Tibet (N), China. H= 6-6.5, D_m=?, D_c= 7.08, IIC 09. Yu & Chou, 1979. Am. Min. 65(1980), 1068 (Abst.). See also Osarsite.

Rucklidgeite, $(Bi,Pb)_3Te_4$, [A] Rhom. Metallic white. Kochkar deposit, Ural Mts., USSR. H= 2, D_m= 7.739, D_c= 8.06, IIA 15. Zav'yalov & Begizov, 1977. Am. Min. 63(1978), 599 (Abst.).

Ruizite, $Ca_2Mn_2Si_4O_{11}(OH)_4 \cdot 2H_2O$, [A] Mon. Translucent orange, brown. Christmas mine, Christmas, Gila Co., Arizona, USA. H= 5, D_m= 2.9, D_c= 2.89, VIIIB 19. Williams & Duggan, 1977. Min. Mag. 41 (1977), 429. Am. Min. 70 (1985), 171.

Rusakovite, $(Fe,Al)_5(VO_4,PO_4)_2(OH)_9 \cdot 3H_2O$, [Q] Syst=? Dull yellow-orange, reddish-yellow. Balasauskandyk, Kara Tau, Kazakhstan, USSR. H= 1.5-2, D_m= 2.8, D_c=?, VIIC 24. Ankinovich, 1960. Am. Min. 45(1960), 1316 (Abst.).

Russellite, Bi_2WO_6, [G] Tet. Translucent pale yellow, greenish. Castle on Dinos, St. Ausdell, Cornwall, England. H= 3.5, D_m= 7.35, D_c= 7.40, IVD 14b. Schaller, 1914. Min. Mag. 37 (1970), 705.

Rustenburgite, Pt_3Sn, [A] Cub. Metallic creamy. Rustenburg mine, Merensky Reef, Transvaal, South Africa. H= 4.5, D_m=?, D_c= 15.08, IA 07. Mihalik et al., 1975. Can. Min. 13 (1975), 146.

Rustumite, $Ca_{10}(Si_2O_7)_2(SiO_4)(OH)_2Cl_2$, [A] Mon. Transparent colorless. Ardnamurchan, Kilchoan, Scotland. H=?, D_m=?, D_c= 2.835, VIIIB 27. Agrell, 1965. Am. Min. 50(1965), 2104 (Abst.). Am. Min. 64(1979), 659 (Abst.).

Ruthenarsenite, (Ru,Ni)As, [A] Orth. Metallic orange-brown/brownish-grey. New Guinea. H= 6, D_m=?, D_c= 10.0, IIB 09d. Harris, 1974. Can. Min. 12 (1974), 280.

Rutheniridosmine, (Os,Ir,Ru), [A] Hex. Metallic white. Japan. H=?, D_m=?, D_c= 20.49, IA 08. Strunz, 1966. CIM SV 23 (1981), 91. See also Iridosmine.

Ruthenium, Ru, [A] Hex. Metallic white. Horokanai placer, Hokkaido, Japan. H= 6.5, D_m= 12.2, D_c= 12.368, IA 08. Urashima et al., 1974. Am. Min. 61(1976), 177 (Abst.).

Ruthenosmiridium, (Ir,Os,Ru), [A] Cub. Metallic white. Japan. H=?, D_m=?, D_c= 21.17, IA 07. Aoyama, 1936. CIM SV23 (1981), 83. See also Osmiridium.

Rutherfordine, $(UO_2)CO_3$, [G] Orth. Dull pale brownish-yellow, yellow. Morogoro, Uluguru Mountains, Tanzania. H=?, D_m= 4.82, D_c= 5.72, VbD 04. Marckwald, 1906. Am. Min. 41 (1956), 127. Science 121 (1955), 472.

Rutile, TiO_2, [G] Tet. Metallic/adamantine red, brown, yellowish, etc. H= 6-6.5, D_m= 4.23, D_c= 4.250, IVD 02. Werner, 1803. Dana, 7th ed.(1944), v.1, 554. Can. Min. 17 (1979), 77. See also Anatase, Brookite.

Rynersonite, $Ca(Ta,Nb)_2O_6$, [A] Orth. Earthy reddish-pink. Himalaya mine, Mesa Grande Dist., San Diego Co., California, USA. H= 4.5, D_m= 6.40, D_c= 6.394, IVD 11. Foord & Mrose, 1978. Am. Min. 63, (1978), 709. See also Aeschynite-(Ce), Vigezzite.

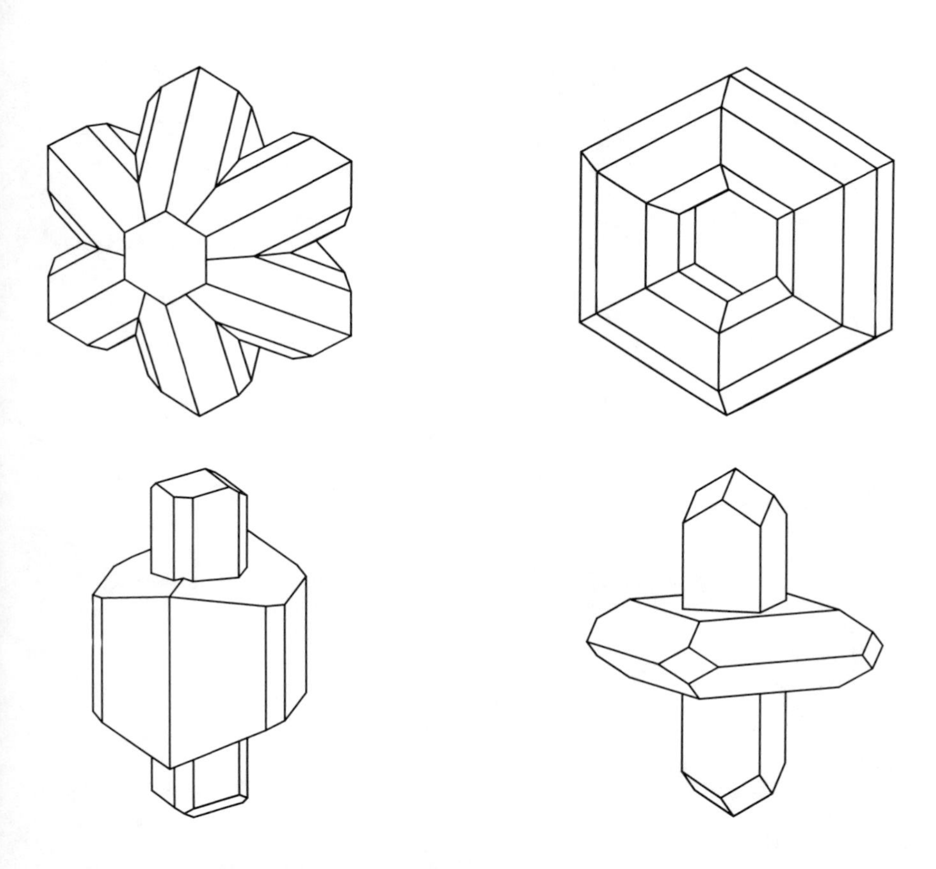

(Upper Left) Crysoberyl sixling twin showing the forms {111}, {001}, {121} and {010}.

(Upper Right) Rutile doughnut sixling twin showing the forms {100} and {110}.

(Lower Left) Augite enclosed by a hornblende. The augite shows the forms {110}, {100}, {010}, {001} and {111} and the hornblende {110}, {010}, {011} and {120}.

(Lower Right) Xenotime enclosing zircon — from North Carolina — showing the forms {100} and {111} for both members of the intergrowth. (After Dana, 1892)

S

Sabatierite, Cu_6TlSe_4, [A] Orth. Metallic grey-blue/yellow-brown. Bukov, Moravia, Czechoslovakia. H=?, D_m=?, D_c= 6.78, IIA 09. Johan et al., 1978. Bull. Min. 101 (1978), 557. Zts. Krist. 181 (1987), 241.

Sabieite, $NH_4Fe(SO_4)_2$, [A] Rhom. Translucent white. Lone Creek Fall Cave, Sabie, Transvaal (E), South Africa. H=?, D_m=?, D_c= 2.361, VIA 03. Martini, 1983. Am. Min. 71(1986), 229 (Abst.).

Sabinaite, $Na_4TiZr_2O_4(CO_3)_4$, [A] Mon. Vitreous/silky/chalky white. Francon quarry, St.-Michel dist., Montreal, Québec, Canada. H=?, D_m= 3.36, D_c= 3.48, VbA 07. Jambor et al., 1980. Can. Min. 23 (1985), 17.

Sabugalite, $HAl(UO_2)_4(PO_4)_4 \cdot 16H_2O$, [G] Mon. Translucent yellow. Minho, Kariz, Portugal; Mina de Quarta Seira, Sabugal, Portugal. H= 2.5, D_m= 3.20, D_c= 3.150, VIID 20c. Frondel, 1951. J.Phys.Chem.Min. 9 (1983), 23.

Sacrofanite, $(Na,Ca)_9(Si,Al)_{12}O_{24}(OH,SO_4)_4 \cdot nH_2O$, [A] Hex. Transparent colorless. Sacrofano, Lazio (Latium), Roma, Italy. H= 5.5-6, D_m= 2.423, D_c= 2.446, VIIIF 05. Burragato et al., 1980. Am. Min. 66(1981), 1100 (Abst.).

Sadanagaite, $(K,Na)Ca_2(Fe,Mg)_5(Si,Al)_8O_{22}(OH)_2$, [A] Mon. Vitreous dark brown, black. Yuge Island and Myojin Island, Ehime, Japan. H= 6, D_m=?, D_c= 3.30, VIIID 05b. Shimazaki et al., 1984. Min. Mag. 53 (1989), 99. See also Magnesio-sadanagaite.

Safflorite, $CoAs_2$, [G] Orth. Metallic white. Nordmark, Vårmland, Sweden. H= 4.5-5, D_m= 7.2, D_c= 7.471, IIC 07. Breithaupt, 1835. Am. Min. 53 (1968), 1856. See also Clinosafflorite.

Sahamalite-(Ce), $(Ce,La,Nd)_2(Mg,Fe)(CO_3)_4$, [G] Mon. Translucent colorless. Mountain Pass mine, San Bernardino Co., California, USA. H=?, D_m= 4.30, D_c= 4.323, VbA 05. Jaffe et al., 1953. Am. Min. 38 (1953), 741. Min. Abstr. 84M/2545.

Sahlinite, $Pb_{14}O_9(AsO_4)_2Cl_4$, [G] Mon. Translucent yellow. Långban mine, Filipstad (near), Värmland, Sweden. H= 2-3, D_m= 8.00, D_c= 8.096, IIID 03. Aminoff, 1934. Am. Min. 71(1986), 231 (Abst.).

Sainfeldite, $Ca_5(AsO_4)_2(AsO_3OH)_2 \cdot 4H_2O$, [A] Mon. Transparent colorless, light pink. Gabe Gottes vein, Rauenthal, Ste.-Marie-aux-Mines, Vosges Mtns., Haut-Rhin, Alsace, France. H=?, D_m= 3.04, D_c= 3.027, VIIC 02. Pierrot, 1964. Am. Min. 50(1965), 806 (Abst.). Bull. Min. 95 (1972), 33.

Sakhaite, $Ca_3Mg(BO_3)_2(CO_3) \cdot nH_2O$, [A] Cub. Vitreous grey, greyish-white, colorless. Ural Mts. (N), Siberia, USSR. H= 5, D_m= 2.82, D_c= 2.8, Vc 01e. Ostrovskaya et al., 1966. Min. Mag. 54 (1990), 105. Min. Abstr. 81-1239. See also Harkerite.

Sakharovaite, $(Pb,Fe)(Bi,Sb)_2S_4$, [Q] Syst=? Metallic grey; white in reflected light. Ustarasaisk deposits, Tyan-Shan (W), USSR. H=?, D_m=?, D_c=?, IID 09e. Kostov, 1959. Am. Min.45(1960), 1134 (Abst.).

Sakuraiite, $(Cu,Zn,Fe,In,Sn)S$, [A] Cub. Metallic grey. Ikuno mine, Hyogo, Kanagase, Senju-Hon, Honshu, Japan. H= 4, D_m=?, D_c= 4.45, IIB 01a. Kato, 1965. Can. Min. 24 (1986), 679.

Sal ammoniac, α–NH_4Cl, [G] Cub. Vitreous colorless, white, grey, yellow. H=?, D_m= 1.532, D_c= 1.535, IIIA 03. Agricola, 1546. Dana, 7th ed.(1951), 16.

Saléeite, $Mg(UO_2)_2(PO_4)_2 \cdot 10H_2O$, [G] Mon. Transparent yellow. Shinkolobwe, Shaba, Zaïre. H= 2-3, D_m= 3.27, D_c= 3.21, VIID 20c. Thoreau & Vaes, 1932. Bull. Min. 103 (1980), 630. Zts. Krist. 177 (1986), 247.

Salesite, $CuIO_3(OH)$, [G] Orth. Transparent bluish-green. Chuquicamata, Chile. H= 3, D_m= 4.77, D_c= 4.900, IVG 01. Palache & Jarrell, 1939. Am. Min. 24 (1939), 388. Am. Min. 63 (1978), 172.

Salmonsite, D A mixture of hureaulite and jahnsite. Min. Mag. 42 (1978), 309. Schaller, 1912.

Samarskite-(Y), $(Y,Ce,U,Fe)_3(Nb,Ta,Ti)_5O_{16}$, G Orth. Vitreous/resinous black. Miask, Ural Mts., USSR. H= 5-6, D_m= 5.69, D_c=?, IVD 10b. Rose, 1847. Am. Min. 70 (1985), 856.

Sampleite, $NaCaCu_5(PO_4)_4Cl \cdot 5H_2O$, G Orth. Pearly light blue, blue-green. Chuquicamata, Antofagasta, Chile. H= 4, D_m= 3.20, D_c= 3.26, VIID 16. Hurlbut, 1942. Min. Mag. 42 (1978), 369. See also Lavendulan.

Samsonite, $Ag_4MnSb_2S_6$, G Mon. Metallic steel-black; bluish-white in reflected light. Samson mine, St. Andreasberg, Harz, Germany. H= 2.5, D_m= 5.50, D_c= 5.51, IID 01d. Werner & Fraatz, 1910. Dana, 7th ed.(1944), v.1, 393. Zts. Krist. 140 (1974), 87.

Samuelsonite, $(Ca,Ba)_9(Mn,Fe)_4Al_2(PO_4)_{10}(OH)_2$, A Mon. Sub-adamantine colorless. Palermo #1 mine, North Groton, Grafton Co., New Hampshire, USA. H= 5, D_m= 3.353, D_c= 3.355, VIIB 20. Moore et al., 1975. Am. Min. 60 (1975), 957. Am. Min. 62 (1977), 229.

Sanbornite, $BaSi_2O_5$, G Orth. Pearly white. Trumbull Peak, Mariposa Co., California, USA. H= 5, D_m= 3.77, D_c= 3.774, VIIIE 03. Rogers, 1932. Am. Min. 43 (1958), 517. Zts. Krist. 153 (1980), 33.

Sanderite, $MgSO_4 \cdot 2H_2O$, Q Syst=? Wathlingen, Harz, Germany. H=?, D_m=?, D_c=?, VIC 03a. Berdesinski, 1952. Am. Min. 37(1952), 1072 (Abst.).

Saneroite, $Na_2Mn_{10}VSi_{11}O_{34}(OH)_4$, A Tric. Resinous/greasy orange. Gumbatesa mine, Val Greveglia, Liguria, Italy. H=?, D_m= 3.47, D_c= 3.464, VIIID 20. Cortesogno et al., 1979. Am. Min. 66(1981), 1277 (Abst.). Am. Min. 66(1981), 1277 (Abst.).

Sangarite, D Inadequate data. Min. Mag. 36 (1967), 133. Drits & Kossovskaya, 1963.

Sanidine, $(K,Na)(Si,Al)_4O_8$, G Mon. Vitreous colorless, white. H= 6, D_m= 2.6, D_c= 2.54, VIIIF 03. Nose, 1808. Am. Min. 63 (1978), 1264. Am. Min. 72 (1987), 973.

Sanjuanite, $Al_2(PO_4)(SO_4)(OH) \cdot 9H_2O$, G Tric. Chalky/silky white. San Juan, Argentina. H= 3, D_m= 1.94, D_c= 1.96, VIID 04. Abeledo et al., 1968. Min. Mag. 53 (1989), 385.

Sanmartinite, $(Zn,Fe)WO_4$, G Mon. Resinous dark brown, brownish-black. San Martin (7 km W of), Los Corrillos, San Luis, Argentina. H=?, D_m= 6.70, D_c= 7.87, IVD 08. Angelelli & Gordon, 1948. NBS Monogr. 2 (1963), 40. See also Ferberite, Hübnerite.

Santaclaraite, $CaMn_4Si_5O_{14}(OH)_2 \cdot H_2O$, A Tric. Vitreous pale pink, reddish-orange. Diablo Range, Santa Clara Co., California, USA. H= 6.5, D_m= 3.31, D_c= 3.398, VIIID 13. Erd & Ohashi, 1984. Am. Min. 69 (1984), 200. Am. Min. 66 (1981), 154.

Santafeite, $(Ca,Sr,Na)_3(Mn,Mg,Al,Fe)_4(VO_4)_4(OH)_5 \cdot 2H_2O$, G Orth. Sub-adamantine black. Grants dist., McKinley Co., New Mexico, USA. H=?, D_m= 3.379, D_c= 3.34, VIID 12. Sun & Weber, 1958. Min. Mag. 50 (1986), 299.

Santanaite, $Pb_{11}CrO_{16}$, A Hex. Adamantine yellow. Santa Ana mine, Caracoles, Sierra Gorda, Chile. H= 4, D_m=?, D_c= 9.155, IVB 03. Mücke, 1972. Am. Min. 58(1973), 966 (Abst.).

Santite, $KB_5O_6(OH)_4 \cdot 2H_2O$, A Orth. Translucent colorless. Larderello, Val di Cecina, Pisa, Toscana, Italy. H=?, D_m=?, D_c= 1.740, Vc 08. Merlino & Sartori, 1970. Am. Min. 56(1971), 636 (Abst.).

Saponite, $(Ca,Na)_{0.3}(Mg,Fe)_3(Si,Al)_4O_{10}(OH)_2 \cdot 4H_2O$, G Mon. Greasy white, yellowish, greyish, green bluish, reddish. Lizard Point, Cornwall, England. H=?, D_m= 2.50, D_c= 2.26, VIIIE 08b. Svanberg, 1840. JCPDS 13-305. Min. Abstr. 78-2716.

Sapphirine-2M, $Mg_7Al_{18}Si_3O_{40}$, G Mon. Translucent pale/dark blue, green. Fiskenäs, Nuuk area, Greenland (SW). H= 7.5, D_m= 3.45, D_c= 3.492, VIIIA 15. Giesecke, 1819. Am. Min. 73 (1988), 1134. Am. Min. 54 (1969), 31.

Sapphirine-1Tc, $Mg_7Al_{18}Si_3O_{40}$, P Tric. Wilson Lake, Labrador, Canada. H=?, D_m=?, D_c= 3.310, VIIIA 15. Merlino, 1973. Am. Min. 59 (1974), 632 (Abst.).

Sarabauite, $CaSb_{10}O_{10}S_6$, A Mon. Resinous red. Sarabau mine, Kuching (near), Sarawak, Malaysia. H= 4, D_m= 4.8, D_c= 4.99, IIE 02. Nakai et al., 1978. Am. Min. 63 (1978), 715.

Sarcolite, $(Ca,Na)_8Al_4Si_6O_{23}(PO_4,CO_3,SO_4,F,Cl,OH,H_2O)_2$, G Tet. Vitreous pink, red, reddish-white. Monte Somma, Mt. Vesuvius, Napoli, Campania, Italy. H= 6, D_m= 2.545, D_c= 2.93, VIIIF 09. Thompson, 1807. Can. Min. 25 (1987), 731. Am. Min. 64(1979), 245 (Abst.).

Sarcopside, $(Fe,Mn,Mg)_3(PO_4)_2$, G Mon. Silky flesh-red, reddish-brown, dark brown. Góry Sowie (Michelsdorf), Silesia, Poland. H= 4, D_m= 3.7, D_c= 3.94, VIIA 03. Websky, 1868. Am. Min. 50 (1965), 1608. Am. Min. 57 (1972), 24.

Sarkinite, $Mn_2AsO_4(OH)$, G Mon. Greasy red, reddish-yellow, yellow. Harstigen mine, Pajsberg, Värmland, Sweden. H= 4-5, D_m= 4.15, D_c= 4.208, VIIB 03b. Sjögren, 1885. Min. Mag. 43 (1980), 681. Struct. Repts. 41A (1975), 338.

Sarmientite, $Fe_2(AsO_4)(SO_4)(OH)\cdot5H_2O$, G Mon. Translucent pale yellow-orange. Santa Elena mine, La Alcaparrosa, between San Juan and Calingasta, Barreal Dept., Argentina. H=?, D_m= 2.58, D_c= 2.58, VIID 04. Angelelli & Gordon, 1941. Am. Min. 53 (1968). 2077. See also Diadochite.

Sartorite, $PbAs_2S_4$, G Mon. Metallic grey; white in reflected light. Lengenbach quarry, Binntal, Valais (Wallis), Switzerland. H= 3, D_m= 5.10, D_c= 4.97, IID 15. Dana, 1868. Dana, 7th ed.(1944), v.1, 478. Acta Cryst. 14 (1961), 1291.

Saryarkite-(Y), $Ca(Y,Th)Al_5(SiO_4)_2(PO_4)_2(OH)_7\cdot6H_2O$, G Tet. Dull/greasy white. Unspecified locality, Kazakhstan(?), USSR. H= 3.5-4, D_m= 3.1, D_c= 3.35, VIID 12. Krol et al., 1964. Am. Min. 49(1964), 1775 (Abst.).

Sasaite, $(Al,Fe)_6(PO_4,SO_4)_5(OH)_3\cdot36H_2O$, A Orth. Chalky white. West Driefontein Cave, Transvaal, South Africa. H=?, D_m= 1.75, D_c= 1.780, VIID 06. Martini, 1978. Can. Min. 21 (1983), 489.

Sassolite, $B(OH)_3$, G Tric. Pearly white/grey. Sasso, Toscana, Italy. H= 1, D_m= 1.46, D_c= 1.50, IVF 01. Karsten, 1800. Am. Min. 42 (1957), 56. Acta Cryst. 7 (1954), 305.

Satimolite, $KNa_2Al_4(B_2O_5)_3Cl_3\cdot13H_2O$, A Orth. Translucent white. Unspecified locality, Kazakhstan(?), USSR. H=?, D_m= 1.70, D_c= 1.70, Vc 05. Bocharov et al., 1969. Am. Min. 55(1970), 1069 (Abst.).

Satpaevite, $Al_{12}V_8O_{37}\cdot30H_2O$(?), Q Syst=? Pearly/dull yellow. Kurumsak and Balasauskandyk, Kara Tau, Kazakhstan, USSR. H= 1.5, D_m= 2.4, D_c=?, VIIC 24. Ankinovich, 1959. Am. Min. 44(1959), 1325 (Abst.).

Satterlyite, $(Fe,Mg)_2PO_4(OH)$, A Rhom. Vitreous pale yellow, pale brown. Big Fish River, Yukon Territory, Canada. H=?, D_m= 3.68, D_c= 3.60, VIIB 03c. Mandarino et al., 1978. Can. Min. 16 (1978), 411. See also Wolfeite, Holtedahlite.

Sauconite, $Na_{0.3}Zn_3(Si,Al)_4O_{10}(OH)_2\cdot4H_2O$, G Mon. Greasy white, bluish, brownish-yellow, yellow. Saucon–Lehigh Valley, Pennsylvania, USA. H=?, D_m=?, D_c=?, VIIIE 08b. Roepper, 1875. JCPDS 8-445.

Sayrite, $Pb_2(UO_2)_5O_6(OH)_2\cdot4H_2O$, A Mon. Translucent yellow-orange, red-orange. Shinkolobwe, Shaba, Zaïre. H=?, D_m=?, D_c= 6.76, IVF 13. Piret et al., 1983. Am. Min. 69(1984), 568 (Abst.). Bull. Min. 106 (1983), 299.

Sazhinite-(Ce), $Na_2CeSi_6O_{14}(OH)\cdot6H_2O$, A Orth. Vitreous/pearly white, grey, cream. Mt. Karnasurt, Lovozero massif, Kola Peninsula, USSR. H= 2.3, D_m= 2.61, D_c= 2.71, VIIID 11. Malinko, 1974. Am. Min. 60(1975), 162 (Abst.). Sov.Phys.Cryst. 25 (1980), 419.

Sborgite, $NaB_5O_6(OH)_4 \cdot 3H_2O$, [G] Mon. Larderello, Val di Cecina, Pisa, Toscana, Italy. H=?, D_m= 1.713, D_c= 1.711, Vc 08. Cipriani, 1957. Am. Min. 43(1958), 378 (Abst.). Acta Cryst. B28 (1972), 3559.

Scacchite, $MnCl_2$, [G] Rhom. Translucent colorless, red, brown. Mt. Vesuvius, Napoli, Campania, Italy. H=?, D_m= 2.98, D_c= 3.04, IIIA 06. Adam, 1869. Dana, 7th ed.(1951), v.2, 40.

Scapolite, A group name for framework silicates with additional anions, and the general formula $(Na, Ca)_4(Si, Al)_{12}O_{24}(Cl, CO_3, SO_4)$. See Marialite, Meionite, Mizzonite.

Scarbroite, $Al_5CO_3(OH)_{13} \cdot 5H_2O$, [G] Mon. White. South Bay, Scarborough coast, Yorkshire, England. H=?, D_m= 2.21, D_c= 2.28, IVF 02. Vernon, 1829. Min. Mag. 32 (1960), 353.

Scawtite, $Ca_7(Si_3O_9)_2(CO_3) \cdot 2H_2O$, [G] Mon. Vitreous colorless. Scawt Hill, Antrim Co., Northern Ireland. H= 4.5-5, D_m= 2.77, D_c= 2.765, VIIIC 02. Tilley, 1929. Am. Min. 40 (1955), 505. Acta Cryst. B29 (1973), 73.

Schachnerite, β-$Ag_{1.1}Hg_{0.9}$, [A] Hex. Metallic grey. Vertrauen auf Gott mine and Moschellandsberg, Obermoschel, Rheinland-Pfalz, Germany. H= 3.5, D_m=?, D_c= 13.52, IA 06. Seeliger & Mücke, 1972. N.Jb.Min.Abh. 117 (1972), 1.

Schafarzikite, $FeSb_2O_4$, [G] Tet. Metallic red, red-brown. Pernek mine, Pezinok, Západoslovenský kraj, Slovensko (Slovakia), Czechoslovakia. H= 3.5, D_m= 4.3, D_c= 5.531, VIIA 15. Krenner, 1921. Am. Min. 66 (1981), 1073. Struct. Repts. 41A (1975), 437. See also Trippkeite.

Schairerite, $Na_{21}(SO_4)_7ClF_6$, [G] Rhom. Vitreous colorless. Well G 75, Searles Lake, San Bernardino Co., California, USA. H= 3.5, D_m= 2.616, D_c= 2.619, VIB 04. Foshag, 1931. Am. Min. 56 (1971), 174. Min. Mag. 40 (1975), 131.

Schallerite, $(Mn, Fe)_{16}As_3Si_{12}O_{36}(OH)_{17}$, [G] Rhom. Vitreous/waxy light brown. Franklin, Sussex Co., New Jersey, USA. H= 4.5-5, D_m= 3.365, D_c= 3.45, VIIIE 12. Gage, 1924. Am. Min. 66 (1981), 1054. See also Nelenite.

Schapbachite, $AgBiS_2$, [G] Cub. Schapbach, Baden, Germany. H=?, D_m= 7.02, D_c= 7.04, IID 07. Kenngott, 1853. Min. Mag. 46 (1982), 513. Acta Cryst. 12 (1959), 46. See also Matildite.

Schaurteite, $Ca_3Ge(SO_4)_2(OH)_6 \cdot 3H_2O$, [A] Hex. Silky white. Tsumeb, Namibia. H=?, D_m= 2.65, D_c= 2.64, VID 06. Strunz & Tennyson, 1965. Am. Min. 53(1968), 507 (Abst.). See also Despujolsite, Fleischerite.

Scheelite, $CaWO_4$, [G] Tet. Vitreous/adamantine colorless, white, yellow, greenish, etc. Sweden. H= 4.5-5, D_m= 6.10, D_c= 6.12, VIF 01. Leonhard, 1821. Dana, 7th ed.(1951), v.2, 1074. Struct. Repts. 29 (1964), 331. See also Powellite.

Schertelite, $(NH_4)_2Mg(PO_3OH)_2 \cdot 4H_2O$, [G] Orth. Skipton Caves, Ballarat, Victoria, Australia. H=?, D_m= 1.82, D_c= 1.84, VIIC 14. Mac Ivor, 1887. Am. Min. 48 (1963), 635. Acta Cryst. B28 (1972), 683.

Scheteligite, [D] Discredited; may = betafite. Am. Min. 62 (1977), 403.

Schieffelinite, $Pb(Te, S)O_4 \cdot H_2O$, [A] Orth. Adamantine colorless, white. Joe mine, Tombstone, Cochise Co., Arizona, USA. H= 2, D_m= 4.98, D_c= 5.15, VIG 07. Williams, 1980. Min. Mag. 43 (1980), 771.

Schirmerite, $Ag_3Pb_{3-6}Bi_{7-9}S_{18}$, [G] Orth. Metallic grey; white in reflected light. Treasury mine, Geneva dist., Park Co., Colorado, USA. H= 3.5, D_m= 6.74, D_c= 7.58, IID 09b. Genth, 1874. Can. Min. 11 (1973), 952.

Schlossmacherite, $(H_3O, Ca)Al_3(SO_4, AsO_4)_2(OH)_6$, [A] Rhom. Translucent green, greyish-green. Emma Luisa mine, Guanaco, Chile. H=?, D_m=?, D_c= 3.00, VIB 03b. Schmetzer et al., 1980. N.Jb.Min.Mh. (1980), 215.

Schmiederite, $Cu_2Pb_2(SeO_3)(SeO_4)(OH)_4$, [G] Mon. Condor mine, La Rioja, Sierra de Cacheuta, Argentina. H=?, D_m=?, D_c=?, VIB 06. Olsacher, 1962. Am. Min. 49(1964), 1498 (Abst.). Min. Abstr. 87M/3984. See also Linarite.

Schmitterite, $(UO_2)TeO_3$, [A] Orth. Pearly pale yellow. Moctezuma mine, Moctezuma, Sonora, Mexico. H= 1, D_m= 6.878, D_c= 6.91, VIG 11. Gaines, 1971. Am. Min. 56 (1971), 411. Acta Cryst. B29 (1973), 1251.

Schneiderhöhnite, $Fe_4As_5O_{13}$, [A] Tric. Metallic/adamantine dark brown. Tsumeb, Namibia. H= 3, D_m= 4.3, D_c= 4.238, VIIA 15. Ottemann et al., 1973. Am. Min. 59(1974), 1139 (Abst,). Can. Min. 23 (1985), 675.

Schoderite, $Al_2(PO_4)(VO_4) \cdot 8H_2O$, [G] Mon. Translucent yellow, orange-yellow. Eureka (S), Fish Creek Range, Eureka Co., Nevada, USA. H= 2, D_m= 1.92, D_c= 1.931, VIIC 24. Hausen, 1962. Am. Min. 64 (1979), 713.

Schoenfliesite, $MgSn(OH)_6$, [A] Cub. Brooks Mt., Seward Peninsula, Alaska, USA. H=?, D_m=?, D_c= 3.483, IVF 06. Faust & Schaller, 1971. Am. Min. 57(1972), 1557 (Abst.).

Schoepite, $UO_3 \cdot 2H_2O$, [G] Orth. Adamantine yellow, amber brown, golden brown. Shinkolobwe, Shaba, Zaïre. H= 2-3, D_m= 4.8, D_c= 4.83, IVF 13. Walker, 1923. Am. Min. 45 (1960), 1026.

Schöllhornite, $Na_{0.3}CrS_2 \cdot H_2O$, [A] Rhom. Metallic grey. Norton Co., Kansas, USA. H= 2, D_m= 2.70, D_c= 2.74, IIB 17. Okada et al., 1985. Am. Min. 70 (1985), 638.

Scholzite, $CaZn_2(PO_4)_2 \cdot 2H_2O$, [G] Orth. Translucent colorless, whitish-grey. Pleystein (near), Hagendorf-Nord pegmatite, Oberpfalz, Bayern (Bavaria), Germany. H=?, D_m=?, D_c= 3.110, VIIC 03. Strunz, 1948. Am. Min. 36(1951), 382 (Abst.). Am. Min. 60 (1975), 1019. See also Parascholzite.

Schoonerite, $ZnMnFe_3(PO_4)_3(OH)_2 \cdot 9H_2O$, [A] Orth. Translucent pale tan/brown, reddish-brown. Palermo #1 Pegmatite, North Groton, Grafton Co., New Hampshire, USA. H= 4, D_m= 2.87, D_c= 2.79, VIID 05. Moore & Kampf, 1977. Am. Min. 62 (1977), 246. Am. Min. 62 (1977), 250.

Schörl, $NaFe_3Al_6(BO_3)_3Si_6O_{18}(OH)_4$, [G] Rhom. Vitreous black. H= 7, D_m= 3.20, D_c= 3.175, VIIIC 08. Matthesius, 1524. Can. Min. 13 (1975), 173. Am. Min. 74 (1989), 422. See also Dravite.

Schorlomite, $Ca_3(Ti,Fe)_2[(Si,Fe)O_4]_3$, [G] Cub. Vitreous/metallic/bituminous black, brown-black. Magnet Cove, Hot Spring Co., Arkansas, USA. H= 7-7.5, D_m= 3.761, D_c= 3.77, VIIIA 06a. Shepard, 1846. Dana/Ford (1932), 596. Phys.Chem.Min. 13 (1986), 198. See also Andradite.

Schreibersite, $(Fe,Ni,Cr)_3P$, [G] Tet. Metallic white. Copiapo (Deesa) Meteorite, Atacama, Chile. H= 6.5-7, D_m= 7.2, D_c= 7.128, IC 05. Haidinger, 1847. Geoch.Cos.Acta 28 (1964), 971. Struct. Repts. 35A (1970), 71.

Schreyerite, $V_2Ti_3O_9$, [A] Mon. Reddish-brown in reflected light. Kwale dist., Voi (S), Kenya. H= 7, D_m=?, D_c= 4.480, IVD 05. Medenbach & Schmetzer, 1978. Am. Min. 63 (1978), 1182. See also Kyzylkumite.

Schröckingerite, $NaCa_3(UO_2)(SO_4)(CO_3)_3F \cdot 10H_2O$, [G] Tric. Vitreous greenish-yellow. Jáchymov (St. Joachimsthal), Západočeský kraj, Čechy (Bohemia), Czechoslovakia. H= 2.5, D_m= 2.51, D_c= 2.561, VbD 04. Schrauf, 1873. Am. Min. 44 (1959), 1020. Min. Abstr. 86M/4306.

Schubnelite, $FeVO_4 \cdot H_2O$, [A] Tric. Brilliant black, translucent yellowish/greenish-brown. Mounana mine, Franceville, Haut-Ogooué, Gabon. H=?, D_m= 3.28, D_c= 3.406, VIIC 04. Cesbron, 1970. Am. Min. 57(1972), 1556 (Abst.).

Schuetteite, $Hg_3O_2(SO_4)$, [G] Hex. Translucent yellow. Oceanic mine dump, San Luis Obispo Co., California, USA. H= 3, D_m= 8.18, D_c= 8.36, VIB 10. Bailey et al., 1959. Am. Min. 44 (1959), 1026.

Schuilingite-(Nd), $CuPb(Nd,Gd,Sm,Y)(CO_3)_3(OH) \cdot 1.5H_2O$, [G] Orth. Translucent turquoise, azure blue. Kalompe, Shaba, Zaïre. H= 3-4, D_m= 5.2, D_c= 4.74, VbD 03. Vaes, 1947. Bull.Min. 105 (1982), 225.

Schulenbergite, $(Cu,Zn)_7(SO_4,CO_3)_2(OH)_{10}\cdot 3H_2O$, A Rhom. Pearly light green-blue. Glücksrad mine, Oberschulenberg, Harz, Germany. H= 2, D_m= 3.28, D_c= 3.38, VID 01. v. Hodenberg et al., 1984. Am. Min. 70(1985), 438 (Abst.).

Schultenite, $Pb(AsO_3OH)$, G Mon. Vitreous/adamantine colorless. Otavi, Tsumeb, Namibia. H= 2.5, D_m= 6.07, D_c= 6.079, VIIA 09. Spencer, 1926. Min. Mag. 49 (1985), 65. Min. Abstr. 87M/2149.

Schumacherite, $Bi_3O(VO_4,AsO_4,PO_4)_2(OH)$, A Tric. Adamantine yellow, yellow-brown. Schneeberg, Sachsen (Saxony), Germany. H= 3, D_m=?, D_c= 6.90, VIIB 17. Walenta et al., 1983. Am. Min. 70(1985), 438 (Abst.). See also Preisingerite.

Schwartzembergite, $Pb_3O(IO_3)Cl_2(OH)$, G Tet. Adamantine yellow, reddish brown. Cachinal, Atacama desert, Chile. H= 2-2.5, D_m= 7.39, D_c= 7.77, IIID 06. Dana, 1868. Am. Min. 55 (1970), 1814.

Sclarite, $(Zn,Mg,Mn)_4Zn_3(CO_3)_2(OH)_{10}$, A Mon. Vitreous colorless. Franklin, Sussex Co., New Jersey, USA. H= 3-4, D_m= 3.51, D_c= 3.547, VbB 11. Grice & Dunn, 1989. Am. Min. 74 (1989), 1355. Am. Min. 74 (1989), 1355.

Scolecite, $Ca(Si_3Al_2)O_{10}\cdot 3H_2O$, G Mon. Vitreous/silky colorless. H= 5-5.5, D_m= 2.2, D_c= 2.284, VIIIF 10. Gehlen & Fuchs, 1813. DHZ (1963), v.4, 358. Zts. Krist. 166 (1984), 219.

Scorodite, $FeAsO_4\cdot 2H_2O$, G Orth. Vitreous/sub-resinous pale green greyish-green, brown, etc. H= 3.5-4, D_m= 3.28, D_c= 3.339, VIIC 05. Breithaupt, 1818. Dana, 7th ed.(1971), v.2, 763. Acta Cryst. B32 (1976), 2891. See also Mansfieldite.

Scorzalite, $(Fe,Mg)Al_2(PO_4)_2(OH)_2$, G Mon. Vitreous blue, bluish-green. Corrego Frio Pegmatite, Minas Gerais, Brazil. H= 5.5-6, D_m= 3.38, D_c= 3.32, VIIB 05. Pecora & Fahey, 1947. Am. Min. 67 (1982), 610. Acta Cryst. 12 (1959), 695. See also Barbosalite, Lazulite.

Scotlandite, $PbSO_3$, A Mon. Adamantine/pearly pale yellow, greyish-white, colorless. Susanna vein, Leadhills, Lanarkshire, Scotland. H= 2, D_m= 6.37, D_c= 6.456, VIA 08. Paar et al., 1984. Min. Mag. 48 (1984), 283. Min. Abstr. 86M/2937. See also Molybdomenite.

Scrutinyite, α–PbO_2, A Orth. Sub-metallic dark reddish-brown. Sunshine #1 mine, Bingham, Hansonberg Dist., Socorro Co., New Mexico, USA. H=?, D_m=?, D_c= 9.867, IVD 07. Taggart et al., 1988. Can. Min. 26 (1988), 905. Struct. Repts. 16 (1952), 224.

Seamanite, $Mn_3[B(OH)_4](PO_4)(OH)_2$, G Orth. Transparent pale yellow, wine-yellow. Chicagoan mine, Iron River (near), Iron Co., Michigan, USA. H= 4, D_m= 3.08, D_c= 3.132, Vc 03c. Kraus et al., 1930. Dana, 7th ed.(1951), v.2, 388. Am. Min. 56 (1971), 1527.

Searlesite, $NaBSi_2O_5(OH)_2$, G Mon. Translucent white. Searles Lake, San Bernardino Co., California, USA. H=?, D_m= 2.46, D_c= 2.46, VIIIE 03. Larsen & Hicks, 1914. Dana/Ford (1932), 687. Am. Min. 61 (1976), 123.

Sederholmite, β–NiSe, G Hex. Metallic orange/orange-yellow. Kuusamo, Finland (NE). H=?, D_m=?, D_c= 7.06, IIB 09a. Vuorelainen et al., 1964. Am. Min. 50(1965), 519 (Abst.).

Sedovite, $U(MoO_4)_2$, G Orth. Translucent brown, reddish-brown. Unspecified locality, USSR. H= 3.3, D_m= 4.2, D_c= 5.11, VIF 02. Skvortsova & Sidorenko, 1965. Am. Min. 51(1966), 530 (Abst.).

Seeligerite, $Pb_3O(IO_3)Cl_3$, A Orth. Translucent yellow. Santa Ana mine, Caracoles, Sierra Gorda, Chile. H=?, D_m= 6.83, D_c= 7.052, IVG 01. Mücke, 1971. Am. Min. 57(1972), 327 (Abst.).

Segelerite, $CaMgFe(PO_4)_2(OH)\cdot 4H_2O$, A Orth. Translucent pale yellow-green, colorless. Tip Top Pegmatite, Custer, Custer Co., South Dakota, USA. H= 4,

D_m= 2.67, D_c= 2.61, VIID 25. Moore, 1974. Am. Min. 59 (1974), 48. Am. Min. 62 (1977), 692.

Seidozerite, $(Na,Ca)_4MnTiZr(Si_2O_7)_2O_2(F,OH)_2$, G Mon. Vitreous brownish-red, reddish-yellow. Lake Seidozero, Lovozero tundra, Kola Peninsula, USSR. H= 4-5, D_m= 3.47, D_c= 3.53, VIIIB 09. Semenov et al., 1958. Am. Min. 44(1959), 467 (Abst.). Struct. Repts. 23 (1959), 472.

Seinäjokite, $(Fe,Ni)(Sb,As)_2$, A Orth. Metallic light grey. Seinäjoki, Vaasa, Finland. H= 4.5, D_m=?, D_c= 7.938, IIC 07. Mozgova et al., 1976. Am. Min. 62(1977), 1059 (Abst.).

Sekaninaite, $(Fe,Mg)_2Al_4Si_5O_{18}$, A Orth. Vitreous blue, violet-blue. Dolní Bory, Moravia, Czechoslovakia. H= 7-7.5, D_m=?, D_c= 2.78, VIIIC 06. Stanek & Miskovsky, 1975. Am. Min. 62(1977), 395 (Abst.). See also Cordierite.

Selenium, γ-Se, G Rhom. Metallic grey/red. H= 2, D_m= 4.80, D_c= 4.809, IB 03. Am. Min. 41 (1956), 156. See also Selen-tellurium.

Selenostephanite, $Ag_5Sb(Se,S)_4$, A Orth. Metallic greyish white, olive tint. Chukotskiy (Tushukotka), USSR. H= 3, D_m=?, D_c= 7.5, IID 01d. Botova et al., 1985. Am. Min. 72 (1987), 225 (Abst.). See also Stephanite.

Selen-tellurium, [Se, Te](?), Q Syst=? Metallic blackish-grey. El Plomo mine, Ojojona dist., Tegucigalpa, Honduras. H= 2-2.5, D_m=?, D_c=?, IB 04. Dana & Wells, 1890. Dana, 6th ed. (1892), 11. See also Selenium, Tellurium.

Seligmannite, $CuPbAsS_3$, G Orth. Metallic grey/black; rose-white in reflected light. Lengenbach quarry, Binntal, Valais (Wallis), Switzerland. H= 3, D_m= 5.4, D_c= 5.41, IID 01e. Baumhauer, 1901. Min. Abstr. 81-2424. Zts. Krist. 131 (1970), 397. See also Bournonite, Součekite.

Sellaite, MgF_2, G Tet. Vitreous colorless, white. Les Allues, Gebroulaz glacier, Moutiers (near), Savoie (Savoy), France. H= 5, D_m= 3.15, D_c= 3.08, IIIA 05. Strüver, 1868. Dana, 7th ed.(1951), v.2, 37. See also Rutile.

Semenovite-(Ce), $(Na,Ca)_9Fe(Ce,La)_2(Si,Be)_{20}(O,OH)_{48}$, A Orth. Transparent colorless. Taseq Slope, Ilímaussaq Intrusion, Greenland (S). H= 3.5-4, D_m= 3.140, D_c= 3.35, VIIIE 22. Petersen & Ronsbo, 1972. Am. Min. 58(1973), 1114 (Abst.). Am. Min. 64 (1979), 202.

Semseyite, $Pb_9Sb_8S_{21}$, G Mon. Metallic grey/black; white in reflected light. Baia Sprie (Felsöbánya), Romania. H= 2.5, D_m= 6.03, D_c= 6.08, IID 09a. Krenner, 1881. Min. Mag. 37 (1969), 442. Acta Cryst. B30 (1974), 2935. See also Rayite.

Senaite, $Pb(Ti,Fe,Mn)_{21}O_{38}$, G Rhom. Sub-metallic black. Diamantina (Diamond-bearing sands of), Minas Gerais, Brazil. H= 6+, D_m= 5.301, D_c= 4.59, IVC 06. Hussak & Prior, 1898. Am. Min. 53 (1968), 869. Acta Cryst. B32 (1976), 1509.

Senandorite, $AgPbSb_3S_6$, Q Orth. Greyish-white in reflected light. Les Borderies, France. H=?, D_m=?, D_c= 5.456, IID 09b. Moëlo et al., 1984. Am. Min. 70(1985), 219 (Abst.). Zts. Krist. 180 (1987), 141.

Senarmontite, Sb_2O_3, G Cub. Resinous colorless, greyish-white. Djebel Haminate mine, Ain Beida, Qacentina (Constantine), Algeria. H= 2-2.5, D_m= 5.50, D_c= 5.584, IVC 02. Dana, 1851. Dana, 7th ed.(1944), v.1, 544. Acta Cryst. B31 (1975), 2016. See also Valentinite, Arsenolite.

Senegalite, $Al_2PO_4(OH)_3 \cdot H_2O$, A Orth. Vitreous colorless, pale yellow. Kouroudaiko, Senegal. H= 5.5, D_m= 2.552, D_c= 2.229, VIID 10. Johan, 1976. Am. Min. 62(1977), 595 (Abst.). Am. Min. 64 (1979), 1243.

Sengierite, $Cu_2(UO_2)_2(VO_4)_2(OH)_2 \cdot 6H_2O$, G Mon. Vitreous/adamantine green. Luiswishi, Shaba, Zaïre. H= 2.5, D_m= 4.05, D_c= 4.10, VIID 23. Vaes & Kerr, 1949. Am. Min. 66(1981), 220 (Abst.). Bull. Min. 103 (1980), 176.

Sepiolite, $Mg_4Si_6O_{15}(OH)_2 \cdot 6H_2O$, G Orth. Greyish-white, white, bluish, green. Baldissero Canavese, Piemonte, Italy. H= 2-2.5, D_m= 2, D_c= 2.257, VIIIE 13.

Glocker, 1847. Rev. Min. 19 (1988), 631. Clays Cl. Mins. 22 (1974), 285. See also Falcondoite.

Serandite, $Na(Mn,Ca)_2Si_3O_8(OH)$, G Tric. Translucent reddish-pink. Roma Island, Los Islands, Guinea. H=?, D_m= 3.215, D_c= 3.46, VIIID 08. Lacroix, 1931. Am. Min. 63 (1978), 274. Am. Min. 61 (1976), 229. See also Pectolite.

Serendibite, $Ca_2(Mg,Al)_6(Si,Al,B)_6O_{20}$, G Tric. Translucent blue. Sri Lanka. H= 6.7, D_m= 3.4, D_c= 3.67, VIIID 07. Prior & Coomaraswamy, 1903. Can. Min. 15 (1977), 108.

Sergeevite, $Ca_2Mg_{11}(CO_3)_9(HCO_3)_4(OH)_4 \cdot 6H_2O$, A Rhom. Dull white. Tyrnyauz deposit, Caucasus, USSR. H= 3.5, D_m= 2.3, D_c= 2.64, VbD 01. Yakhontova et al., 1980. ZVMO 109 (1980), 217. See also Huntite.

Serpentine, A group name for sheet silicates with the general formula $(Mg,Al,Fe,Mn,Ni,Zn)_{2-3}(Si,Al,Fe)_2O_5(OH)_4$.

Serpierite, $Ca(Cu,Zn)_4(SO_4)_2(OH)_6 \cdot 3H_2O$, G Mon. Vitreous blue. Lávrion (Laurium), Attikí, Greece. H=?, D_m= 3.07, D_c= 3.08, VID 14. Des Cloiseaux, 1881. Am. Min. 54(1969), 328 (Abst.). Acta Cryst. B24 (1968), 1214. See also Orthoserpierite.

Shabaite-(Nd), $Ca(Nd,Sm,Y)_2(UO_2)(CO_3)_4(OH)_2 \cdot 6H_2O$, A Mon. Pearly yellowish-white. Kamoto Est, Kolwezi Mining Dist., Shaba (S), Zaïre. H= 2.5, D_m= 3.13, D_c= 3.18, VbD 04. Deliens & Piret, 1989. Europ.Jour.Min. 1 (1989), 85.

Shabynite, $Mg_5BO_3(OH)_5(Cl,OH)_2 \cdot 4H_2O$, A Mon. Translucent white. Korshunov, Irkutsk, Siberia, USSR. H= 3, D_m= 2.32, D_c=?, Vc 01f. Pertsev et al., 1980. Am. Min. 66(1981), 1101 (Abst.).

Shachialite, D Inadequate data. Min. Abstr. 79-1659.

Shadlunite, $(Fe,Cu)_8(Pb,Cd)S_8$, A Cub. Metallic greyish-yellow. Mayak mine, Talnakh, Noril'sk (near), Siberia (N), USSR. H= 3.5-4, D_m=?, D_c= 4.61, IIA 07. Evstigneeva et al., 1973. ZVMO 102 (1973), 63.

Shafranovskite, $(Na,K)_6(Mn,Fe)_3Si_9O_{24} \cdot 6H_2O$, A Rhom. Vitreous dark/olive/yellowish green. Khibina massif and Lovozero massif, Kola Peninsula, USSR. H= 2-3, D_m= 2.78, D_c= 2.78, VIIIC 07. Khomyakov et al., 1982. Am. Min. 68(1983), 644 (Abst.).

Shakhovite, $Hg_4SbO_3(OH)_3$, A Mon. Adamantine green. Kelyan deposit, Buryat, and Khaidarkan deposit, Kirgizia, USSR. H= 3-3.5, D_m= 8.38, D_c= 8.606, VIIB 17. Tillmanns et al., 1982. Am. Min. 73(1988), 1499 (Abst.). Am. Min. 68(1983), 1041 (Abst.).

Shandite, $Ni_3Pb_2S_2$, G Rhom. Metallic white. Trial Harbour, Tasmania, Australia. H= 4, D_m= 8.72, D_c= 8.865, IIA 05. Ramdohr, 1950. Am. Min. 59 (1974), 296. N.Jb.Min.Mh. (1978), 256. See also Parkerite, Rhodplumsite.

Sharpite, $Ca(UO_2)_6(CO_3)_5(OH)_4 \cdot 6H_2O$, G Orth. Translucent greenish-yellow. Shinkolobwe, Shaba, Zaïre. H= 2.5, D_m= 4.45, D_c= 4.51, VbD 04. Melon, 1938. Am. Min. 70(1985), 220 (Abst.).

Shattuckite, $Cu_5(SiO_3)_4(OH)_2$, G Orth. Translucent dark blue. Shattuck mine, Bisbee, Cochise Co., Arizona, USA. H= 3.5, D_m= 4.11, D_c= 4.128, VIIID 04. Schaller, 1915. Am. Min. 51(1966), 266 (Abst.). Am. Min. 62 (1977), 491.

Shcherbakovite, $NaK(Ba,K)Ti_2(Si_2O_7)_2$, G Orth. Vitreous dark brown. Khibina tundra, Kola Peninsula, USSR. H= 6.5, D_m= 2.968, D_c= 2.97, VIIID 15. Es'kova & Kazakova, 1954. JCPDS 8-101.

Shcherbinaite, V_2O_5, A Orth. Vitreous yellow-green. Mt. Rasvumchorr, Beizȳmyannīvolcano, Kamchatka, USSR. H= 3-3.5, D_m= 3.2, D_c= 3.37, IVE 01. Borisenko, 1972. Am. Min. 58(1973), 560 (Abst.). Acta Cryst. C42 (1986), 1467.

Shentulite, D Name also written shen-t'u-shih; = thorite or thorogummite. Min. Mag. 33 (1962), 261. Peng, 1959.

Sherwoodite, $Ca_{4.5}AlV_{14}O_{40} \cdot 28H_2O$, G Tet. Vitreous/earthy dark blue-black. Peanut mine, Montrose Co., Colorado, USA. H= 2, D_m= 2.8, D_c= 2.56, VIIC 23. Thompson et al., 1958. Am. Min. 43 (1958), 749. Am. Min. 63 (1978), 863.

Shigaite, $Mn_7Al_4(SO_4)_2(OH)_{22} \cdot 8H_2O$, A Rhom. Translucent light yellow. Ioi mine, Shiga, Honshu, Japan. H= 2, D_m= 2.32, D_c= 2.35, IVF 03e. Peacor et al., 1985. Am. Min. 71(1986), 1546 (Abst.). See also Lawsonbauerite, Mooreite, Torreyite.

Shortite, $Na_2Ca_2(CO_3)_3$, G Orth. Vitreous colorless, pale yellow. West Vaca, Green River (W), Sweetwater Co., Wyoming, USA. H= 3, D_m= 2.63, D_c= 2.621, VbA 05. Fahey, 1939. Am. Min. 24 (1939), 514. Struct. Repts. 37A (1971), 281.

Shubnikovite, $Ca_2Cu_8(AsO_4)_6Cl(OH) \cdot 7H_2O$(?), Q Orth. Translucent light blue. Unspecified locality, USSR. H= 2, D_m=?, D_c=?, VIID 16. Nefedov, 1953. Am. Min. 40(1955), 552 (Abst.).

Shuiskite, $Ca_2MgCr_2(SiO_4)(Si_2O_7)(OH)_2 \cdot H_2O$, A Mon. Vitreous dark brown. Bisersk deposit, Ural Mts., USSR. H= 6, D_m= 3.24, D_c= 3.46, VIIIB 26. Ivanov et al., 1981. Am. Min. 67(1982), 869 (Abst.).

Sibirskite, $CaHBO_3$, Q Orth. Unspecified locality, Siberia(?), USSR. H=?, D_m=?, D_c=?, Vc 01g. Vasilkova, 1962. Am. Min. 48(1963), 433 (Abst.).

Sicklerite, $Li(Mn,Fe)PO_4$, G Orth. Sub-translucent yellowish/dark brown. Naylor-Vanderburg mine, Pala (near), San Diego Co., California, USA. H= 4, D_m= 3.36, D_c= 3.50, VIIA 02. Schaller, 1912. JCPDS 33-802. See also Ferrisicklerite.

Siderazot, $Fe_{2.5}N$, G Hex. Metallic white. Mount Etna, Italy. H=?, D_m= 3.15, D_c= 3.080, IC 05. Silvestri, 1876. Dana, 7th ed. (1944), v.1, 126. Zts. Krist. 74 (1930), 511.

Siderite, $FeCO_3$, G Rhom. Vitreous yellowish-brown, grey, pale green, white. H= 4, D_m= 3.96, D_c= 3.936, VbA 02. Haidinger, 1845. Dana, 7th ed.(1951), v.2, 166. Zts. Krist. 156 (1981), 233. See also Magnesite, Rhodochrosite.

Sideronatrite, $Na_2Fe(SO_4)_2(OH) \cdot 3H_2O$, G Orth. Translucent yellow, orange, yellow-brown. San Simon mine, Huantajaya, Tarapacá, Chile. H= 1.5-2.5, D_m= 2.28, D_c= 2.276, VID 13. Raimondi, 1878. Per. Mineral. 54 (1985), 15.

Siderophyllite, $K_2(Fe,Al)_6(Si_5Al_3)O_{20}(OH)_4$, G Mon. Black, dark brown. Pike's Peak, Park Co., Colorado, USA. H= 2.5-3, D_m= 3.0, D_c= 3.27, VIIIE 05b. Lewis, 1880. Jour. Petrol. 14 (1973), 159.

Siderotil, $(Fe,Cu)SO_4 \cdot 5H_2O$, G Tric. Vitreous white, yellowish, pale green. Idrija (Idria), Slovenija, Yugoslavia. H=?, D_m= 2.1, D_c= 2.212, VIC 03a. Schrauf, 1891. Can. Min. 7 (1963), 751.

Sidorenkite, $Na_3Mn(PO_4)(CO_3)$, A Mon. Vitreous/pearly pale rose. Mt. Alluaiv region, Lovozero massif, Kola Peninsula, USSR. H= 2, D_m= 2.90, D_c= 2.96, VbB 06. Khomyakov et al., 1979. Am. Min. 64(1979), 1331 (Abst.). Min. Abstr. 81-3827. See also Bonshtedtite, Bradleyite.

Sidwillite, $MoO_3 \cdot 2H_2O$, A Mon. Resinous yellow. Lake Como, Hinsdale Co., Colorado, USA. H= 2.5, D_m= 3.12, D_c= 3.11, IVF 10. Cesbron & Ginderow, 1985. Am. Min. 71(1986), 1546 (Abst.). Bull. Min. 108 (1985), 813.

Siegenite, $CoNi_2S_4$, G Cub. Metallic grey; white in reflected light. Siegen dist., Westphalia, Germany. H= 4.5-5.5, D_m= 4.6, D_c= 4.83, IIC 01. Dana, 1850. Min. Abstr. 80-4873.

Sieleckiite, $Cu_3Al_4(PO_4)_2(OH)_{12} \cdot 2H_2O$, A Tric. Pearly deep blue. Mt. Oxide mine, Mt. Isa (N), Queensland, Australia. H= 3, D_m= 3.02, D_c= 2.94, VIID 36. Birch & Pring, 1988. Min. Mag. 52 (1988), 515.

Sigloite, $FeAl_2(PO_4)_2(OH)_3 \cdot 7H_2O$, G Tric. Translucent straw-yellow, light brown. Llallagua, Potosí, Bolivia. H= 3, D_m= 2.35, D_c= 2.500, VIID 03. Hurlbut & Honea, 1962. Am. Min. 47 (1962), 1. Min. Petrol. 38 (1988), 201.

Silhydrite, $Si_3O_6 \cdot H_2O$, [A] Orth. Earthy white. Bonanza King Quad., Trinity Center (E), Trinity Co., California, USA. H=?, D_m= 2.141, D_c= 2.116, VIIIE 28. Gude & Sheppard, 1972. Am. Min. 57 (1972), 1053.

Silicate-wiikite, [D] A mixture of yttropyrochlore and other minerals. Am. Min. 62 (1977), 403. Strunz, 1957.

Silicomanganberzeliite, [D] Unnecesssary name for manganoan silicatian berzeliite. Min. Mag. 36 (1968), 1144. Kayupova, 1963.

Silicomonazite, [D] Unnecessary name for silicatian monazite. Min. Mag. 43 (1980), 1055. Nekrasov, 1972.

Silicorhabdophane, [D] Unnecessary name for silicatian rhabdophane. Min. Mag. 36 (1967), 133. Semenov, 1959.

Sillénite, γ–$Bi_{12}SiO_{20}$, [G] Cub. Waxy/earthy olive, green, yellowish-green. Durango, Mexico. H=?, D_m= 9.30, D_c= 8.98, IVC 02. Frondel, 1943. JCPDS 29-235.

Sillimanite, Al_2SiO_5, [G] Orth. Translucent colorless, white, yellow, brown, green. Chester, Connecticut, USA and/or Moldau and Schüttenhofen, Czechoslovakia. H= 6.5-7.5, D_m= 3.25, D_c= 3.239, VIIIA 14. Bowen, 1824. DHZ, 2nd ed.(1982), v.1A, 719. Am. Min. 64 (1979), 573. See also Andalusite, Kyanite.

Silver-2H, Ag, [P] Hex. Metallic white. Unspecified locality, USSR. H=?, D_m=?, D_c= 10.11, IA 01b. Novgorodova et al., 1979. ZVMO 108 (1979), 552.

Silver-3C, Ag, [P] Cub. Metallic white. H= 2.5-3, D_m= 10.487, D_c= 10.500, IA 01b. Dana, 7th ed.(1944), v.1, 96. See also Gold.

Silver-4H, Ag, [P] Hex. Metallic white. Unspecified locality, USSR. H=?, D_m=?, D_c= 9.53, IA 01b. Novgorodova et al., 1979. Am. Min. 65(1980), 1069 (Abst.).

Simonellite, $C_{19}H_{24}$, [G] Orth. Fognano, Toscana, Italy. H=?, D_m=?, D_c= 1.104, IXB 01. Boeris, 1919. Am. Min. 55(1970), 1818 (Abst.).

Simonite, $TlHgAs_3S_6$, [A] Mon. Alšar (Allchar), Roždën (near), Makedonija (Macedonia), Yugoslavia. H=?, D_m=?, D_c= 5.036, IID 05c. Engel & Nowacki, 1982. Am. Min. 69(1984), 211 (Abst.). Zts. Krist. 161 (1982), 159.

Simonkolleite, $Zn_5(OH)_8Cl_2 \cdot H_2O$, [A] Rhom. Vitreous colorless. Richelsdorfer Hütte, Richelsdorf, Hessen, Germany. H= 1.5, D_m= 3.20, D_c= 3.35, IIIC 02. Schmetzer et al., 1985. Am. Min. 73(1988), 194 (Abst.).

Simplotite, $CaV_4O_9 \cdot 5H_2O$, [G] Mon. Vitreous black, greenish-black, yellowish-green. Peanut mine, Montrose Co., Colorado, USA. H= 1, D_m= 2.64, D_c= 2.65, VIIC 22. Thompson et al., 1958. Am. Min. 43 (1958), 16.

Simpsonite, $Al_4(Ta,Nb)_3O_{13}(F,OH)$, [G] Hex. Transparent colorless, cream. Tabba Tabba, Western Australia, Australia. H=?, D_m= 6.67, D_c= 6.83, IVD 15. Bowley, 1938. Min. Mag. 33 (1963), 458. Struct. Repts. 27 (1962), 540.

Sincosite, $Ca(VO)_2(PO_4)_2 \cdot 5H_2O$, [G] Tet. Vitreous/sub-metallic green, brownish-green, yellow-green. Sincos, Peru. H=?, D_m= 2.98, D_c= 2.970, VIIC 22. Schaller, 1924. Am. Min. 70 (1985), 409.

Sinhalite, $MgAlBO_4$, [G] Orth. Vitreous brown, yellow, greenish-brown. Sri Lanka. H=?, D_m= 3.49, D_c= 3.452, Vc 03b. Claringbull & Hey, 1952. Min. Mag. 29 (1952), 841. Min. Mag. 35 (1965), 196.

Sinjarite, $CaCl_2 \cdot 2H_2O$, [A] Tet. Vitreous/resinous pink. Sinjar town, Mosul (near), Iraq. H= 1.5, D_m= 1.66, D_c= 1.96, IIIC 10b. Aljubouri & Aldabbagh, 1980. Min. Mag. 43 (1980), 643.

Sinkankasite, $MnAl(PO_3OH)_2(OH) \cdot 6H_2O$, [A] Tric. Vitreous colorless. Barker Pegmatite, Keystone (near), Pennington Co., South Dakota, USA; Palermo mine, North Groton, Grafton, New Hampshire, USA. H= 4, D_m= 2.27, D_c= 2.25, VIID 07. Peacor et al., 1984. Amer. Mineral. 69 (1984), 380.

Sinnerite, $Cu_6As_4S_9$, [G] Tric. Metallic grey. Lengenbach quarry, Binntal, Valais (Wallis), Switzerland. H=?, D_m=?, D_c= 4.47, IID 11. Marumo & Nowacki, 1964. Am. Min. 57 (1972), 824. Am. Min. 60 (1975), 998.

Sinoite, Si_2N_2O, G Orth. Translucent. Jajh deh Kot Lalu meteorite, Sind Province, Pakistan. H=?, D_m=?, D_c= 2.84, IC 05. Anderson et al., 1964. Science 146 (1964), 256.

Sjögrenite, $Mg_6Fe_2CO_3(OH)_{16}\cdot 4H_2O$, G Hex. Pearly creamy-white, yellowish. Sjo or Långban mine, Filipstad (near), Värmland, Sweden. H= 2.5, D_m= 2.11, D_c= 2.096, IVF 03c. Frondel, 1941. Am. Min. 73(1988), 199 (Abst.). N.Jb.Min.Mh. (1966), 161. See also Pyroaurite.

Skinnerite, Cu_3SbS_3, A Mon. Light bluish-grey in reflected light. Ilímaussaq Intrusion, Greenland (S). H= 3.3, D_m=?, D_c= 5.10, IID 01a. Karup-Møller & Makovicky, 1974. Am. Min. 59 (1974), 889.

Skippenite, Bi_2Se_2Te, A Rhom. Metallic white. Otish Mts, Québec, Canada. H= 2.2, D_m=?, D_c= 7.94, IIC 03. Johan et al., 1987. Can. Min. 25 (1987), 625.

Sklodowskite, $(H_3O)_2Mg(UO_2)_2(SiO_4)_2\cdot 4H_2O$, G Mon. Translucent pale yellow. Shinkolobwe, Shaba, Zaïre. H=?, D_m= 3.54, D_c= 3.662, VIIIA 25. Schoep, 1924. Am. Min. 66 (1981), 610. Struct. Repts. 43A (1977), 323. See also Cuprosklodowskite.

Skutterudite, $CoAs_3$, G Cub. Metallic white/grey. Skutterud, Norway. H= 5.5-6, D_m= 6.5, D_c= 6.82, IIC 12. Haidinger, 1845. Dana, 7th ed.(1944), v.1, 342. Struct. Repts. 40A (1974), 6. See also Nickel-skutterudite.

Slavíkite, $NaMg_2Fe_5(SO_4)_7(OH)_6\cdot 33H_2O$, G Rhom. Vitreous greenish-yellow. Czechoslovakia. H=?, D_m= 1.89, D_c= 1.90, VID 05. Jirovksy & Ulrich, 1926. Bull. Min. 87 (1964), 622. Struct. Repts. 41A (1975), 351.

Slawsonite, $(Sr,Ca)Al_2Si_2O_8$, A Mon. Translucent colorless, grey. Martin Bridge formation, Wallowa Co., Oregon, USA. H= 5.5-6.5, D_m=?, D_c= 3.12, VIIIF 04. Griffen et al., 1977. Am. Min. 72(1987), 225 (Abst.). Am. Min. 62 (1977), 31.

Smectite, A group name for sheet silicates with exchangeable cations and the general formula $(Ca,Na,Li)_{0-1}(Mg,Fe,Al,Li,Ni,Cr,Zn)_{2-3}(Si,Al)_4O_{10}\cdot nH_2O$.

Smirnite, Bi_2TeO_5, A Orth. Translucent colorless, light grey, light yellow. Zod deposit, Armenia; Kazakhstan and Zakarpatsk, USSR. H= 3.3-4, D_m= 7.78, D_c= 7.72, VIG 05. Spiridonov et al., 1984. Am. Min. 70(1985), 876.

Smithite, $AgAsS_2$, G Mon. Adamantine red. Lengenbach quarry, Binntal, Valais (Wallis), Switzerland. H= 1.5-2, D_m= 4.88, D_c= 4.93, IID 07. Solly, 1905. Dana, 7th ed.(1944), v.1, 430. Struct. Repts. 29 (1964), 25. See also Trechmannite.

Smithsonite, $ZnCO_3$, G Rhom. Vitreous greyish-white, green blue, brown, etc. H= 4-4.5, D_m= 4.2, D_c= 4.434, VbA 02. Beudant, 1832. Am. Min. 39 (1954), 47. Zts. Krist. 156 (1981), 233.

Smolianinovite, $(Co,Ni,Mg,Ca)_3(Fe^{+3},Al)_2(AsO_4)_4\cdot 11H_2O$, G Orth. Earthy yellow. Bou Azzer, Morocco. H= 2, D_m= 2.1, D_c= 2.2, VIIC 26. Yakhontova, 1956. Am. Min. 59(1974), 1141 (Abst.).

Smythite, $(Fe,Ni)_9S_{11}$, G Hex. Metallic bronze-yellow. Bloomington Crushed Stone Quarry, Monroe Co., Indiana, USA. H=?, D_m= 4.06, D_c= 4.09, IIB 09b. Erd et al., 1957. Am. Min. 55 (1970), 1650. Struct. Repts. 21 (1957), 142.

Sobolevite, $Na_{14}Ca_2Ti_3MnO_4(Si_2O_7)_2(PO_4)_4$, A Mon. Metallic/pearly/resinous brown. Mt. Alluaiv, Lovozero massif (NW), Kola Peninsula, USSR. H= 4.5-5, D_m= 3.03, D_c= 3.00, VIIIB 11. Khomyakov et al., 1983. Am. Min. 69(1984), 813 (Abstr.).

Sobolevskite, PdBi, A Hex. Metallic greyish-white. Oktyabr deposit, Talnakh, Noril'sk (near), Siberia (N), USSR. H= 4, D_m=?, D_c= 11.88, IIB 09a. Evstigneeva et al., 1975. Am. Min.61(1976), 1054 (Abst.).

Sodalite, $Na_4(Si_3Al_3)O_{12}Cl$, G Cub. Translucent pink, grey, yellow, blue, green, white. H= 5.5-6, D_m= 2.2, D_c= 2.30, VIIIF 07. Thomson, 1811. DHZ (1963), v.4, 289. Acta Cryst. B40 (1984), 6.

Soddyite, $(UO_2)_2SiO_4 \cdot 2H_2O$, G Orth. Translucent greenish-yellow, yellow, amber-yellow. Shinkolobwe, Shaba, Zaïre. H= 3-4, D_m= 4.70, D_c= 4.69, VIIIA 25. Schoep, 1922. Am. Min. 66 (1981), 610.

Sodium phlogopite, D = wonesite. Am. Min. 66(1981), 639 (Abst.). Schreyer et al., 1980.

Sodium alum, $NaAl(SO_4)_2 \cdot 12H_2O$, G Cub. Vitreous colorless. H= 3, D_m= 1.67, D_c= 1.671, VIC 08. Dana, 7th ed.(1951), v.2, 474. Acta Cryst. 22 (1967), 182.

Sodium-anthophyllite, $Na(Mg,Fe)_7(Si_7Al)O_{22}(OH)_2$, A Orth. H=?, D_m=?, D_c=?, VIIID 06. Leake et al., 1978. Am. Min. 63 (1978), 1023. See also Anthophyllite.

Sodium autunite, $Na_2(UO_2)_2(PO_4)_2 \cdot 8H_2O$, G Tet. Pearly/vitreous yellow, greenish-yellow. Unspecified locality, USSR. H= 2-2.5, D_m= 3.584, D_c= 3.89, VIID 20b. Chernikov et al., 1957. Am. Min. 43(1958), 383 (Abst.).

Sodium betpakdalite, $Na_2CaFe_2^{3+}(As_2O_4)(MoO_4)_6 \cdot 15H_2O$, A Mon. Dull yellow. Unspecified locality, USSR. H=?, D_m= 2.02, D_c= 2.84, VIID 15b. Skvortsova et al., 1971. Am. Min. 57(1972),1312 (Abst.).

Sodium boltwoodite, $(H_3O)(Na,K)(UO_2)SiO_4 \cdot H_2O$, G Orth. Powdery pale yellow. Unspecified arid regions, USSR. H=?, D_m= 4.1, D_c= 4.4, VIIIA 25. Chernikov et al., 1975. Am. Min. 66 (1981), 610.

Sodium dachiardite, $Na_4(Al_4Si_{20})O_{48} \cdot 13H_2O$, Q Mon. Alpe di Suisi, Italy; Tsugawa, Japan (NE). H=?, D_m= 2.16, D_c= 2.141, VIIIF 12. Yoshimura & Wakabayashi, 1977. Am. Min. 64(1979), 244 (Abst.).

Sodium-gedrite, $Na(Mg,Fe)_6Al(Si_6Al_2)O_{22}(OH)_2$, A Mon. H=?, D_m=?, D_c=?, VIIID 06. Leake et al., 1978. Am. Min. 63 (1978), 1023. See also Gedrite.

Sodium pharmacosiderite, $(Na,K)_2Fe_4(AsO_4)_3(OH)_5 \cdot 7H_2O$, A Cub. Vitreous pale green. Marda, Western Australia, Australia. H= 3, D_m= 2.79, D_c= 2.90, VIID 14b. Peacor & Dunn, 1985. Min. Rec. 16 (1985), 121. See also Alumopharmacosiderite, Pharmacosiderite.

Sodium-uranospinite, $(Na_2,Ca)(UO_2)_2(AsO_4)_2 \cdot 5H_2O$, G Tet. Vitreous/pearly yellow-green, yellow. Unspecified Locality, USSR. H= 2.5, D_m= 3.846, D_c= 3.65, VIID 20b. Kopchenova & Skvortsova, 1957. Am. Min. 43(1958), 383 (Abst.). See also Uranospinite, Metauranospinite, Meta-Na-uranospinite.

Sodium-zippeite, $Na_4(UO_2)_6(SO_4)_3(OH)_{10} \cdot 4H_2O$, A Orth. Translucent yellow. Numerous localities, Utah, USA. H= 2, D_m=?, D_c=?, VID 08. Frondel et al., 1976. Can. Min. 14 (1976), 429. See also Zippeite, Nickel-zippeite, Zinc-zippeite, Magnesium-zippeite, Cobalt-zippeite.

Sogdianite, $KNa(Li,Al)_3(Fe,Ti,Zr)Si_{12}O_{30}$, G Hex. Vitreous violet. Alai Mts., Turan dist., Tadzhikistan, USSR. H= 7, D_m= 2.90, D_c= 2.75, VIIIC 10. Dusmatov et al., 1968. Am. Min. 54(1969), 1221 (Abst.). Sov.Phys.Cryst. 19 (1974), 460.

Söhngeite, $Ga(OH)_3$, A Cub. Translucent light brown. Tsumeb, Namibia. H= 4-4.5, D_m= 3.84, D_c= 3.847, IVF 06. Strunz, 1965. Am. Min. 51(1966), 1815 (Abst.). See also Dzhalindite.

Sokolovite, D Inadequate data; may be goyazite. Min. Mag. 33 (1962), 261. Sharova & Gladovskii, 1958.

Solongoite, $Ca_2B_3O_4(OH)_4Cl$, A Mon. Vitreous colorless. Solongo deposit, Buryat, Ural Mts., USSR. H= 3.5, D_m= 2.514, D_c= 2.57, Vc 05. Malinko, 1974. Am. Min. 60(1975), 162 (Abst.). Min. Abstr. 75-3060.

Sonolite, $Mn_9(SiO_4)_4(OH,F)_2$, G Mon. Translucent pinkish-brown, reddish-orange, dark brown. Sono mine and Hanawa mine, Japan. H= 5.5, D_m= 3.82, D_c= 4.08, VIIIA 16. Yoshinaga, 1963. Am. Min. 54 (1969), 1392. N.Jb.Min.Mh. (1989), 410. See also Jerrygibbsite.

Sonoraite, $FeTeO_3(OH) \cdot H_2O$, [A] Mon. Transparent yellowish-green. Moctezuma mine, Moctezuma, Sonora, Mexico. H= 3, D_m= 3.95, D_c= 4.179, VIG 02. Gaines et al., 1968. Am. Min. 53 (1968), 1828. Struct. Repts. 40A (1974), 311.

Sopcheite, $Ag_4Pd_3Te_4$, [A] Orth. Metallic brownish-grey. Sopcha massif, Monchegorsk pluton, USSR. H= 3.5, D_m=?, D_c= 9.95, IIA 14. Orsoev et al., 1982. ZVMO 111 (1982), 114.

Sophiite, $Zn_2(SeO_3)Cl_2$, [A] Orth. Tolbachik volcano, Kamchatka, USSR. H= 2, D_m=?, D_c= 3.64, VIG 13. Vergasova et al., 1989. ZVMO 118 (1989), 65. ZVMO 118 (1989), 65.

Sorbyite, $Pb_{19}(Sb,As)_{20}S_{49}$, [A] Mon. Metallic grey/black; white in reflected light. Taylor pit, Madoc, Huntingdon Twp., Hastings Co., Ontario, Canada. H= 3.5, D_m=?, D_c= 5.52, IID 10. Jambor, 1967. Min. Rec. 13 (1982), 93.

Sörensenite, $Na_4Be_2Sn(Si_3O_9)_2 \cdot 2H_2O$, [A] Mon. Silky colorless, pinkish. Nakalaq, Ilímaussaq Intrusion, Greenland (S). H= 5.5, D_m= 2.9, D_c= 2.92, VIIID 24. Semenov et al., 1965. Am. Min. 51(1966), 1547 (Abst.). Acta Cryst. B32 (1976), 2553.

Sosedkoite, $(K,Na)_5Al_2(Ta,Nb)_{22}O_{60}$, [A] Orth. Adamantine colorless. Unspecified locality, Kola Peninsula, USSR. H= 6, D_m=?, D_c= 6.90, IVE 02. Voloshin et al., 1982. Am. Min 68 (1983), 644 (Abst.).

Součekite, $CuPbBi(S,Se)_3$, [A] Orth. Metallic lead-grey. Oldřichov, Tachov (near), Čechy (Bohemia), Czechoslovakia. H= 3.5, D_m=?, D_c= 7.60, IID 01e. Čech & Vavrin, 1979. Am. Min. 65(1980), 209 (Abst.). See also Bournonite, Seligmannite.

Souzalite, $(Mg,Fe)_3(Al,Fe)_4(PO_4)_4(OH)_6 \cdot 2H_2O$, [G] Tric. Vitreous dark green, blue-green. Corrego Frio Pegmatite, Minas Gerais, Brazil. H= 5.5-6, D_m= 3.09, D_c= 3.087, VIID 28. Pecora & Fahey, 1949. Can. Min. 19 (1981),381. See also Gormanite.

Spadaite, $MgSiO_2(OH)_2 \cdot H_2O$(?), [Q] Syst=? Reddish. Capo di Bove, Roma (near), Italy. H= 2.5, D_m=?, D_c=?, VIIIE 08b. Kobell,1843. Dana/Ford (1932), 679.

Spangolite, $Cu_6AlSO_4(OH)_{12}Cl \cdot 3H_2O$, [G] Hex. Vitreous dark green, bluish green. Globe Dist., Tombstone (near), Cochise Co., Arizona, USA. H= 2, D_m= 3.14, D_c= 3.14, IVF 03e. Penfield, 1890. Am. Min. 34 (1949), 181.

Spencerite, $Zn_4(PO_4)_2(OH)_2 \cdot 3H_2O$, [G] Mon. Pearly/vitreous white. Hudson Bay zinc mine, Nelson (near), West Kootenay Dist., British Columbia, Canada. H= 3, D_m= 3.14, D_c= 3.242, VIID 02. Walker, 1916. Dana, 7th ed.(1951), v.2, 931. Min. Mag. 38 (1972), 687.

Sperrylite, $PtAs_2$, [G] Cub. Metallic white. Vermilion mine, Denison Twp., Sudbury Dist., Ontario, Canada. H= 6-7, D_m= 10.58, D_c= 10.806, IIC 05. Wells, 1889. Dana, 7th ed.(1944), v.1, 292. Can. Min. 17 (1979), 117.

Spertiniite, $Cu(OH)_2$, [A] Orth. Vitreous blue, blue-green. Jeffrey mine, Shipton Twp., Richmond Co., Québec, Canada. H=?, D_m= 3.93, D_c= 3.94, IVF 03a. Grice & Gasparrini, 1981. Can. Min. 19 (1981), 337.

Spessartine, $Mn_3Al_2(SiO_4)_3$, [G] Cub. Vitreous/resinous dark red, brownish-red. Aschaffenburg, Spessart (near), Germany. H= 6.5-7.5, D_m= 4.18, D_c= 4.21, VIIIA 06a. Beudant, 1832. DHZ, 2nd ed.(1982), v.1A, 468. Am. Min. 56 (1971), 791. See also Almandine.

Sphaerocobaltite, $CoCO_3$, [A] Rhom. Vitreous rose-red. Schneeberg, Sachsen (Saxony), Germany. H= 4, D_m= 4.13, D_c= 4.208, VbA 02. Weisbach, 1877. Dana, 7th ed.(1951), v.2, 175. Acta Cryst. C42 (1986), 4.

Sphalerite, α–ZnS, [G] Cub. Resinous/adamantine colorless, brown, etc. H= 3.5, D_m= 4.08, D_c= 4.096, IIB 01a. Glocker, 1847. Dana, 7th ed.(1944), v.1, 210. Acta Cryst. A36 (1980), 482. See also Matraite, Wurtzite.

Spheniscidite, $(NH_4,K)(Fe,Al)_2(PO_4)_2(OH) \cdot 2H_2O$, [A] Mon. Earthy brown. Elephant Island, British Antarctic Territory, Antarctica. H=?, D_m=?, D_c= 2.71,

VIID 14a. Wilson & Bain, 1986. Min. Mag. 50 (1986), 291. See also Leucophosphite, Tinsleyite.

Spinel, $MgAl_2O_4$, G Cub. Vitreous red, blue, green, brown, etc. H= 7.5-8, D_m=?, D_c= 3.55, IVB 01. Agricola, 1546. Dana, 7th ed.(1944), v.1, 689. Acta Cryst. B40 (1984), 96. See also Magnesiochromite, Gahnite, Hercynite.

Spionkopite, $Cu_{39}S_{28}$, A Rhom. Metallic blue-grey. Yarrow Creek – Spionkop Creek deposit, Alberta (SW), Canada. H= 2.5, D_m=?, D_c= 5.13, IIA 01a. Goble, 1980. Can. Min. 18 (1980), 511. Can. Min. 23 (1985), 61.

Spiroffite, $(Mn, Zn)_2Te_3O_8$, G Mon. Adamantine red, purple. Moctezuma, Sonora, Mexico. H= 3.5, D_m= 5.01, D_c= 4.97, VIG 06. Mandarino et al., 1962. Am. Min. 47(1962), 196 (Abst.).

Spodiosite, Ca_2PO_4F, Q Orth. Porcelaneous, vitreous ash/grey, brown. Nyttsta Krangrūva, Värmland, Sweden. H= 5, D_m= 2.94, D_c=?, VIIB 03c. Tiberg, 1872. Dana, 7th ed.(1951), v.2, 848.

Spodumene, $LiAlSi_2O_6$, G Mon. Vitreous colorless, greyish-white, lilac, pale green, etc. H= 6.5-7, D_m= 3.1, D_c= 3.184, VIIID 01a. d'Andrada, 1800. DHZ, 2nd ed.(1978), v.2A, 527. MSA Spec. Paper 2 (1969), 31.

Spurrite, $Ca_5(SiO_4)_2(CO_3)$, G Mon. Translucent pale grey. Velardeña, Durango, Mexico. H= 5, D_m= 3.01, D_c= 3.02, VIIIA 10. Wright, 1908. Dana/Ford (1932), 687. Acta Cryst. 13 (1960), 451. See also Paraspurrite.

Srebrodolskite, $Ca_2Fe_2O_5$, A Orth. Adamantine/metallic black, brownish-red. Kopeysk, Chelyabinsk coal basin, Ural Mts. (S), USSR. H= 5.5, D_m= 4.04, D_c= 4.03, IVA 08. Chesnokov & Bazhenova, 1985. Am. Min. 71(1986), 1279 (Abst.).

Srilankite, $(Ti, Zr)O_2$, A Orth. Sub-metallic/adamantine black. Sabaragamuva, Rakwana, Sri Lanka. H= 6.5, D_m=?, D_c= 4.765, IVD 07. Willgallis et al., 1983. Am. Min. 69(1984), 212 (Abst.). Zts. Krist. 164 (1983), 59.

Stanfieldite, $Ca_4(Mg, Fe, Mn)_5(PO_4)_6$, A Mon. Transparent reddish, amber. Estherville meteorite, Emmet Co., Iowa, USA. H= 4-5, D_m= 3.15, D_c= 3.15, VIIA 04. Fuchs, 1967. Am. Min. 53(1968), 508 (Abst.).

Stanleyite, $VOSO_4 \cdot 6H_2O$, A Orth. Translucent blue. Minasragra, Cerro de Pasco (near), Junin, Peru. H= 1-1.5, D_m= 1.95, D_c= 2.01, VID 12. Livingstone, 1982. Min. Mag. 45 (1982), 163.

Stannite, Cu_2FeSnS_4, G Tet. Metallic grey/black; olive-grey in reflected light. Wheal Rock, Cornwall, England. H= 4, D_m= 4.4, D_c= 4.44, IIB 03a. Beudant, 1832. Can. Min. 17 (1979), 125. Can. Min. 16 (1978), 131.

Stannoidite, $Cu_8(Fe, Zn)_3Sn_2S_{12}$, A Orth. Metallic brass brown. Konjo mine, Aida, Mito-cho, Okayama, Honshu, Japan. H= 4, D_m=?, D_c= 4.29, IIB 03b. Kato, 1969. Can. Min. 17 (1979), 125. Zts. Krist. 144 (1976), 145.

Stannoluzonite, D Superfluous name for a stannian luzonite. Min. Mag. 36 (1967), 133. Moh & Ottemann, 1962.

Stannomicrolite, $(Sn, Fe, Mn)_2(Ta, Nb, Sn)_2(O, OH, F)_7$, A Cub. Translucent yellowish-brown. Sukula, Tammela, Finland (SW). H=?, D_m= 8.34, D_c=?, IVC 09a. Hogarth, 1977. Am. Min. 62 (1977), 403.

Stannopalladinite, Pd_3Sn_2, G Hex. Metallic brown-rose. Monchegorsk, USSR. H= 4.5-5, D_m=?, D_c= 9.821, IA 10. Maslenitzky et al., 1947. Can. Min. 19 (1981), 599.

Staringite, $(Fe, Mn)(Sn, Ti)_9(Ta, Nb)_2O_{24}$, A Tet. Dark grey in reflected light. Seridózinho and Pedra Lavreda, Paraiba, Brazil. H=?, D_m=?, D_c= 7.17, IVD 04. Burke et al., 1969. Am. Min. 55(1970), 1446 (Abst.). See also Manganotapiolite, Tapiolite.

Starkeyite, $MgSO_4 \cdot 4H_2O$, A Mon. Dull white. Starkey mine, Madison Co., Missouri, USA. H=?, D_m= 2.01, D_c= 2.007, VIC 02. Grawe, 1956. Min. Rec. 6 (1975), 144. Acta Cryst. 17 (1964), 863.

Staurolite, $(Fe,Mg)_4Al_{17}(Si,Al)_8O_{44}(OH)_4$, G Mon. Translucent dark/reddish/yellow-brown, pale yellow. H= 7.5, D_m= 3.8, D_c= 3.75, VIIIA 14. Delamétherie, 1792. Am. Min. 67 (1982), 292. Am. Min. 74 (1989), 610.

Steacyite, $K_{0.3}(Na,Ca)_2ThSi_8O_{20}$, A Tet. Transparent brown. Mont Saint-Hilaire, Rouville Co., Québec, Canada. H= 4.9, D_m= 2.95, D_c= 3.32, VIIIE 17. Perrault & Szymanski, 1982. Can. Min. 20 (1982), 59. Acta Cryst. B28 (1972), 1794. See also Ekanite, Iraqite.

Steenstrupine-(Ce), $Na_{14}Ce_6Mn_2Fe_2Zr(PO_4)_7Si_{12}O_{36}(OH)_2 \cdot 3H_2O$, G Rhom. Dark brown, black. Kangerdluarsuk, Ilímaussaq, Greenland (S). H= 4, D_m= 3.4, D_c= 3.631, VIIIA 21. Lorenzen, 1881. N.Jb.Min.Abh. 140 (1981), 301. Am. Min. 69(1984), 215 (Abst.).

Steigerite, $AlVO_4 \cdot 3H_2O$, G Mon. Waxy yellow, greenish-yellow, olive-green. Sullivan Brothers Claim, Gypsum Valley (N side), San Miguel Co., Colorado, USA. H= 2.5-3, D_m= 2.55, D_c= 2.58, VIIC 24. Henderson, 1935. ZVMO 116 (1987), 100.

Stellerite, $Ca(Si_7Al_2)O_{18} \cdot 7H_2O$, G Orth. Transparent colorless. Komandor Islands, Bering Sea. H= 3.5-4, D_m= 2.15, D_c=?, VIIIF 13. Morozewicz, 1909. Bull. Min. 101 (1978), 368. Bull Min. 98 (1975), 11. See also Barrerite.

Stenhuggarite, $CaFeSbAs_2O_7$, A Tet. Translucent orange. Långban mine, Filipstad (near), Värmland, Sweden. H= 4, D_m= 4.63, D_c= 4.56, VIIA 15. Moore, 1970. Am. Min. 56(1971), 636 (Abst.). Acta Cryst. B33 (1977), 1807.

Stenonite, $(Sr,Ba,Na)_2AlCO_3F_5$, G Mon. Vitreous colorless, white. Ivigtut, Greenland (SW). H= 3.5, D_m= 3.86, D_c= 3.847, IIID 02. Pauly, 1962. Am. Min. 48(1963), 1178 (Abst.). Can. Min. 22 (1984), 245.

Stepanovite, $NaMgFe(C_2O_4)_3 \cdot 8\text{–}9H_2O$, G Rhom. Vitreous greenish. Talnakh coal deposit, Lena river, USSR. H= 2, D_m= 1.69, D_c= 1.69, IXA 01. Nefedov, 1953. Am. Min. 49(1964), 442 (Abst.).

Stephanite, Ag_5SbS_4, G Orth. Metallic black; grey in reflected light. Freiberg, Sachsen (Saxony), Germany. H= 2-2.5, D_m= 6.26, D_c= 6.28, IID 01d. Haidinger, 1845. Dana, 7th ed.(1944), v.1, 358. Acta Cryst. B26 (1970), 201. See also Selenostephanite.

Stercorite, $(NH_4)Na(PO_3OH) \cdot 4H_2O$, G Tric. Vitreous white, yellowish, brownish. Ichaboe Island, Namibia. H= 2, D_m= 1.574, D_c= 1.570, VIIC 14. Herepath, 1850. Dana, 7th ed.(1951), v.2, 698. Acta Cryst. B30 (1974), 504.

Sterlinghillite, $Mn_3(AsO_4)_2 \cdot 4H_2O$, A Syst=? Silky white, light pink. Sterling Hill mine, Ogdensburg, Sussex Co., New Jersey, USA. H= 3, D_m= 2.95, D_c=?, VIIC 07. Dunn, 1981. Am. Min. 66 (1981), 182.

Sternbergite, $AgFe_2S_3$, G Orth. Metallic brown. Jáchymov (St. Joachimsthal), Západočeský kraj, Čechy (Bohemia), Czechoslovakia. H= 1-1.5, D_m= 4.2, D_c= 4.32, IIB 08. Haidinger, 1827. Am. Min. 54 (1969), 1198. N.Jb.Min.Mh. (1987), 458. See also Argentopyrite.

Sterryite, $(Ag,Cu)_2Pb_{10}(Sb,As)_{12}S_{29}$, A Orth. Metallic black; white/grey in reflected light. Taylor pit, Huntingdon Twp., Hastings Co., Ontario, Canada. H=?, D_m=?, D_c= 6.088, IID 10. Jambor, 1967. Min. Rec. 13 (1982), 93.

Stetefeldtite, $Ag_2Sb_2(O,OH)_7$(?), Q Cub. Black, brown. Combination claim, Belmont dist., Tonopah, Nye Co., Nevada, USA. H= 3.5-4.5, D_m= 4.6, D_c= 6.63, IVC 08. Riotte, 1867. Min. Mag. 30 (1953), 100.

Stevensite, $Na_{0.15}Mg_3Si_4O_{10}(OH)_4$, G Mon. Greasy pink, brown, buff. Bergen Hill, New Jersey, USA. H=?, D_m=?, D_c=?, VIIIE 08b. Leeds, 1873. Bull. Min. 103 (1980), 579.

Stewartite, $MnFe_2(PO_4)_2(OH)_2 \cdot 8H_2O$, G Tric. Translucent brownish-yellow. Stewart mine, Pala, San Diego Co., California, USA. H=?, D_m= 2.94, D_c= 2.483, VIID 03. Schaller, 1912. Dana, 7th ed.(1951), v.2, 730. Am. Min. 59 (1974), 1272. See also Laueite, Strunzite.

Stibarsen, SbAs, [A] Rhom. Metallic white. Varuträsk, Sweden (N). H= 3-4, D_m= 6.0, D_c= 6.277, IB 01. Wretblad, 1941. Am. Min. 59 (1974), 1331.

Stibiconite, $Sb_3O_6(OH)$, [G] Cub. Pearly/earthy yellow, yellowish-white, reddish-white. Goldkronach, Bayern (Bavaria), Germany. H= 4-5.5, D_m= 5.58, D_c= 5.88, IVC 08. Brush, 1862. Am. Min. 37 (1952), 982.

Stibiobetafite, $(Ca,Sb)_2(Ti,Nb,Ta)_2O_6(O,OH)$, [A] Cub. Vitreous dark brown. Věžná (Vezna), Moravia, Czechoslovakia. H= 5, D_m= 5.30, D_c= 5.19, IVC 09a. Černy et al., 1979. Can. Min. 17 (1979), 583.

Stibiocolumbite, $SbNbO_4$, [G] Orth. Resinous/adamantine brown, reddish-yellow, greenish-yellow. Hunalaya mine, Mesa Grande, San Diego Co., California, USA. H= 5.5, D_m= 5.98, D_c= 5.728, IVD 14a. Schaller, 1915. Am. Min. 48 (1963), 1348. See also Stibiotantalite, Bismutotantalite.

Stibiodufrenoysite, [D] Inadequate data; may be veenite. Min. Mag. 38 (1971), 103. Nowacki, 1964.

Stibiomicrolite, $(Ca,Sb,Na)_2(Ta,Nb)_2(O,OH,F)_7$, [A] Cub. Greenish-white/white. Varuträsk, Sweden. H= < 5.5, D_m=?, D_c= 6.0, IVC 09a. Quensel & Berggren, 1938. Am. Min. 73(1988), 1499 (Abst.).

Stibiopalladinite, Pd_5Sb_2, [G] Hex. Metallic white/grey; white in reflected light. Bushveld, Transvaal, South Africa. H= 4-5, D_m= 9.5, D_c= 10.8, IIG 04. Adam, 1927. Am. Min. 61 (1976), 1249.

Stibiotantalite, $SbTaO_4$, [G] Orth. Resinous/adamantine brown, reddish-yellow, greenish-yellow. Greenbushes, Western Australia, Australia. H= 5.5, D_m= 7.34, D_c= 7.583, IVD 14a. Goyder, 1893. Am. Min. 48 (1963), 1348. See also Stibiocolumbite, Bismutotantalite.

Stibivanite, Sb_2VO_5, [A] Mon. Adamantine yellow-green. Lake George antimony deposit, York Co., New Brunswick, Canada. H= 4.3, D_m=?, D_c= 5.267, IVD 05. Kaiman et al., 1980. Can. Min. 18 (1980), 329. Can. Min. 18 (1980), 333.

Stibivanite-2O, Sb_2VO_5, [P] Orth. Adamantine emerald-green. Buca della Vena mine, Apuan Alps, Italy. H= 4, D_m=?, D_c= 5.260, IVD 05. Merlino et al., 1989. Can. Min. 27 (1989), 625. Can. Min. 27 (1989), 625.

Stibnite, Sb_2S_3, [G] Orth. Metallic grey; white in polished section. H= 2, D_m= 4.63, D_c= 4.63, IIC 02. Pliny, 77. Min. Abstr. 77-1483. Zts. Krist. 142 (1976), 447. See also Metastibnite, Bismuthinite, Guanajuatite.

Stichtite, $Mg_6Cr_2CO_3(OH)_{16} \cdot 4H_2O$, [G] Rhom. Waxy/greasy lilac, rose-pink. Adelaide mine, Stichtite Hill, Dundas, Tasmania, Australia. H= 1.5-2, D_m= 2.16, D_c= 2.11, IVF 03b. Petterd, 1910. Am. Min. 26 (1941), 295. Min. Mag. 39 (1973), 377. See also Barbertonite.

Stilbite, $NaCa_4(Si_{27}Al_9)O_{72} \cdot 30H_2O$, [G] Mon. Transparent colorless, red. H= 3.5-4, D_m= 2.19, D_c= 2.23, VIIIF 13. Haüy, 1796. Natural Zeolites (1985), 284. Acta Cryst. B27 (1971), 833.

Stilleite, ZnSe, [G] Cub. Metallic grey. Shaba, Zaïre. H= 5, D_m= 5.3, D_c= 5.267, IIB 01a. Ramdohr, 1956. Am. Min. 42(1957), 584 (Abst.). Acta Cryst. A36 (1980), 482.

Stillwaterite, Pd_8As_3, [A] Hex. Metallic creamy-grey in reflected light. Beartooth Mts. (N), Stillwater Complex, Sweetwater Co., Montana, USA. H= 4.7, D_m= 10.4, D_c= 10.96, IIG 04. Cabri et al., 1975. Can. Min. 13 (1975), 321.

Stillwellite-(Ce), $(Ce,La,Ca)BSiO_5$, [G] Rhom. Translucent colorless. Mary Kathleen Lease, Mount Isa (E of), Queensland, Australia. H=?, D_m= 4.70, D_c= 4.74, VIIIA 23. McAndrew & Scott, 1955. Nature 176 (1955), 509. Sov.Phys.Cryst. 12(1967), 214.

Stilpnomelane, $(K,Ca,Na)(Fe,Mg,Al)_{12}(Si,Al)_{16}(O,OH)_{54} \cdot nH_2O$, [G] Tric. Golden brown, deep reddish-brown, black. H= 3-4, D_m= 2.8, D_c= 2.62, VIIIE

07c. Glocker, 1827. Min. Mag. 42 (1978), 361. Min. Mag. 38 (1972), 693. See also Lennilenapeite.

Stipoverite, D = stishovite. Min. Mag. 36 (1967), 133. Grigoriev, 1962.

Stishovite, SiO_2, G Tet. Vitreous colorless. Canyon Diablo meteorite, Meteor Crater, Coconino Co., Arizona, USA. H=?, D_m= 4.28, D_c= 4.291, IVD 01. Chao et al., 1962. Am. Min. 47(1962), 807 (Abst.). Nature 272 (1978), 714. See also Quartz, Tridymite-Cristobalite, Coesite.

Stistaite, SnSb, A Rhom. Metallic grey. Elkiaidan river, Uzbekistan, USSR. H= 3, D_m= 6.91, D_c= 6.933, IIB 12. Nikolaeva et al., 1970. Min. Abstr. 81-4376.

Stoiberite, $Cu_5O_2(VO_4)_2$, A Mon. Metallic black; light grey in reflected light. Izalco Volcano, El Salvador. H=?, D_m=?, D_c= 4.96, VIIA 12. Birnie & Hughes, 1979. Am. Min. 64 (1979), 941. Acta Cryst. B29 (1973), 1338.

Stokesite, $CaSnSi_3O_9 \cdot 2H_2O$, G Orth. Translucent colorless. Roscommon Cliffs, St. Just, Cornwall, England. H= 6, D_m= 3.2, D_c= 3.192, VIIID 16. Hutchinson, 1899. Dana/Ford (1932), 633. Min. Mag. 33 (1963), 615.

Stolzite, β–$PbWO_4$, G Tet. Resinous/sub-adamantine brown, yellow, red, etc. Cínovec (Zinnwald), Krusné Hory (Erzgebirge), Čechy (Bohemia), Czechoslovakia. H= 2.5-3, D_m= 8.34, D_c= 8.41, VIF 01. Haidinger, 1845. NBS Circ. 539, v.7 (1957), 24. See also Raspite, Wulfenite.

Stottite, $FeGe(OH)_6$, G Tet. Greasy brown. Tsumeb, Namibia. H= 4.5, D_m= 3.59, D_c= 3.545, IVF 06. Strunz et al., 1958. Am. Min. 43(1958), 1006 (Abst.). Am. Min. 73 (1988), 657.

Straczekite, $(Ca,K,Ba)V_8O_{20} \cdot 3H_2O$, A Mon. Greasy greenish-black. Union Carbide mine, Wilson Springs, Garland Co., Arkansas, USA. H=?, D_m= 3.1, D_c= 3.21, IVF 09. Evans et al., 1984. Min. Mag. 48(1984), 289.

Stranskiite, $CuZn_2(AsO_4)_2$, G Tric. Translucent blue. Tsumeb, Namibia. H=?, D_m= 5.3, D_c= 5.1, VIIA 08. Strunz, 1960. Am. Min. 63 (1978), 213. Min. Abstr. 81-1243.

Strashimirite, $Cu_4(AsO_4)_2(OH)_2 \cdot 2.5H_2O$, A Mon. Pearly/greasy white, pale green. Zapachista deposit, Stara-Planina, Bulgaria. H=?, D_m=?, D_c= 3.81, VIID 02. Mincheva-Stefanova, 1968. Am. Min. 54(1969), 1221 (Abst.).

Strätlingite, $Ca_2Al_2SiO_7 \cdot 8H_2O$, A Rhom. Translucent colorless/light green. Ettringer Bellerberg, Mayen, Eifel, Rheinland-Pfalz, Germany. H=?, D_m= 1.9, D_c= 1.95, VIIIF 15. Hentschel & Kuzel, 1976. Am. Min. 62(1977), 395 (Abst.).

Strelkinite, $Na_2(UO_2)_2(VO_4)_2 \cdot 6H_2O$, A Orth. Silky/pearly yellow. Unspecified locality, USSR. H= 2-2.5, D_m= 4.1, D_c= 4.22, VIID 23. Alekseeva et al., 1974. Am. Min. 60(1975), 488 (Abst.).

Strengite, $FePO_4 \cdot 2H_2O$, G Orth. Vitreous red, violet. Eleonore mine, Giessen (near), Hessen, Germany. H= 3.5-4.5, D_m= 2.87, D_c= 2.90, VIIC 05. Nies, 1877. Dana, 7th ed.(1951), v.2, 756. See also Phosphosiderite, Variscite.

Stringhamite, $CaCuSiO_4 \cdot H_2O$, A Mon. Translucent deep blue. Bawana mine, Beaver Co., Utah, USA. H=?, D_m= 3.17, D_c= 3.359, VIIIA 12. Hindman, 1976. Am. Min. 61 (1976), 189. Min. Abstr. 85M/3792.

Stromeyerite, CuAgS, G Orth. Metallic grey; greyish-white in reflected light. Smeinogorsk mine, Kolyvan (near), Siberia, USSR. H= 2.5-3, D_m= 6.2, D_c= 6.26, IIA 04. Beudant, 1832. Dana, 7th ed.(1944), v.1, 190. Struct. Repts. 19 (1953), 412.

Stronalsite, $Na_2SrAl_4Si_4O_{16}$, A Orth. Vitreous white. Kochi, Rendai, Japan. H= 6.5, D_m= 2.95, D_c= 2.943, VIIIF 02. Hori et al., 1987. Min. Jour. 13 (1987), 368.

Strontianite, $SrCO_3$, G Orth. Vitreous/resinous colorless, grey, yellowish, greenish. Strontian, Argyllshire, Scotland. H= 3.5, D_m= 3.76, D_c= 3.804, VbA 04. Sulzer, 1790. Am. Min. 61 (1976), 1001. Bull. Min. 111 (1988), 139.

Strontioborite, $SrB_8O_{11}(OH)_4$, G Mon. Caspian region, USSR. H=?, D_m= 2.40, D_c= 2.38, Vc 12b. Lobanova, 1960. Am. Min. 46(1961), 768 (Abst.). Sov.Phys.Cryst. 20(1975), 563.

Strontio-chevkinite, $(Sr, Ce, La)_4Fe(Ti, Zr)_4O_8(Si_2O_7)_2$, A Mon. Sub-metallic opaque; grey in reflected light. Concepcion and Amambay, Sarambi Carbontite Complex, Paraguay. H=?, D_m=?, D_c= 5.44, VIIIB 16. Haggerty & Mariano, 1983. Am. Min. 69(1984), 1192 (Abst.). See also Chevkinite.

Strontiodresserite, $(Sr, Ca)Al_2(CO_3)_2(OH)_4 \cdot H_2O$, A Orth. Vitreous/silky white. Francon quarry, St.-Michel dist., Montreal Island, Québec, Canada. H=?, D_m= 2.71, D_c= 2.73, VbD 02. Jambor et al., 1977. Min. Abstr. 80-0189. See also Dresserite.

Strontioginorite, $SrCaB_{14}O_{23} \cdot 8H_2O$, Q Mon. Göttingen (near), Reyershausen, Hanover, Niedersachsen, Germany. H= 3-4, D_m=?, D_c= 2.258, Vc 08. Nefedov, 1953. Strunz (1970), 260. See also Ginorite.

Strontiohilgardite, D Name discarded in favor of strontian tyretskite-1Tc. Am. Min. 69(1984), 214 (Abst.). Braitsch, 1959.

Strontiohilgardite-1Tc, D = strontian hilgardite-1Tc. Am. Min. 70 (1985), 636. Braitsch, 1959.

Strontiojoaquinite, $(Na, Fe)_2Ba_2Sr_2Ti_2(SiO_3)_8(O, OH)_2 \cdot H_2O$, A Mon. Translucent green, yellow-green, yellow-brown. Numero Uno mine, San Benito Co., California, USA. H= 5.5, D_m=?, D_c= 3.68, VIIIC 05. Wise, 1982. Am. Min. 67 (1982), 809. See also Strontio-orthojoaquinite.

Strontio-orthojoaquinite, $Na_2Ba_2Sr_2Ti_2(SiO_3)_8(O, OH)_2 \cdot H_2O$, A Orth. Translucent yellow. San Benito Co., California, USA. H= 5.5, D_m= 3.62, D_c= 3.87, VIIIC 05. Wise, 1982. Am. Min. 67 (1982), 809. See also Strontiojoaquinite, Orthojoaquinite.

Strontiopyrochlore, $Sr_{0.6}Nb_2(O, OH)_7$, Q Cub. Translucent yellow. Yeniseĭ Ridge, USSR. H= 4, D_m= 3.80, D_c= 3.99, IVC 09a. Lapin et al., 1986. Am. Min. 73(1988), 930 (Abst.).

Strontium-apatite, $(Sr, Ca)_5(PO_4)_3F$, G Hex. Vitreous, colorless/pale green. Inaglya massif, Yakutiya (S), USSR. H= 5, D_m= 3.84, D_c= 3.95, VIIB 16. Efimov et al., 1962. Am. Min. 47(1962), 808 (Abst.). Sov.Phys.Cryst. 32 (1987), 524. See also Belovite, Apatite.

Strontiumthomsonite, D Unnecessary name for strontian thomsonite. Min. Mag. 36 (1968), 1144. Efimov et al., 1963.

Strunzite, $MnFe_2(PO_4)_2(OH)_2 \cdot 6H_2O$, G Tric. Translucent yellow, brownish-yellow. Hagendorf pegmatite, Oberpfalz, Bayern (Bavaria), Germany. H=?, D_m= 2.52, D_c= 2.581, VIID 05. Frondel, 1958. Am. Min. 43(1958), 793 (Abst.). Min. Abstr. 81-1246. See also Laueite, Stewartite, Ferrostrunzite.

Strüverite, $(Ti, Ta, Fe)O_2$, Q Tet. Black. Craveggia, Piemonte, Italy. H=?, D_m= 5.25, D_c= 5.69, IVD 02. Zambonini, 1907. JCPDS 17-543. See also Ilmenorutile.

Struvite, $(NH_4)MgPO_4 \cdot 6H_2O$, G Orth. Vitreous colorless, yellowish, brown. Hamburg, Germany. H= 2, D_m= 1.711, D_c= 1.70, VIIC 14. Ulex, 1846. Dana, 7th ed,(1951), v.2, 715. Acta Cryst. B42 (1986), 253.

Studtite, $UO_4 \cdot 4H_2O$, G Mon. Translucent yellow. Shinkolobwe, Shaba, Zaïre. H=?, D_m= 3.58, D_c= 3.64, IVF 11. Vaes, 1947. Am. Min. 59 (1974), 166.

Stumpflite, $Pt(Sb, Bi)$, A Hex. Metallic cream. Driekop, Transvaal, South Africa. H= 4.9, D_m=?, D_c= 13.52, IIB 09a. Johan & Picot, 1972. Am. Min. 59(1974), 211 (Abst.).

Sturmanite, $Ca_6Fe_2(SO_4)_2[B(OH)_4](OH)_{12} \cdot 25H_2O$, A Rhom. Vitreous bright yellow. Kuruman (near), Cape Province, South Africa. H= 2.5, D_m= 1.847, D_c= 1.855, VID 07. Peacor et al., 1983. Can. Min. 21 (1983), 705.

Sturtite, $(Mn,Al,Fe,Ca)_3Si_4O_{10}(OH)_3 \cdot H_2O$, Q Amor. Broken Hill, New South Wales, Australia. H=?, D_m=?, D_c=?, VIIIE 07a. Hodge-Smith, 1930. Am. Min. 69(1984), 215 (Abst.).

Stützite, $Ag_{5-x}Te_3$, G Hex. Metallic grey. Săcărâmb (Nagyág), Transylvania, Romania. H= 3.5, D_m= 8.00, D_c= 8.18, IIA 03. Schrauf, 1878. Am. Min. 50 (1965), 795.

Suanite, $Mg_2B_2O_5$, G Mon. Silky/pearly white. Suan, Korea. H= 5.5, D_m= 2.91, D_c= 2.91, Vc 02. Watanabe, 1953. Am. Min. 40(1955), 941 (Abst.). Acta Cryst. 5 (1952), 574.

Sudburyite, (Pd, Ni)Sb, A Hex. Metallic yellowish-white. Copper Cliff South mine, Sudbury dist, Ontario, Canada. H= 4-4.5, D_m=?, D_c= 9.41, IIB 09a. Cabri & Laflamme, 1974. Can. Min. 12 (1974), 275. Min. Abstr. 80-1317.

Sudoite, $Mg_2(Al,Fe)_3(Si_3Al)O_{10}(OH)_8$, G Mon. Knollenbergel-Keuper, Lützelbach, Plochingen, Germany. H=?, D_m=?, D_c= 2.63, VIIIE 09b. v. Engelhardt et al., 1962. Clays Cl. Mins. 37 (1989), 193. Min. Journ. 8 (1976), 158.

Suessite, Fe_3Si, A Cub. Metallic creamy-white. North Haig meteorite, Sleeper Camp (N), Haig(N), Western Australia, Australia. H=?, D_m=?, D_c= 7.08, IC 03. Keil et al., 1982. Am. Min. 67 (1982), 126.

Sugilite, $KNa_2Li_3(Fe,Mn,Al)_2Si_{12}O_{30}$, A Hex. Vitreous light brownish-yellow. Ehime, Iwagi Island, Shikoku, Japan. H= 6-6.5, D_m= 2.74, D_c= 2.80, VIIIC 10. Murakami et al., 1976. Can. Min. 18 (1980), 37. Am. Min. 73 (1988), 595.

Sulfoborite, $Mg_3[B(OH)_4]_2(SO_4)(OH,F)_2$, G Orth. Transparent colorless. Bücking, Westeregeln, Prussia, Germany. H= 4-4.5, D_m= 2.4, D_c= 2.425, Vc 03c. Naupert & Wense, 1893. Min. Abstr. 78-253. Am. Min. 68 (1983), 255.

Sulphate-monazite, D Unnecessary name for sulfatic monazite. Min. Mag. 36 (1967), 133. Kukharenko et al., 1961.

Sulphohalite, $Na_6(SO_4)_2ClF$, G Cub. Vitreous/greasy colorless, pale greenish, yellow, grey. Searles Lake, San Bernardino Co., California, USA. H= 3.5, D_m= 2.50, D_c= 2.50, VIB 04. Hidden & Mackintosh, 1888. Min. Mag. 40 (1975), 131. Struct. Repts. 33A (1968), 377.

Sulphotsumoite, Bi_3Te_2S, A Rhom. Metallic grey-white. Magadan and Egerlyakh deposits, Yakutiya, USSR. H= 2.2, D_m=?, D_c= 8.13, IIB 18. Zav'yalov & Begizov, 1982. Am. Min. 68(1983), 1250 (Abst.). See also Tsumoite.

Sulphur, α–S, G Orth. Translucent resinous yellow. H= 1.5-2.5, D_m= 2.07, D_c= 2.066, IB 04. Dana, 7th ed.(1944), v.1, 140. Acta Cryst. C43 (1987), 2260. See also Rosickyite.

Sulunite, D Inadequate data. Min. Mag. 33 (1962), 261. Nyrkov, 1959.

Sulvanite, Cu_3VS_4, G Cub. Metallic cream-gold. Burra (near), South Australia, Australia. H= 3.5, D_m= 3.9, D_c= 3.918, IIB 04. Goyder, 1900. Am. Min. 59 (1974), 307. Am. Min. 51 (1966), 890. See also Arsenosulvanite.

Sundiusite, D Hypothetical amphibole end-member. Min. Mag. 36 (1968), 1144. Phillips & Layton, 1964.

Sundiusite, $Pb_{10}(SO_4)O_8Cl_2$, A Mon. Adamantine colorless, white. Långban mine, Filipstad (near), Värmland, Sweden. H= 3, D_m= 7.0, D_c= 7.20, IIID 04. Dunn & Rouse, 1980. Am. Min. 65 (1980), 506.

Sungulite, D A mixture of lizardite and sepiolite. Am. Min. 59 (1974), 212 (Abst.). Sokolov, 1925.

Suolunite, $Ca_2Si_2O_5(OH)_2 \cdot H_2O$, G Orth. Vitreous/resinous white. Suolun, Inner Mongolia, China. H=?, D_m=?, D_c= 5.39, VIIIB 06. Huang, 1965. Am. Min. 53(1968), 349 (Abst.). Min. Abstr. 75-871.

Surinamite, $(Mg,Fe)_3BeAl_4Si_3O_{16}$, A Mon. Translucent blue-green. Bakhuis Mts., Surinam (S). H=?, D_m= 3.58, D_c= 3.43, VIIIA 15. de Roever et al., 1976. Am. Min. 61 (1976), 193. Am. Min. 68 (1983), 804.

Surite, $Pb_2Al_2(Si,Al)_4O_{10}(CO_3)_2(OH)_2$, [A] Mon. Glossy white. Cruz del Sur mine, Rio Negro, Argentina. H= 2-3, D_m= 4.0, D_c= 3.91, VIIIE 23. Hayase et al., 1978. Am. Min. 63 (1978), 1175.

Sursassite, $Mn_2Al_3(SiO_4)(Si_2O_7)(OH)_3$, [G] Mon. Translucent copper-red. Alp Parsettens, Sursass (Oberhalbstein), Tinizong, (Tinzen), Grischun (Graubünden), Switzerland. H=?, D_m= 3.252, D_c= 3.57, VIIIB 19. Jakob, 1926. Min. Abstr. 3 (1927), 272. Am. Min. 70(1985), 221 (Abst.). See also MacFallite.

Susannite, $Pb_4(SO_4)(CO_3)_2(OH)_2$, [G] Rhom. Translucent colorless, greenish, yellowish. Susanna mine, Leadhills, Lanarkshire, Scotland. H=?, D_m=?, D_c= 6.534, VbB 03. Brooke, 1827. Min. Mag. 49 (1985), 759. See also Leadhillite, Macphersonite.

Sussexite, $MnBO_2(OH)$, [G] Orth. Silky/earthy white, buff, straw-yellow. Franklin Hill mine, Sussex Co., New Jersey, USA. H= 3-3.5, D_m= 3.30, D_c= 3.43, Vc 02. Brush, 1868. SMPM 39 (1959), 85. See also Szaibelyite.

Suzukiite, $BaVSi_2O_7$, [A] Orth. Vitreous green. Mogurazawa mine, Gumma, Japan. H= 4-4.5, D_m= 4.0, D_c= 4.03, VIIID 21. Matsubara et al., 1982. Am. Min. 68(1983), 282 (Abst.). See also Haradaite.

Svabite, $Ca_5(AsO_4)_3F$, [G] Hex. Vitreous/sub-resinous colorless, yellowish-white, grey. Harstigen mine, Pajsberg, Värmland, Sweden. H= 4-5, D_m= 3.7, D_c= 3.708, VIIB 16. Sjögren, 1891. Strunz (1970), 327.

Svanbergite, $SrAl_3(SO_4)(PO_4)(OH)_6$, [G] Rhom. Vitreous/adamantine colorless, yellow, rose, reddish-brown. Horrsjöberg, Värmland, Sweden. H= 5, D_m= 3.22, D_c= 3.280, VIB 03b. Igelström, 1854. Dana, 7th ed.(1951), v.2, 1005. Min. Jour. 8 (1977), 419.

Sveite, $KAl_7(NO_3)_4(OH)_{16}Cl_2 \cdot 8H_2O$, [A] Mon. Translucent white. Autana Cave, Amazonas, Venezuela. H=?, D_m= 2.0, D_c= 2.185, Va 04. Martini, 1980. Am. Min. 67(1982), 1076 (Abst.).

Sverigeite, $NaBe_2(Mn,Mg)_2SnSi_3O_{12}(OH)$, [A] Orth. Vitreous yellow. Långban mine, Filipstad (near), Värmland, Sweden. H= 6.5, D_m= 3.60, D_c= 3.61, VIIID 24. Dunn et al., 1984. Am. Min. 70(1985), 1332 (Abst.). Am. Min. 74 (1989), 1343.

Svetlozarite, $(Ca,K,Na)_3(Si,Al)_{24}O_{48} \cdot 12H_2O$, [A] Mon. Vitreous/pearly colorless, white. Zvezdel, Rhodope Mts., Bulgaria. H= 4, D_m= 2.166, D_c= 1.99, VIIIF 12. Maleev, 1976. Am. Min. 62(1977), 1060 (Abst.). Min. Mag. 45 (1982), 157.

Svyatoslavite, $CaAl_2Si_2O_8$, [A] Orth. Vitreous colorless. Chelyabinsk, Ural Mts. (S), USSR. H= 6, D_m= 2.695, D_c= 2.687, VIIIF 02. Chesnokov et al., 1989. ZVMO 118 (1989), 111.

Svyazhinite, $(Mg,Mn)(Al,Fe)(SO_4)_2F \cdot 14H_2O$, [A] Tric. Translucent yellowish, colorless. Ilmen Mts., Ural Mts., USSR. H=?, D_m=?, D_c= 1.69, VID 09. Chesnokov et al., 1984. Am. Min. 70 (1985), 877. See also Aubertite.

Swamboite, $H_6U(UO_2)_6(SiO_4)_6 \cdot 30H_2O$, [A] Mon. Translucent pale yellow. Swambo deposit, Shaba, Zaïre. H=?, D_m= 4.0, D_c= 4.064, VIIIA 25. Deliens & Piret, 1981. Can. Min. 19 (1981), 553.

Swartzite, $CaMg(UO_2)(CO_3)_3 \cdot 12H_2O$, [G] Mon. Dull whitish-yellow. Hillside mine, Bagdad, Yavapai Co., Arizona, USA. H=?, D_m= 2.3, D_c= 2.356, VbD 04. Axelrod et al., 1951. Am. Min. 36 (1951), 1. Min. Abstr. 87M /2145.

Swedenborgite, $NaBe_4SbO_7$, [G] Hex. Transparent colorless, yellow. Långban mine, Filipstad (near), Värmland, Sweden. H= 8, D_m= 4.28, D_c= 4.28, IVB 04. Aminoff, 1924. JCPDS 23-656.

Sweetite, $Zn(OH)_2$, [A] Tet. Translucent white. Milltown (NW of), Ashover (near), Derbyshire, England. H=?, D_m= 3.33, D_c= 3.41, IVF 03a. Clark et al., 1984. Min. Mag. 48 (1984), 267. See also Wülfingite.

Swinefordite, $(Li,Ca)(Al,Li,Mg)_5Si_8O_{20}(OH,F)_4$, [A] Mon. Translucent light greenish-grey, greyish-olive. Foote Mineral Company spodumene mine, Kings Mt.

(near), Cleveland Co., North Carolina, USA. H= 1, D_m=?, D_c= 2.02, VIIIE 08a. Tien et al., 1975. Am. Min. 60 (1975), 540.

Switzerite, $(Mn,Fe)_3(PO_4)_2 \cdot 7H_2O$, A Mon. Vitreous/pearly pale pink. Foote Mineral Company spodumene mine, Kings Mt. (near), Cleveland Co., North Carolina, USA. H=?, D_m= 2.535, D_c= 2.562, VIIC 07. Leavens & White, 1967. Am. Min. 71 (1986), 1221. Am. Min. 71 (1986), 1224.

Sylvanite, $AgAuTe_4$, G Mon. Metallic grey/white; creamy-white in reflected light. Baia de Arieş (Offenbánya) and Săcărâmb (Nagyág), Transylvania, Romania. H= 1.5-2, D_m= 8.16, D_c= 8.17, IIC 04. Necker, 1835. Dana, 7th ed.(1944), v.1, 338. Min. Abstr. 85M/2406. See also Kostovite.

Sylvite, KCl, G Cub. Vitreous colorless, white, grey, yellow, red. Mt. Vesuvius, Napoli, Campania, Italy. H= 2, D_m= 1.993, D_c= 1.99, IIIA 02. Beudant, 1832. Min. Mag. 29 (1951), 667.

Symplesite, $Fe_3(AsO_4)_2 \cdot 8H_2O$, G Mon. Vitreous/pearly green, greenish-black, blue. Lobenstein, Germany. H= 2.5, D_m= 3.01, D_c= 3.09, VIIC 10. Breithaupt, 1837. Dana, 7th ed.(1951), v.2, 752. Struct. Repts. 13 (1950), 307. See also Parasymplesite, Metaköttigite, Metavivianite.

Synadelphite, $(Mn,Mg,Ca)_9(AsO_4)_2(AsO_3)(OH)_9 \cdot 2H_2O$, G Orth. Vitreous colorless, red, brown. Moss mine, Nordmark, Sweden. H= 4.5, D_m= 3.57, D_c= 3.59, VIID 31. Sjögren, 1884. Dana, 7th ed.(1951), v.2, 780. Am. Min. 55 (1970), 2023.

Synchysite-(Ce), $Ca(Ce,La)(CO_3)_2F$, G Orth. Vitreous/greasy yellow, brown. Narsarsuk, Greenland (S). H= 4.5, D_m= 4.1, D_c= 3.99, VbB 04. Flink, 1900. Am. Min. 38 (1953), 932. Am. Min. 38 (1953), 932.

Synchysite-(Nd), $Ca(Nd,La)(CO_3)_2F$, A Orth. Dull translucent light greyish-blue. Grebnik Bauxite deposit, Srbija (Serbia), Yugoslavia. H= 1, D_m=?, D_c= 4.14, VbB 04. Scharm & Kühn, 1983. Min. Abstr. 85M/0850.

Synchysite-(Y), $Ca(Y,Ce)(CO_3)_2F$, A Orth. Translucent brownish-red. Scrub Oaks mine, Dover, New Jersey, USA. H=?, D_m=?, D_c= 3.40, VbB 04. Levinson, 1966. Am. Min. 51 (1966), 152.

Syngenite, $K_2Ca(SO_4)_2 \cdot H_2O$, G Mon. Vitreous colorless, pale yellow, white. Kalusz, Galicia, Poland. H= 2.5, D_m= 2.60, D_c= 2.606, VIC 15. Zepharovich, 1872. Dana, 7th ed.(1951), v.2, 4421. Sov.Phys.Cryst. 23(1978), 271. See also Koktaite.

Szaibelyite, $MgBO_2(OH)$, G Mon. Silky/earthy white, buff, straw-yellow. Băiţa Bihorului (Rézbánya), Romania. H= 3-3.5, D_m= 2.62, D_c= 2.738, Vc 02. Peters, 1861. Dana, 7th ed.(1951), v.2, 375. Am. Min. 60 (1975), 273. See also Sussexite.

Szmikite, $MnSO_4 \cdot H_2O$, G Mon. Translucent dirty white, reddish, rose-red. Baia Mare (Felsöbánya), Romania. H= 1.5, D_m= 3.15, D_c= 2.943, VIC 01. Schröckinger, 1877. Per. Mineral. 54 (1985), 32.

Szomolnokite, $FeSO_4 \cdot H_2O$, G Mon. Vitreous yellow, reddish-brown, blue, colorless. Smolnik (Szomolnok), Slovensko (Slovakia), Czechoslovakia. H= 2.5, D_m= 3.05, D_c= 3.08, VIC 01. Krenner, 1891. Per. Min. 54 (1985), 32.

Sztrokayite, Bi_3TeS_2, Q Syst=? Metallic. Nagyborzsony, Hungary (N). H=?, D_m=?, D_c=?, IIB 18. Nagy, 1983. Am. Min. 72(1987), 1027 (Abst.).

T

Taaffeite-8H, $Mg_3BeAl_8O_{16}$, G Hex. Transparent mauve. Unknown locality, Sri Lanka (?). H=?, D_m= 3.613, D_c= 3.579, IVC 05c. Anderson & Claringbull, 1951. Am. Min. 37(1952), 360 (Abst.). Min. Abstr. 85M/0172.

Tacharanite, $Ca_{12}Al_2Si_{18}O_{33}(OH)_{36}$, G Mon. Translucent white. Portree, Isle of Skye, Scotland. H=?, D_m= 2.36, D_c= 2.28, VIIID 10. Sweet, 1961. Min. Mag. 40 (1975), 113.

Tachyhydrite, $CaMg_2Cl_6 \cdot 12H_2O$, G Rhom. Vitreous colorless, yellow. Stassfurt salt mines, Germany. H= 2, D_m= 1.667, D_c= 1.673, IIIB 08. Rammelsberg, 1856. Dana, 7th ed.(1951), v.2, 95. Acta Cryst. B36 (1980), 2736.

Tadzhikite-(Ce), $Ca_3(Ce,Y)_2(Ti,Al,Fe)B_4Si_4O_{22}$, A Mon. Vitreous greyish-brown, dark brown. Turkestana, Tadzhikistan, USSR. H= 6, D_m= 3.72, D_c= 3.77, VIIID 23. Efimov et al., 1970. Am. Min. 56(1971), 1838 (Abst.). Min. Abstr. 83M/4211.

Taeniolite, $KLiMg_2Si_4O_{10}F_2$, G Mon. Vitreous greenish-brown. Narsarsuk, Greenland (S). H= 3.5, D_m= 2.85, D_c= 2.84, VIIIE 05b. Flink, 1900. Min. Rec. 14 (1983), 39. Zts. Krist. 146 (1983), 39.

Taenite, γ-(Ni, Fe), G Cub. Metallic white. Gorge River, South Island, New Zealand. H=?, D_m=?, D_c= 8.19, IA 05. Reichenbach, 1861. Am. Min. 51 (1966), 37. See also Kamacite, Tetrataenite.

Tagilite, D = pseudomalachite. Am. Min. 73(1988), 935 (Abst.). Breithaupt, 1841.

Taikanite, $(Sr,Ba)_4Mn_2(Si_2O_7)_2$, A Mon. Vitreous/greasy greenish-black. Taikan Mts., USSR (Far E). H= 6.5, D_m= 4.72, D_c= 4.81, VIIIB 10. Kalinin et al., 1985. Am. Min. 72(1987), 226 (Abst.).

Taimyrite I, $(Pd,Cu,Pt)_3Sn$, Q Orth. Metallic bronze-grey. Talnakh, Noril'sk (near), Taimyr Peninsula, Siberia (N), USSR. H= 5, D_m=?, D_c=?, IA 10. Begizov et al., 1982. ZVMO 111 (1982), 78. See also Taimyrite II.

Taimyrite II, $(Pd,Cu,Pt)_3Sn$, Q Orth. Metallic bronze-grey. Talnakh , Noril'sk (near), Taimyr Peninsula, Siberia (N), USSR. H= 5, D_m=?, D_c=?, IA 10. Begizov et al., 1982. ZVMO 111 (1982), 78. See also Taimyrite I.

Takanelite, $(Mn,Ca)Mn_4O_9 \cdot H_2O$, A Hex. Sub-metallic/dull grey, black. Nomura mine, Ehime, Japan. H= 5, D_m= 3.41, D_c= 3.78, IVF 05b. Nambu & Tanida, 1971. Am. Min. 56(1971), 1487 (Abst.). See also Ranciéite.

Takeuchiite, $(Mg,Mn)_3BO_5$, A Orth. Metallic black. Långban mine, Filipstad (near), Värmland, Sweden. H= 6, D_m=?, D_c= 3.93, Vc 01c. Bovin & O'Keeffe, 1980. Am. Min. 65 (1980), 1130. Zts. Krist. 181 (1987), 135. See also Fredrikssonite, Orthopinakiolite, Pinakiolite.

Takovite, $Ni_6Al_2CO_3(OH)_{16} \cdot 4H_2O$, A Rhom. Translucent yellowish-green, bluish-green. Takovo, Srbija (Serbia), Yugoslavia. H=?, D_m= 2.80, D_c= 2.947, IVF 03b. Maksimovic, 1957. Am. Min. 62 (1977), 458.

Talc, $Mg_3Si_4O_{10}(OH)_2$, G Tric. Translucent colorless, white, pale green, brown. H= 1, D_m= 2.7, D_c= 2.798, VIIIE 04. DHZ (1962), v.3, 121. Zts. Krist. 156 (1981), 177. See also Minnesotaite, Willemseite.

Talmessite, $Ca_2Mg(AsO_4)_2 \cdot 2H_2O$, A Tric. Translucent pale green. Talmessi mine, Anarak, Iran. H=?, D_m= 3.57, D_c= 3.42, VIIC 12. Bariand & Herpin, 1960. Am. Min. 50(1965), 813 (Abst.). Bull. Min. 100 (1977), 230. See also Gaitite.

Talnakhite, $Cu_9(Fe,Ni)_8S_{16}$, A Cub. Metallic yellow. Talnakh, Noril'sk (near), Siberia (N), USSR. H=?, D_m= 4.24, D_c= 4.36, IIB 02. Bud'ko & Kulagov, 1968. Am. Min. 55(1970), 2135 (Abst.). Can. Min. 13 (1975), 168.

Tamarugite, $NaAl(SO_4)_2 \cdot 6H_2O$, [G] Mon. Vitreous colorless. Cerro Pintados, Iquique, Tarapacá, Chile. H= 3, D_m= 2.06, D_c= 2.066, VIC 07. Schulze, 1889. N. Jb. Min. Mh. (1987), 171. Am. Min. 54 (1969), 19.

Tancoite, $HNa_2LiAl(PO_4)_2(OH)$, [A] Orth. Vitreous colorless, pale pink. Tanco mine, Bernic Lake, Lac-du-Bonnet (ENE), Manitoba, Canada. H= 4-4.5, D_m= 2.752, D_c= 2.777, VIIB 02. Ramik et al., 1980. Can. Min. 18 (1980), 185. Am. Min. 69(1984), 215 (Abst.).

Taneyamalite, $(Na,Ca)(Mn,Mg)_{12}(Si,Al)_{12}(O,OH)_{44}$, [A] Tric. Vitreous greenish grey-yellow. Iwaizawa mine, Saitama, Kumanto, Toyo, Japan. H= 5, D_m=?, D_c= 3.30, VIIID 20. Matsubara, 1981. Min. Journ. 10 (1981), 385. See also Howieite.

Tangenite, [D] Mixture. Am. Min. 62 (1977), 403. Gagarin & Cuomo, 1949.

Tantal-aeschynite-(Y), $(Y,Ce)(Ta,Ti,Nb)_2O_6$, [A] Orth. Resinous brownish-black, black. Raposa pegmatite, São José do Sabugí, Paraíba, Brazil. H= 5.5-6, D_m= 5.9, D_c=?, IVD 11. Adusumilli et al., 1974. Min. Mag. 39 (1974), 571. See also Aeschynite-(Y).

Tantalcarbide, TaC, [G] Cub. Metallic bronze. Nizhniĭ Tagilsk, Ural Mts., USSR. H= 6-7, D_m= 14.5, D_c= 14.47, IC 01. Strunz, 1966. Am. Min. 47 (1962), 786. Struct. Repts. 26 (1961), 104.

Tantalite, $(Fe,Mn)Ta_2O_6$, [G] Orth. Tammela, Finland. H=?, D_m=?, D_c=?, IVD 10a. Ekeberg, 1802. Can. Min. 14 (1976), 540. See also Ferrotantalite, Manganotantalite.

Tantalo-obruchevite, [D] Natural occurrence not proven. Am. Min. 62 (1977), 403. Van der Veen, 1963.

Tantalum, [D] Natural occurrence not substantiated. Am. Min. 47 (1962), 786. Walther, 1909.

Tanteuxenite-(Y), $(Y,Ce,Ca)(Ta,Nb,Ti)_2(O,OH)_6$, [G] Orth. Resinous brownish-black. Pilbara, Western Australia, Australia. H= 5-6, D_m= 5.7, D_c=?, IVD 10b. Simpson, 1928. Strunz (1970), 206. See also Euxenite-(Y), Yttrocrasite-(Y).

Tantite, Ta_2O_5, [A] Tric. Adamantine colorless. Kola Peninsula, USSR. H= 7, D_m=?, D_c= 8.45, IVE 01. Voloshin et al., 1983. Am. Min. 69(1984), 1193 (Abst.).

Tapiolite, $(Fe,Mn)(Ta,Nb)_2O_6$, [A] Tet. Kulmala farm (near), Sukula, Tammela, Finland. H=?, D_m=?, D_c=?, IVD 04. Nordenskiöld, 1863. Am. Min. 70(1985), 217 (Abst.). See also Ferrotantalite, Manganotapiolite, Staringite.

Taramellite, $Ba_4(Fe,Ti)_4B_2(Si_8O_{27})O_2Cl_x$, [G] Orth. Translucent reddish-brown. Candoglia, Novara, Piemonte, Italy. H= 5.5, D_m= 3.9, D_c= 4.08, VIIIC 04. Tacconi, 1908. Am. Min. 44(1959), 469 (Abst.). Am. Min. 65 (1980), 123. See also Titantaramellite, Nagashimalite.

Taramite, $Na_2Ca(Mg,Fe)_3Fe_2(Si_6Al_2)O_{22}(OH)_2$, [A] Mon. Walli–Tarama Valley, Mariupol (near), Ukraine, USSR. H=?, D_m=?, D_c=?, VIIID 05c. Morozewicz, 1923. Am. Min. 63 (1978), 1023. See also Magnesiotaramite, Alumino-taramite.

Taranakite, $H_6K_3(Al,Fe)_5(PO_4)_8 \cdot 18H_2O$(?), [G] Rhom. Powdery white. Sugar Loaves, Taranaki, New Zealand. H=?, D_m= 2.06, D_c= 2.11, VIIC 01. Hector & Skey, 1865. Am. Min. 60 (1975), 331. Am. Min. 61 (1976), 329.

Tarapacáite, K_2CrO_4, [G] Orth. Transparent yellow. Tarapacá, Chile. H=?, D_m= 2.74, D_c= 2.736, VIE 01. Raimondi, 1878. Dana, 7th ed.(1951), v.2, 644. Acta Cryst. 28B (1972), 2845.

Tarasovite, [D] Irregular interstratification of mica and smectite. Am. Min. 67(1982), 397 (Abst.). Lazarenko & Korolev, 1970.

Tarbuttite, $Zn_2PO_4(OH)$, [G] Tric. Vitreous colorless, pale yellow, brown, red, green. Kabwe (Broken Hill), Central province, Zambia. H= 3.7, D_m= 4.19, D_c= 4.21, VIIB 04. Spencer, 1907. Dana, 7th ed.(1951), v.2, 869. Zts. Krist. 123 (1966), 321.

Tatarskite, $Ca_6Mg_2(SO_4)_2(CO_3)_2Cl_4(OH)_4 \cdot 7H_2O$, G Syst=? Vitreous colorless, yellowish. Tatarske River, Siberia, USSR. H= 2.5, D_m= 2.341, D_c=?, VID 09. Lobanova, 1963. Am. Min. 49(1964), 1151 (Abst.).

Tausonite, $SrTiO_3$, A Cub. Adamantine red, reddish-brown, grey. Alden Muranski massif, Olekminsk (near), Yakutiya, USSR. H= 6-6.5, D_m= 4.88, D_c= 4.85, IVC 07. Vorob'ev et al., 1984. Am. Min. 70(1985), 218 (Abst.).

Tavorite, $LiFePO_4(OH)$, G Tric. Translucent greenish-yellow. Sapucaia pegmatite, Galileia, Minas Gerais, Brazil. H=?, D_m= 3.29, D_c= 3.33, VIIB 02. Lindberg & Pecora, 1955. Am. Min. 40 (1955), 952. Min. Abstr. 85M/0191.

Tazheranite, $(Zr,Ti,Ca)_4O_6$, A Cub. Adamantine/greasy yellowish-orange, reddish-orange. Tazheranskii Alkaline Massif, Lake Baikal, Eastern Siberia, USSR. H= 7.5, D_m= 5.01, D_c= 4.83, IVD 16a. Konev et al., 1969. Am. Min. 55(1970), 318 (Abst.).

Teallite, $PbSnS_2$, G Orth. Metallic greyish-black; white in reflected light. Santa Rosa mine, Bolivia. H= 1.5, D_m= 6.36, D_c= 6.570, IIB 12. Prior, 1904. USGS Prof.Pap. 750 (1961), 347.

Teepleite, $Na_2B(OH)_4Cl$, G Tet. Vitreous colorless, white. Searles Lake, Lake Co., California, USA. H= 3-3.5, D_m= 2.076, D_c= 2.070, Vc 03a. Gale et al., 1938. Dana, 7th ed.(1951), v.2, 372. Acta Cryst. B38 (1982), 82.

Teineite, $CuTeO_3 \cdot 2H_2O$, G Orth. Translucent blue. Teine mine, Japan. H= 2.5, D_m= 3.80, D_c= 3.864, VIG 01. Yoshimura, 1939. Am. Min. 46(1961), 466 (Abst.). Min. Abstr. 78-1502. See also Chalcomenite.

Telargpalite, $(Pd,Ag)_3Te$, A Cub. Metallic grey. Oktyabr deposit, Talnakh, Noril'sk (near), Siberia (N), USSR. H= 2-2.5, D_m=?, D_c= 12.05, IIA 14. Kovalenker et al., 1974. Am. Min. 66(1981), 1103 (Abst.).

Tellurantimony, Sb_2Te_3, A Rhom. Metallic pink/cream. Mattagami Lake mines, Mattagami, Galinée Twp., Abitibi Co., Québec, Canada. H= 2, D_m=?, D_c= 6.51, IIC 03. Thorpe & Harris, 1973. Can. Min. 12 (1973), 55. Min. Abstr. 88M/1826. See also Tellurobismuthite.

Tellurite, TeO_2, G Orth. Sub-adamantine white, yellow. H= 2, D_m= 5.90, D_c= 5.749, IVD 06. Nicol, 1849. Dana, 7th ed.(1944), v.1, 593. Zts. Krist. 124 (1967), 228. See also Paratellurite.

Tellurium, Te, G Rhom. Metallic white. Faťa Băii (Faczebaja), Transylvania, Romania. H= 2-2.5, D_m= 6.2, D_c= 6.236, IB 03. Haüy, 1822. Dana, 7th ed.(1944), v.1, 138. Acta Cryst. 23 (1967), 670. See also Selen-tellurium.

Tellurobismuthite, Bi_2Te_3, G Rhom. Metallic grey; white in reflected light. Field's vein, Dahlonega, Lumpkin Co., Georgia and/or Little Mildred mine, Hildalgo Co., New Mexico, USA. H= 1.5-2, D_m= 7.81, D_c= 7.882, IIC 03. Balch, 1863. Dana, 7th ed.(1944), v.1, 160. Min. Abstr. 88M/1826. See also Tellurantimony.

Tellurohauchecornite, Ni_9BiTeS_8, A Tet. Metallic bronze. Strathcona mine, Sudbury dist., Ontario, Canada. H= 4-6, D_m= 6.50, D_c= 6.393, IIC 13. Gait & Harris, 1980. Min. Mag. 43 (1980), 877.

Telluropalladinite, Pd_9Te_4, A Mon. Metallic cream. Stillwater Complex, Montana, USA. H= 4-4.5, D_m= 10.25, D_c= 10.62, IIA 14. Cabri et al., 1979. Can. Min. 17 (1979), 589.

Temagamite, Pd_3HgTe_3, A Orth. Metallic white. Temagami mine, Temagami Island, Nipissing Dist., Ontario, Canada. H= 2.5, D_m= 9.5, D_c= 9.45, IIA 14. Cabri et al., 1973. Can. Min. 12 (1973), 193.

Tengchongite, $Ca(UO_2)_6(MoO_4)_2O_5 \cdot 12H_2O$, A Orth. Vitreous yellow. Tengchong Co., Yunnan, China. H= 2-2.5, D_m= 4.25, D_c= 4.24, VIF 02. Chen et al., 1986. Am. Min. 73(1988), 195 (Abst.).

Tengerite-(Y), $Y_2(CO_3)_3 \cdot 2H_2O$, [G] Orth. Dull, chalky white. Ytterby, Stockholm (near), Sweden. H=?, D_m=?, D_c= 3.110, VbC 03b. Dana, 1868. Min. Abstr. 75-3580. IMA 1986, Abstr. p. 173.

Tennantite, $(Cu,Fe)_{12}As_4S_{13}$, [G] Cub. Metallic grey/black; grey/olive-brown in reflected light. H= 3-4.5, D_m= 4.6, D_c= 4.61, IID 01a. Phillips & Phillips, 1819. Dana, 7th ed.(1944), v.1, 374. Zts. Krist. 123 (1966), 1. See also Tetrahedrite.

Tenorite, CuO, [G] Mon. Metallic grey/black; greyish-white in reflected light. Långban mine, Filipstad (near), Värmland, Sweden. H= 3.5, D_m= 6.4, D_c= 6.515, IVA 05. Semmola, 1841. Dana, 7th ed.(1944), v.1, 507. Acta Cryst. B26 (1970), 8.

Tephroite, Mn_2SiO_4, [G] Orth. Translucent olive-green, bluish-green, grey. Sterling Hill, Sussex Co., New Jersey, USA. H= 6, D_m= 3.9, D_c= 4.06, VIIIA 03. Breithaupt, 1823. DHZ, 2nd ed.(1982), v.1A, 337. Am. Min. 65 (1980), 1263. See also Fayalite.

Teremkovite, [D] = owyheeite. Min. Mag. 38 (1971), 103. Timofeevskii, 1967.

Terlinguaite, Hg_2OCl, [G] Mon. Adamantine yellow, greenish-yellow, brown. Terlingua, Brewster Co., Texas, USA. H= 2.5, D_m= 9.27, D_c= 9.31, IIIC 03. Turner, 1900. Dana, 7th ed.(1951), v.2, 52. Acta Cryst. 9 (1956), 956.

Terskite, $Na_4ZrSi_6O_{15}(OH)_2 \cdot H_2O$, [A] Orth. Vitreous pale lilac. Mt. Alluaiv and Mt. Karnasurt, Lovozero massif, Kola Peninsula, USSR. H= 5, D_m= 2.71, D_c= 2.74, VIIIG 01. Khomyakov et al., 1983. Am. Min. 69(1984), 212 (Abst.).

Tertschite, $Ca_4B_{10}O_{19} \cdot 20H_2O$, [G] Mon. Silky white. Kurtpinari mine, Sultancayin, Faras, Anatolia (W), Turkey. H=?, D_m=?, D_c=?, Vc 08. Meixner, 1953. Am. Min. 39(1954), 849 (Abst.).

Teruggite, $Ca_4Mg[AsB_6O_{11}(OH)_6]_2 \cdot 14H_2O$, [A] Mon. Vitreous colorless. Loma Blanca borate deposit, Jujuy, Argentina. H= 2.5, D_m= 2.20, D_c= 2.192, Vc 11. Aristarain & Hurlbut, 1968. Am. Min. 53 (1968), 1815. Am. Min. 58 (1973), 1034.

Teschemacherite, $(NH_4)HCO_3$, [G] Orth. Transparent colorless, white, yellowish. Patagonia region, Chile or Argentina(?). H= 1.5, D_m= 1.57, D_c= 1.552, VbA 01. Dana, 1868. Am. Min. 57 (1972), 1304. Min. Abstr. 84M/3850.

Testibiopalladite, Pd(Sb,Bi)Te, [Q] Cub. Metallic white. China (SW and NE). H= 3.5-4, D_m=?, D_c= 9.00, IIC 06b. Am. Min. 61(1976), 182 (Abst.). See also Maslovite, Michenerite.

Tetraauricupride, CuAu, [A] Tet. Metallic yellow. Sardala, Marneshi Co., Xinjiang Region, China. H= 4.5, D_m=?, D_c= 14.67, IA 01c. Chen et al., 1982. Am. Min.68(1983), 1250 (Abst.). See also Auricupride.

Tetradymite, $Bi_{14}Te_{13}S_8$, [G] Rhom. Metallic grey; white in reflected light. Narverud, Telemark, Norway. H= 1.5-2, D_m= 7.3, D_c= 7.208, IIC 03. Haidinger, 1831. Dana, 7th ed.(1944), v.1, 161. Am. Min. 60 (1975), 994.

Tetraferroplatinum, PtFe, [A] Tet. Metallic. Tulameen River, British Columbia, Canada. H=?, D_m= 14.3, D_c= 15.22, IA 12. Cabri & Feather, 1975. Can. Min. 13 (1975), 117.

Tetrahedrite, $(Cu,Fe)_{12}Sb_4S_{13}$, [G] Cub. Metallic grey/black; grey/olive-brown in reflected light. H= 3-4.5, D_m= 4.97, D_c= 4.99, IID 01a. Haidinger, 1845. Am. Min. 73 (1988), 389. Min. Mag. 50 (1986), 717. See also Tennantite, Freibergite.

Tetranatrolite, $(Na,K)_2(Si,Al)_5O_{10} \cdot 2H_2O$, [A] Tet. Vitreous/dull white. Ilímaussaq, Greenland and Mont Saint-Hilaire, Rouville Co., Québec, Canada. H=?, D_m= 2.272, D_c= 2.230, VIIIF 10. Chen & Chao, 1980. Can. Min. 18 (1980), 77. Zts. Krist. 189 (1989), 191. See also Natrolite.

Tetrataenite, FeNi, [A] Tet. Metallic creamy-white. Estherville, Santa Catharina, and 16 other meteorites. H= 3.5, D_m=?, D_c= 8.275, IA 05. Clarke & Scott, 1980. Am. Min. 65 (1980), 624. See also Kamacite, Taenite.

Tetrawickmanite, $MnSn(OH)_6$, [A] Tet. Translucent honey-yellow, brownish-orange. Foote Mineral Company spodumene mine, Kings Mt. (near),

Cleveland Co., North Carolina, USA. H=?, D_m= 3.65, D_c= 3.79, IVF 06. White & Nelen, 1973. Am. Min. 58(1973), 966 (Abst.). See also Wickmanite.

Texasite, D Natural occurrence not proven. Am. Min. 67 (1982), 156. Crook, 1977.

Thadeuite, $(Ca, Mn)(Mg, Fe)_3(PO_4)_2(OH, F)_2$, A Orth. Vitreous yellow-orange. Panasqueira mine, Panasqueria, Beira Baixa, Portugal. H= 4, D_m= 3.25, D_c= 3.21, VIIB 03c. Isaacs et al., 1979. Am. Min. 64 (1979), 359. Am. Min. 67 (1982), 120.

Thalcusite, $Cu_3FeTl_2S_4$, A Tet. Metallic grey. Talnakh deposit, Noril'sk, USSR. H= 2.5, D_m=?, D_c= 6.54, IIA 10. Kovalenker et al., 1976. N.Jb.Min.Abh. 138 (1980), 122. See also Bukovite, Murunskite.

Thalenite-(Y), $Y_3Si_3O_{10}(OH)$, G Mon. Translucent flesh-red. Österby, Dalarne, Sweden. H= 6.5, D_m= 4.2, D_c= 4.29, VIIIB 28. Benedicks, 1898. Am. Min. 71 (1986), 188. Sov.Phys.Cryst. 33 (1988), 356.

Thalfenisite, $Tl_6(Fe, Ni, Cu)_{25}S_{26}Cl$, A Cub. Brown in reflected light. Oktyabr deposit, Talnakh, Noril'sk (near), Siberia (N), USSR. H= 3.3, D_m=?, D_c= 5.26, IIF 02. Rudashevskii et al., 1979. Am. Min. 66(1981), 219 (Abst.). See also Djerfisherite.

Thaumasite, $Ca_3Si(OH)_6(CO_3)(SO_4) \cdot 12H_2O$, G Hex. Translucent/earthy white. Bjelkesgruvan, Åreskuta, Jämtland, Sweden. H= 3.5, D_m= 1.881, D_c= 1.876, VID 07. Nordenskiöld, 1878. Klockmann, 16th ed.(1978), 683. N.Jb.Min.Mh. (1983), 60.

Theisite, $Cu_5Zn_5(As, Sb)_2O_8(OH)_{14}$, A Orth. Translucent pale blue-green. Durango (near), La Plata Co., Colorado, USA. H= 1.5, D_m= 4.3, D_c= 4.45, VIIB 10a. Bevins et al., 1982. Min. Mag. 46 (1982), 49.

Thenardite, α-Na_2SO_4, G Orth. Vitreous/pearly colorless, white. Aranjuez, Madrid, Spain. H=?, D_m= 2.664, D_c= 2.659, VIA 05. Casaseca, 1826. Dana, 7th ed.(1951), v.2, 404. Can. Min. 13 (1975), 181.

Theophrastite, $Ni(OH)_2$, A Rhom. Vitreous green. Vermion, Makedhonia (Macedonia), Greece. H= 2, D_m= 4.00, D_c= 3.95, IVF 03a. Marcopoulos & Economou, 1981. Am. Min. 66 (1981), 1021.

Thermonatrite, $Na_2CO_3 \cdot H_2O$, G Orth. Vitreous colorless, white, greyish, yellowish. H= 1-1.5, D_m= 2.255, D_c= 2.256, VbC 02. Haidinger, 1845. Dana, 7th ed.(1951), v.2, 224. Acta Cryst. B31 (1975), 890.

Thometzekite, $Pb(Cu, Zn)_2(AsO_4)_2 \cdot 2H_2O$, A Syst=? Earthy bluish-green, green. Tsumeb, Namibia. H=?, D_m=?, D_c=?, VIIC 17. Schmetzer et al., 1985. Am. Min. 73(1988), 931 (Abst.). See also Helmutwinklerite, Tsumcorite.

Thomsenolite, $NaCaAlF_6 \cdot H_2O$, G Mon. Vitreous colorless, white. Ivigtut, Greenland (SW). H= 2, D_m= 2.981, D_c=?, IIIC 09. Dana, 1868. Acta Cryst. 23 (1967), 162. Struct. Repts. 52A (1985), 119. See also Pachnolite.

Thomsonite, $NaCa_2(Al_5Si_5)O_{20} \cdot 6H_2O$, G Orth. Transparent colorless, yellowish. Kilpatrick, Dumbartonshire, Scotland. H= 5, D_m= 2.4, D_c= 2.36, VIIIF 16. Brooke, 1820. Natural Zeolites (1985), 57. Min. Abstr. 86M/1429.

Thorbastnäsite, $Th(Ca, Ce)(CO_3)_2F_2 \cdot 3H_2O$, G Hex. Translucent brown. Siberia (E), USSR. H=?, D_m= 4.04, D_c= 5.70, VbB 04. Pavlenko et al., 1965. Am. Min. 50(1965), 1505 (Abst.).

Thoreaulite, $SnTa_2O_6$, G Mon. Resinous/adamantine brown. Shaba, Zaïre. H= 6, D_m= 7.7, D_c= 7.5, IVD 14c. Buttgenbach, 1933. Am. Min. 59 (1974), 1026. Am. Min. 55 (1970), 367. See also Foordite.

Thorgadolinite, D Unnecessary name for thorian gadolinite. Min. Mag. 43 (1980), 1055. Zubkov et al., 1970.

Thorianite, ThO_2, G Cub. Horny/sub-metallic dark grey, brownish-black, black. Sri Lanka. H= 6.5, D_m= 9.7, D_c= 9.99, IVD 16b. Dunstan, 1904. NBS Circ. 539, v.1 (1953), 57. See also Uraninite, Cerianite.

Thorikosite, $Pb_3O_3(Sb, As)(OH)Cl_2$, A Tet. Vitreous light yellow. Lávrion (Laurium), Attikí, Greece. H=?, D_m=?, D_c= 7.24, IIID 03. Dunn & Rouse, 1985. Am. Min. 70 (1985), 845.

Thorite, $ThSiO_4$, [G] Tet. Black, orange-yellow. Lövö, Brevik (near), Langesundfjord, Norway. H= 4.5-5, D_m= 6.63, D_c= 6.70, VIIIA 07. Berzelius, 1829. Min. Mag. 43 (1980), 1031. Acta Cryst. B34 (1978), 1074. See also Huttonite.

Thornasite, $(Na,K)ThSi_{11}(O,H_2O,F,Cl)_{33}$, [A] Rhom. Vitreous/waxy colorless, pale green. De Mix Quarry, Mont Saint-Hilaire, Rouville Co., Québec, Canada. H=?, D_m= 2.62, D_c= 2.627, VIIIE 17. Ansell & Chao, 1987. Can. Min. 25 (1987), 181.

Thoro-aeschynite, [D] Unnecessary name for thorian aeschynite. Min. Mag. 36 (1968), 1144. Es'kova et al., 1964.

Thorogummite, $(Th,U)[(SiO_4),(OH)_4]$, [G] Tet. Earthy yellow, white, grey-green. Brevik (near), Langesundfjord, Norway. H=?, D_m= 4.54, D_c= 5.05, VIIIA 07. Hidden & Mackintosh, 1889. Am. Min. 38 (1953), 1007. See also Coffinite.

Thorosteenstrupine, $(Ca,Th,Mn)_3Si_4O_{11}F \cdot 6H_2O$, [G] Amor. Greasy/vitreous dark brown. Siberia (E), USSR. H= 4, D_m= 3.02, D_c=?, VIIIA 21. Kupriyanova et al., 1962. Am. Min. 48(1963), 433 (Abst.).

Thortveitite, $(Sc,Y)_2Si_2O_7$, [G] Mon. Greyish-green, black. Iveland, Setersdalen, Norway. H= 6-7, D_m= 3.57, D_c= 3.35, VIIIB 01. Schetelig, 1911. Dana/Ford (1932), 620. Am. Min. 73 (1988), 601. See also Keiviite-(Y), Keiviite-(Yb).

Thorutite, $(Th,U,Ca)Ti_2(O,OH)_6$, [G] Mon. Resinous black. Unspecified locality, USSR(?). H=?, D_m= 6.0, D_c= 6.08, IVD 12. Gotman & Khapaev, 1958. Am. Min. 43(1958), 1007 (Abst.). Acta Cryst. 21 (1966), 974. See also Brannerite.

Threadgoldite, $Al(UO_2)_2(PO_4)_2(OH) \cdot 8H_2O$, [A] Mon. Translucent greenish-yellow. Kobokobo, Kivu, Zaïre. H=?, D_m= 3.4, D_c= 3.33, VIID 20c. Deliens & Piret, 1979. Am. Min. 65(1980), 209 (Abst.). Acta Cryst. B35 (1979), 3017.

Tibiscumite, $(Na,Ca)(Al,Mg,Fe)_2Si_4O_{10}(OH) \cdot H_2O$, [Q] Mon. Unspecified locality, Caransebes Basin, Romania. H=?, D_m=?, D_c= 2.50, VIIIE 08c. Ghergari & Nicolescu, 1987. Min. Abstr. 89M/0178.

Tiemannite, HgSe, [G] Cub. Metallic grey; light brownish-grey in reflected light. Harz, Germany. H= 2.5, D_m= 8.19, D_c= 8.266, IIB 01a. Naumann, 1855. Dana, 7th ed.(1944), v.1, 217. Struct. Repts. 20 (1956), 152. See also Metacinnabar.

Tienshanite, $KNa_9Ba_6Ca_2Mn_6Ti_6B_{12}Si_{36}O_{123}(OH)_2$, [A] Hex. Vitreous green. Turkestan-Alai, Tienshan (S), USSR. H= 6-6.5, D_m= 3.29, D_c= 3.23, VIIIC 07. Dusmatov et al., 1967. Am. Min. 53(1968), 1426 (Abst.). Min. Abstr. 79-2103.

Tikhonenkovite, $SrAlF_4(OH) \cdot H_2O$, [G] Mon. Vitreous colorless, rosy. Karasug, Tannu-Ola, Tuva, USSR. H= 3.5, D_m= 3.26, D_c= 3.30, IIIC 09. Khomyakov et al., 1964. Am. Min. 49(1964), 1774 (Abst.).

Tilasite, $CaMgAsO_4F$, [G] Mon. Resinous/vitreous grey, violet-grey, olive-green. Långban mine, Filipstad (near), Värmland, Sweden. H= 5, D_m= 3.77, D_c= 3.722, VIIB 11a. Sjögren, 1895. Min. Rec. 9 (1978), 385. Am. Min. 57 (1972), 1880.

Tilleyite, $Ca_5Si_2O_7(CO_3)_2$, [G] Mon. Translucent white. Crestmore, Riverside Co., California, USA. H=?, D_m= 2.838, D_c= 2.88, VIIIB 06. Larsen & Dunham, 1933. Min. Abstr. 5 (1934), 387. Zts. Krist. 132 (1970), 288.

Tin, β-Sn, [G] Tet. Metallic white. H= 1.5-1.8, D_m= 7.30, D_c= 7.281, IA 04. Dana, 7th ed.(1944), v.1, 126.

Tin-tantalite, [D] Unnecessary name for stannian manganotantalite. Min. Mag. 36 (1967), 133. Matias, 1961.

Tinaksite, $K_2Na(Ca,Mn)_2TiSi_7O_{19}(OH)$, [G] Tric. Vitreous pale yellow. Murun massif, Olekminsk (near), Yakutiya, USSR. H= 6, D_m= 2.82, D_c= 2.321, VIIID 11. Rogov et al., 1965. Am. Min. 50(1965), 2098 (Abst.). Acta Cryst. B36 (1980), 259.

Tincalconite, $Na_2B_4O_5(OH)_4 \cdot 3H_2O$, [G] Rhom. Vitreous colorless. Searles Lake, San Bernardino Co., California, USA. H=?, D_m= 1.88, D_c= 1.94, Vc 07a. Shepard, 1878. Dana, 7th ed.(1951), v.2, 337. Am. Min. 58 (1973), 523.

Tinsleyite, $KAl_2(PO_4)_2(OH) \cdot 2H_2O$, [A] Mon. Vitreous magenta-red. Tip Top Pegmatite, Custer, Custer Co., South Dakota, USA. H= 5, D_m= 2.69, D_c= 2.62, VIID 14a. Dunn et al., 1984. Am. Min. 69 (1984), 374. See also Leucophosphite, Spheniscidite.

Tinticite, $Fe_4(PO_4)_3(OH)_3 \cdot 5H_2O$, [G] Mon. Earthy creamy, brown. Tintic Mining Dist., Utah, USA. H=?, D_m= 2.94, D_c= 2.97, VIID 05. Stringham, 1946. N.Jb.Min.Mh. (1988), 446.

Tintinaite, $Cu_2Pb_{11}Sb_{15}S_{35}$, [A] Orth. Metallic grey. Tintina silver mines, Watson Lake (NW), Whitehorse Div., Yukon Territory, Canada. H=?, D_m=?, D_c= 5.48, IID 08. Harris et al., 1968. Can. Min. 22 (1984), 219. See also Kobellite.

Tinzenite, $(Ca, Mn, Fe^{3+})_3Al_2BSi_4O_{15}(OH)$, [G] Tric. Translucent yellow, orange-red. Alp Parsettens, Sursass (Oberhalbstein), Tinizong (Tinzen), Grischun (Graubünden), Switzerland. H=?, D_m= 3.29, D_c= 3.37, VIIIC 04. Jakob, 1923. JCPDS 6-444. See also Manganaxinite.

Tiptopite, $K_2(Li, Na, Ca)_6Be_6(PO_4)_6(OH)_2 \cdot 1.3H_2O$, [A] Hex. Clear colorless. Tip Top Pegmatite, Custer (near), Custer Co., South Dakota, USA. H=?, D_m= 2.65, D_c= 2.52, VIIB 21. Grice et al., 1985. Can. Min. 23 (1985), 43. Am. Min. 72 (1987), 816.

Tiragalloite, $Mn_4AsSi_3O_{12}(OH)$, [A] Mon. Sub-adamantine orange. Molinello mine, Val Graveglia, Liguria, Italy. H=?, D_m= 3.84, D_c= 3.86, VIIIB 23. Gramaccioli et al., 1980. Min. Abstr. 81-2383. Acta Cryst. B35 (1979), 2287.

Tirodite, $Mn_2(Mg, Fe)_5Si_8O_{22}(OH)_2$, [G] Mon. Vitreous yellow. Tirodi, India (central). H= 6.5, D_m= 3.24, D_c= 3.25, VIIID 05a. Dunn & Roy, 1938. Am. Min. 25 (1940), 380. Can. Min. 15 (1977), 309. See also Dannemorite.

Tisinalite, $H_3Na_3(Mn, Ca, Fe)TiSi_6(O, OH)_{18} \cdot 2H_2O$, [A] Rhom. Vitreous yellow-orange. Mt. Koashva, Khibina massif, Kola Peninsula, USSR. H= 5, D_m= 2.68, D_c= 2.682, VIIIC 07. Kapustin et al., 1980. Am. Min. 66(1981), 219 (Abst.).

Titanclinohumite, $(Mg, Fe, Ti)_9(SiO_4)_4(O, OH)_2$, [Q] Mon. Vitreous brown, yellow. H=?, D_m= 3.364, D_c= 3.35, VIIIA 16. Machatschki, 1930. ZVMO 117 (1988), 675. Am. Min. 58 (1973), 43. See also Clinohumite.

Titanite, $CaTiSiO_5$, [G] Mon. Adamantine/resinous brown, grey, yellow, green, red, etc. H= 5-5.5, D_m= 3.5, D_c= 3.52, VIIIA 20. Klaproth, 1795. Am. Min. 61 (1976), 878. Am. Min. 61 (1976), 238. See also Malayaite.

Titanmicrolite, [D] Natural occurrence not proven. Am. Min. 62 (1977), 403. Strunz, 1966.

Titano-aeschynite, [D] Superfluous name for aeschynite. Min. Mag. 36 (1967), 133. Zhabin et al., 1960.

Titanomaghemite, $Fe(Fe, Ti)_2O_4$, [Q] Cub. Bon Accord, Transvaal, South Africa. H=?, D_m=?, D_c= 4.712, IVB 01. Basta, 1959. Min. Mag. 53 (1989), 299. Am. Min. 73 (1988), 153. See also Maghemite.

Titanopyrochlore, [D] A hypothetical Ti equivalent of pyrochlore. Am. Min. 62 (1977), 403. Machatschki, 1932.

Titanorhabdophane, [D] = tundrite-(Ce). Min. Mag. 36 (1967), 133. Semenov, 1959.

Titantaramellite, $Ba_4(Ti, Fe, Mg)_4B_2Si_8O_{27}O_2Cl_x$, [A] Orth. Vitreous deep red. Candoglia, Piemonte, Italy. H= 6, D_m= 4.03, D_c= 4.02, VIIIC 04. Alfors & Pabst, 1984. Am. Min. 69 (1984), 358. See also Nagashimalite, Taramellite.

Tivanite, $TiVO_3(OH)$, [A] Mon. Metallic/sub-metallic black; grey in reflected light. Kalgoorlie, Western Australia, Australia. H= 5.5, D_m=?, D_c= 4.17, IVD 05. Grey & Nickel, 1981. Am. Min. 66 (1981), 866. Am. Min. 66 (1981), 866.

Tlalocite, $Cu_{10}Zn_6Te_3O_{11}Cl(OH)_{25} \cdot 27H_2O$, [A] Orth. Velvety blue. Bambollita mine, Moctezuma, Sonora, Mexico. H= 1, D_m= 4.55, D_c= 4.58, VIG 09. Williams, 1975. Min. Abstr. 80-0755.

Tlapallite, $H_6(Ca,Pb)_2(Cu,Zn)_3SO_4(TeO_3)_4TeO_6$, [A] Mon. Translucent green. Bambollita mine, Moctezuma, Sonora, Mexico. H= 3, D_m= 5.38, D_c= 5.465, VIG 08. Williams & Duggan, 1978. Min. Mag. 42 (1978), 183.

Tobelite, $(NH_4,K)Al_2(Si_3Al)O_{10}(OH)_2$, [A] Mon. Clayey white, yellowish-green. Ohgidani deposit, Tobe; Ehime and Horo deposits, Toyosaka, Hiroshima, Japan. H=?, D_m= 2.58, D_c= 2.617, VIIIE 05a. Higashi, 1982. Am. Min. 68(1983), 850 (Abst.).

Tobermorite, $Ca_5Si_6O_{16}(OH)_2 \cdot xH_2O$, [G] Orth. Translucent pale pinkish-white. Tobermore Island, Hall, Scotland. H=?, D_m= 2.423, D_c= 2.58, VIIID 10. Heddle, 1880. JCPDS 19-1364. Zts. Krist. 182 (1988), 114.

Tochilinite, $6(Fe_{0.9}S) \cdot 5[(Mg,Fe)(OH)_2)]$, [A] Tric. Metallic bronze-black. Lower Maman intrusive, Voronezh, USSR. H= 1.5-2, D_m= 2.97, D_c= 3.03, IIE 01. Organova et al., 1971. Am. Min. 57(1972), 1552 (Abst.). Sov.Phys.Cryst. 17 (1973), 667. See also Haapalaite, Valleriite, Yushkinite.

Tocornalite, Ag, Hg, I, [Q] Syst=? Translucent yellow. Chañarcillo, Copiapó, Atacama, Chile. H=?, D_m=?, D_c=?, IIIA 02. Domeyko, 1867. Am. Min. 58(1973), 348 (Abst.).

Toddite, [D] A mixture of columbite and samarskite. Am. Min. 47 (1962), 1363. Ellsworth, 1926.

Todorokite, $(Na,Ca,K,Ba,Sr)_{1-x}(Mn,Mg,Al)_6O_{12} \cdot 3\text{-}4H_2O$, [G] Mon. Metallic black. Todoroki mine, Hokkaido, Japan. H=?, D_m= 3.67, D_c=?, IVF 05a. Yoshimura, 1934. Am. Min. 68 (1983), 972. Am. Min. 73 (1988), 861. See also Woodruffite.

Tokkoite, $K_2Ca_4Si_7O_{17}(O,OH,F)_4$, [A] Tric. Vitreous, colorless. Murun massif, between the Charo and Tokko rivers, Yakutiya (SW), USSR. H= 4-5, D_m= 2.76, D_c= 2.77, VIIID 11. Lazebnik et al., 1986. Am. Min. 73(1988), 196 (Abst.). Zts. Krist. 189 (1989), 195.

Tolbachite, $CuCl_2$, [A] Mon. Translucent brown. Tolbachik volcano, Kamchatka, USSR. H=?, D_m=?, D_c= 3.42, IIIA 06. Bergasova & Filatov, 1983. Am. Min. 69(1984), 408 (Abst.).

Tolovkite, IrSbS, [A] Orth. Metallic grey. Ust'-Bel'skiĭ massif, Tolovka river basin, USSR (NE). H= 5-6, D_m=?, D_c= 10.50, IIC 06b. Razin et al., 1981. Am. Min. 67(1982), 1076 (Abst.). Am. Min. 74 (1989), 1168.

Tombarthite-(Y), $Y_4[SiO_4,(OH)_4]_3$, [A] Mon. Dull brownish-black. Høgetveit, Evje, Norway (S). H= 5-6, D_m= 3.65, D_c= 3.64, VIIIA 08. Neumann & Nilssen, 1968. Am. Min. 54(1969), 327 (Abst.).

Tomichite, $(V,Fe)_4Ti_3AsO_{13}(OH)$, [A] Mon. Opaque black. Kalgoorlie, Western Australia, Australia. H= 6, D_m= 4.16, D_c= 4.42, IVC 10. Nickel & Grey, 1979. Min. Mag. 43 (1979), 469. Am. Min. 72 (1987), 201. See also Derbylite.

Tongbaite, Cr_3C_2, [A] Orth. Metallic pale brownish yellow. Liu Zhuang, Tongbai Co., Henan, China. H= 8.5, D_m=?, D_c= 6.66, IC 02. Tian et al., 1983. Am. Min. 70(1985), 218 (Abst.). Struct. Repts. 34A (1969), 56.

Topaz, $Al_2SiO_4(F,OH)_2$, [G] Orth. Translucent colorless, white, yellow, grey, green, red, blue. H= 8, D_m= 3.5, D_c= 3.56, VIIIA 14. de Boodt, 1636. Min. Mag. 43 (1980), 943. Am. Min. 56 (1971), 24.

Torbernite, $Cu(UO_2)_2(PO_4)_2 \cdot 10H_2O$, [G] Tet. Vitreous/sub-adamantine green. Jáchymov (St. Joachimsthal), Západočeský kraj, Čechy (Bohemia), Czechoslovakia. H= 2-2.5, D_m= 3.22, D_c= 3.16, VIID 20a. Werner, 1786. Dana, 7th ed.(1951), v.2, 981.

Törnebohmite-(Ce), $(Ce,La,Nd)_2Al(SiO_4)_2(OH)$, [A] Syst=? Bastnäs, Riddarhyttan, Sweden. H=?, D_m=?, D_c=?, VIIIA 26. Geijer, 1921. Am. Min. 51 (1966), 152.

Törnebohmite-(La), $(La,Ce)_2Al(SiO_4)_2(OH)$, [G] Mon. Translucent green, olive. Mt. Nepkha, Lovozero tundra, Kola Peninsula, USSR. H= 4.5, D_m= 4.94,

D_c= 5.12, VIIIA 26. Geijer, 1921. Am. Min. 51 (1966), 152. Am. Min. 67 (1982), 1021.

Torreyite, $(Mg,Mn)_9Zn_4(SO_4)_2(OH)_{22} \cdot 8H_2O$, G Mon. Dull/vitreous white, colorless. Sterling Hill, Sussex Co., New Jersey, USA. H=?, D_m= 2.66, D_c= 2.65, IVF 03e. Prewitt-Hopkins, 1949. Am. Min. 64 (1979), 949. See also Lawsonbauerite, Mooreite, Shigaite.

Tosalite, D Intermediate member of bementite-greenalite series. Min. Mag. 43 (1980), 1055. Yoshimura, 1967.

Tosudite, $Na_{0.5}(Al,Mg)_6(Si,Al)_8O_{18}(OH)_{12} \cdot 5H_2O$, G Orth. Dark blue. Crimea peninsula, USSR. H=?, D_m=?, D_c= 2.46, VIIIE 08c. Frank-Kamenetskii et al., 1963. JCPDS 22-956. See also Chlorite, Smectite.

Tourmaline, A group name for ring silicates with the general formula $(Na,K,Ca)(Mg,Fe,Mn,Li,Al)_3(Al,Fe,Cr,V)_6Si_6O_{18}(BO_3)_3(O,OH,F)_4$.

Trabzonite, $Ca_4Si_3O_{10} \cdot 2H_2O$, A Mon. Vitreous colorless. Varda Yaylasi, Ikizdere-Rise, Trabzon, Turkey. H=?, D_m= 2.9, D_c= 3.08, VIIID 08. Sarp & Burri, 1986. Am. Min. 73(1988), 1497 (Abst.).

Tranquillityite, $Fe_8Ti_3(Zr,Y)_2Si_3O_{24}$, A Hex. Nearly opaque. Tranquillity Base, Moon. H=?, D_m=?, D_c= 4.7, VIIIA 21. Lovering et al., 1971. Am. Min. 58(1973), 140 (Abst.).

Traskite, $Ba_{12}Fe_2Ti_6Si_{12}O_{54}Cl_3 \cdot 7H_2O$, A Hex. Vitreous brownish-red. Big Creek and Rush Creek, Fresno Co., California, USA. H=?, D_m= 3.71, D_c= 3.52, VIIIC 07. Alfors et al., 1965. Am. Min. 50 (1965), 1500. Min. Abstr. 78-202.

Treasurite, $Ag_7Pb_6Bi_{15}S_{32}$, A Mon. Metallic. Treasury mine, Geneva dist., Park Co., Colorado, USA. H=?, D_m=?, D_c= 7.25, IID 09b. Makovicky & Karup-Møller, 1977. Am. Min. 64(1979), 243 (Abst.).

Trechmannite, $AgAsS_2$, G Rhom. Adamantine red. Lengenbach quarry, Binntal, Valais (Wallis), Switzerland. H= 1.5-2, D_m=?, D_c= 4.78, IID 07. Solly, 1904. Am. Min. 53 (1968), 1212. Zts. Krist. 129 (1969), 163. See also Smithite.

Tremolite, $Ca_2(Mg,Fe)_5Si_8O_{22}(OH)_2$, G Mon. Vitreous colorless, grey. Tremola Valley, St. Gotthard, Switzerland. H= 5-6, D_m= 3.0, D_c= 2.964, VIIID 05b. Höpfner, 1790. DHZ (1963), v.2, 249. Can. Min. 14 (1976), 334. See also Actinolite, Ferro-actinolite.

Trevorite, $NiFe_2O_4$, G Cub. Metallic/sub-metallic brown. Barberton, Transvaal, South Africa. H= 5, D_m= 5.164, D_c= 5.20, IVB 01. Crosse, 1921. Dana, 7th ed.(1944), v.1, 698.

Triangulite, $Al_3(UO_2)_4(PO_4)_4(OH)_5 \cdot 5H_2O$, A Tric. Translucent bright yellow. Kobokobo, Kivu, Zaïre. H=?, D_m= 3.7, D_c= 3.68, VIID 21. Deliens & Piret, 1982. Am. Min. 69(1984), 212 (Abst.).

Tridymite, SiO_2, G Mon. Vitreous colorless, white. Cerro San Cristóbal, Pachuca, Mexico. H= 7, D_m= 2.26, D_c= 2.27, IVD 01. vom Rath, 1868. Dana, 7th ed.(1962), v.3, 259. Acta Cryst. B32 (1976), 2486. See also Coesite, Cristobalite, Quartz, Stishovite.

Trigonite, $Pb_3Mn(AsO_3)_2(AsO_2OH)$, G Mon. Vitreous/adamantine yellow, yellowish-brown, dark brown. Långban mine, Filipstad (near), Värmland, Sweden. H= 2-3, D_m= 6.6, D_c= 6.347, VIID 31. Flink, 1920. Dana, 7th ed.(1951), v.2, 1032. Min. Abstr. 89M/1661.

Trikalsilite, $K_2NaAl_3(SiO_4)_3$, G Hex. Kabfumu, Kivu (N), Zaïre. H=?, D_m=?, D_c= 2.636, VIIIF 01. Sahama & Smith, 1957. Am. Min. 42 (1957), 287. N.Jb.Min.Mh. (1988), 559. See also Kaliophilite, Kalsilite, Panunzite.

Trimerite, $CaBe_3Mn_2(SiO_4)_3$, G Mon. Translucent salmon-pink, colorless. Harstigen mine, Pajsberg, Värmland, Sweden. H= 6-7, D_m= 3.474, D_c= 3.47, VIIIA 02. Flink, 1890. Dana/Ford (1932), 600. Zts. Krist. 145 (1977), 46.

Triphylite, $LiFePO_4$, G Orth. Vitreous/sub-resinous bluish-grey, greenish-grey. Hühnerkobel, Rabenstein (near), Bayern (Bavaria), Germany. H= 4-5, D_m= 3.565, D_c= 3.59, VIIA 02. Fuchs, 1834. Dana, 7th ed.(1951), v.2, 665. Min. Abstr. 79-2143. See also Lithiophilite.

Triplite, $(Mn,Fe,Mg,Ca)_2PO_4(F,OH)$, G Mon. Vitreous/resinous dark brown, reddish-brown, pink. Chanteloube, Limoges, Haute-Vienne, France. H= 5-5.5, D_m= 3.7, D_c= 3.61, VIIB 03a. Hausmann, 1813. Am. Min. 36 (1951), 256. Zts. Krist. 130 (1969), 1. See also Zwieselite, Magniotriplite.

Triploidite, $(Mn,Fe)_2PO_4(OH)$, G Mon. Vitreous/greasy/adamantine pinkish, yellow, yellowish-brown. Branchville, Fairfield Co., Connecticut, USA. H= 4.5-5, D_m= 3.697, D_c= 3.80, VIIB 03b. Brush & Dana, 1878. Dana, 7th ed.(1951), v.2, 853. Zts. Krist. 131 (1970), 1. See also Wolfeite.

Trippkeite, $CuAs_2O_4$, G Tet. Brilliant greenish-blue. Copiapó, Chile. H=?, D_m= 4.8, D_c= 4.482, VIIA 15. vom Rath, 1880. Dana, 7th ed.(1951), v.2, 1034. Struct. Repts. 41A (1975), 340. See also Schafarzikite.

Tripuhyite, $FeSb_2O_6$, G Tet. Dull greenish yellow to dark brown. Tripuhy, Ouro Preto, Minas Gerais, Brazil. H= 7, D_m= 5.6, D_c= 6.70, IVD 04. Hussak & Prior, 1897. Min. Mag. 30 (1953), 100.

Tristramite, $(Ca,U,Fe)(PO_4,SO_4)\cdot 2H_2O$, A Hex. Lustrous yellow, greenish-yellow. Wheal Trewavas, Cornwall, England. H=?, D_m= 4.0, D_c= 4.18, VIIC 19. Atkin et al., 1983. Min. Mag. 47 (1983), 393.

Tritomite-(Ce), $Ca_2(Ce,La)_3(Si_2B)O_{13}$, G Hex. Resinous dark brown. Låven Island, Brevik (near), Langesundfjord, Norway. H= 5.5, D_m= 4.2, D_c= 4.94, VIIIA 19. Weibye, 1849. Am. Min. 47 (1962), 9.

Tritomite-(Y), $(Y,Ca,Ce)_5(Si,B,Al)_3(O,OH)_{13}$, A Hex. Vitreous dark greenish-black. Cranberry Lake, Sussex Co., New Jersey, USA. H= 3.5, D_m= 3.40, D_c=?, VIIIA 19. Jaffe & Molinski, 1962. Am. Min. 47 (1962), 9.

Trögerite, $(H_3O)_2(UO_2)_2(AsO_4)_2\cdot 6H_2O$, G Tet. Weisser Hirsch mine, Neustädtel, Schneeberg (near), Sachsen (Saxony), Germany. H=?, D_m=?, D_c= 3.55, VIID 20b. Weisbach, 1871. Min. Abstr. 76-874.

Trogtalite, $CoSe_2$, G Cub. Metallic rose-violet. Trogtal Quarry, Lautenthal (near), Harz, Niedersachsen, Germany. H= 7, D_m=?, D_c= 7.09, IIC 05. Ramdohr & Schmitt, 1955. Am. Min. 41(1956), 164 (Abst.). Struct. Repts. 19 (1955), 123. See also Hastite.

Troilite, FeS, G Hex. Metallic bronze. Albareto Meteorite, Albareto, Modena, Italy. H= 4, D_m= 4.67, D_c= 4.83, IIB 09b. Haidinger, 1863. Acta Cryst. B38 (1982), 1877. Struct. Repts. 35A (1970), 140.

Trolleite, $Al_4(PO_4)_3(OH)_3$, G Mon. Vitreous pale green. Vestanå mine, Nästum, Skåne, Sweden. H= 5.5-6, D_m= 3.09, D_c= 3.08, VIIB 05. Blomstrand, 1868. Am. Min. 64 (1979), 1175. Am. Min. 59 (1974), 974.

Trona, $Na_3(HCO_3)(CO_3)\cdot 2H_2O$, G Mon. Vitreous colorless, grey, yellowish-white. H= 2.5-3, D_m= 2.14, D_c= 2.112, VbC 02. Bagge, 1773. Am. Min. 44 (1959), 724. Struct. Repts. 52A (1985), 258.

Trudellite, D A mixture of natroalunite and chloraluminite. Am. Min. 57 (1972), 1317 (Abst.). Gordon, 1926.

Truscottite, $(Ca,Mn)_{14}Si_{24}O_{58}(OH)_8\cdot 2H_2O$, G Rhom. Pearly white. Lebong Donok mine, Benkulen, Sumatra, Indonesia. H=?, D_m=?, D_c= 2.51, VIIIE 14. Grutterink, 1925. Min. Mag. 43 (1979), 333. See also Minehillite, Reyerite.

Trüstedtite, Ni_3Se_4, G Cub. Metallic yellow. Kuusamo, Finland (NE). H= 3, D_m=?, D_c= 6.62, IIC 01. Vuorelainen et al., 1964. Am. Min. 50(1965), 520 (Abst.). See also Wilkmanite.

Tschermakite, $Ca_2(Mg,Fe)_5(Si_6Al_2)O_{22}(OH)_2$, A Mon. H=?, D_m=?, D_c=?, VIIID 05b. Winchell, 1945. Am. Min. 63 (1978), 1023. See also Ferrotschermakite, Alumino-tschermakite.

Tschermigite, $NH_4Al(SO_4)_2 \cdot 12H_2O$, G Cub. Vitreous/silky colorless, white. Čermíky (Tschermig), Čechy (Bohemia), Czechoslovakia. H= 1.5, D_m= 1.645, D_c= 1.64, VIC 08. Kobell, 1853. N. Jb. Min. Mh. (1987), 171. Zts. Krist. 157 (1982), 147.

Tsumcorite, $Pb(Zn,Fe)_2(AsO_4)_2 \cdot (OH,H_2O)_2$, A Mon. Translucent reddish-brown. Tsumeb, Namibia. H= 4-5, D_m= 5.2, D_c= 5.39, VIIC 17. Geier et al., 1971. N.Jb.Min.Mh. (1985), 446. Acta Cryst. B29 (1973), 2789. See also Helmutwinklerite, Thometzekite.

Tsumebite, $CuPb_2(PO_4)(SO_4)(OH)$, G Mon. Vitreous green. Tsumeb, Otavi, Namibia. H= 3.5, D_m= 6.13, D_c= 6.18, VIIB 14. Busz, 1912. Am. Min. 51(1966), 259 (Abst.). Am. Min. 51(1966), 267 (Abst.). See also Arsentsumebite.

Tsumoite, BiTe, A Rhom. Metallic creamy-white. Tsumo mine, Hiroshima (NW), Shimane, Japan. H= 2-2.5, D_m= 8.16, D_c= 8.23, IIB 18. Shimazaki & Ozawa, 1978. Am. Min. 63 (1978), 1162. See also Sulphotsumoite.

Tucanite, D Probably a partly dehydrated scarbroite. Min. Mag. 36 (1968), 1144. Karsulin, 1964.

Tučekite, $Ni_9Sb_2S_8$, A Tet. Metallic pale yellow. Kanowna, Western Australia, Australia. H= 5-6, D_m=?, D_c= 6.15, IIC 13. Just & Feather, 1978. Min. Mag. 42 (1978), 278.

Tugarinovite, MoO_2, A Mon. Greasy/metallic lilac-brown. Siberia (E), USSR. H= 4.6, D_m=?, D_c= 6.58, IVD 07. Kruglova et al., 1980. Am. Min. 66(1981), 438 (Abst.).

Tugtupite, $Na_4BeAlSi_4O_{12}Cl$, G Tet. Translucent white, pink to deep red, pale blue. Ilímaussaq Intrusion, Greenland (S). H= 5.5-6.5, D_m= 2.30, D_c= 2.36, VIIIF 07. Sörensen, 1962. Am. Min. 48(1963), 1178 (Abst.). Acta Cryst. 20 (1966), 812.

Tuhualite, $(Na,K)Fe_2Si_6O_{15}$, G Orth. Translucent violet. Mayor Island, Opo Bay, North Island, New Zealand. H= 3-4, D_m= 2.89, D_c= 3.03, VIIID 14. Marshall, 1932. Am. Min. 41(1956), 959 (Abst.). Science 166 (1969), 1399. See also Zektzerite.

Tulameenite, Pt_2FeCu, A Tet. Metallic white. Similkameen River and Tulameen River, Similkameen Dist., British Columbia, Canada. H= 5, D_m=?, D_c= 15.6, IA 12. Cabri et al., 1973. Can. Min. 12 (1973), 21. See also Ferronickelplatinum.

Tundrite-(Ce), $Na_2Ce_2TiO_2SiO_4(CO_3)_2$, G Tric. Vitreous brownish-yellow, greenish-yellow. Mt. Nepkha, Lovozero tundra, Kola Peninsula, USSR. H= 3, D_m= 3.70, D_c= 4.25, VbB 08. Semenov, 1963. Am. Min. 53(1968), 1780 (Abst.). Struct. Repts. 42A (1976), 418.

Tundrite-(Nd), $Na_2Nd_2TiO_2(SiO_4)(CO_3)_2$, G Syst=? Ilímaussaq Intrusion, Greenland (S). H=?, D_m=?, D_c=?, VbB 08. Semenov et al., 1967. Am. Min. 53(1968), 1780 (Abst.).

Tunellite, $SrB_6O_9(OH)_2 \cdot 3H_2O$, G Mon. Sub-vitreous/pearly colorless. U. S. Borax Corp open pit, Boron, Kern Co., California, USA. H= 2.5, D_m= 2.40, D_c= 2.381, Vc 12a. Erd et al., 1961. Am. Min. 47(1962), 416 (Abst.). Am. Min. 49 (1964), 1549.

Tungstenite-2H, WS_2, G Hex. Metallic grey; white in reflected light. Emma mine, Little Cottonwood dist., Salt Lake Co., Utah, USA. H= 2.5, D_m= 7.45, D_c= 7.73, IIC 10. Wells & Butler, 1917. NBS Circ. 539, v.8 (1959), 65. See also Drysdallite, Molybdenite.

Tungstenite-3R, WS_2, P Rhom. Metallic white. Emma mine, Little Cottonwood dist., Salt Lake Co., Utah, USA. H=?, D_m=?, D_c= 7.80, IIC 10. Gait & Mandarino, 1970. JCPDS 35-651. See also Molybdenite-3R.

Tungstite, $WO_3 \cdot H_2O$, G Orth. Resinous yellow, yellowish-green. Lanes' mine, Monroe, Connecticut, USA. H= 2.5, D_m= 5.52, D_c= 5.75, IVF 10. Dana, 1868. Am. Min. 29 (1944), 192. Can. Min. 22 (1984), 681.

Tungusite, $Ca_4Fe_2Si_6O_{15}(OH)_6$, A Syst=? Pearly green, yellowish-green. Lower Tunguska River, Siberia, USSR. H= 2, D_m= 2.59, D_c=?, VIIIE 14. Kudryashova, 1966. Am. Min. 52(1967), 927 (Abst.).

Tunisite, $NaCa_2Al_4(CO_3)_4(OH)_8Cl$, A Tet. Translucent colorless, white. Sakiet Sidi Youssef deposit, El Kef, Tunisia. H= 4.5, D_m= 2.51, D_c= 2.512, VbB 02. Johan et. al., 1969. Am. Min. 54 (1969), 1. Am. Min. 67(1982), 418 (Abst.).

Tuperssuatsiaite, $NaFe_3Si_8O_{20}(OH)_2 \cdot 5H_2O$, A Mon. Vitreous reddish-brown. Tuperssuatsiat Bay, Ilímaussaq Intrusion, Greenland (S). H=?, D_m=?, D_c= 2.468, VIIIE 13. Karup-Møller & Petersen, 1984. Am. Min. 70(1985), 1332 (Abst.). See also Palygorskite, Yofortierite.

Turanite, $Cu_5(VO_4)_2(OH)_4$(?), Q Syst=? Translucent olive-green. Tyuya Muyun, Fergana, Uzbekistan, USSR. H= 5, D_m=?, D_c=?, VIIA 12. Nenadkevich, 1909. Dana, 7th ed.(1951), v.2, 818.

Turite, D Unnecessary name for cerian götzenite. Min. Mag. 36 (1968), 1144. Kukharenko et al., 1965.

Turneaureite, $Ca_5(AsO_4, PO_4)_3Cl$, A Hex. Vitreous/greasy colorless. Långban mine, Filipstad (near), Värmland, Sweden; Balmat, St. Lawrence Co., New York, USA; Franklin, Sussex Co., New Jersey, USA. H= 5, D_m= 3.60, D_c= 3.63, VIIB 16. Dunn et al., 1985. Can. Min. 23 (1985), 251.

Turquoise, $CuAl_6(PO_4)_4(OH)_8 \cdot 4H_2O$, G Tric. Vitreous blue, bluish-green, green. Virginia, USA. H= 5-6, D_m= 2.84, D_c= 3.189, VIID 08. Am. Min. 38 (1953), 964. Zts. Krist. 121 (1965), 87. See also Chalcosiderite.

Tuscanite, $KCa_6(Si, Al)_{10}O_{22}(SO_4, CO_3)_2(OH) \cdot H_2O$, A Mon. Transparent colorless. Pitigliano, Grosseto, Toscana, Italy. H= 5.5-6, D_m= 2.83, D_c= 2.77, VIIIE 20. Orlandi et al., 1977. Am. Min. 62 (1977), 1110. Am. Min. 62 (1977), 1114. See also Latiumite.

Tusionite, $MnSn(BO_3)_2$, A Rhom. Vitreous colorless, yellow-brown. Tusion River Valley, Pamir Mts. (SW), USSR. H= 5-7.5, D_m= 4.73, D_c= 4.85, Vc 01b. Konovalenko et al., 1983. Am. Min. 69(1984), 1193 (Abst.). See also Nordenskiöldine, Dolomite.

Tvalchrelidzeite, $Hg_{12}(As, Sb)_8S_{12}$, A Tric. Adamantine reddish-black. Gomi deposit, Georgia (N), USSR. H= 3.5, D_m= 7.38, D_c= 7.234, IID 09b. Gruzdev et al., 1975. Dokl.Earth Sci. 290(1986), 184.

Tveitite-(Y), $Ca_{14}Y_5F_{43}$, A Rhom. Translucent greasy white/pale yellow. Hoydalen, Telemark, Norway. H=?, D_m=?, D_c= 3.892, IIIA 07. Bergstøl et al., 1977. Am. Min. 62(1977), 1060 (Abst.). J.Sol.St.Chem. 44 (1982), 75.

Twinnite, $Pb(Sb, As)_2S_4$, A Orth. Metallic black; white in reflected light. Taylor pit, Huntingdon Twp., Hastings Co., Ontario, Canada. H= 3.5, D_m=?, D_c= 5.323, IID 15. Jambor, 1967. Can. Min. 9 (1967), 191. See also Guettardite.

Tychite, $Na_6Mg_2(CO_3)_4SO_4$, G Cub. Vitreous white. Searles Lake, San Bernardino Co., California, USA. H= 3.5-4, D_m= 2.549, D_c= 2.586, VbB 03. Penfield & Jamieson, 1905. Am. Min. 54 (1969), 302. See also Ferrotychite, Northupite.

Tynite, D Inadequate data. Min. Mag. 36 (1967), 133. Ovchinnikov, 1962.

Tyretskite-1Tc, $Ca_2B_5O_9(OH) \cdot H_2O$, A Tric. Tyret, Siberia (E), USSR. H=?, D_m= 2.189, D_c= 2.571, Vc 13. Kondrat'eva, 1964. Am. Min. 70 (1985), 636. See also Hilgardite-1Tc.

Tyrolite, $CaCu_5(AsO_4)_2(CO_3)(OH)_4 \cdot 6H_2O$, G Orth. Vitreous/pearly pale apple-green, verdigris-green, blue. Galkenstein or Schwatz, Tyrol, Austria. H= 2, D_m= 3.1, D_c= 3.606, VIID 09. Haidinger, 1845. Bull. Min. 79 (1956), 7. See also Clinotyrolite.

Tyrrellite, $(Cu, Co, Ni)_3Se_4$, G Cub. Metallic brassy-bronze. Beaver Lodge Lake, Goldfields district, Saskatchewan, Canada. H= 3.5, D_m= 6.6, D_c= 7.06, IIC 01. Robinson, post-1952. Am. Min. 37 (1952), 542.

Tyuyamunite, $Ca(UO_2)_2(VO_4)_2 \cdot 5\text{–}8H_2O$, G Orth. Translucent yellow, greenish-yellow. Tyuya Muyun, Fergana, Uzbekistan, USSR. H= 2, D_m=?, D_c= 3.4, VIID 23. Nenadkevich, 1912. Am. Min. 41 (1956), 187. See also Carnotite, Margaritasite.

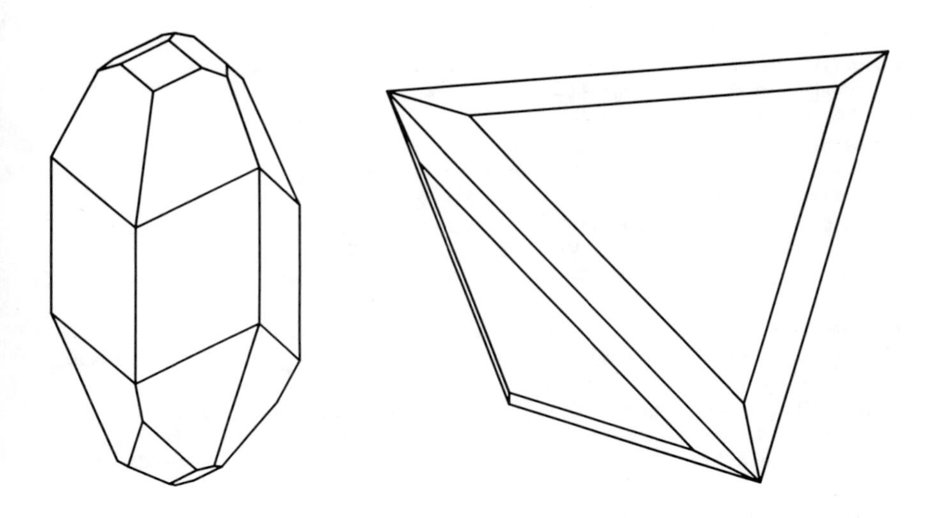

(Left) Proustite from Chañarcillo, Chile showing the forms {101}, {10$\bar{1}$}, {012}, {01$\bar{2}$}, {211}, {12$\bar{1}$} and {110} (after Miers & Prior, 1887).

(Right) Tetrahedrite from Dillenberg, Germany showing the forms {111} and {112} (after Kretschmer, 1911).

U

Uchucchacuaite, $AgMnPb_3Sb_5S_{12}$, [A] Orth. Metallic grey. Uchucchacua mine, Oyon, Cajatambo, Peru. H= 3.5, D_m=?, D_c= 5.61, IID 09b. Moëlo et al., 1984. Am. Min. 70(1985), 1332 (Abst.). See also Andorite.

Udokanite, [D] Inadequate data; probably = antlerite. Min. Mag. 43 (1980), 1055. Yurgenson et al., 1968.

Uduminelite, [D] Inadequate data. Min. Mag. 39 (1974), 929.

Uhligite, $Ca_3(Ti, Al, Zr)_9O_{20}$(?), [Q] Cub. Metallic black. Lake Magadi, Rift Valley, Kenya. H= 5.5, D_m= 4.15, D_c=?, IVC 07. Hauser, 1909. Dana, 7th ed.(1944), v.1, 735.

Uigite, [D] A mixture of thomsonite and gyrolite. Min. Mag. 33 (1962), 262. Heddle, 1856.

Uklonskovite, $NaMgSO_4(OH, F) \cdot 2H_2O$, [G] Mon. Vitreous colorless. Amu-Darya River, Kara-Kalpakii, Kara Tau, Kazakhstan, USSR. H=?, D_m= 2.42, D_c= 2.41, VID 05. Slyusareva et al., 1964. Am. Min. 50(1965), 520 (Abst.). Struct. Repts. 30A (1965), 377.

Ulexite, $NaCaB_5O_6(OH)_6 \cdot 5H_2O$, [G] Tric. Silky, vitreous colorless, white. Iquique, Tarapacá, Chile. H= 2.5, D_m= 1.955, D_c= 1.955, Vc 08. Dana, 1850. Am. Min. 44 (1959), 712. Am. Min. 63 (1978), 160.

Ullmannite, NiSbS, [G] Cub. Metallic grey; white in reflected light. Freusburg, Siegen (Near), Westphalia, Germany. H= 5-5.5, D_m= 6.65, D_c= 6.793, IIC 06b. Frobel, 1850. Can. Min. 24 (1986), 27. Am. Min. 65 (1980), 154. See also Willyamite.

Ulrichite, $CaCu(UO_2)(PO_4)_2 \cdot 4H_2O$, [A] Mon. Vitreous pale apple/lime-green. Lake Boga, Victoria, Australia. H= 3.5, D_m=?, D_c= 3.71, VIID 20c. Birch et al., 1988. Austral. Min. 3 (1988), 125. Austral. Min. 3 (1988), 125.

Ulvöspinel, Fe_2TiO_4, [G] Cub. Metallic brown in reflected light. Ulvö Islands, Sweden. H=?, D_m=?, D_c= 4.820, IVB 01. Mogensen, 1943. Can. Min. 18 (1980), 339. J. Appl. Cryst. 11 (1978), 121.

Umangite, Cu_3Se_2, [G] Tet. Metallic red/violet-blue; red-violet in reflected light. Sierra de Umango, La Rioja Province, Argentina. H= 3, D_m= 6.4, D_c= 6.590, IIA 13. Klockmann, 1891. Am. Min. 35 (1950), 354. Struct. Repts. 42A (1976), 75.

Umbite, $K_2ZrSi_3O_9 \cdot H_2O$, [A] Orth. Vitreous colorless, yellowish. Lake Umba (near), Khibina massif, Kola Peninsula, USSR. H= 4.5, D_m= 2.79, D_c= 2.79, VIIID 09. Khomyakov et al., 1983. ZVMO 112 (1983), 461. Am. Min. 67(1984), 416 (Abst.). See also Kostylevite.

Umbozerite, $Na_3Sr_4ThSi_8(O, OH)_{24}$, [A] Amor. Vitreous green, greenish-brown. Umbozero, Kola Peninsula, USSR. H= 5, D_m= 3.60, D_c=?, VIIIE 17. Es'kova et al., 1974. Am. Min. 60(1975), 341 (Abst.).

Umohoite, $(UO_2)MoO_4 \cdot 4H_2O$, [G] Mon. Splendent blue-black, dark green. Freedom #2 mine, Marysville, Piute Co., Utah, USA. H=?, D_m= 4.53, D_c= 4.53, VIF 02. Brophy & Kerr, 1953. Am. Min. 42 (1957), 657. Struct. Repts. 28 (1963), 225.

Ungemachite, $K_3Na_8Fe(SO_4)_6(NO_3)_2 \cdot 6H_2O$, [G] Rhom. Vitreous colorless/pale yellow. Chile. H= 2.5, D_m= 2.287, D_c= 2.259, VID 16. Peacock & Bandy, 1936. Dana, 7th ed.(1951), v.2, 596. Am. Min. 71 (1986), 826.

Ungursaite, [D] Published without CNMMN approval. Am. Min. 71(1986), 1546 (Abst.). Voloshin et al., 1985.

Upalite, $Al(UO_2)_3(PO_4)_2O(OH) \cdot 7H_2O$, [A] Mon. Translucent amber-yellow. Kobokobo, Kivu, Zaïre. H=?, D_m= 3.5, D_c= 3.94, VIID 21. Deliens & Piret, 1979. Am. Min. 65(1980), 208 (Abst.). Bull. Min. 106 (1983), 383.

Uralborite, $CaB_2O_2(OH)_4$, [G] Mon. Transparent colorless. Turinsk, Ural Mts., USSR. H= 4, D_m= 2.60, D_c= 2.58, Vc 04c. Malinko, 1961. Am. Min. 47(1962), 1482 (Abst.). Min. Abstr. 70-3027. See also Vimsite.

Uralite, [D] = Actinolite pseudomorphous after a pyroxene. Am. Min. 63 (1978), 1023. Origin not known.

Uralolite, $Ca_2Be_4(PO_4)_3(OH)_3 \cdot 5H_2O$, [G] Mon. Silky/vitreous colorless, white. Ural Mts., USSR. H= 2.5, D_m= 2.1, D_c= 1.52, VIID 01. Grigor'ev, 1964. Min. Rec. 9 (1978), 99.

Uramphite, $NH_4(UO_2)PO_4 \cdot 3H_2O$, [G] Tet. Vitreous green. Unspecified locality, USSR. H=?, D_m= 3.70, D_c= 3.26, VIID 20b. Nekrasova, 1957. JCPDS 29-121.

Urancalcarite, $Ca(UO_2)_3CO_3(OH)_6 \cdot 3H_2O$, [A] Orth. Translucent yellow. Shinkolobwe, Shaba, Zaïre. H= 2-3, D_m= 4.03, D_c= 4.10, VbD 04. Deliens & Piret, 1984. Am. Min. 70(1985), 438 (Abst.).

Uraninite, UO_2, [G] Cub. Sub-metallic/pitchy black, brownish-black, greyish, greenish. Jáchymov (St. Joachimsthal), Západočeský kraj, Čechy (Bohemia), Czechoslovakia. H= 5-6, D_m= 10.6, D_c= 10.88, IVD 16b. Born, 1772. Min. Rec. 5 (1974), 79. See also Thorianite, Cerianite.

Uranmicrolite, $(U,Ca,Ce)_2(Ta,Nb)_2O_6(OH,F)$, [A] Cub. Translucent yellowish-brown, greenish-brown, brownish-black. Posse farm, Conçeicao Co., Minas Gerais, Brazil. H= 5.5, D_m= 5.8, D_c=?, IVC 09a. Strunz, 1957. Am. Min. 62 (1977), 403.

Uranoanatase, [D] Probably a uranian anatase. Min. Mag. 36 (1968), 1144. Vuorelainen et al., 1964.

Uranocircite, $Ba(UO_2)_2(PO_4)_2 \cdot 10H_2O$, [G] Tet. Pearly yellow-green. Bergen, Falkenstein (near), Voigtland, Sachsen (Saxony), Germany. H= 2-2.5, D_m= 3.53, D_c= 3.46, VIID 20a. Weisbach, 1877. Strunz (1970), 351. See also Meta-uranocircite.

Uranophane, $Ca(UO_2)_2(SiO_3OH)_2 \cdot 5H_2O$, [G] Mon. Translucent yellow. Miedzianka (Kupferberg), Silesia, Poland. H= 2-3, D_m= 3.9, D_c= 3.78, VIIIA 25. Websky, 1853. Am. Min. 66 (1981), 610. Acta Cryst. C44 (1988), 421. See also Beta-Uranophane.

Uranopilite, $(UO_2)_6SO_4(OH)_{10} \cdot 12H_2O$, [G] Mon. Silky yellow. Jáchymov (St. Joachimsthal), Západočeský kraj, Čechy (Bohemia), Czechoslovakia. H=?, D_m= 3.96, D_c=?, VID 08. Weisbach, 1882. Am. Min. 37 (1952), 950. See also Meta-uranopilite.

Uranosilite, $(UO_2)Si_7O_{15}$, [A] Orth. Vitreous yellowish-white. Menzenschwand, Schwarzwald (S), Germany. H=?, D_m=?, D_c= 3.25, VIIIA 25. Walenta, 1983. Am. Min. 69(1984), 408 (Abst).

Uranospathite, $HAl(UO_2)_4(PO_4)_4 \cdot 40H_2O$, [G] Tet. Translucent yellow. Redruth, Cornwall, England. H= 2-2.5, D_m= 2.50, D_c= 2.49, VIID 20a. Hallimond, 1915. Min. Mag. 42 (1978), 117.

Uranosphaerite, $Bi_2U_2O_9 \cdot 3H_2O$, [G] Orth. Greasy orange-yellow, brick-red. Weisser Hirsch mine, Neustädtel, Schneeberg (near), Sachsen (Saxony), Germany. H= 2-3, D_m= 6.36, D_c=?, IVF 14. Weisbach, 1873. Am. Min. 42 (1957), 905.

Uranospinite, $Ca(UO_2)_2(AsO_4)_2 \cdot 10H_2O$, [G] Tet. Translucent yellow. Neustädtel, Schneeberg (near), Sachsen (Saxony), Germany. H= 2-3, D_m= 3.45, D_c= 3.34, VIID 20a. Weisbach, 1873. Strunz (1970), 352. See also Meta-uranospinite.

Uranotungstite, $(Fe,Ba,Pb)(UO_2)_2WO_4(OH)_4 \cdot 12H_2O$, [A] Orth. Translucent yellow, orange, brownish. Menzenschwand, Schwarzwald (S) and Clara mine (Grube Clara), Wolfach, Schwarzwald, Baden-Württemberg, Germany. H= 2, D_m=?, D_c= 4.27, VIF 02. Walenta, 1985. Am. Min. 71(1986), 1547 (Abst.).

Uranpyrochlore, $(U,Ca,Ce)_2(Nb,Ta)_2O_6(OH,F)$, [A] Cub. Adamantine red, red-brown. Mitchell Co. mica mines, North Carolina, USA. H= 4-5, D_m= 4.509, D_c=?, IVC 09a. Hogarth, 1977. Am. Min. 62 (1977), 403. See also Ishikawaite.

Urea, $CO(NH_2)_2$, [A] Tet. Translucent pale yellow/brown. Toppin Hill, Western Australia, Australia. H=?, D_m=?, D_c= 1.33, IXE 01. Bridge, 1973. Min. Mag. 39 (1973), 346.

Uricite, $C_5H_4N_4O_3$, [A] Mon. Dingo Donga Cave, Western Australia, Australia. H=?, D_m= 1.851, D_c= 1.851, IXE 01. Klaproth, 1807. Min. Mag. 39 (1974), 889. Acta Cryst. 19 (1965), 286.

Ursilite, $(Mg,Ca)_4(UO_2)_4(Si_2O_5)_{5.5}(OH)_5 \cdot 13H_2O$, [G] Orth. Silky yellow. Unspecified locality, USSR. H= 2-3, D_m= 3.18, D_c= 3.23, VIIIA 25. Chernikov et al., 1957. ZVMO 106 (1977), 553.

Urvantsevite, $Pd(Bi,Pb)_2$, [A] Hex. Metallic greyish-white. Talnakh, Noril'sk (near), Siberia (N), USSR. H= 2, D_m=?, D_c= 9.6, IIC 07. Rudashevskii et al., 1976. Am. Min. 62(1977), 1260 (Abst.).

Ushkovite, $MgFe_2(PO_4)_2(OH)_2 \cdot 8H_2O$, [A] Tric. Vitreous/pearly/greasy pale yellow, orange-yellow, brown. Ilmen Mts., Ural Mts., USSR. H= 3.5, D_m= 2.38, D_c= 2.47, VIID 03. Chesnokov et al., 1983. Am. Min. 69(1984), 212 (Abst.).

Usovite, $CaBa_2MgAl_2F_{12}$, [A] Mon. Vitreous/greasy brown. Noiby river, Yeniseĭ, Siberia, USSR. H= 3.5, D_m= 4.18, D_c= 4.26, IIIB 04. Nozhkin et al., 1967. Am. Min. 60(1975), 739 (Abst.).

Ussingite, $Na_2AlSi_3O_8(OH)$, [G] Tric. Translucent reddish-violet. Kangerdluarsuk, Ilímaussaq, Greenland (S). H= 6-7, D_m= 2.49, D_c= 2.51, VIIIF 21. Bøggild, 1915. Dana/Ford (1932), 551. Am. Min. 59 (1974), 335.

Ustarasite, $Pb(Bi,Sb)_6S_{10}$, [G] Syst=? Metallic grey; white in reflected light. Ustarasaisk deposit, Tan-shan (W), Siberia, USSR. H= 2.5, D_m=?, D_c=?, IID 09e. Sakharova, 1955. Am. Min. 41(1956), 814 (Abst.).

Uvanite, $(UO_2)_2V_6O_{17} \cdot 15H_2O$(?), [Q] Orth. Translucent brownish-yellow. San Rafael Swell, Temple Rock, Emery Co., Utah, USA. H=?, D_m=?, D_c=?, VIID 23. Hess & Schaller, 1914. Dana, 7th ed.(1951), v.2, 1056.

Uvarovite, $Ca_3Cr_2(SiO_4)_3$, [G] Cub. Vitreous/resinous green. Saransk (Near), USSR. H= 7.5, D_m= 3.5, D_c= 3.82, VIIIA 06a. Hess, 1832. DHZ, 2nd ed.(1982), v.1A, 468. Am. Min. 56 (1971), 791. See also Grossular.

Uvite, $Ca(Mg,Fe)_3Al_5Mg(BO_3)_3Si_6O_{18}(OH)_4$, [G] Rhom. Vitreous black, brown, colorless, light green. Sri Lanka. H= 7.5, D_m= 2.97, D_c= 3.01, VIIIC 08. Kunitz, 1929. DHZ, 2nd ed.(1986), v.1B, 559.

Uytenbogaardtite, Ag_3AuS_2, [A] Tet. Metallic greyish-white. Tambang Sawah, Benkulen Dist., Sumatra, Indonesia; Comstock Lode, Virginia City, Storey Co., Nevada, USA; Smeinogorsk (Schlangenberg), Altai Mts., USSR. H= 2, D_m=?, D_c= 8.34, IIA 12. Barton et al., 1978. Can. Min. 16 (1978), 651.

Uzonite, As_4S_5, [A] Mon. Pearly yellow. Uson caldera, Kamchatka, USSR. H= 1.5, D_m=?, D_c= 3.38, IIB 19. Popova & Poljakov, 1985. Am. Min. 71(1986), 1280 (Abst.).

V

Vaesite, NiS_2, G Cub. Metallic grey. Kasompi mine, Shaba, Zaïre. H=?, D_m= 4.45, D_c= 4.435, IIC 05. Kerr, 1945. Am. Min. 74 (1989), 1168. Struct. Repts. 24 (1960), 231. See also Cattierite.

Valentinite, Sb_2O_3, G Orth. Adamantine colorless, white, yellowish, reddish. Chalanches, Allemont (near), Dauphiné, France. H= 2.5-3, D_m= 5.76, D_c= 5.412, IVC 02. Haidinger, 1845. Dana, 7th ed.(1944), v.1, 547. Acta Cryst. B30 (1974), 458. See also Senarmontite.

Vallachite, D Mixed-layer clay mineral. Min. Mag. 38 (1971), 103. Gita & Gita, 1962.

Valleriite, $2[(Fe, Cu)S] \cdot 1.53[(Mg, Al)(OH)_2]$, G Rhom. Metallic bronze; creamy bronze/purple in reflected light. Aurora mine, Nya-Kopparberg, Sweden. H= 1.5, D_m= 3.1, D_c= 3.21, IIE 01. Blomstrand, 1870. Am. Min. 57 (1972), 1037. Zts. Krist. 127 (1968), 73. See also Haapalaite, Tochilinite, Yushkinite.

Vanadinite, $Pb_5(VO_4)_3Cl$, G Hex. Sub-resinous/sub-adamantine orange, red, brown, yellow, etc. Zimapan, Hidalgo, Mexico. H= 2.7-3, D_m= 6.86, D_c= 6.953, VIIB 16. Kobell, 1838. Dana, 7th ed.(1951), v.2, 895. Can. Min. 27 (1989), 189.

Vanalite, $NaAl_8V_{10}O_{38} \cdot 30H_2O$, G Mon. Waxy/vitreous bright orange-yellow. Kara Tau, Dzhambul (near), Kazakhstan, USSR. H=?, D_m= 2.3, D_c= 2.150, VIIC 24. Ankinovich, 1962. Am. Min. 57(1972), 597 (Abst.).

Vandenbrandeite, $CuUO_2(OH)_4$, G Tric. Translucent dark green. Kolongwe, Shinkolobwe, Shaba, Zaïre. H= 4, D_m= 5.03, D_c= 5.26, IVF 12. Schoep, 1932. Am. Min. 36 (1951), 394.

Vandendriesscheite, $PbU_7O_{22} \cdot 12H_2O$, G Orth. Translucent yellowish-orange, orange. Kolongwe, Shinkolobwe, Shaba, Zaïre. H=?, D_m= 5.45, D_c= 5.86, IVF 13. Vaes, 1947. Am. Min. 45 (1960), 1026. See also Metavandendriesscheite.

Vanmeersscheite, $U(UO_2)_3(PO_4)_2(OH)_6 \cdot 4H_2O$, A Orth. Translucent yellow. Kobokobo, Kivu, Zaïre. H=?, D_m=?, D_c= 4.49, VIID 21. Piret & Deliens, 1982. Am. Min. 67(1982), 1077 (Abst.). Bull. Min. 105 (1982), 125.

Vanoxite, $V_6O_{13} \cdot 8H_2O$(?), Q Syst=? Opaque black. Bill Bryan Claim, Wild Steer Canyon, Paradox Valley (S), Colorado, USA. H=?, D_m=?, D_c=?, IVF 07. Hess, 1924. Dana, 7th ed.(1944), v.1, 601.

Vantasselite, $Al_4(PO_4)_3(OH)_3 \cdot 9H_2O$, A Orth. Pearly white. Stavelot massif, Bihain (near), Belgium. H= 2-2.5, D_m= 2.30, D_c= 2.312, VIIID 06. Fransolet, 1987. Am. Min. 73(1988), 931 (Abst.).

Vanthoffite, $Na_6Mg(SO_4)_4$, G Mon. Vitreous/pearly colorless. Wilhelmshall, Stassfurt, Sachsen (Saxony), Germany. H= 3.5, D_m= 2.694, D_c= 2.67, VIA 02. Kubierschky, 1902 . Dana, 7th ed.(1951), v.2, 430. Acta Cryst. 17 (1964), 1613.

Vanuralite, $Al(UO_2)_2(VO_4)_2(OH) \cdot 11H_2O$, G Mon. Translucent yellow. Mounana mine, Franceville, Haut-Ogooué, Gabon. H= 2, D_m= 3.62, D_c= 3.16, VIID 23. Branche et al., 1963. Am. Min. 56(1971), 639 (Abst.).

Vanuranylite, D Inadequate data. Min. Mag. 36 (1968), 1144. Buryanova et al., 1965.

Variscite, $AlPO_4 \cdot 2H_2O$, G Orth. Vitreous pale/emerald green. Variscia, Voitland, Germany. H= 3.5-4.5, D_m= 2.57, D_c= 2.59, VIIC 05. Breithaupt, 1837. Am. Min. 57 (1972), 36. Acta Cryst. B33 (1977), 263. See also Metavariscite, Strengite.

Varlamoffite, $(Sn, Fe)(O, OH)_2$, Q Tet. Earthy, yellow; vitreous/greasy orange-red. Mesaraba, Kivu, Zaïre. H=?, D_m= 2.6, D_c= 3.26, IVD 02. De Dycker, 1946. Min. Abstr. 83M/3628.

Varulite, $NaCaMn_3(PO_4)_3$, G Mon. Vitreous olive-green. Varuträsk, Sweden (N). H= 5, D_m= 3.5, D_c= 3.89, VIIA 05b. Quensel, 1937. Min. Mag. 43 (1979), 227. See also Hagendorfite.

Vashegyite, $Al_{11}(PO_4)_9(OH)_6 \cdot 38H_2O$, G Orth. Dull white, pale green/yellow, brownish. Vashegy mine, Szirk (near), Gömör, Hungary. H= 2-3, D_m= 1.93, D_c= 1.934, VIID 06. Zimanyi, 1909. Can. Min. 21 (1983),489.

Vaterite, $CaCO_3$, G Hex. Translucent colorless. Ballycraigy, Larne, Northern Ireland. H=?, D_m= 2.54, D_c= 2.659, VbA 02. Meigen, 1911. Min. Mag. 32 (1960), 533. Zts. Krist. 128 (1969), 182. See also Aragonite, Calcite.

Vaughanite, $TlHgSb_4S_7$, A Tric. Metallic; greenish/bluish grey in reflected light. Golden Giant orebody, Hemlo mine, Marathon, Ontario (E), Canada. H= 3-3.5, D_m=?, D_c= 5.62, IID 05c. Harris et al., 1989. Min. Mag. 53 (1989), 79.

Vauquelinite, $CuPb_2(CrO_4)(PO_4)(OH)$, G Mon. Adamantine/resinous green, brown. Berezov, Sverdlovsk, Ural Mts., USSR. H= 2.5-3, D_m= 6.16, D_c= 6.16, VIE 02. Berzelius, 1818. Bull. Min. 103 (1980), 469. Zts. Krist. 126 (1968), 433. See also Fornacite, Molybdofornacite.

Vauxite, $FeAl_2(PO_4)_2(OH)_2 \cdot 6H_2O$, G Tric. Vitreous blue. Siglo mine, Llallagua, Potosí, Bolivia. H= 3.5, D_m= 2.39, D_c= 2.41, VIID 03. Gordon, 1922. Dana, 7th ed.(1951), v.2, 974. Am. Min. 53 (1968), 1025. See also Gordonite.

Väyrynenite, $BeMnPO_4(OH,F)$, G Mon. Vitreous rose-red. Viitaniemi, Eräjärvi, Orivesi, Finland. H= 5, D_m= 3.22, D_c= 3.23, VIIB 01. Volborth, 1954. Am. Min. 41(1956), 371 (Abst.). Zts. Krist. 112 (1959), 275.

Veatchite, $Sr_2[B_5O_8(OH)]_2B(OH)_3 \cdot H_2O$, G Mon. Vitreous/silky colorless, white. Tick Canyon, Lang (near), Los Angeles Co., California, USA. H= 2, D_m= 2.66, D_c= 2.664, Vc 10. Switzer, 1938. Sov.Phys.Cryst. 16(1971), 236. Am. Min. 56 (1971), 1934. See also p-Veatchite, Veatchite-A.

Veatchite-A, $Sr_2[B_5O_8(OH)]_2B(OH)_3 \cdot H_2O$, A Tric. Pearly colorless. Emet colemanite deposit, Kiyahya, Turkey. H=?, D_m= 2.73, D_c= 2.77, Vc 10. Kumbasar, 1979. Am. Min. 64 (1979), 362. See also Veatchite, p-Veatchite.

Veenite, $Pb_2(Sb,As)_2S_5$, G Orth. Metallic grey; white in reflected light. Taylor pit, Madoc, Huntingdon Twp., Hastings Co., Ontario, Canada. H= 3.5, D_m= 5.92, D_c= 5.96, IID 08. Jambor, 1967. Can. Min. 9 (1967), 7. See also Cosalite.

Velikite, D Incomplete description. Am. Min. 62(1977), 1260 (Abst.). Kaplunnik et al., 1977.

Vermiculite, $Mg_{0.7}(Mg,Fe,Al)_6(Si,Al)_8O_{22}(OH)_2 \cdot 8H_2O$, G Mon. Greyish, brownish. Milbury, Worchester Co., Massachusetts, USA. H= 1-2, D_m= 2.756, D_c=?, VIIIE 08e. Webb, 1824. N.Jb.Min.Mh. (1988), 297. Am. Min. 51 (1966), 1124.

Vernadite, δ-$(Mn,Fe,Ca,Na)(O,OH)_2 \cdot nH_2O$, Q Hex. Black. USSR. H=?, D_m=?, D_c=?, IVF 05b. Betechtin, 1944. Am. Min. 64(1979), 1334 (Abst.). See also Nsutite, Pyrolusite, Ramsdellite.

Verplanckite, $Ba_4Mn_2Si_4O_{12}(OH,H_2O)_3Cl_3$, A Hex. Vitreous brownish-orange, light brownish-yellow. Big Creek and Rush Creek, Fresno Co., California, USA. H= 2.5-3, D_m= 3.52, D_c= 3.33, VIIIC 08. Alfors et al., 1965. Am. Min. 50 (1965), 1500. Acta Cryst. B29 (1973), 2019.

Versiliaite, $Fe_6Sb_6O_{16}S$, A Orth. Metallic black. Buca della Vena mine, Stazzema, Alpe Apuane, Toscana, Italy. H= 4.5, D_m= 5.12, D_c= 5.20, VIIA 15. Mellini et al., 1979. Am. Min. 64 (1979), 1230. Am. Min. 64 (1979), 1235.

Vertumnite, $Ca_4Al_4Si_4O_6(OH)_{24} \cdot 3H_2O$, A Mon. Vitreous colorless. Campo Morto, Montalto di Castro, Viterbo, Toscana, Italy. H= 5, D_m= 2.15, D_c= 2.15, VIIIE 21. Passaglia & Galli, 1977. Am. Min. 62(1977), 1061 (Abst.). Min. Abstr. 81-1207.

Vésignéite, $Cu_3Ba(VO_4)_2(OH)_2$, G Mon. Vitreous yellow-green, dark olive-green. Perm, Ural Mts. and Agalik, Uzbekistan, USSR; Freidrichsroda,

Thüringen, Germany. H= 3-4, D_m= 4.05, D_c= 4.765, VIIB 14. Guillemin, 1955. Min. Abstr. 80-753. Min. Abstr. 78-897.

Vesuvianite, $Ca_{19}Fe(Mg, Al)_8Al_4(SiO_4)_{10}(Si_2O_7)_4(OH)_{10}$, G Tet. Vitreous/resinous brown, green, yellow, pale blue. Mt. Vesuvius, Napoli, Campania, Italy. H= 6.5, D_m= 3.4, D_c= 3.42, VIIIB 17. Werner, 1795. Contr.Min.Petr. 89 (1985), 205. Min. Jour. 13 (1986), 1.

Veszelyite, $(Cu, Zn)_3PO_4(OH)_3 \cdot 2H_2O$, G Mon. Vitreous greenish-blue, dark blue. Moravicza (Vaskö), Banat, Romania. H=?, D_m= 3.4, D_c= 3.42, VIID 09. Schrauf, 1874. Dana, 7th ed.(1951), v.2, 916. Am. Min. 59 (1974), 573.

Vigezzite, $(Ca, Ce)(Nb, Ta, Ti)_2O_6$, A Orth. Translucent orange-yellow. Alpe Rosso and Pizzo Marcio, Orcesco (near), Valle Vigezzo, Novara, Piemonte, Italy. H= 4.5-5, D_m=?, D_c= 5.54, IVD 11. Graeser et al., 1979. Min. Mag. 43 (1979), 459. See also Aeschynite-(Ce), Rynersonite.

Viitaniemiite, $Na(Ca, Mn)AlPO_4(F, OH)_3$, A Mon. Vitreous grey, white. Viitaniemi, Eräjärvi, Orivesi, Finland. H= 5, D_m= 3.245, D_c= 3.242, VIIB 02. Lahti, 1981. Am. Min. 66(1981), 1102 (Abst.). Am. Min. 69 (1984), 961.

Vikingite, $Ag_5Pb_8Bi_{13}S_{30}$, A Mon. Ivigtut, Greenland (SW). H= 3.5, D_m=?, D_c= 6.940, IID 09b. Makovicky & Karup-Møller, 1977. Am. Min. 64(1979), 243 (Abst.).

Villamaninite, $(Cu, Ni, Co, Fe)S_2$, G Cub. Metallic grey. Villamanín, León, Spain. H= 4.5, D_m= 4.5, D_c= 4.50, IIC 05. Schoeller & Powell, 1919. Am. Min. 74 (1989), 1168.

Villiaumite, NaF, G Cub. Vitreous red, colorless. Roma Island, Los Islands, Guinea. H= 2-2.5, D_m= 2.78, D_c= 2.81, IIIA 02. Lacroix, 1908. Dana, 7th ed.(1951), v.2, 10.

Villyaellenite, $(Mn, Ca, Zn)_5(AsO_3OH)_2(AsO_4)_2 \cdot 4H_2O$, A Mon. Vitreous colorless, light pink. Ste.-Marie-aux-Mines, Vosges Mtns., Haut-Rhin, Alsace, France. H= 4, D_m= 3.69, D_c= 3.72, VIIC 02. Sarp, 1984. Am. Min. 71(1986), 1547 (Abst.). Am. Min. 73 (1988), 1172.

Vimsite, $CaB_2O_2(OH)_4$, A Mon. Vitreous colorless. Ural Mts., USSR. H= 4, D_m= 2.54, D_c= 2.56, Vc 04c. Shashkin et al., 1968. Am. Min. 54(1969), 1219 (Abst.). Min. Abstr. 77-2701. See also Uralborite.

Vincentite, $(Pd, Pt)_3(As, Sb, Te)$, A Syst=? Metallic light brownish-grey in reflected light. Riam Kanan River, Borneo (SE). H= 5, D_m=?, D_c=?, IIG 04. Stumpfl and Tarkian, 1974. Min. Mag. 39 (1974), 525.

Vinciennite, $Cu_{10}Fe_4Sn(As, Sb)S_{16}$, A Tet. Metallic orange. Chizeuil, Bourbon-Lancy, Saône-et-Loire, France. H= 4, D_m=?, D_c= 4.29, IIB 03b. Cesbron et al., 1985. Am. Min. 71(1986), 1280 (Abst.).

Vinogradovite, $(Na, Ca)_4Ti_4Si_8O_{26} \cdot (H_2O, K_3)$, G Mon. Vitreous colorless, white. Kola Peninsula, USSR. H= 4, D_m= 2.88, D_c= 2.97, VIIID 01b. Semenov et al., 1956. Am. Min. 42(1957), 308 (Abst.). Sov.Phys.Cryst. 29(1984), 403.

Violarite, $FeNi_2S_4$, G Cub. Metallic violet-grey. Vermilion mine, Sudbury (near), Ontario, Canada. H= 4.5-5.5, D_m= 4.6, D_c= 4.79, IIC 01. Lindgren & Davy, 1924. Min. Mag. 43 (1980), 733.

Virgilite, $LiAlSi_2O_6$, A Hex. Translucent colorless. Macusani, Peru. H= 5.5-6, D_m= 2.395, D_c= 2.399, VIIID 01b. French et al., 1978. Am. Min. 63 (1978), 461. Zts. Krist. 127 (1968), 327.

Viséite, $Ca_{10}Al_{24}(PO_4)_{14}(SiO_4)_6F_3O_{13} \cdot 72H_2O$, G Cub. Chalky white, blue. Visé, Liège (near), Belgium. H= 3-4, D_m= 2.2, D_c= 2.17, VIID 13. Melon, 1942. Min. Mag. 41 (1977), 437.

Vishnevite, $(Na, K, Ca)_8(Si_6Al_6)O_{24}(SO_4) \cdot 2H_2O$, G Hex. Translucent colorless, white, light blue. Vishnevye Gory, Ural Mts. (S), USSR. H= 5-6, D_m= 2.4, D_c= 2.37, VIIIF 05. Belyankin, 1931. ZVMO 118 (1989), 78. Can. Min. 22 (1984), 333.

Vismirnovite, $ZnSn(OH)_6$, [A] Cub. Vitreous pale yellow. Trudov deposit and Mushiston deposit, Tadzhikistan, USSR. H= 3.9, D_m=?, D_c= 4.073, IVF 06. Marshukova et al., 1981. Am. Min. 67(1982), 1077 (Abst.).

Vitusite-(Ce), $Na_3(Ce,La,Nd)(PO_4)_2$, [A] Orth. Vitreous pale pink, white, pale green. Ilímaussaq Intrusion, Greenland (S); Mt. Karnasurt and Mt. Sengistchorr, Kola Peninsula, USSR. H= 4.5, D_m= 3.60, D_c= 3.86, VIIA 04. Rønsbo et al., 1979. Am. Min. 65(1980), 812 (Abst.). Sov.Phys.Cryst. 25(1980), 650.

Vivianite, $Fe_3(PO_4)_2 \cdot 8H_2O$, [G] Mon. Vitreous/earthy colorless, pale blue, greenish-blue. St. Agnes, Cornwall, England. H= 1.5-2, D_m= 2.68, D_c= 2.696, VIIC 10. Werner, 1817. Dana, 7th ed.(1951), v.2, 742. Bull. Min. 103 (1980), 135.

Vladimirite, $Ca_5(AsO_4)_2(AsO_3OH)_2 \cdot 5H_2O$, [G] Mon. Translucent pale rose, colorless, white. Kovouaksi, Tuva, USSR. H=?, D_m= 3.14, D_c= 3.17, VIIC 02. Nefedov, 1953. Bull. Min. 87 (1964), 169. Zts. Krist. 157 (1981), 296.

Vlasovite (triclinic), $Na_2ZrSi_4O_{11}$, [G] Tric. Vitreous/pearly/greasy colorless. Ascension Island, Atlantic Ocean (S). H= 6, D_m= 2.97, D_c= 3.07, VIIID 18. Tikhonenkova & Kazakova, 1961. Min. Mag. 36 (1967), 233. Sov.Phys.Cryst. 19 (1974), 152.

Vlasovite (monoclinic), $Na_2ZrSi_4O_{11}$, [G] Mon. Colorless. Mt. Varnbed (near), Lovozero massif, Kola Peninsula, USSR. H= 6, D_m=?, D_c= 3.029, VIIID 18. Tikhonenkova & Kazakova, 1961. Powd. Diff. 2 (1987), 176.

Vochtenite, $(Fe^{2+},Mg)Fe^{3+}(UO_2)_4(PO_4)_4(OH) \cdot 12\text{–}13H_2O$, [A] Mon. Bronzy brown. Basset mine, Redruth, Cornwall, England. H= 2.5, D_m=?, D_c= 3.663, VIID 20c. Zwaan et al., 1989. Min. Mag. 53 (1989), 473.

Voglite, $Ca_2Cu(UO_2)(CO_3)_4 \cdot 6H_2O$, [G] Mon. Pearly green. Eliáš mine, Jáchymov (St. Joachimsthal), Západočeský kraj, Čechy (Bohemia), Czechoslovakia. H=?, D_m= 2.8, D_c= 3.06, VbD 04. Haidinger, 1853. Dana, 7th ed.(1951), v.2, 237. J. Appl. Cryst. 12 (1979), 616.

Volborthite, $Cu_3V_2O_7(OH)_2 \cdot 2H_2O$, [G] Mon. Vitreous/pearly dark olive-green, green, yellowish-green. Sisersk and Nizhniĭ Tagilsk, Ural Mts., USSR. H= 3.5, D_m= 3.42, D_c= 3.52, VIIC 04. Hess, 1837. Am. Min. 59 (1974), 372. N.Jb.Min.Mh. (1988), 385.

Volfsonite, [D] Am. Min. 73(1988), 441 (Abst.). Kovalenker et al., 1986.

Volkonskoite, $Ca_{0.3}(Cr,Mg)_2(Si,Al)_4O_{10}(OH)_2 \cdot 4H_2O$, [A] Mon. Translucent green, bluish-green. Mt. Efimiatsk, Ural Mts., Siberia, USSR. H=?, D_m=?, D_c=?, VIIIE 08a. Kämmerer, 1831. Clays Cl. Mins. 35(1987) 139.

Volkovite, $Sr_2B_{14}O_{17}(OH)_{12} \cdot 2H_2O$, [Q] Mon. Königshall-Hindenburg mine, Reyershausen, Hanover, Niedersachsen, Germany. H= 3-4, D_m=?, D_c= 2.380, Vc 08. Nefedov, 1953. Strunz (1970), 260.

Volkovskite, $Ca[B_3O_4(OH)_2]_2 \cdot H_2O$, [G] Mon. Vitreous colorless. Inder, Kazakhstan, USSR (?). H=?, D_m= 2.31, D_c= 2.39, Vc 06. Kondrat'eva et al., 1966. Am. Min. 51(1966), 1550 (Abst.).

Voltaite, $K_2Fe_8Al(SO_4)_{12} \cdot 18H_2O$, [G] Cub. Resinous greenish-black, black. Pozzuoli, Napoli (Naples), Italy. H= 3, D_m= 2.645, D_c= 2.662, VIC 14. Scacchi, 1841. Dana, 7th ed.(1951), v.2, 464. Struct. Repts. 39A (1973), 314.

Volynskite, $AgBiTe_2$, [G] Rhom. Metallic, pale purplish in reflected light. Armenia, USSR. H= 2-2.5, D_m=?, D_c= 8.014, IID 07. Bezsmertnaya & Soboleva, 1965. Am. Min. 51(1966), 531 (Abst.). See also Bohdanowiczite, Matildite.

Vonsenite, $(Fe,Mg)_2FeO_2(BO_3)$, [G] Orth. Silky black/ greenish-black. Old City quarry, Riverside, Riverside Co., California, USA. H=?, D_m= 4.7, D_c= 4.795, Vc 01c. Eakle, 1920. N.Jb.Min.Mh. H.3/4 (1974), 95. Am. Min. 68 (1983), 827. See also Ludwigite.

Vozhminite, $(Ni, Co)_4(As, Sb)S_2$, [A] Hex. Metallic brownish-yellow. Vozhmin massif, Karelia (NE), USSR. H= 4-5, D_m=?, D_c= 6.2, IIA 05. Rudashevskii et al., 1982. ZVMO 111 (1982), 480.

Vrbaite, $Hg_3Tl_4As_8Sb_2S_{20}$, [G] Orth. Submetallic grey-black; bluish-white in reflected light. Alšar (Allchar), Roždеn (near), Makedonija (Macedonia), Yugoslavia. H= 3.5, D_m= 5.30, D_c= 5.451, IID 05c. Ježek, 1912. Am. Min. 53(1968), 351 (Abst.). Zts. Krist. 134 (1971), 360.

Vuagnatite, $CaAlSiO_4(OH)$, [A] Orth. Vitreous white. Bögürtlencik Tepe, Doğanbaba, Burdur, Taurus Mts., Turkey (SW). H=?, D_m= 3.22, D_c= 3.42, VIIIA 26. Sarp et al., 1976. Am. Min. 61 (1976), 825. Am. Min. 61 (1976), 831.

Vulcanite, CuTe, [G] Orth. Metallic yellow/blue-grey. Good Hope mine, Vulcan, Gunnison Co., Colorado, USA. H= < 2, D_m=?, D_c= 7.11, IIB 05. Cameron & Threadgold, 1961. Am. Min. 46 (1961), 258.

Vuonnemite, $Na_5TiNb_2(Si_2O_7)_2O_2F_2 \cdot 2Na_3PO_4$, [A] Tric. Vitreous light yellow. Vuonnem River Valley, Khibina massif, Kola Peninsula, USSR. H= 2-3, D_m= 3.13, D_c= 3.18, VIIIB 11. Bussen et al., 1973. Am. Min. 69(1984), 569 (Abst.). GAC-MAC (1987) Abstracts, 41.

Vuorelainenite, $(Mn, Fe)(V, Cr)_2O_4$, [A] Cub. Brownish-grey in reflected light. Stäta (Doverstorp) deposit, Sweden (central). H= 6.3, D_m=?, D_c= 4.64, IVB 01. Zakrzewski et al., 1982. Can. Min. 20 (1982), 281.

Vyacheslavite, $UPO_4(OH) \cdot 2.5H_2O$, [A] Orth. Translucent green, dark green. Unspecified locality, USSR. H=?, D_m=?, D_c= 5.01, VIID 19. Belova et al., 1984. Am. Min. 70(1985), 878 (Abst.).

Vysotskite, (Pd, Ni)S, [G] Tet. Metallic greyish-white. Noril'sk, Siberia (N), USSR. H=?, D_m= 8.4, D_c= 6.728, IIB 16. Genkin et al., 1962. Am. Min. 63 (1978), 832. Acta Cryst. C41 (1985), 1829. See also Braggite.

Vyuntspakhkite-(Y), $Y_4Al_3Si_5O_{18}(OH)_5$, [A] Mon. Adamantine colorless. Kola Peninsula, USSR. H= 6-7, D_m= 4.02, D_c= 4.07, VIIIA 11. Voloshin et al., 1983. Am. Min. 69(1984), 1193 (Abst.). Sov.Phys.Cryst. 29(1984), 141.

W

Wadeite, $K_2ZrSi_3O_9$, G Hex. Translucent colorless. Walgidee Hills, Fitzroy Basin, Kimberley Div., Western Australia, Australia. H=?, D_m= 3.10, D_c= 3.123, VIIIC 01. Prider, 1939. Min. Mag. 30 (1955), 585. Sov.Phys.Cryst. 22 (1977), 31.

Wadsleyite, β–$(Mg, Fe)_2SiO_4$, A Orth. Transparent pale fawn. Peace River meteorite. H=?, D_m=?, D_c= 3.84, VIIIA 03. Price et al., 1983. Can. Min. 21 (1983), 29. See also Forsterite, Ringwoodite.

Wagnerite, $(Mg, Fe)_2PO_4F$, G Mon. Vitreous yellow, greyish, red, greenish. Höllengraben, Werfen (near), Salzburg, Austria. H= 5-5.5, D_m= 3.15, D_c= 3.11, VIIB 03a. Fuchs, 1821. Dana, 7th ed.(1951), v.2, 845. Struct. Repts. 32A (1967), 362.

Wairakite, $Ca(Al_2Si_4)O_{12} \cdot 2H_2O$, G Mon. Vitreous/dull colorless, white. Lake Taupo (near), Wairakei, New Zealand. H= 5.5-6, D_m= 2.26, D_c= 2.278, VIIIF 11. Steiner, 1955. Am. Min. 65 (1980), 1212. Am. Min. 64 (1979), 993.

Wairauite, CoFe, G Cub. Metallic white. Red Hill, Wairau Valley, South Island, New Zealand. H= 4.5, D_m=?, D_c= 8.43, IA 05. Challis & Long, 1964. Min. Mag. 33 (1964), 942.

Wakabayashilite, $SbAs_{10}S_{18}$, A Mon. Silky/resinous yellow. White Caps mine, Manhatten, Nye Co., Nevada, USA. H= 1.5, D_m= 3.96, D_c= 4.06, IIB 19. Kato et al., 1970. Can. Min. 13 (1975), 418.

Wakefieldite-(Ce), $(Ce, Pb)VO_4$, A Tet. Black; yellow in transmitted light. Kusu deposit, Kinshasa (SW), Zaïre. H= 4.5, D_m=?, D_c= 5.30, VIIA 10. Deliens & Piret, 1977. Bull. Min. 109 (1986), 305. See also Wakefieldite-(Y).

Wakefieldite-(Y), YVO_4, A Tet. Translucent yellow. Evans-Lou mine, St-Pierre-de-Wakefield (near), Portland Twp., Papineau Co., Québec, Canada. H= 5, D_m=?, D_c= 4.25, VIIA 10. Miles et al., 1971. Am. Min. 56 (1971), 395. See also Chernovite-(Y), Xenotime-(Y), Wakefieldite-(Ce).

Walentaite, $H_4Ca_4Fe_{12}(AsO_4)_{10}(PO_4)_6 \cdot 28H_2O$, A Orth. Translucent greenish-yellow. White Elephant mine, Pringle, Custer Co., South Dakota, USA. H=?, D_m= 2.72, D_c= 2.74, VIID 15b. Dunn et al., 1984. Austral. Min. 2 (1987) (1), 9.

Wallisite, $CuPbTlAs_2S_5$, G Tric. Lengenbach quarry, Binntal, Valais (Wallis), Switzerland. H=?, D_m=?, D_c= 5.71, IID 05c. Nowacki, 1965. Am. Min. 54(1969), 1497 (Abst.). Zts. Krist. 127 (1968), 349. See also Hatchite.

Wallkilldellite, $Ca_4Mn_6(AsO_4)_4(OH)_8 \cdot 18H_2O$, A Hex. Vitreous dark red. Sterling Hill mine, Ogdensburg, Sussex Co., New Jersey, USA. H= 3, D_m= 2.85, D_c= 2.90, VIID 12. Dunn & Peacor, 1983. Am. Min. 68 (1983),1029. See also Kittatinnyite.

Walpurgite, $Bi_4O_4(UO_2)(AsO_4)_2 \cdot 2H_2O$, G Tric. Adamantine/greasy yellow. Weisser Hirsch mine, Neustädtel, Schneeberg (near), Sachsen (Saxony), Germany. H= 3.5, D_m=?, D_c= 6.69, VIID 19. Weisbach, 1871. Dana, 7th ed.(1951), v.2, 796. Min. Abstr. 83M/1226.

Walstromite, $BaCa_2Si_3O_9$, A Tric. Sub-vitreous/pearly white, colorless. Big Creek and Rush Creek, Fresno Co., California, USA. H=?, D_m= 3.67, D_c= 3.73, VIIIC 02. Alfors et al., 1965. Am. Min. 50 (1965), 314. Am. Min. 53 (1968), 9.

Wardite, $NaAl_3(PO_4)_2(OH)_4 \cdot 2H_2O$, G Tet. Vitreous blue-green, pale green, colorless. Clay Canyon, Fairfield, Utah Co., Utah, USA. H= 5, D_m= 2.81, D_c= 2.805, VIID 13. Davison, 1896. Am. Min. 37 (1952), 849. Min. Mag. 37 (1970), 598. See also Cyrilovite.

Wardsmithite, $Ca_5Mg(B_4O_7)_6 \cdot 30H_2O$, A Hex. Vitreous white. Hard Scrabble Claim, Death Valley National Monument, Inyo Co., California, USA. H= 2.5, D_m= 1.88, D_c=?, Vc 07b. Erd et al., 1970. Am. Min. 53 (1970), 349.

Warikahnite, $Zn_3(AsO_4)_2 \cdot 2H_2O$, [A] Tric. Translucent pale yellow, colorless. Tsumeb, Namibia. H=?, D_m= 4.24, D_c= 4.29, VIIC 03. Keller et al., 1979. Am. Min. 65(1980), 408 (Abst.). Struct. Repts. 46A (1980), 341.

Warthaite, [D] A mixture of cosalite and galena. Am. Min. 49 (1964), 1501 (Abst.). Krenner, 1909.

Warwickite, $(Mg, Ti, Fe, Al)_2O(BO_3)$, [G] Orth. Dull/sub-metallic/pearly brown, black. Warwick, Orange Co., New York, USA. H= 3.5-4, D_m= 3.35, D_c= 3.402, Vc 01c. Shepard, 1838. Min. Abstr. 79-1141. Am. Min. 59 (1974), 985.

Watkinsonite, $Cu_2PbBi_4(Se, S, Te)_8$, [A] Mon. Metallic black; bluish-white in reflected light. Otish Mts., Québec, Canada. H= 3.5, D_m=?, D_c= 7.82, IID 16. Johan et al., 1987. Can. Min. 25 (1987), 625.

Wattevillite, $Na_2Ca(SO_4)_2 \cdot 4H_2O$(?), [Q] Syst=? Silky white. Bauersberg, Bischofsheim, Bayern (Bavaria), Germany. H=?, D_m= 1.81, D_c=?, VIC 12. Singer, 1879. Dana, 7th ed., vol. 2, 452.

Wavellite, $Al_3(PO_4)_2(OH, F)_3 \cdot 5H_2O$, [G] Orth. Vitreous/pearly/resinous green, yellow, white, brown, etc. Barnstable, Devonshire, England. H= 3.2-4, D_m= 2.36, D_c= 2.325, VIID 06. Babington, 1805. Dana, 7th ed.(1951), v.2, 962. Zts. Krist. 127 (1968), 21.

Wawayandaite, $Ca_{12}Be_{18}Mn_4B_2Si_{12}O_{46}(OH, Cl)_{30}$, [A] Mon. Pearly/dull colorless, white. Franklin mine, Franklin, Sussex Co., New Jersey, USA. H= 1, D_m= 3.0, D_c= 2.98, VIIIA 02. Dunn et al., 1990. Am. Min. 75 (1990), 405.

Waylandite, $(Bi, Ca)Al_3(PO_4, SiO_4)_2(OH)_6$, [A] Rhom. Vitreous white. Wampiro Hill, Busiro Co., Buganda, Uganda. H= 4-5, D_m=?, D_c= 4.08, VIIB 15b. Von Knorring & Mrose, 1962. Min. Mag. 50 (1986), 730.

Weberite, Na_2MgAlF_7, [G] Orth. Vitreous light grey. Ivigtut, Greenland (SW). H= 3.5, D_m= 2.955, D_c= 2.948, IIIB 05. Bøgvad, 1938. Am. Min. 34 (1949), 383. TMPM 25 (1978), 57.

Weddellite, $CaC_2O_4 \cdot 2H_2O$, [G] Tet. Transparent colorless, white. Weddell Sea, Antarctica. H= 4, D_m= 1.94, D_c= 1.936, IXA 01. Frondel & Prien, 1942. Acta Cryst. 18 (1965), 917. Am. Min. 65 (1980), 327.

Weeksite, $K_2(UO_2)_2Si_6O_{15} \cdot 4H_2O$, [G] Orth. Waxy/silky yellow. Autunite #8 Claim, Thomas Range, Juab Co., Utah, USA. H=?, D_m= 4.1, D_c= 4.02, VIIIA 25. Outerbridge et al., 1960. ZVMO 106 (1977), 553. Am. Min. 66 (1981), 610.

Wegscheiderite, $Na_5(HCO_3)_3(CO_3)$, [G] Tric. Vitreous colorless. Perkins Well #1, Sweetwater Co., Wyoming, USA. H= 2.5-3, D_m= 2.34, D_c= 2.334, VbA 01. Fahey & Yorks, 1963. Am. Min. 48 (1963), 404.

Wehrlite, [D] A mixture of hessite and a bismuth telluride. Am. Min. 69 (1984), 215 (Abst.). Huot, 1841.

Weibullite, $Ag_{0.3}Pb_{5.3}Bi_{8.3}Se_6S_{12}$, [A] Orth. Metallic grey; light grey in reflected light. Falun, Kopparberg, Sweden. H= 2-2.5, D_m= 7.0, D_c= 7.28, IID 09e. Flink, 1910. Am. Min. 65 (1980), 789. Can. Min. 18 (1980), 1.

Weibyeite, [D] A mixture of bastnäsite and ancylite. Am. Min. 49 (1964), 1154 (Abst.). Brøgger, 1890.

Weilerite, [D] Inadequate compositional data. Min. Mag. 36 (1967), 133. Walenta & Wimmenauer, 1961.

Weilite, $Ca(AsO_3OH)$, [A] Tric. Porcelanous/greasy white. Schneeberg, Sachsen (Saxony), Germany. H=?, D_m= 3.50, D_c= 3.541, VIIA 09. Herpin & Pierrot, 1963. Am. Min. 49(1964), 816 (Abst.). Acta Cryst. B26 (1970), 354. See also Monetite.

Weishanite, $(Au, Ag)_3Hg_2$, [A] Hex. Metallic light yellow. Poshan Mining Dist., Tongbai, Hunan, China. H= 2.4, D_m=?, D_c= 18.17, IA 06. Li et al., 1984. Am. Min. 73(1988), 196 (Abst.).

Weissbergite, $TlSbS_2$, [A] Tric. Metallic grey; creamy-white in reflected light. Carlin mine, Elko (NW), Eureka Co., Nevada, USA. H= 1.5, D_m= 5.79, D_c= 6.1, IID 14. Dickson & Radtke, 1978. Am. Min. 63 (1978), 720.

Weissite, $Cu_{2-x}Te$, [G] Hex. Metallic bluish-black. Good Hope mine, Vulcan, Gunnison Co., Colorado, USA. H= 3, D_m= 6, D_c= 8.913, IIA 01b. Trolle-Wachtmeister, 1828. Dana, 7th ed.(1944), v.1, 199.

Welinite, $Mn_3(W,Mg)_{0.7}(SiO_4)(O,OH)_3$, [A] Hex. Resinous red-brown, reddish-black. Långban mine, Filipstad (near), Värmland, Sweden. H= 4, D_m= 4.47, D_c= 4.41, VIIIA 17. Moore, 1967. Am. Min. 71 (1986), 1522. Struct. Repts. 34A (1969), 387.

Wellsite, $(Ba,Ca,K_2)(Si_6Al_2)O_{16}\cdot 6H_2O$, [Q] Orth. Translucent colorless, white. Cullakanee mine, Buck Creek, Clay Co., North Carolina, USA. H=?, D_m=?, D_c=?, VIIIF 14. Pratt & Foote, 1897. Bull. Min. 111 (1988), 671. See also Harmotome, Phillipsite.

Weloganite, $Na_2(Sr,Ca)_3Zr(CO_3)_6\cdot 3H_2O$, [A] Tric. Vitreous yellow, amber. Francon quarry, St.-Michel, Montreal Island, Québec, Canada. H= 3.5, D_m= 3.22, D_c= 3.49, VbC 03a. Sabina et al., 1968. Can. Min. 9 (1968), 468. Can. Min. 13 (1975), 209. See also Donnayite-(Y), Mckelveyite-(Y).

Welshite, $Ca_2Mg_4Be_2FeSbSi_4O_{20}$, [A] Tric. Sub-adamantine reddish-brown/black. Långban mine, Filipstad (near), Värmland, Sweden. H= 5, D_m= 3.77, D_c= 3.71, VIIID 07. Moore, 1978. Min. Mag. 42 (1978), 129.

Wendwilsonite, $Ca_2(Mg,Co)(AsO_4)_2\cdot 2H_2O$, [A] Mon. Vitreous pink, red. Bou Azzer, Morocco and Sterling Hill mine, Ogdensburg, New Jersey, USA. Also Coahuila, Mexico. H= 3-4, D_m= 3.52, D_c= 3.57, VIIC 12. Dunn et al., 1987. Am. Min. 72 (1987), 217.

Wenkite, $Ba_4Ca_6(Si,Al)_{20}O_{39}(OH)_2(SO_4)_3\cdot nH_2O$, [G] Hex. Translucent grey. Candoglia, Novara, Piemonte, Italy. H= 6, D_m= 3.10, D_c= 3.16, VIIIF 05. Papageorgakis, 1962. Am. Min. 48(1963), 213 (Abst.). Zts. Krist. 137 (1973), 113.

Werdingite, $(Mg,Fe)_2Al_{14}Si_4B_4O_{37}$, [A] Tric. Vitreous brownish-yellow. Bok se Puts, Namaqualand, South Africa. H= 7, D_m= 3.04, D_c= 3.07, VIIIA 24. Moore et al., 1990. Am. Min. 75 (1990), 415.

Wermlandite, $CaMg_7(Al,Fe)(SO_4)_2(OH)_{18}\cdot 12H_2O$, [A] Rhom. Translucent pale greenish-grey. Långban mine, Filipstad (near), Värmland, Sweden. H= 1.5, D_m= 1.932, D_c= 1.96, IVF 03e. Moore, 1971. Am. Min. 57(1972), 327 (Abst.). Zts. Krist. 168 (1984), 133. See also Hydrotalcite.

Westerveldite, $(Fe,Ni,Co)As$, [G] Orth. Metallic brownish-white/grey. La Gallego, Spain. H=?, D_m=?, D_c= 8.13, IIB 09d. Oen et al., 1972. Am. Min. 57 (1972), 354.

Wheatleyite, $Na_2Cu(C_2O_4)_2\cdot 2H_2O$, [A] Tet. Translucent bright blue. Wheatley mine, Phoenixville (near), Chester Co., Pennsylvania, USA; Nishimomaki mine, Gumma, Japan. H= 1-2, D_m= 2.27, D_c= 2.250, IXA 01. Rouse et al., 1986. Am. Min. 71 (1986), 1240. Acta Cryst. B36 (1980), 2145.

Wherryite, $CuPb_4O(SO_4)_2(CO_3)(OH,Cl)_2$, [G] Mon. Translucent pale yellow, yellowish-green, light green. Mammoth mine, Tiger, Pinal Co., Arizona, USA. H=?, D_m= 6.45, D_c= 7.22, VIB 11. Fahey et al., 1950. Am. Min. 55 (1970), 505.

Whewellite, $CaC_2O_4\cdot H_2O$, [G] Mon. Vitreous/pearly colorless, yellowish, brownish. Havre (near), Montana, USA. H= 2.5-3, D_m= 2.23, D_c= 2.218, IXA 01. Brooke & Miller, 1852. Am. Min. 53 (1968), 455. Am. Min. 65 (1980), 327.

Whiteite-(CaFeMg), $Ca(Fe,Mn)Mg_2Al_2(PO_4)_4(OH)_2\cdot 8H_2O$, [A] Mon. Translucent tan. Ilhade Tequaral, Minas Gerais, Brazil. H= 3-4, D_m= 2.58, D_c= 2.51, VIID 24. Moore & Ito, 1978. Min. Mag. 42 (1978), 309.

Whiteite-(MnFeMg), $MnFeMg_2Al_2(PO_4)_4(OH)_2 \cdot 8H_2O$, [A] Mon. Translucent brown. Ilhade Tequaral, Minas Gerais, Brazil. H= 3-4, D_m= 2.67, D_c= 2.62, VIID 24. Moore & Ito, 1978. Min. Mag. 42 (1978), 309.

Whiteite-(CaMnMg), $CaMnMg_2Al_2(PO_4)_4(OH)_2 \cdot 8H_2O$, [A] Mon. Transparent yellow, greenish-yellow, pink, pale lavender. Tip Top pegmatite, Custer, Custer Co., South Dakota, USA. H= 3.5, D_m= 2.63, D_c= 2.64, VIID 24. Grice et al., 1989. Can. Min. 27 (1989), 699.

Whitlockite, $Ca_{18}(Mg,Fe)_2(H_2,Ca)(PO_4)_{14}$, [G] Rhom. Vitreous/sub-resinous colorless, white, grey, yellowish. Palermo #1 Pegmatite, North Groton, Grafton Co., New Hampshire, USA. H= 5, D_m= 3.12, D_c= 3.102, VIIA 09. Frondel, 1940. Dana, 7th ed.(1951), v.2, 684. Am. Min. 60 (1975), 120.

Whitmoreite, $Fe_3(PO_4)_2(OH)_2 \cdot 4H_2O$, [A] Mon. Vitreous/sub-adamantine pale tan, dark brown, greenish-brown. Palermo #1 mine, North Groton, Grafton Co., New Hampshire, USA. H= 3, D_m= 2.87, D_c= 2.85, VIID 27. Moore et al., 1974. Am. Min. 59 (1974), 900. Am. Min. 59 (1974), 900. See also Arthurite, Earlshannonite, Ojuelaite.

Wickenburgite, $CaPb_3Al_2Si_{10}O_{24}(OH)_6$, [A] Hex. Transparent colorless, pink. Potter–Cramer claim, Wickenburg (near), Maricopa Co., Arizona, USA. H= 5, D_m= 3.85, D_c= 3.88, VIIIE 23. Williams, 1968. Am. Min. 53 (1968), 1433.

Wickmanite, $MnSn(OH)_6$, [A] Cub. Translucent brownish-yellow. Långban mine, Filipstad (near), Värmland, Sweden. H=?, D_m= 3.89, D_c= 3.82, IVF 06. Moore & Smith, 1967. Am. Min. 53(1968), 1063 (Abst.). See also Tetrawickmanite.

Wicksite, $NaCa_2MgFe(Fe,Mn)_4(PO_4)_6 \cdot 2H_2O$, [A] Orth. Sub-metallic dark blue. Big Fish River, Yukon Territory, Canada. H= 4.5-5, D_m= 3.54, D_c= 3.58, VIIC 17. Sturman et al., 1981. Can. Min. 19 (1981), 377.

Widenmannite, $Pb_2UO_2(CO_3)_3$, [G] Orth. Pearly/silky yellow. Michael mine, Weiler, Lehr, Schwarzwald, Baden-Würtemberg, Germany. H=?, D_m=?, D_c= 6.89, VbD 04. Walenta & Wimmenauer, 1961. Min. Abstr. 77-2184.

Wightmanite, $Mg_5O(BO_3)(OH)_5 \cdot 2H_2O$, [G] Mon. Vitreous colorless. Crestmore, Riverside Co., California, USA. H= 5.5, D_m= 2.59, D_c= 2.780, Vc 01f. Murdoch, 1962. Am. Min. 47 (1962), 718. Am. Min. 59 (1974), 985.

Wiikite, [D] A mixture of yttropyrochlore, euxenite and other minerals. Am. Min. 62 (1977), 403. Ramsay, 1899.

Wilcoxite, $MgAl(SO_4)_2F \cdot 18H_2O$, [A] Tric. Translucent colorless, white. Lone Pine mine, Silver City, Grant Co., New Mexico, USA. H= 2, D_m= 1.58, D_c= 1.67, VID 09. Williams & Cesbron, 1983. Min. Mag. 47 (1983), 37.

Wilhelmvierlingite, $CaMnFe(PO_4)_2(OH) \cdot 2H_2O$, [A] Orth. Translucent pale yellow, brownish. Hagendorf pegmatite, Oberpfalz, Bayern (Bavaria), Germany. H= 4, D_m= 2.58, D_c= 2.60, VIID 25. Mücke, 1983. Am. Min. 69 (1984), 568 (Abst.).

Wilkinsonite, $NaFe^{2+}_2Fe^{3+}Si_3O_{10}$, [A] Tric. Black with brown streak and vitreous luster. Warrumbungle Volcano, New South Wales, Australia. H= 5, D_m=?, D_c= 3.89, Duggan, 1990. Am. Min. 75 (1990), 694.

Wilkmanite, Ni_3Se_4, [G] Mon. Metallic greyish-yellow. Kuusamo, Finland (NE). H=?, D_m=?, D_c= 6.96, IIB 09c. Vuorelainen et al., 1964. Am. Min. 50(1965), 519 (Abst.). See also Trüstedtite.

Willemite, Zn_2SiO_4, [G] Rhom. Vitreo-resinous white, greenish-yellow, green, red, etc. Franklin Furnace or Sterling Hill, Sussex Co., New Jersey, USA. H= 5.5, D_m= 4.1, D_c= 4.17, VIIIA 01. Lévy, 1830. Dana/Ford (1932), 601. Sov.Phys.Cryst. 15(1970), 314.

Willemseite, $(Ni,Mg)_3Si_4O_{10}(OH)_2$, [G] Mon. Translucent light green. Scotia Talc mine, Bon Accord, Barberton Dist., Transvaal, South Africa. H= 2, D_m= 3.3, D_c= 3.348, VIIIE 04. De Waal, 1970. Am. Min. 55 (1970), 31. See also Talc, Minnesotaite.

Willhendersonite, $KCa(Al_3Si_3)O_{12} \cdot 5H_2O$, [A] Tric. Vitreous colorless. San Venanzo quarry, Perugia, Umbria, Italy. H= 3, D_m= 2.18, D_c= 2.173, VIIIF 15. Peacor et al., 1984. Am. Min. 69 (1984), 186. N.Jb.Min.Mh. (1985), 547.

Willyamite, (Co, Ni)SbS, [A] Cub. Metallic white. A.B.H. Consols mine, Broken Hill, New South Wales, Australia. H= 5.5, D_m=?, D_c= 7.05, IIC 06b. Pittman, 1893. Am. Min. 56(1971), 361 (Abst.). See also Ullmannite.

Winchite, $NaCa(Mg, Fe)_5(Si, Al)_8O_{22}(OH)_2$, [A] Mon. Translucent blue. India (central). H= 5-6, D_m= 2.97, D_c= 2.96, VIIID 05c. Fermor, 1906. JCPDS 20-1390. See also Ferrowinchite, Alumino-winchite.

Winebergite, $Al_4SO_4(OH)_{10} \cdot 7H_2O$(?), [Q] Syst=? White. Löwmühl, Passau (Near), Bavaria, Germany. H=?, D_m=?, D_c=?, IVF 02. Gümbel, 1868. Dana, 7th ed., vol. 2, 586.

Winstanleyite, $TiTe_3O_8$, [A] Cub. Translucent yellow, tan, cream. Grand Central mine, Tombstone, Cochise Co., Arizona, USA. H= 4, D_m= 5.57, D_c= 5.632, VIG 06. Williams, 1979. Min. Mag. 43 (1979), 453.

Wiserite, $(Mn, Mg)_{14}(B_2O_5)_4(OH)_8 \cdot (Si, Mg)(O, OH)_4Cl$, [G] Tet. Silky/vitreous white, pinkish-brown, reddish. Naus mine, Sargans (near), Gonzen (near), Bergwerk, St. Gall, Switzerland. H= 2.5, D_m= 3.54, D_c= 3.57, Vc 02. Haidinger, 1845. Am. Min. 74 (1989), 1374. Am. Min. 74 (1989), 1351.

Witherite, $BaCO_3$, [G] Orth. Vitreous/resinous colorless, white, greyish. Alston Moor, Cumberland, England. H= 3-3.5, D_m= 4.291, D_c= 4.314, VbA 04. Werner, 1789. Am. Min. 64 (1979), 742. Am. Min. 56 (1971), 758.

Wittichenite, Cu_3BiS_3, [G] Orth. Grey/white; white/creamy-white in reflected light. Wittichen (near), Schwarzwald, Baden-Württemberg, Germany. H= 3.5, D_m= 6.01, D_c= 6.11, IID 06a. Kobell, 1853. Min. Mag. 43 (1979), 109. Acta Cryst. B29 (1973), 2528.

Wittite, $Pb_{0.35}Bi_{0.44}(S, Se)$, [G] Mon. Metallic grey. Falun, Kopparberg, Sweden. H= 2-2.5, D_m= 7.12, D_c=?, IID 09e. Johansson, 1924. Am. Min. 65 (1980), 789.

Wodginite, $Mn(Sn, Ta)Ta_2O_8$, [G] Mon. Resinous reddish-brown, dark brown. Wodgina, Western Australia, Australia. H=?, D_m= 7.19, D_c= 7.80, IVD 09. Nickel et al., 1963. Can. Min. 7 (1963), 390. Can. Min. 14 (1976), 550.

Wöhlerite, $Na_2Ca_4ZrNb(Si_2O_7)_2(O, F)_4$, [G] Mon. Vitreous/resinous pale yellowish, white, brownish, greyish. Langesundfjord, Norway. H= 5-6, D_m= 3.316, D_c= 3.192, VIIIB 08. Scheerer, 1843. Min. Abstr. 75-460. Min. Abstr. 81-1197.

Wolfeite, $(Fe, Mn)_2PO_4(OH)$, [G] Mon. Vitreous/greasy/adamantine reddish-brown, dark brown. Palermo mine, North Groton, New Hampshire, USA. H= 4.5-5, D_m= 3.82, D_c= 3.90, VIIB 03b. Frondel, 1949. Powd. Diff. 4 (1989), 34. See also Satterlyite, Triploidite.

Wolframite, (Fe, Mn)WO4, [G] Mon. Sub-metallic/adamantine greyish-black. H= 4-4.5, D_m= 7.3, D_c= 7.46, IVD 08. Wallerius, 1747. Dana, 7th ed.(1951), v.2, 1064. Zts. Krist. 144 (1976), 238. See also Hübnerite, Ferberite.

Wolframo-ixiolite, [D] Inadequate data; probably tungstenian ixiolite. Min. Mag. 43 (1980), 1055. Borneman-Starynkevich, 1976.

Wollastonite-1T, $CaSiO_3$, [G] Tric. Translucent white, colorless. H= 4.5-5, D_m= 3.0, D_c= 2.911, VIIID 08. Leman, 1818. Min. Jour. 7 (1973), 180. Am. Min. 63 (1978), 274.

Wollastonite-2M, $CaSiO_3$, [P] Mon. Translucent white, colorless. H= 4.5-5, D_m= 3.0, D_c= 2.92, VIIID 08. Peacock, 1935. Min. Abstr. 69-1046. Zts. Krist. 168 (1984), 93.

Wollastonite-3T, $CaSiO_3$, [P] Tric. Kushiro, Hiroshima, Japan. H=?, D_m=?, D_c= 2.93, VIIID 08. Henmi et al., 1983. Am. Min. 68 (1983), 156.

Wollastonite-4T, $CaSiO_3$, [P] Tric. Kushiro, Hiroshima, Japan. H=?, D_m=?, D_c=?, VIIID 08. Henmi et al., 1983. Am. Min. 68 (1983), 156.

Wollastonite-5T, $CaSiO_3$, P Tric. Kushiro, Hiroshima, Japan. H=?, D_m=?, D_c= 2.95, VIIID 08. Henmi et al., 1983. Am. Min. 68 (1983), 156.

Wollastonite-7T, $CaSiO_3$, P Tric. Fuka, Okayama, Japan. H=?, D_m=?, D_c= 2.90, VIIID 08. Henmi et al., 1978. Am. Min. 64(1979), 658 (Abst.).

Wölsendorfite, $(Pb, Ca)U_2O_7 \cdot 2H_2O$, G Orth. Translucent orange-red. Wölsendorf, Bayern (Bavaria), Germany. H=?, D_m= 6.8, D_c= 6.82, IVF 12. Protas, 1957. Am. Min. 42(1957), 919 (Abst.).

Wonesite, $(Na, K)(Mg, Fe, Al)_6(Si, Al)_8O_{20}(OH, F)_4$, A Mon. Similar to phlogopite. Post Pond Volcanics, Mt. Cube Quadrangle (SW), Vermont, USA. H=?, D_m=?, D_c= 2.875, VIIIE 05b. Spear et al., 1981. Am. Min. 68 (1983), 554.

Woodhouseite, $CaAl_3(SO_4)(PO_4)(OH)_6$, G Rhom. Vitreous colorless, flesh-colored, white. Champion Andalusite mine, Mono Co., California, USA. H= 4.5, D_m= 3.012, D_c= 2.972, VIB 03b. Lemmon, 1937. Dana, 7th ed.(1951), v.2, 1006. Min. Abstr. 77-4077.

Woodruffite, $(Zn, Mn)Mn_3O_7 \cdot 1\text{–}2H_2O$, G Tet. Dull black. Sterling Hill, Sussex Co., New Jersey, USA. H= 6-7.5, D_m= 4.01, D_c= 3.98, IVF 05a. Frondel, 1953. Min. Mag. 33 (1963), 506. See also Torodokite.

Woodwardite, $(Cu, Al)_8SO_4(OH)_{16} \cdot nH_2O$, G Rhom. Translucent greenish-blue, turquoise-blue. Cornwall, England. H=?, D_m= 2.38, D_c=?, IVF 03d. Church, 1866. Min. Mag. 40 (1976), 644.

Wroewolfeite, $Cu_4SO_4(OH)_6 \cdot 2H_2O$, A Mon. Vitreous deep greenish-blue. Loudville lead mine, Loudville, Hampden Co., Massachusetts, USA. H= 2.5, D_m= 3.27, D_c= 3.320, VID 01. Dunn et al., 1975. Min. Mag. 40 (1975), 1. Am. Min. 70 (1985), 1050. See also Langite.

Wulfenite, $PbMoO_4$, G Tet. Resinous/adamantine orange-yellow, yellow, grey, brown, etc. Bleiberg, Schwarzenbach, Carinthia, Austria. H= 2.7-3, D_m= 6.78, D_c= 6.828, VIF 01. Haidinger, 1845. Dana, 7th ed.(1951), v.2, 1081. Zts. Krist. 121 (1965), 158. See also Stolzite.

Wülfingite, $Zn(OH)_2$, A Orth. Translucent colorless. Richelsdorfer Hütte, Richelsdorf, Hessen, Germany. H= 3, D_m= 3.05, D_c= 3.06, IVF 03a. Schmetzer et al., 1985. Am. Min. 73(1988), 196 (Abst.). See also Sweetite.

Wurtzite-2H, β-ZnS, G Hex. Resinous brownish-black. H= 3.5-4, D_m= 3.98, D_c= 4.10, IIB 06. Friedel, 1861. Dana, 7th ed.(1944), v.1, 226. Am. Min. 62 (1977), 540. See also Matraite, Sphalerite, Cadmoselite, Greenockite.

Wüstite, FeO, G Cub. Metallic grey in reflected light. Scharnhaüsen, Württemberg, Germany. H= 5, D_m= 5.74, D_c= 5.97, IVA 04. Schenck & Dingmann, 1927. JCPDS 6-615. Acta Cryst. B38 (1982), 1451.

Wyartite, $Ca_3U(UO_2)_6(CO_3)_2(OH)_{18} \cdot 4H_2O$, G Orth. Dull/vitreous black, violet-black. Shinkolobwe, Shaba, Zaïre. H= 3-4, D_m= 4.69, D_c= 4.81, VbD 04. Guillemin & Protas, 1959. Am. Min. 45 (1960), 200.

Wyllieite, $(Na, Ca, Mn)(Mn, Fe)(Fe, Mg)Al(PO_4)_3$, A Mon. Vitreous/sub-metallic green. Old Mike Pegmatite, Custer (near), Custer Co., South Dakota, USA. H=?, D_m=?, D_c=?, VIIA 05a. Moore & Ito, 1973. Min. Mag. 43 (1979), 227. Am. Min. 59 (1974), 280. See also Ferrowyllieite, Rosemaryite, Quingheiite.

X

Xanthiosite, $Ni_3(AsO_4)_2$, G Mon. Translucent golden-yellow. South Terras mine, St. Stephen-in-Brannel, Cornwall, England. H=?, D_m= 5.4, D_c= 5.388, VIIA 08. Adam, 1869. Min. Mag. 35 (1965), 72.

Xanthoconite, Ag_3AsS_3, G Mon. Adamantine red/orange/clove-brown. Himmelsfürst mine, Freiberg, Sachsen (Saxony), Germany. H= 2-3, D_m= 5.54, D_c= 5.53, IID 04. Breithaupt, 1840. Min. Mag. 29 (1950), 346. Acta Cryst. B24 (1968), 77. See also Proustite.

Xanthoxenite, $Ca_4Fe_2(PO_4)_4(OH)_2 \cdot 3H_2O$, A Tric. Waxy/pulverulent pale yellow, brownish-yellow. Rabenstein, Bayern (Bavaria), Germany. H= 2.5, D_m= 2.97, D_c= 3.38, VIID 12. Laubmann & Steinmetz, 1920. Min. Mag. 42 (1978), 309.

Xenotime-(Y), YPO_4, G Tet. Vitreous/resinous yellowish/reddish brown, yellow, red, etc. H= 4-5, D_m= 4.8, D_c= 4.277, VIIA 10. Beudant, 1832. Min. Mag. 39 (1973), 145. Zts. Krist. 121 (1965), 315. See also Chernovite, Wakefieldite.

Xiangjiangite, $(Fe, Al)(UO_2)_4(PO_4)_2(SO_4)_2(OH) \cdot 22H_2O$, A Tet. Silky/earthy yellow. Xiangjiang (Hsiang) River, China. H= 1-2, D_m= 3.0, D_c= 2.87, VIID 20a. Am. Min. 64(1979), 466 (Abst.).

Xifengite, Fe_5Si_3, A Hex. Metallic steel-grey. Yanshan, China. H= 5, D_m=?, D_c= 7.15, IC 03. Yu, 1984. Am. Min. 71(1986), 228 (Abst.). Struct. Repts. 10 (1945), 63.

Xilingolite, $Pb_{3.3}Bi_{1.8}S_6$, A Mon. Metallic grey; white in reflected light. Chaobuleng dist., Xilingola League, Inner Mongolia, China. H= 3, D_m= 7.08, D_c= 7.23, IID 09b. Hong et al., 1982. N.Jb.Min.Abh. 160 (1989), 269.

Xingzhongite, $(Pb, Cu, Fe)(Ir, Pt, Rh)_2S_4$, Q Cub. Metallic grey. Unspecified locality, China. H= 6, D_m=?, D_c= 9.06, IIC 01. Yu et al, 1974. Am. Min. 74(1989), 1220 (Abst.).

Xitieshanite, $FeSO_4Cl \cdot 6H_2O$, A Mon. Translucent green (yellow tint). Xitieshan deposit, Qaidam Basin, Qinghai, China. H= 2.7, D_m= 1.99, D_c= 2.025, VID 02. Li et al., 1983. Scient. Geol. Sin. (1989), 106. Am. Min. 74(1989), 1404 (Abst.).

Xocomecatlite, $Cu_3TeO_4(OH)_4$, A Orth. Translucent green. Bambollita mine, Moctezuma, Sonora, Mexico. H= 4, D_m= 4.65, D_c= 4.42, VIG 09. Williams, 1975. Min. Mag. 40 (1975), 221.

Xonotlite, $Ca_6Si_6O_{17}(OH)_2$, G Mon. Translucent white, grey, pale pink. Tetela de Xonotla, Mexico. H= 6.5, D_m= 2.7, D_c= 2.7, VIIID 11. Rammelsberg, 1866. Dana/Ford (1932), 641. Zts. Krist. 154 (1981), 271.

Y

Yafsoanite, $(Zn,Ca,Pb)_3TeO_6$, [A] Cub. Vitreous light/dark brown. Unspecified locality, Aldan (central), Yakutiya, USSR. H= 5.7, D_m=?, D_c= 5.537, VIG 10. Kim et al., 1982. Am. Min. 68(1983), 282 (Abst.). Min. Abstr. 85M/0176.

Yagiite, $(Na,K)_{1.5}Mg_2(Al,Mg,Fe)_3(Si,Al)_{12}O_{30}$, [A] Hex. Translucent colorless. Colomera iron meteorite. H=?, D_m=?, D_c= 2.70, VIIIC 10. Bunch & Fuchs, 1969. Am. Min. 54 (1969), 14.

Yakhontovite, $(Ca,Na,K)_{0.2}(Cu,Fe,Mg)_2Si_4O_{10}(OH)_2 \cdot 3H_2O$, [A] Mon. Dull green. Komsomolsk region, USSR (E). H= 2-3, D_m=?, D_c=?, VIIIE 08b. Postnikova et al., 1986. Min. Abstr. 88M/1097.

Yamatoite, [D] Not proven to occur naturally. Min. Mag. 36 (1967), 133. Yoshimura & Momoi, 1964.

Yaroslavite, $Ca_3Al_2F_{10}(OH)_2 \cdot H_2O$, [G] Orth. Vitreous white. Siberia, USSR. H= 4, D_m= 3.09, D_c= 3.15, IIIC 09. Novikova et al., 1966. Am. Min. 51(1966), 1546 (Abst.).

Yarrowite, Cu_9S_8, [A] Rhom. Metallic bluish-grey. Yarrow Creek, Yarrow Creek – Spionkop Creek deposit, Alberta (SW), Canada. H= 2.5, D_m=?, D_c= 4.89, IIA 01a. Goble, 1980. Can. Min. 23 (1985), 61.

Yavapaiite, $KFe(SO_4)_2$, [G] Mon. Vitreous pale pink. United Verde mine, Jerome, Yavapai Co., Arizona, USA. H= 2.5-3, D_m= 2.88, D_c= 2.891, VIA 03. Hutton, 1959. Am. Min. 44 (1959), 1105. Am. Min. 56 (1971), 1917.

Yeatmanite, $Mn_7Zn_8Sb_2^{+5}Si_4O_{28}$, [A] Tric. Translucent brown. Franklin, Sussex Co., New Jersey, USA. H= 4, D_m= 5.02, D_c= 5.079, VIIIA 18. Palache et al., 1938. Am. Min. 65 (1980), 196. Min. Jour. 13 (1986), 53.

Yecoraite, $Fe_3Bi_5O_9(TeO_3)(TeO_4)_2 \cdot 9H_2O$, [A] Syst=? Pitchy/resinous yellow. Maria Elena mine, Yecora, Sonora, Mexico. H= 3, D_m= 5.59, D_c=?, VIG 08. Williams & Cesbron, 1985. Am. Min. 71(1986), 1547 (Abst.).

Yedlinite, $Pb_6CrCl_6(O,OH,H_2O)_8$, [A] Rhom. Translucent red-violet. Mammoth mine, Tiger, Pinal Co., Arizona, USA. H= 2.5, D_m= 5.85, D_c= 5.80, IIIC 13. McLean et al., 1974. Am. Min. 59 (1974), 1157. Am. Min. 59 (1974), 1160.

Ye'elimite, $Ca_4Al_6O_{12}SO_4$, [A] Cub. Translucent colorless. Hatrurim formation, Israel. H=?, D_m=?, D_c= 2.61, IVC 05a. Gross, 1985. Am. Min. 72(1987), 226 (Abst.).

Yftisite, [D] Incomplete chemical analysis; not approved by CNMMN. Am. Min. 62(1977), 396 (Abst.). Pletneva et al., 1971.

Yimengite, $K(Cr,Ti,Fe,Mg)_{12}O_{19}$, [A] Hex. Metallic black; grey in reflected light. Yimengshan area, Shandong, China. H= 4.1, D_m= 4.34, D_c= 4.35, IVC 05a. Dong et al., 1983. Am. Min. 70(1985), 218 (Abst.). See also Hibonite, Magnetoplumbite.

Yixunite, PtIn, [Q] Cub. Metallic white. Unspecified locality, Dao dist., China. H= 3, D_m=?, D_c=?, IA 11. Yu et al., 1974. Am. Min. 65(1980), 408 (Abst.). See also Platinum.

Yoderite, $(Mg,Al)_8O_2(SiO_4)_4(OH)_2$, [G] Mon. Translucent purple. Mautia Hill, Kongwa, Tanzania. H= 6, D_m= 3.39, D_c= 3.36, VIIIA 14. McKie & Radford, 1959. Min. Mag. 32 (1959), 282. Am. Min. 67 (1982), 76.

Yofortierite, $(Mn,Mg)_5Si_8O_{20}(OH)_2 \cdot 8\text{–}9H_2O$, [A] Mon. Transparent maroon. De-Mix Quarry, Mont Saint-Hilaire, Rouville Co., Québec, Canada. H= 2.5, D_m= 2.18, D_c= 2.81, VIIIE 13. Perrault et al., 1975. Can. Min. 13 (1975), 68. See also Palygorskite.

Yoshimuraite, $Ba_2TiMn_2(SiO_4)_2(SO_4,PO_4)(OH,Cl)$, [G] Tric. Noda-Tamagawa mine, Misago Ore Body, Iwate, Honshu, Japan. H= 4.5, D_m= 4.13, D_c= 4.21, VIIIB 12. Watanabe et al., 1961. Am. Min. 46(1961), 1515 (Abst.).

Yoshiokaite, $Ca_3Al_6Si_2O_{16}$, A Rhom. Colorless, devitrified glassy spheres. Apollo 14 Base, Moon. H=?, D_m=?, D_c=?, Vaniman & Bish, 1990. Am. Min. 75 (1990), 676.

Yttrialite-(Y), α-$(Y, Th)_2Si_2O_7$, G Mon. Vitreous/greasy olive-green, orange-yellow. Barringer Hill, Blufton (near), Llano Co., Texas, USA. H= 5-5.5, D_m= 4.575, D_c=?, VIIIB 01. Hidden & Mackintosh, 1889. Min. Abstr. 75-303. Sov.Phys.Cryst. 16(1971), 786.

Yttrobetafite-(Y), $(Y, U, Ce)_2(Ti, Nb, Ta)_2O_6(OH)$, A Amor. Greasy greenish. Alakurtti region, Karelia (NW), USSR. H=?, D_m= 4, D_c=?, IVC 09a. Kalita et al., 1962. Am. Min. 49(1964), 440 (Abst.). See also Betafite.

Yttroceberysite-(Y), $(Y, Ce)BeSiO_4(OH)$, Q Mon. Vitreous white, light yellow, light green. Greater Khingan area, Manchuria, China. H= 5-5.5, D_m= 4.57, D_c= 4.53, VIIIA 23. Ding et al., 1981. Am. Min. 73(1988), 442 (Abst.).

Yttrocolumbite, $(Fe, Mn, Y)(Nb, Ta)_2O_6$, Q Orth. Vitreous black. Mozambique. H= 5, D_m= 5.49, D_c=?, IVD 10b. Lepierre, 1937. Min. Abstr. 7-470. See also Yttrotantalite-(Y).

Yttrocrasite-(Y), $(Y, Th, Ca, U)(Ti, Fe)_2(O, OH)_6$, Q Syst=? Resinous/pitchy black. Burnet Co., Texas, USA. H= 5.5-6, D_m= 4.80, D_c=?, IVD 10b. Hidden & Warren, 1906. See also Euxenite-(Y), Tanteuxenite-(Y).

Yttromicrolite, D Mixture. Am. Min. 67 (1982), 156. Crook, 1979.

Yttropyrochlore-(Y), $(Y, Ce, Nd, Th)(Nb, Ta)_2(O, OH)_7$, A Amor. Vitreous/adamantine brown. Alakurtti region, Karelia (NW), USSR. H= 4.5-5, D_m= 3.7, D_c=?, IVC 09a. Hogarth, 1977. Am. Min. 43(1958), 797 (Abst.). See also Pyrochlore.

Yttrotantalite-(Y), $(Y, U, Ca)(Ta, Fe)_2(O, OH)_6$, G Orth. Sub-metallic/vitreous black, brown. Gl. Ytterby, Sweden. H= 5-5.5, D_m= 5.7, D_c=?, IVD 10b. Ekeberg, 1802. Dana, 7th ed.(1944), v.1, 763. See also Yttrocolumbite-(Y), Tantalite.

Yttrotungstite-(Y), $(Y, La, Ca)(W, Fe, Si, Al, Ti)_2(O, OH, H_2O)_9$, G Mon. Earthy yellow. Kramat Pulai mine, Kinta dist., Perak, Malaysia. H=?, D_m= 5.96, D_c=?, IVF 15. Beard, 1950. Min. Mag. 38 (1971), 261. Min. Mag. 38 (1971), 261. See also Cerotungstite-(Ce), Tungstite.

Yugawaralite, $Ca(Al_2Si_6)O_{16} \cdot 4H_2O$, G Mon. Vitreous colorless, white. Kanagawa, Yugawara hot spring, Hudo-No-Taki waterfall, Honshu, Japan. H= 4.5, D_m= 2.20, D_c= 2.236, VIIIF 17. Sakurai & Hayashi, 1952. Am. Min. 38(1953), 426 (Abst.). Zts. Krist. 174 (1986), 265.

Yukonite, $Ca_2Fe_5(AsO_4)_4(OH)_7 \cdot 7H_2O$, Q Syst=? Waxy reddish-brown, dark brown. Windy Arm (W), Tagish Lake, Yukon Territory, Canada. H=?, D_m=?, D_c=?, VIID 13. Tyrell & Graham, 1913. Min. Mag. 46 (1982), 261.

Yuksporite, $(K, Ba)NaCa_2(Si, Ti)_4O_{11}(F, OH) \cdot H_2O$, G Orth. Opaque yellowish-rose. Yukspor Mts., Khibina massif, Kola Peninsula, USSR. H= 5, D_m= 3.05, D_c= 2.79, VIIID 11. Fersman, 1922. Min. Zhurn. 7 (4) (1985), 74.

Yushkinite, $V_{1-x}S \cdot n(Mg, Al)(OH)_2$, A Hex. Metallic pinkish-violet. Pay-Khoĭ, Silova-Yakha River, Ural Mts., USSR. H= 1, D_m= 2.94, D_c= 3.00, IIE 01. Makeev et al., 1984. Am. Min. 71(1986), 846 (Abst.). See also Haapalaite, Tochilinite, Valleriite.

Z

Zabuyelite, Li_2CO_3, [A] Mon. Vitreous colorless. Nagri, Tibet, China. H= 3, D_m= 2.09, D_c= 2.096, VbA 06. Zheng & Liu, 1987. Am. Min. 75(1990), 243 (Abst.).

Zaherite, $Al_{12}(SO_4)_5(OH)_{26}\cdot 20H_2O$, [A] Tric. Pearly/earthy white. Salt Range, Pakistan. H= 3.5, D_m= 2.007, D_c= 2.006, VID 03. Ruotsala & Babcock, 1977. Min. Mag. 49 (1985), 145.

Zairite, $Bi(Fe,Al)_3(PO_4)_2(OH)_6$, [A] Rhom. Translucent greenish. Eta-Eta, Kivu (N), Zaïre. H= 4.5, D_m= 4.37, D_c= 4.42, VIIB 15a. Van Wambeke, 1975. Am. Min. 62(1977), 174 (Abst.).

Zakharovite, $Na_4Mn_5Si_{10}O_{24}(OH)_6\cdot 6H_2O$, [A] Rhom. Pearly/waxy yellow. Mt. Karnasurt, Lovozero massif; also Yukspor Mts. and Koashua Mts., Khibina, Kola Peninsula, USSR. H= 2, D_m= 2.6, D_c= 2.67, VIIIC 07. Khomyakov et al., 1982. Am. Min. 68(1983), 1040 (Abst.).

Zapatalite, $Cu_3Al_4(PO_4)_3(OH)_9\cdot 4H_2O$, [A] Tet. Translucent pale blue. Naco, Cerro Morita, Sonora, Mexico. H= 1.5, D_m= 3.016, D_c= 3.017, VIID 36. Williams, 1972. Min. Mag. 38 (1972), 541.

Zaratite, $Ni_3CO_3(OH)_4\cdot 4H_2O$, [Q] Cub. Vitreous/greasy green. Texas, Lancaster Co., Pennsylvania, USA. H= 3.5, D_m= 2.6, D_c= 2.67, IVF 03b. Casares, 1851. Min. Mag. 33 (1963), 663.

Zavaritskite, BiOF, [G] Tet. Semi-metallic/greasy grey. Sherlova Gory, Transbaikal (E), USSR. H=?, D_m= 9.0, D_c= 9.21, IIIC 16. Dolomanova et al., 1962. Am. Min. 48(1963), 210 (Abst.). See also Bismoclite, Daubréeite.

Zeiringite, [D] A mixture of aragonite and aurichalcite. Am. Min. 48 (1963), 1184. Pantz, 1811.

Zektzerite, $NaLiZrSi_6O_{15}$, [A] Orth. Vitreous colorless, pink. Washington Pass, Okanogan Co., Washington, USA. H= 6, D_m= 2.79, D_c= 2.80, VIIID 14. Dunn et al., 1977. Am. Min. 62 (1977), 416. Am. Min. 63 (1978), 304. See also Tuhualite.

Zellerite, $Ca(UO_2)(CO_3)_2\cdot 5H_2O$, [A] Orth. Translucent yellow. Lucky MC mine, Gas Hills deposit, Fremont Co., Wyoming, USA. H= 2, D_m= 3.25, D_c= 3.24, VbD 04. Coleman et al., 1966. Am. Min. 51 (1966), 1567.

Zemannite, $(H,Na)_2(Zn,Fe)_2(TeO_3)_3\cdot nH_2O$, [A] Hex. Adamantine light/dark brown. Moctezuma mine, Moctezuma, Sonora, Mexico. H=?, D_m=?, D_c= 4.34, VIG 04. Mandarino & Williams, 1961. Can. Min. 14 (1976), 387. See also Kinichilite.

Zeolite, A group name for framework silicates with exchangeable cations.

Zeophyllite, $Ca_{13}Si_{10}O_{28}(OH)_2F_8\cdot 6H_2O$, [G] Rhom. Translucent white. Gross-Priesen, Čechy (Bohemia), Czechoslovakia. H= 3, D_m= 2.8, D_c= 2.79, VIIIE 14. Pelikan, 1902. Min. Mag. 47 (1983), 397. Acta Cryst. B28 (1968), 2726.

Zeunerite, $Cu(UO_2)_2(AsO_4)_2\cdot 16H_2O$, [G] Tet. Transparent green. Weisser Hirsch mine, Neustädtel, Schneeberg (near), Sachsen (Saxony), Germany. H=?, D_m= 3.47, D_c= 3.58, VIID 20a. Weisbach, 1872. Am. Min. 42 (1957), 905.

Zhanghengite, CuZn, [A] Cub. Metallic yellow. Meteorite, Xiaoyanzhuang, Boxian Co., Anhui, China. H= 5, D_m=?, D_c= 8.30, IA 02. Wang, 1986. Acta Min. Sin. 6(3)(1986), 220.

Zharchikhite, $AlF(OH)_2$, [A] Mon. Transparent colorless. Zharchinskoye deposit, Zabaykalye, USSR. H= 4.5, D_m= 2.81, D_c= 2.82, IIIC 17. Bolokhontseva et al., 1988. Am. Min. 74(1989), 504 (Abst.).

Zhemchuzhnikovite, $NaMg(Al,Fe)(C_2O_4)_3\cdot 8H_2O$, [G] Rhom. Vitreous smoky-green. Chaitumusuk deposits, Lena River, Siberia, USSR. H= 2, D_m= 1.69, D_c= 1.66, IXA 01. Knipovich et al., 1963. Am. Min. 49(1964), 442 (Abst.).

Zhonghuacerite-(Ce), $Ba_2Ce(CO_3)_3F$, [Q] Rhom. Vitreous/resinous yellow. Bayan Obo, Inner Mongolia, China. H= 4.6, D_m= 4.3, D_c= 4.66, VbB 04. Zhang & Tao, 1981. Am. Min. 67(1982), 1078 (Abst.).

Ziesite, β–$Cu_2V_2O_7$, [A] Mon. Metallic black; white in reflected light. Izalco Volcano, El Salvador. H=?, D_m= 3.86, D_c= 3.869, VIIA 12. Hughes & Birnie, 1980. Am. Min. 65 (1980), 1146. N. Jb. Min. Mh. (1989), 41.

Zimbabweite, $Na(PbNa_{0.5}K_{0.5})_2(Ta,Nb,Ti)_4As_4O_{18}$, [A] Orth. Adamantine yellow-brown. St. Ann's mine, Karoi Dist., Miami (SE of), Zimbabwe. H= 5-5.5, D_m= 6.20, D_c= 6.16, IVC 16. Foord et al., 1986. Bull. Min. 109 (1986), 331. Am. Min. 73 (1988), 1186.

Zinalsite, $Zn_7Al_4(SiO_4)_6(OH)_2 \cdot 9H_2O$(?), [Q] Syst=? White, rose, reddish-brown. Akdzhal and Achisai deposits, Kazakhstan, USSR. H= 2.5-3, D_m= 3.007, D_c=?, VIIIE 10a. Chukhrov, 1956. Am. Min. 44(1959), 208 (Abst.). See also Fraipontite.

Zinc, Zn, [G] Hex. Metallic white. H= 2, D_m=?, D_c= 7.140, IA 03. Dana, 7th ed.(1944), v.1, 127.

Zincaluminite, $Zn_6Al_6(SO_4)_2(OH)_{26} \cdot 5H_2O$(?), [Q] Syst=? Translucent white, bluish-white, pale blue. Lávrion (Laurium), Attikí, Greece. H= 2.5-3, D_m= 2.26, D_c=?, IVF 03e. Bertrand & Damour, 1881. Dana, 7th ed.(1951), v.2, 579.

Zincalunite, [D] Inadequate data. Min. Mag. 36 (1967), 133. Omori & Kerr, 1963.

Zincblödite, $Na_2Zn(SO_4)_2 \cdot 4H_2O$, [Q] Mon. H=?, D_m= 2.511, D_c= 2.51, VIC 10. John, 1821. Per. Mineral. 54 (1985), 12. Acta Cryst. 11 (1958), 789. See also Blödite.

Zincite, (Zn, Mn)O, [G] Hex. Sub-adamantine orange, red, yellow. Sterling Hill and Franklin Furnace, Sussex Co., New Jersey, USA. H= 4, D_m= 5.67, D_c= 5.699, IVA 03. Haidinger, 1845. Dana, 7th ed.(1944), v.1, 504. See also Bromellite.

Zinclavendulan, $NaCa(Zn,Cu)_5(AsO_4)_4Cl \cdot 4$–$5H_2O$, [Q] Orth. Tsumeb, Namibia. H=?, D_m=?, D_c=?, VIID 16. Strunz, 1959. Strunz (1970), 349.

Zinc-melanterite, $(Zn,Cu,Fe)SO_4 \cdot 7H_2O$, [G] Syst=? Vitreous pale greenish-blue. Good Hope mine and Vulcan mine, Vulcan, Gunnison Co., Colorado, USA. H= 2, D_m= 2.02, D_c=?, VIC 03c. Larsen & Glenn, 1920. Dana, 7th ed.(1951), v.2, 508. See also Melanterite.

Zincobotryogen, $(Zn,Mg,Mn)Fe(SO_4)_2(OH) \cdot 7H_2O$, [G] Mon. Vitreous/greasy bright orange-red. Tsadam Basin (N), China. H= 2.5, D_m= 2.201, D_c= 2.24, VID 04a. Kuang et al., 1964. Am. Min. 49(1964), 1776 (Abst.). See also Botryogen.

Zincochromite, $ZnCr_2O_4$, [A] Cub. Semi-metallic brown; brownish-grey in reflected light. Onega trough, South Karelia, USSR. H= 5.8, D_m=?, D_c= 5.434, IVB 01. Nesterov & Rumjantzeva, 1987. Am. Min. 73(1988), 931 (Abst.).

Zincocopiapite, $ZnFe_4(SO_4)_6(OH)_2 \cdot 20H_2O$, [G] Tric. Vitreous yellowish-green. Chaidamupendi (Tsadam Basin), Qinghai (Chinghai), China. H= 2, D_m= 2.181, D_c= 2.09, VID 04b. Ti, 1964. Can. Min. 23 (1985), 53.

Zincovoltaite, $K_2Zn_5Fe_4(SO_4)_{12} \cdot 18H_2O$, [A] Cub. Pitchy/resinous/vitreous green-black, oil-green. Xitieshan mine, Qinghai, China. H= 3, D_m= 2.756, D_c= 2.767, VIC 14. Li et al., 1987. Am. Min. 75(1990), 244 (Abst.).

Zincrosasite, $(Zn,Cu)_2CO_3(OH)_2$, [G] Syst=? Tsumeb, Namibia. H=?, D_m=?, D_c=?, VbB 01. Strunz, 1959. Am. Min. 44(1959), 1323 (Abst.).

Zincroselite, $Ca_2Zn(AsO_4)_2 \cdot 2H_2O$, [A] Mon. Translucent colorless, white. Tsumeb mine, Tsumeb, Namibia. H= 3, D_m= 3.75, D_c= 3.77, VIIC 12. Keller et al., 1986. Am. Min. 73(1988), 932 (Abst.).

Zincsilite, $Zn_3Si_4O_{10}(OH)_2 \cdot 4H_2O$(?), [Q] Syst=? Translucent white, bluish. Batystau, Kazakhstan, USSR. H= 1.5-2, D_m= 2.7, D_c=?, VIIIE 08b. Smol'yaninova et al., 1960. Am. Min. 46(1961), 241 (Abst.).

Zinc-zippeite, $Zn_2(UO_2)_6(SO_4)_3(OH)_{10} \cdot 16H_2O$, [A] Syst=? Translucent yellow. Hillside mine, Bagdad, Yavapai Co., Arizona, USA. H= 2, D_m=?, D_c=?, VID 08. Frondel et al., 1976. Can. Min. 14 (1976), 429. See also Zippeite, Sodium-zippeite, Nickel-zippeite, Magnesium-zippeite, Cobalt-zippeite.

Zinkenite, $Pb_9Sb_{22}S_{42}$, [G] Hex. Metallic grey; grey-white in reflected light. Wolfsberg, Harz, Germany. H= 3-3.5, D_m= 5.36, D_c= 5.16, IID 05b. Rose, 1826. Am. Min. 71 (1986), 194. Zts. Krist. 141 (1975), 79.

Zinkosite, $ZnSO_4$, [Q] Orth. Translucent. Barranco Jaroso mine, Sierra Almagrera, Spain. H=?, D_m= 4.33, D_c= 3.87, VIA 01. Breithaupt, 1852. Dana, 7th ed.(1951), v.2, 428. Min. Petrol. 39 (1988), 201.

Zinnwaldite, $K(Al, Fe, Li)_3(Si, Al)_4O_{10}(OH)F$, [G] Mon. Translucent greyish-brown, yellowish-brown, pale violet. Cínovec (Zinnwald), Krusné Hory (Erzgebirge), Čechy (Bohemia), Czechoslovakia. H= 2.5-4, D_m= 3.0, D_c= 2.99, VIIIE 05b. Haidinger, 1845. DHZ (1962), v.3, 92. Am. Min. 62 (1977), 1158.

Zippeite, $K_4(UO_2)_6(SO_4)_3(OH)_{10} \cdot 4H_2O$, [G] Mon. Translucent golden yellow. Jáchymov (St. Joachimsthal), Západočeský kraj, Čechy (Bohemia), Czechoslovakia. H= 2, D_m= 3.66, D_c= 3.68, VID 08. Haidinger, 1845. Am. Min. 37 (1952), 394. See also Sodium-zippeite, Nickel-zippeite, Zinc-zippeite, Magnesium-zippeite Cobalt-zippeite.

Zircon, $ZrSiO_4$, [G] Tet. Adamantine colorless, yellow, grey, reddish-brown, green. H= 7.5, D_m= 4.7, D_c= 4.714, VIIIA 07. Werner, 1783. DHZ, 2nd ed.(1982), v.1B, 418. Am. Min. 56 (1971), 782. See also Hafnon.

Zirconolite-2M, $CaZrTi_2O_7$, [A] Mon. Brown, black. Afrikanda massif, Kola Peninsula, USSR. H=?, D_m= 4.2, D_c= 4.395, IVC 09b. Borodin et al., 1956. Min. Mag. 53 (1989), 565. Acta Cryst. B37 (1981), 306.

Zirconolite-3O, $(Ca, Fe, Y, Th)_2Fe(Ti, Nb)_3Zr_2O_{14}$, [A] Orth. Submetallic/metallic black. Fredricksvärn, Norway. H= 6.5, D_m= 4.8, D_c=?, IVC 09b. Berzelius, 1828. Min. Mag. 53 (1989), 565. Am. Min. 68 (1983), 262. See also Calciobetafite, Zirkelite.

Zirconolite-3T, $CaZrTi_2O_7$, [A] Tet. Submetallic/resinous black, brownish-black. Bambarabotuwa, Sri Lanka. H= 5.5-6, D_m= 4.8, D_c= 4.88, IVC 09b. Blake & Smith, 1913. Min. Mag. 53 (1989), 565. Am. Min. 68 (1983), 262.

Zircophyllite, $(K, Na)_3(Mn, Fe)_7Zr_2Si_8O_{27}(OH, F)_4$, [A] Tric. Vitreous/adamantine dark brown. Korgeredabinsk, Tuva, USSR. H= 4-4.5, D_m= 3.34, D_c=?, VIIID 25. Kapustin, 1972. Am. Min. 58(1973), 967 (Abst.).

Zircosulfate, $Zr(SO_4)_2 \cdot 4H_2O$, [G] Orth. Translucent dull colorless/white. Korgeredabin massif, Tuva (SE), USSR. H= 2.5-3, D_m= 2.85, D_c= 2.833, VIC 16. Kapustin, 1965. Am. Min. 51(1966), 529 (Abst.).

Zirkelite, $(Ti, Ca, Zr)O_{2-x}$, [A] Cub. Resinous black. Jacupiranga, São Paulo, Brazil. H= 5.5, D_m= 4.741, D_c=?, IVC 09b. Hussak & Prior, 1895. Min. Mag. 53 (1989), 565. See also Calciobetafite, Pyrochlore, Polymignite.

Zirklerite, $(Fe, Mg)_9Al_4Cl_{18}(OH)_{12} \cdot 14H_2O(?)$, [Q] Rhom. Translucent. Adolfsglück mine, Hope, Hannover, Germany. H= 3.5, D_m= 2.6, D_c=?, IIIC 12. Harbort, 1928. Dana, 7th ed.(1951), v.2, 87.

Zirsinalite, $Na_6(Ca, Mn, Fe)ZrSi_6O_{18}$, [A] Rhom. Vitreous colorless, yellowish-grey. Mt. Koashva, Khibina massif, Kola Peninsula, USSR. H= 5.5, D_m= 2.98, D_c= 3.08, VIIIB 09. Kapustin et al., 1974. Dokl.Earth Sci. 237(1977), 208.

Zirsite, [D] Inadequate data. Min. Mag. 36 (1967), 133. Dorfman, 1962.

Zodacite, $Ca_4MnFe_4(PO_4)_6(OH)_4 \cdot 12H_2O$, [A] Mon. Vitreous yellow. Mangualde and Mesquitella (between), Portugal. H= 4, D_m= 2.68, D_c= 2.65, VIID 35. Dunn et al., 1988. Am. Min. 73 (1988), 1179.

Zoisite, $Ca_2Al_3(Si_2O_7)(SiO_4)(O, OH)_2$, [G] Orth. Vitreous/pearly greyish-white, grey, yellowish-brown, etc. Saualpe, Carinthia, Austria. H= 6-6.5, D_m= 3.3,

D_c= 3.329, VIIIB 15. Werner, 1805. DHZ, 2nd ed.(1986), v.1B, 4. Am. Min. 53 (1968), 1882. See also Clinozoisite.

Zorite, $Na_6Ti_5Si_{12}O_{34}(O,OH)_5 \cdot 11H_2O$, [A] Orth. Vitreous rosy. Lovozero tundra, Kola Peninsula, USSR. H= 3-4, D_m= 2.3, D_c= 2.23, VIIIF 18. Mer'kov et al., 1973. Am. Min. 58(1973), 1113 (Abst.). Sov.Phys.Cryst. 24 (1980), 686.

Zoubekite, $AgPb_4Sb_4S_{10}$, [A] Orth. Metallic grey; yellowish white in reflected light. Příbram (near), Stȩedočský kraj, Čechy (Bohemia), Czechoslovakia. H= 3.4, D_m=?, D_c= 5.15, IID 07. Megarskaya & Rykl, 1986. N. Jb. Min. Mh. (1986), 1.

Zunyite, $Al_{13}Si_5O_{20}(OH,F)_{18}Cl$, [G] Cub. Vitreous colorless. Zuni mine, Silverton, Anvil Mt., San Juan Co., Colorado, USA. H= 7, D_m= 2.875, D_c= 2.89, VIIIB 20. Hillebrand, 1884. Bull. Min. 97 (1974), 271. Acta Cryst. B38 (1982), 390.

Zussmanite, $K(Fe,Mg,Mn)_{13}(Si,Al)_{18}O_{42}(OH)_{14}$, [A] Rhom. Translucent pale green. Laytonville Quarry, Laytonville (S), Mendocino Co., California, USA. H=?, D_m= 3.146, D_c=?, VIIIE 07c. Agrell et al., 1964. Min. Mag. 43 (1980), 605. Min. Mag. 37 (1969), 49.

Zvyagintsevite, $Pd_3(Pb,Sn)$, [A] Cub. Metallic white. Talnakh, Noril'sk, Siberia (N), USSR. H= 4.5, D_m=?, D_c= 12.32, IA 09. Genkin et al., 1966. Can. Min. 8 (1966), 541.

Zwieselite, $(Fe,Mn)_2PO_4(F,OH)$, [G] Mon. Vitreous/resinous dark brown. Rabenstein, Zwiesel, Bayern (Bavaria), Germany. H= 5-5.5, D_m= 3.7, D_c= 4.08, VIIB 03a. Breithaupt, 1841. JCPDS 30-654. See also Triplite, Magniotriplite.

Zykaite, $Fe_4(AsO_4)_3SO_4(OH) \cdot 15H_2O$, [A] Orth. Dull greyish-white. Kaňk, Kutná Hora (Near), Středočeský kraj, Čechy (Bohemia), Czechoslovakiaa. H= 2, D_m= 2.50, D_c= 2.504, VIID 04. Čech et al., 1978. Am. Min. 63(1978), 1284 (Abst.).

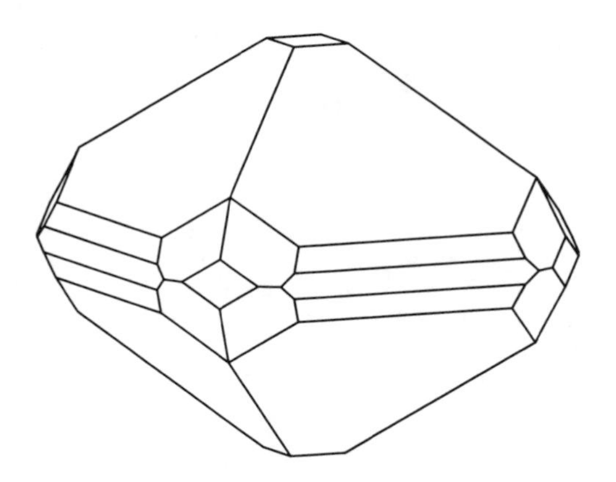

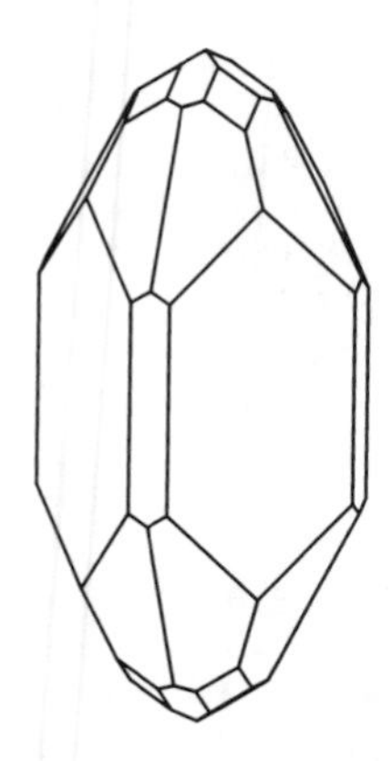

(Left) Zircon from Colorado showing the forms {001}, {111}, {100}, {311}, {110} and {331} (after Hovey).

(Right) Zircon from North Carolina showing the forms {100}, {110}, {111}, {101} and {311} (after Dana, 1892).

Appendix A

MINERAL CLASSIFICATION SCHEME

I: Elements, alloys, carbides, nitrides, phosphides, silicides

- A: Metals and intermetallic alloys
- B: Semi-metals and non-metals
- C: Carbides, nitrides, phosphides, silicides

II: Sulfides, selenides, tellurides, arsenides, antimonides, bismuthinides

- A: Cations:anions $>$ 1
- B: Cations:anions $=$ 1
- C: Cations:anions $<$ 1
- D: Sulfosalts
- E: Oxysulfides and hydroxysulfides
- F: Sulfides, etc. with halogen
- G: Alloys of metals with semimetals As, Sb, Bi

III: Halides

- A: Simple halides
- B: Double halides
- C: Oxyhalides and hydroxyhalides
- D: Halides with other anions

IV: Oxides, hydroxides and iodates

- A: M_2O and MO compounds
- B: M_3O_4 and related compounds
- C: M_2O_3 and related compounds
- D: MO_2 and related compounds
- E: M_xO_y compounds with metal:oxygen < 0.5
- F: Hydroxides
- G: Iodates

Va: Nitrates

Vb: Carbonates

- A: Anhydrous, without additional anions
- B: Anhydrous, with additional anions
- C: Hydrous, without additional anions
- D: Hydrous, with additional anions

Vc: Borates

VI: Sulfates, sulfites, chromates, molybdates, wolframates, selenates, selenites, tellurates, tellurites

- A: Anhydrous sulfates without additional anions
- B: Anhydrous sulfates with additional anions
- C: Hydrous sulfates without additional anions
- D: Hydrous sulfates with additional anions
- E: Chromates
- F: Molybdates and wolframates
- G: Selenates, selenites, tellurates, tellurites, sulfites

VII: Phosphates, arsenates, vanadates, antimonates
- A: Anhydrous, without additional anions
- B: Anhydrous, with additional anions
- C: Hydrous, without additional anions
- D: Hydrous, with additional anions

VIII: Silicates, germanates
- A: Orthosilicates
- B: Sorosilicates
- C: Cyclosilicates
- D: Inosilicates
- E: Phyllosilicates
- F: Tectosilicates
- G: Unclassified silicates
- H: Germanates

IX: Organic minerals
- A: Salts of organic acids
- B: Carbohydrates, etc.
- C: Resins
- D: Porphyrins
- E: Miscellaneous

Appendix B

REFERENCE ABBREVIATIONS

Acta Chem. Scand.	Acta Chemica Scandinavica
Acta Cryst.	Acta Crystallographica
Acta Min. Sin.	Acta Mineralogica Sinica
Am. J. Sci.	American Journal of Science
Am. Min.	American Mineralogist
APMA	Acta Petrologica, Mineralogica et Analytica
Aust. J. Chem.	Australian Journal of Chemistry
Austral. Min.	Australian Mineralogist
Bull. Min.	Bulletin de Minéralogie or Bulletin de la Société francaise de Minéralogie et de Cristallographie
Can. Jour. Chem.	Canadian Journal of Chemistry
Can. Min.	Canadian Mineralogist
Chem. Erde	Chemie der Erde
CIM SV	CIM Special Volume 23, "Platinum-Group Elements: Mineralogy, Geology, Recovery", Cabri, ed.
Clays Cl. Mins.	Clays and Clay Minerals
Contr. Min. Pet.	Contributions to Mineralogy and Petrology
CRASP	Comptes Rendus de l'Academie des Sciences, Paris
Czech. J. Phys.	Czechoslovakian Journal of Physics
DAN USSR	Doklady Akademiia Nauk Ukrainskoi SSR
Dana	"Dana's System of Mineralogy", 7th ed., Wiley, New York
Dana/Ford	"Dana's Textbook of Mineralogy", 4th ed., Wiley, New York
DHZ	"Rock Forming Minerals", Dear, Howie & Zussman, Longmans, London
Dokl. AN SSSR	Doklady Akademiia Nauk, SSSR (USSR)
Dokl. Earth Sci.	Transactions (Doklady) of the USSR Academy of Sciences, Earth Sciences Section
Econ. Geol.	Economic Geology
Embrey & Fuller	"A Manual of New Mineral Names 1892-1978", Embrey & Fuller, British Museum (Nat. History), London
Europ. Jour. Min.	European Journal of Mineralogy
Fleischer	"Glossary of Mineral Species", 5th ed. , Mineralogical Record, Tucson
Fortsch. Min.	Fortschritte der Mineralogie
GAC-MAC	Abstracts of the Joint Annual Meeting of the Geological and Mineralogical Associations of Canada
Geoch.Cos.Acta	Geochimica et Cosmochimica Acta
Geol. Rud. Mest.	Geologiya Rudnykh Mestorozhdenii
Hey	"Chemical Index of Minerals", 2nd ed. and Appendices, Hey, British Museum (Nat. History), London
IMA	Abstracts of General Meeting of the International Mineralogical Association
Inorg. Chem.	Inorganic Chemistry

J. Appl. Cryst.	Journal of Applied Crystallography
J. Chem. Phys.	Journal of Chemical Physics
J. Less Comm. Met.	Journal of the Less-Common Metals
J. Min. Soc. Jap.	Journal of the Mineralogical Society of Japan
J. Sol. St. Chem.	Journal of Solid State Chemistry
JCPDS	JCPDS International Centre for Diffraction Data
Jour. Petrol.	Journal of Petrology
Klockmann	"Lehrbuch der Mineralogie", Ramdohr & Strunz, Enke Verl., Stuttgart
Mem. GSA	Memoirs of the Geological Society of America
Min. Abstr.	Mineralogical Abstracts
Min. Deposita	Mineralium Deposita
Min. Journ.	Mineralogical Journal
Min. Mag.	Mineralogical Magazine
Min. Petrol.	Mineralogy & Petrology
Min. Rec.	Mineralogical Record
Min. Zhurn.	Mineralogicheskii Zhurnal
MSA Spec. Paper	Special Paper of the Mineralogical Society of America
Nat. Phys. Sci.	Nature, Physical Sciences Section
NBS Circ.	Circular of the USA National Bureau of Standards
NBS Monogr.	Monograph of the USA National Bureau of Standards
N.Jb.Min.Abh.	Neues Jahrbuch für Mineralogie, Abhandlungen
N.Jb.Min.Mh.	Neues Jahrbuch für Mineralogie, Monatshefte
Per. Mineral.	Periodico di Mineralogia
Phys. Chem. Min.	Journal of the Physics and Chemistry of Minerals
Powd. Diff.	Powder Diffraction
Ramdohr	"The Ore Minerals and Their Intergrowths", 2nd ed., Ramdohr, Pergamon Press
Rev. Min.	"Reviews in Mineralogy", Mineralogical Society of America
Scott. J. Geol.	Scottish Journal of Geology
SMPM	Schweizerische Mineralogische und Petrographische Mitteilungen
Sov. Phys. Cryst.	Soviet Physics, Crystallography
Sov. Phys. Dokl.	Soviet Physics, Doklady
Struct. Repts.	Structure Reports or Strukturbericht
Strunz	"Mineralogische Tabellen", 5th ed., Strunz and Tennyson Akad. Verlag., Leipzig
TMPM	Tschermaks Mineralogische und Petrographische Mitteilungen
USGS Prof. Pap.	United States Geological Survey Professional Paper
Zts. Krist.	Zeitschrift für Kristallographie
ZVMO	Zapiski Vsesoyuznogo Mineralogicheskogo Obshchestva

Appendix C

SYNONYMY OF NONSPECIES NAMES

Abkhazite .... Tremolite
Abriachanite .... Riebeckite
Absite .... Brannerite
Abukumalite .... Britholite-(Y)
Achroite .... Elbaite
Acmite .... Ægirine
Actinolite .... Ferro-actinolite
Actinote .... Actinolite
Actynolin .... Actinolite
Actynolite .... Actinolite
Adelpholite .... Samarskite-(Y)
Adularia .... Orthoclase
Aeschynite .... Niobo-aeschynite-(Ce)
Agate .... Quartz
Ainalite .... Cassiterite
Alabandine .... Alabandite
Alabaster .... Gypsum
Alexandrite .... Chrysoberyl
Allcharite .... Goethite
Allemontite .... Stibarsen
Allevardite .... Rectorite
Allopalladium .... Stibiopalladinite
Almandite .... Almandine
Almeriite .... Natroalunite
Alpha-catapleite .... Gaidonnayite
Alpha-duftite .... Duftite
Alpha-uranophane .... Uranophane
Altmarkite .... Leadamalgam
Alum .... Sodium Alum
Aluminite-meta .... Meta-aluminite
Alumino-taramite .... Taramite
Alumino-winchite .... Winchite
Aluminum .... Aluminium
Alumobritholite .... Britholite-(Ce)
Amalgam Silver-3C
Amazonite .... Microcline
Amesite-6H Amesite-6R
Amethyst .... Quartz
Ammonia-niter .... Nitrammite
Amosite .... Anthophyllite
Amosite .... Grunerite
Ampangabeite .... Samarskite-(Y)
Amphibole-anthophyllite .... Cummingtonite
Analcite .... Analcime
Anarakite .... Paratacamite
Ancylite .... Calcioancylite-(Ce)
Andorite IV .... Quatrandorite
Andorite VI .... Senandorite
Ankoleite-meta .... Meta-ankoleïte
Anophorite .... Magnesio-arfvedsonite
Anosovite .... Armalcolite
Anthogrammite .... Anthophyllite
Antholite .... Anthophyllite
Antholith .... Anthophyllite
Anthophylline .... Anthophyllite
Anthophyllite .... Magnesio-anthophyllite
Anthophyllite sodium .... Sodium-anthophyllite
Antiglaucophane .... Crossite
Antiglaucophane .... Glaucophane
Antimonite .... Stibnite
Apatite .... Chlorapatite
Apatite .... Fluorapatite
Apatite .... Hydroxylapatite
Apophyllite .... Fluorapophyllite
Apophyllite .... Hydroxyapophyllite
Apophyllite .... Natroapophyllite
Aquamarine .... Beryl
Arfvedsonite .... Magnesio-arfvedsonite
Argentite .... Acanthite
Argentocuproaurite .... Auricupride
Arizonite .... Pseudorutile
Arnimite .... Antlerite
Arsenate-belovite .... Talmessite
Arseniodialyte .... Hausmannite
Ascharite .... Szaibelyite
Ashtonite .... Mordenite
Astochite .... Richterite
Astorite .... Richterite
Astrakanite .... Blödite
Astrakhanite .... Blödite
Astrolite Muscovite-2M1
Astrophyllite .... Magnesium astrophyllite
Attacolite .... Attakolite
Attapulgite .... Palygorskite
Aurocuproite .... Auricupride
Austinite-nickel .... Nickelaustinite
Autunite sodium .... Sodium autunite
Avelinoite .... Cyrilovite
Axinite .... Ferro-axinite
Axinite .... Magnesio-axinite
Bababudanite .... Magnesio-riebeckite
Barioorthojoaquinite .... Bario-orthojoaquinite
Barkevikite .... Alumino-ferro-hornblende
Barrandite .... Strengite
Barrandite .... Variscite
Barroisite .... Ferro-alumino-barroisite
Barroisite .... Ferro-ferri-barroisite
Barsanovite .... Eudialyte
Baryte .... Barite
Belovite .... Talmessite
Bentonite .... Montmorillonite
Bergamaschite .... Hastingsite
Berndtite-C27 .... Berndtite-4H
Beryllium sodalite .... Tugtupite
Beryllosodalite .... Tugtupite
Bessmertnovite .... Bezsmertnovite
Beta-domeykite .... Metadomeykite
Beta-uranopilite .... Meta-uranopilite
Betafite .... Yttrobetafite-(Y)
Betaroselite .... Roselite-beta
Betpakdalite sodium .... Sodium betpakdalite
Bialite .... Wavellite
Bidalotite .... Gedrite
Binnite .... Tennantite
Bisbeeite .... Chrysocolla
Biteplapalladite .... Merenskyite
Biteplatinite .... Moncheite
Blackjack .... Sphalerite

Blanchardite .... Brochantite
Bleiglanz .... Galena
Blende .... Sphalerite
Blockite .... Penroseite
Bloedite .... Blödite
Blodite-nickel .... Nickelblödite
Blodite-zinc .... Zincblödite
Blomstrandine .... Æschynite-(Y)
Blomstrandite .... Uranpyrochlore
Blue Copper .... Azurite
Boltwoodite sodium .... Sodium boltwoodite
Bombiccite .... Hartite
Boodtite Heterogenite-3R
Boric acid .... Sassolite
Borickite .... Delvauxite
Bort .... Diamond
Botryogen-zinc .... Zincobotryogen
Braunite .... Braunite II
Bravoite .... Pyrite
Breunnerite .... Magnesite
Brocenite .... Beta-fergusonite-(Ce)
Bröggerite .... Uraninite
Bromlite .... Alstonite
Bromyrite .... Bromargyrite
Bronzite .... Enstatite
Buetschliite .... Bütschliite
Calafatite .... Alunite
Calamine .... Hemimorphite
Calamite .... Tremolite
Calciocatapleiite .... Calcium catapleiite
Calciocelsian .... Armenite
Calcium-larsenite .... Esperite
Calcium-rinkite .... Götzenite
Calcium-uranospinite .... Meta-uranospinite
Calciumhilgardite .... Hilgardite
Canbyite .... Hisingerite
Caratiite .... Piypite
Carbonado .... Diamond
Carbonate-apatite .... Carbonate-fluorapatite
Carbonate-apatite .... Carbonate-hydroxylapatite
Carborundum Moissanite-6H
Carnelian .... Quartz
Carpathite .... Karpatite
Carphosiderite .... Hydronium jarosite
Caryocerite .... Melanocerite-(Ce)
Cathophorite .... Brabantite
Catoptrite .... Katoptrite
Celestite .... Celestine
Cenosite .... Kainosite-(Y)
Cerargyrite .... Chlorargyrite
Cerphosphorhuttonite .... Huttonite
Ceruranopyrochlore .... Pyrochlore
Cervantite .... Molybdite
Cestibtantite .... Cesstibtantite
Chalcedony .... Quartz
Chalcolamprite .... Pyrochlore
Chalcolite .... Torbernite
Chalcopyrrhotite .... Isocubanite
Chalcosine .... Chalcocite
Chalcosite .... Chalcocite
Chalcotrichite .... Cuprite
Challantite .... Ferricopiapite
Chalybite .... Siderite
Chengbolite .... Moncheite
Chessylite .... Azurite
Chiastolite .... Andalusite
Chile saltpeter .... Nitratine
Chile-loweite .... Humberstonite
Chloanthite .... Nickel-skutterudite
Chlorarsenian .... Allactite
Chlorhastingsite .... Hastingsite
Chlormanasseite .... Chlormagaluminite
Chloropal .... Nontronite
Chlorotile .... Agardite-(Y)
Chromdisthene .... Kyanite
Chrome-tremolite .... Actinolite
Chrome-tremolite .... Tremolite
Chromephlogopite .... Phlogopite
Chrominium .... Phoenicochroite
Chromrutile .... Redledgeite
Chromsteigerite .... Steigerite
Chrysolite .... Fayalite
Chrysolite .... Forsterite
Chrysotile .... Clinochrysotile
Chrysotile .... Orthochrysotile
Chrysotile .... Parachrysotile
Cl-tyretskite Hilgardite-1Tc
Cleavelandite .... Albite
Cliftonite Graphite-2H
Clino-anthophyllite .... Magnesio-cummingtonite
Clinoeulite .... Clinoferrosilite
Clinohypersthene .... Clinoenstatite
Clinohypersthene .... Clinoferrosilite
Clinokupfferite .... Cummingtonite
Clinostrengite .... Phosphosiderite
Clinovariscite .... Metavariscite
Cobalt-frohbergite .... Frohbergite
Cobaltocalcite .... Sphaerocobaltite
Coeruleite .... Ceruléite
Collophane .... Carbonate-fluorapatite
Collophane .... Carbonate-hydroxylapatite
Columbite .... Ferrocolumbite
Columbite .... Magnocolumbite
Columbite .... Manganocolumbite
Columbite .... Yttrocolumbite
Columbomicrolite .... Pyrochlore
Copiapite-zinc .... Zincocopiapite
Copperas .... Melanterite
Coronene .... Karpatite
Cosmochlore .... Kosmochlor
Cossyrite .... Ænigmatite
Coutinite .... Lanthanite-(Nd)
Crasite .... Yttrocrasite-(Y)
Crocidolite .... Riebeckite
Cryptomorphite .... Ginorite
Cubic Chalcopyrite .... Putoranite
Cubic Cubanite .... Isocubanite
Cumengeite .... Cumengite
Cummingtonite .... Magnesio-cummingtonite
Cuproartinite .... Nakauriite
Cuprohydromagnesite .... Nakauriite
Cuprouranite .... Torbernite
Curtisite .... Idrialite
Cyanite .... Kyanite
Cyanophillite .... Cyanophyllite
Cyrtolite .... Zircon
Dachiardite sodium .... Sodium dachiardite
Dahlite .... Carbonate-hydroxylapatite
Dansite .... D'Ansite
Daphnite .... Chamosite

Dehrnite ... Fluorapatite
Delatorreite ... Todorokite
Delorenzite ... Tanteuxenite-(Y)
Delrioite-meta ... Metadelrioite
Delta-mooreite ... Torreyite
Desmine ... Stilbite
Destinezite ... Diadochite
Devillite ... Devilline
Dialogite ... Rhodochrosite
Didymolite ... Plagioclase
Dillnite ... Zunyite
Disthene ... Kyanite
Djalmaite ... Uranmicrolite
Domeykite-beta ... Metadomeykite
Doverite ... Synchysite-(Y)
Droogmansite ... Kasolite
Duftite-alpha ... Duftite
Dysanalyte ... Perovskite
Eardleyite ... Takovite
Ebelmenite ... Cryptomelane
Eckermannite ... Ferro-eckermannite
Eckrite ... Winchite
Edenite ... Ferro-edenite
Eggonite ... Kolbeckite
Eisenrichterite ... Ferro-richterite
Ektropite ... Caryopilite
Electrum ... Gold
Electrum ... Silver-3C
Ellestadite ... Chlorellestadite
Ellestadite ... Fluorellestadite
Ellestadite ... Hydroxylellestadite
Ellsworthite ... Uranpyrochlore
Ellweilerite ... Sodium-uranospinite
Embolite ... Bromargyrite
Embolite ... Chlorargyrite
Emerald ... Beryl
Emery ... Corundum
Endeiolite ... Pyrochlore
Endellionite ... Bournonite
Endlichite ... Vanadinite
Enigmatite ... Ænigmatite
Ephesite-2M1 ... Ephesite
Epidesmine ... Stilbite
Epiianthinite ... Schoepite
Erubescite ... Bornite
Eschynite ... Æschynite
Eschynite ... Æschynite-(Y)
Eucolite ... Eudialyte
Exitele ... Valentinite
Fahlerz ... Tennantite
Fahlerz ... Tetrahedrite
Fahlore ... Tennantite
Fahlore ... Tetrahedrite
Feldspath ... Feldspar
Femaghastingsite ... Magnesio-hastingsite
Femolite ... Molybdenite-2H
Fengluanite ... Isomertieite
Feranthophyllite ... Ferro-anthophyllite
Fergusonite-beta ... Beta-fergusonite
Ferisilicite ... Fersilicite
Ferri-edenite ... Ferro-edenite
Ferrifayalite ... Laihunite
Ferriglaucophane ... Magnesio-riebeckite
Ferrihedrite ... Gedrite
Ferrimolybdite ... Molybdite
Ferripumpellyite ... Julgoldite-(Mg)
Ferririchterite ... Magnesio-arfvedsonite
Ferro-tremolite ... Ferro-actinolite
Ferroalunite ... Alunite
Ferrobabingtonite ... Babingtonite
Ferrocopiapite ... Copiapite
Ferrofillowite ... Johnsomervilleite
Ferrohalotrichite ... Halotrichite
Ferrohastingsite ... Hastingsite
Ferrohornblende ... Alumino-ferro-hornblende
Ferroplatinum ... Isoferroplatinum
Ferroplatinum ... Tetraferroplatinum
Ferropseudobrookite ... Armalcolite
Ferropumpellyite ... Pumpellyite-(Fe^{2+})
Ferroschallerite ... Nelenite
Ferrosilite ... Clinoferrosilite
Ferrostibian ... Långbanite
Ferroszaibelyite ... Szaibelyite
Ferutite ... Davidite-(La)
Feuermineral ... Mawsonite
Fluochlore ... Pyrochlore
Fluorapatite ... Apatite
Fluorene ... Kratochvilite
Flusspat ... Fluorite
Foucherite ... Delvauxite
Fowlerite ... Rhodonite
Francolite ... Carbonate-fluorapatite
Fritzcheite ... Fritzscheite
Fuchsite ... Muscovite-2M1
Fueloeppite ... Fülöppite
Galenite ... Galena
Gamisgradite ... Edenite
Gamisgradite ... Magnesio-hornblende
Gastaldite ... Ferro-glaucophane
Gastaldite ... Glaucophane
Gearksite ... Gearksutite
Gedrite ... Ferro-gedrite
Gedrite ... Magnesio-gedrite
Gedrite sodium ... Sodium-gedrite
Gelzircon ... Zircon
Geneveite ... Theisite
Gersbyite ... Lazulite
Giobertite ... Magnesite
Gips ... Gypsum
Girnarite ... Magnesio-hastingsite
Glaserite ... Aphthitalite
Glauber's salt ... Mirabilite
Glaucophane ... Ferro-glaucophane
Glimmer ... Muscovite-2M1
Glockerite ... Lepidocrocite
Glottalite ... Chabazite
Goergyite ... Görgeyite
Goetzenite ... Götzenite
Goshenite ... Beryl
Goureite ... Narsarsukite
Grammatite ... Tremolite
Granat ... Andradite
Granat ... Grossular
Grandite ... Andradite
Grandite ... Grossular
Griqualandite ... Magnesio-riebeckite
Grossularite ... Grossular
Grothine ... Norbergite
Grovesite ... Pennantite
Grunlingite ... Bismuthinite

Name	Species
Grunlingite	Joséite-A
Guanglinite	Isomertieite
Hackmanite	Sodalite
Haddamite	Microlite
Haeggite	Häggite
Haematite	Hematite
Hagendorfite-ferro	Ferrohagendorfite
Halloysite-10A	Endellite
Hanleite	Uvarovite
Hastingsite	Magnesio-hastingsite
Hatchettolite	Uranpyrochlore
Hauynite	Haüyne
Heikkolite	Crossite
Heikolite	Crossite
Heliodor	Beryl
Hemafibrite	Synadelphite
Henwoodite	Turquoise
Herrengrundite	Devilline
Hexagonite	Tremolite
Hexastibiopalladite	Sudburyite
Hiddenite	Spodumene
Hillangsite	Dannemorite
Hoeferite	Chapmanite
Hoepfnerite	Tremolite
Hoernesite	Hörnesite
Hogtveitite	Thalenite-(Y)
Hohmannite-meta	Metahohmannite
Hokutolite	Barite
Holmquistite	Ferro-holmquistite
Holmquistite	Magnesio-holmquistite
Hornblende	Alumino-ferro-hornblende
Hornblende	Magnesio-hornblende
Hornesite	Manganese-hörnesite
Hörnesite	Manganese-hörnesite
Hortonolite	Fayalite
Hoshiite	Magnesite
Hsing-hua-shih	Hsianghualite
Huangheite	Huanghoite-(Ce)
Hudsonite	Hastingsite
Huebnerite	Hübnerite
Huegelite	Hügelite
Huhnerkobelite	Alluaudite
Huhnerkobelite	Ferroalluaudite
Hyacinth	Zircon
Hyalite	Opal
Hydrargillite	Gibbsite
Hydrated Halloysite	Endellite
Hydrocalcite	Monohydrocalcite
Hydrochlorborate	Hydrochlorborite
Hydrochlore	Pyrochlore
Hydrocyanite	Chalcocyanite
Hydrogen Autunite	Chernikovite
Hydrogrossular	Hibschite
Hydrogrossular	Katoite
Hydrohalloysite	Endellite
Hydroparagonite	Brammallite
Hydropyrochlore	Pyrochlore
Hypochlorite	Bismutoferrite
Idocrase	Vesuvianite
Igdloite	Lueshite
Imerinite	Magnesio-arfvedsonite
Imgreite	Melonite
Imogolite	Allophane
Iodobromite	Bromargyrite
Iodyrite	Iodargyrite
Iolite	Cordierite
Iozite	Wüstite
Iquiquieite	Iquiqueite
Iridisite-beta	Beta-iridisite
Iridosmium	Iridosmine
Iron-anthophyllite	Ferro-anthophyllite
Iron-richterite	Ferro-richterite
Isabellite	Richterite
Isoplatinocopper	Copper
Isoplatinocopper	Hongshiite
Isowolframite	Wolframite
Jade	Actinolite
Jade	Jadeite
Jasper	Quartz
Jefferisite	Vermiculite
Jeffersonite	Augite
Jeffersonite	Ægirine
Jenkinsite	Antigorite
Ježekite	Morinite
Johnstonotite	Spessartine
Josephinite	Awaruite
Josephinite	Kamacite
Josephinite	Taenite
Josephinite	Tetrataenite
Juddite	Arfvedsonite
Juddite	Magnesio-arfvedsonite
K-feldspar	Microcline
K-feldspar	Orthoclase
Kaemmererite	Clinochlore
Kaersutite	Ferro-kaersutite
Kalamite	Tremolite
Kamarezite	Brochantite
Kämmererite	Clinochlore
Kasoite	Celsian
Katophorite	Alumino-katophorite
Katophorite	Ferri-katophorite
Katophorite	Magnesio-alumino-katophorite
Katophorite	Magnesio-ferri-katophorite
Keatite	Quartz
Keilhauite	Titanite
Keivyite	Keiviite
Kennedyite	Armalcolite
Kennedyite	Pseudobrookite
Kerolite	Talc
Khademite	Rostite
Khlopinite	Samarskite-(Y)
Khuniite	Iranite
Kievite	Cummingtonite
Kirchheimerite-meta	Metakirchheimerite
Kirkite	Kirkiite
Kirwanite	Ferri-ferro-hornblende
Klinozoisite	Clinozoisite
Klipsteinite	Neotocite
Knebelite	Fayalite
Knipovichite	Alumohydrocalcite
Koettigite	Köttigite
Kondorite	Konderite
Kongsbergite	Silver-3C
Kongsbergite	Silver-4H
Koppite	Pyrochlore
Kotschubeite	Clinochlore
Kozhanovite	Karnasurtite-(Ce)
Kroehnkite	Kröhnkite
Kruzhanovskite	Kryzhanovskite
Kularite	Monazite

Kunzite .... Spodumene
Kupfferite .... Magnesio-anthophyllite
Kurgantaite Tyretskite-1Tc
Kurskite .... Carbonate-fluorapatite
Kurskite .... Carbonate-hydroxylapatite
Küstelite .... Silver
Kusuite .... Wakefieldite-(Ce)
Kylindrite .... Cylindrite
Lapis lazuli .... Lazurite
Lapparentite .... Rostite
Lavendulan-zinc .... Zinclavendulan
Lavenite .... Låvenite
Lavrovite .... Diopside
Lehiite .... Crandallite
Leonhardite .... Laumontite
Leonhardtite .... Starkeyite
Lepidolite Lepidolite-2M1
Lepidolite-2M Lepidolite-2M1
Lepidomelane .... Biotite
Lesserite .... Inderite
Leuchtenbergite .... Clinochlore
Leucophanite .... Leucophane
Levynite .... Levyne
Lewistonite .... Fluorapatite
Limonite .... Goethite
Lingaitukuang .... Brabantite
Linneite .... Linnaeite
Linosite .... Kaersutite
Lithionglaukophan .... Holmquistite
Lithiophosphatite .... Lithiophosphate
Lithium-amphibole .... Holmquistite
Liujinyinite .... Uytenbogaardtite
Lodevite-meta .... Metalodevite
Lodochnikite .... Brannerite
Loellingite .... Löllingite
Loeweite .... Löweïte
Lomonosovite-beta .... Beta-lomonosovite
Lueneburgite .... Lüneburgite
Lunokite .... Lun'okite
Lusakite .... Staurolite
Macallisterite .... Mcallisterite
Maekinenite .... Mäkinenite
Maganthophyllite .... Magnesio-anthophyllite
Magnesian glaucophane .... Glaucophane
Magnesiotaramite .... Magnesio-alumino-taramite
Magnesium orthite .... Dollaseite-(Ce)
Magnesium szomolnokite .... Szomolnokite
Magnesiumorthite .... Dollaseite-(Ce)
Magnetostibian .... Jacobsite
Magnioborite .... Suanite
Magnodravite .... Dravite
Makedonite .... Macedonite
Malacon .... Zircon
Manganamphibole .... Rhodonite
Mangan crocidolite .... Riebeckite
Mangan krokidolith .... Riebeckite
Mangan tremolite .... Tremolite
Mangan-actinolite .... Actinolite
Manganandalusite .... Andalusite
Mangano-anthophyllite .... Tirodite
Manganomelane .... Romanechite
Manganomossite .... Manganocolumbite
Manganophyllite .... Biotite
Manganosteenstrupine .... Steenstrupine-(Ce)
Mangansеverginite .... Manganaxinite
Mangantapiolite .... Manganotapiolite
Manganuralite .... Arfvedsonite
Marignacite .... Ceriopyrochlore-(Ce)
Mariposite Muscovite-2M1
Marmairolite .... Richterite
Marmatite .... Sphalerite
Martite .... Hematite
Matildite-alpha .... Schapbachite
Mayakite .... Majakite
Mboziite .... Taramite
Mcconnelite .... Mcconnellite
Meerschaum .... Sepiolite
Melaconite .... Tenorite
Melanite .... Andradite
Melilite .... Åkermanite
Melilite .... Gehlenite
Melnikovite .... Greigite
Mendelejevite .... Betafite
Mendelyeevite .... Betafite
Merrillite .... Whitlockite
Meta-uranocircite-17A .... Meta-uranocircite I
Metahalloysite Halloysite-7A
Metalomonosovite .... Beta-lomonosovite
Metasideronatrite II .... Metasideronatrite
Metasimpsonite .... Microlite
Metastrengite .... Phosphosiderite
Mica Muscovite-2M1
Mimetesite .... Mimetite
Mindigite Heterogenite-3R
Minguettite .... Stilpnomelane
Mirupolskite .... Bassanite
Mispickel .... Arsenopyrite
Mizzonite .... Marialite
Mizzonite .... Meionite
Mohsite .... Crichtonite
Moissanite-beta .... Beta-moissanite
Montasite .... Grunerite
Montesite .... Herzenbergite
Mooreite-delta .... Torreyite
Morganite .... Beryl
Morion .... Quartz
Mossite .... Ferrocolumbite
Mossite .... Tantalite
Mossite .... Tapiolite
Muchuanite Molybdenite-2H
Mumbite .... Plumbomicrolite
Munkrudite .... Kyanite
Musgravite-9R .... Musgravite
Musgravite-18R .... Musgravite
Na-dachiardite .... Sodium dachiardite
Namaqualite .... Cyanotrichite
Nanekevite .... Bario-orthojoaquinite
Nasturan .... Uraninite
Natrongrammatit .... Richterite
Natronrichterite .... Richterite
Nemalite .... Brucite
Neodigenite .... Digenite
Neodymium churchite .... Churchite-(Nd)
Neomesselite .... Messelite
Neotantalite .... Microlite
Nephrite .... Actinolite
Nevyanskite .... Iridosmine
Niccolite .... Nickeline
Nickel-iron .... Kamacite
Nickel-iron .... Taenite

Nickel-iron .... Tetrataenite
Nickel-kerolite .... Willemseite
Nickelite .... Nickeline
Nimesite .... Brindleyite
Niobian rutile .... Ilmenorutile
Nioboloparite .... Loparite-(Ce)
Niobozirconolite .... Zirkelite
Niobpyrochlore .... Pyrochlore
Niobtantalpyrochlore .... Microlite
Niobtantalpyrochlore .... Pyrochlore
Nisaite .... Phurcalite
Niter .... Nitratine
Nitre .... Nitratine
Nitroglauberite .... Darapskite
Nordenskioldite .... Tremolite
Noselite .... Nosean
Novacekite-meta .... Metanovacekite
Obruchevite .... Yttropyrochlore-(Y)
Octahedrite .... Anatase
Oligiste .... Hematite
Oligonite .... Siderite
Olivine .... Fayalite
Olivine .... Forsterite
Olovotantalite .... Manganotantalite
Opsimose .... Neotocite
Orizite .... Epistilbite
Orthite .... Allanite
Ortho-armalcolite .... Armalcolite
Orthoferrosilite .... Ferrosilite
Ortholomonosovite .... Lomonosovite
Orthoriebeckite .... Riebeckite
Orthose .... Orthoclase
Orthozoisite .... Zoisite
Oryzite .... Epistilbite
Osannite .... Riebeckite
Osumilite-(K,Mg) .... Osumilite-(Mg)
Oxyferropumpellyite .... Pumpellyite-(Fe^{3+})
Oxyjulgoldite .... Julgoldite-(Fe^{3+})
Paeaekkoenenite .... Pääkkönenite
Paigeite .... Vonsenite
Panabase .... Tetrahedrite
Pandaite .... Bariopyrochlore
Para-armalcolite .... Armalcolite
Paragonite Paragonite-2M
Parahilgardite Hilgardite-3Tc
Parapectolite Pectolite-M2abc
Parawollastonite Wollastonite-2M
Pargasite .... Ferro-pargasite
Partridgeite .... Bixbyite
Paternoite .... Kaliborite
Pendletonite .... Karpatite
Penginite .... Penzhinite
Pennine .... Clinochlore
Penninite .... Clinochlore
Penwithite .... Neotocite
Peridot .... Forsterite
Persian Red .... Hematite
Pharaonite .... Davyne
Pharmacosiderite Na .... Sodium pharmacosiderite
Phengite Muscovite-2M1
Phoenicite .... Magnesium-chlorophoenicite
Phosphochromite .... Variscite
Phosphothorogummite .... Thorogummite
Picotite .... Spinel
Picroamosite .... Anthophyllite
Piedmontite .... Piemontite
Pilinite .... Bavenite
Pimelite .... Willemseite
Pitchblende .... Uraninite
Plazolite .... Hibschite
Pleonaste .... Spinel
Pleonectite .... Hedyphane
Plumbago .... Graphite
Plumbozincocalcite .... Calcite
Plumosite .... Boulangerite
Polianite .... Pyrolusite
Polymignite Zirconolite-30
Polymignyte Zirconolite-30
Potash Alum .... Potassium Alum
Potash feldspar .... Microcline
Potash feldspar .... Orthoclase
Potash feldspar .... Sanidine
Potassium feldspar .... Microcline
Potassium feldspar .... Orthoclase
Potassium feldspar .... Sanidine
Pourayite .... P-Ourayite
Priorite .... Æschynite-(Y)
Prismatic schillerspar .... Anthophyllite
Protoastrakhanite .... Konyaite
Pseudoglaucophane .... Crossite
Pseudoglaucophane .... Glaucophane
Pseudoixiolite .... Ixiolite
Pseudomesolite .... Mesolite
Pseudonatrolite .... Mordenite
Psilomelane .... Romanechite
Ptilolite .... Mordenite
Pyralspite .... Almandine
Pyralspite .... Pyrope
Pyralspite .... Spessartine
Pyrochlore-microlite .... Microlite
Pyrochlore-microlite .... Pyrochlore
Pyrosmalite .... Ferropyrosmalite
Pyrosmalite .... Manganpyrosmalite
Pyrrhite Pyrochlore .... Pyrochlore
Pyrrhoarsenite .... Berzeliite
Ramsayite .... Lorenzenite
Ranquilite .... Haiweeite
Raphilite .... Tremolite
Raphisiderite .... Hematite
Rathite-II .... Liveingite
Realgar .... Realgar (high)
Renardite .... Dewindtite
Retinostibian .... Welinite
Rezhikite .... Magnesio-arfvedsonite
Rezhikite .... Magnesio-riebeckite
Rhabdite .... Schreibersite
Rhodoarsenian .... Rhodonite
Rhodusite .... Magnesio-riebeckite
Rhoenite .... Rhönite
Richterite .... Ferro-richterite
Riebeckite .... Magnesio-riebeckite
Rijkeboerite .... Bariomicrolite
Rinkite .... Mosandrite
Rinkolite .... Mosandrite
Ripido lite .... Clinochlore
Rock salt .... Halite
Roemerite .... Römerite
Roepperite .... Fayalite
Roepperite .... Tephroite
Roesslerite .... Rösslerite

Rogersite	Churchite-(Y)
Rokuehnite	Rokühnite
Roselite-beta	Beta-roselite
Roselite-zinc	Zincroselite
Rossite-meta	Metarossite
Royite	Quartz
Rubellite	Elbaite
Ruby	Corundum
Ruby silver	Proustite
Ruby silver	Pyrargyrite
Rutherfordite	Rutherfordine
Sacchite	Sakhaite
Sahlite	Diopside
Salite	Diopside
Saltpeter	Niter
Samiresite	Uranpyrochlore
Sapphire	Corundum
Sard	Quartz
Schefferite	Ægirine
Scheibeite	Phoenicochroite
Schizolite	Pectolite
Schoenite	Picromerite
Schoepite I	Schoepite
Schoepite II	Metaschoepite
Schoepite III	Paraschoepite
Schönite	Picromerite
Schroeckingerite	Schröckingerite
Schuchardtite	Clinochlore
Schuchardtite	Vermiculite
Schulzenite	Heterogenite
Schwatzite	Tetrahedrite
Schwazite	Tetrahedrite
Scleroclase	Sartorite
Sebesite	Tremolite
Selenite	Gypsum
Selenjoseite	Laitakarite
Selenolite	Olsacherite
Septetalc-chlorite	Baumite
Sericite	Muscovite-2M1
Sericite	Paragonite-2M
Shadlunite	Manganese-shadlunite
Shahovite	Shakhovite
Silbolite	Actinolite
Silfbergite	Dannemorite
Silicomanganberzeliite	Berzeliite
Sillbolite	Actinolite
Simpsonite	Richterite
Sjoegrenite	Sjögrenite
Sjögrufvite	Caryinite
Skutterudite	Nickel-skutterudite
Slavyanskite	Tunisite
Smaltite	Skutterudite
Smaragd	Beryl
Smaragdite	Actinolite
Smaragditic txchermakite	Tschermakite
Smirnovite	Thorutite
Smolyaninovite	Smolianinovite
Soapstone	Talc
Sobotkite	Saponite
Soda asbestos	Magnesio-arfvedsonite
Soda	Natron
Soda feldspar	Albite
Soda hornblende	Arfvedsonite
Soda Niter	Nitratine
Soda richterite	Richterite
Soda tremolite	Richterite
Sodium feldspar	Albite
Sodium phlogopite	Wonesite
Sodium-metauranospinite	Meta-Na-uranospinite
Soehngeite	Söhngeite
Soerensenite	Sörensenite
Soretite	Hastingsite
Specularite	Hematite
Spencite	Tritomite-(Y)
Spessartite	Spessartine
Sphene	Titanite
Spherocobaltite	Sphaerocobaltite
Sterretite	Kolbeckite
Stibiopearceite	Antimonpearceite
Stipoverite	Stishovite
Strahlstein	Actinolite
Stratopeite	Neotocite
Strelite	Actinolite
Strelite	Anthophyllite
Strontiohilgardite	Tyretskite-1Tc
Strontiohilgardite-1Tc	Hilgardite-1Tc
Strontiumthomsonite	Thomsonite
Strunzite-ferro	Ferrostrunzite
Subglaucophane	Crossite
Sukulaite	Stannomicrolite
Sulfur	Sulphur
Sulrhodite	Bowieite
Svetlozarite	Dachiardite
Svidneite	Riebeckite
Svitalskite	Celadonite
Switzerite	Metaswitzerite
Switzerite-meta	Metaswitzerite
Syntagmatite	Hastingsite
Szajbelyite	Szaibelyite
Szechenyiite	Richterite
Taaffeite	Taaffeite-8H
Taaffeite-9R	Musgravite
Tachyhdrite	Tachyhydrite
Tagilite	Pseudomalachite
Taiyite	Æschynite-(Y)
Tangeite	Calciovolborthite
Tantalbetafite	Betafite
Tantalhatchettolite	Uranmicrolite
Tantalpyrochlore	Microlite
Tanzanite	Zoisite
Taprobanite	Taaffeite-8H
Taramite	Alumino-taramite
Taramite	Ferri-taramite
Taramite	Magnesio-ferri-taramite
Tatarkaite	Clinochlore
Tavistockite	Hydroxyl-apatite
Taylorite	Arcanite
Taylorite	Mascagnite
Tellurbismuth	Tellurobismuthite
Tellurite-ferro	Ferrotellurite
Teremkovite	Owyheeite
Ternovskite	Magnesio-riebeckite
Tetra-auricupride	Tetraauricupride
Tetrakalsilite	Panunzite
Thalackerite	Anthophyllite
Thierschite	Whewellite
Thulite	Zoisite
Thuringite	Chamosite
Tibergite	Magnesio-hastingsite
Tin-tantalite	Manganotantalite

Tincal .... Borax
Titanaugite .... Augite
Titanbetafite .... Betafite
Titanclinohumite .... Clinohumite
Titanhornblende .... Ænigmatite
Titano-obruchevite .... Yttrobetafite-(Y)
Titanomaghemite .... Maghemite
Titanorhabdophane .... Tundrite-(Ce)
Tobermorite (14A) .... Tobermorite
Topazolite .... Andradite
Torbernite-meta .... Metatorbernite
Torendrikite .... Magnesio-riebeckite
Transvaalite Heterogenite-3R
Tremolite-glaucophane .... Richterite
Trichalcite .... Tyrolite
Tridamite .... Tridymite
Triphane .... Spodumene
Troegerite .... Trögerite
Troostite .... Willemite
Truestedtite .... Trüstedtite
Tsavolite .... Grossular
Tscheffkinite .... Chevkinite-(Ce)
Tschermakite .... Alumino-tschermakite
Tschermakite .... Ferri-tschermakite
Tschermakite .... Ferro-alumino-tschermakite
Tschermakite .... Ferro-ferri-tschermakite
Tsilaisite .... Elbaite
Turgite .... Hematite
Turite .... Götzenite
Turjite .... Hematite
Turuqois .... Turquoise
Tusiite .... Calciocopiapite
Tyretskite Tyretskite-1Tc
Tysonite .... Fluocerite-(Ce)
Ufertite .... Davidite-(La)
Ultramarine .... Lazurite
Ulvite .... Ulvöspinel
Ulvoespinel .... Ulvöspinel
Uralite .... Actinolite
Uranocircite-meta II .... Meta-uranocircite II
Uranophane-alpha .... Uranophane
Uranophane-beta .... Beta-uranophane
Uranopilite-meta .... Meta-uranopilite
Uranospinite sodium .... Sodium-uranospinite
Uranospinite-meta .... Meta-uranospinite
Uranotile .... Uranophane-alpha
Uranotile-beta .... Uranophane-beta
Ureyite .... Kosmochlor
Usonite .... Uzonite
Uzbekite .... Volborthite
Vaeyrynenite .... Väyrynenite
Valleite .... Anthophyllite
Vandenbrandite .... Vandenbrandeite
Vandendriesscheite II ... Metavandendriesscheïte
Vanmeersscheite-meta ..... Metavanmeersscheite
Vanuralite-meta .... Metavanuralite
Variscite-meta .... Metavariscite
Varlamoffite .... Cassiterite
Vauxite-meta .... Metavauxite
Vegasite .... Plumbojarosite
Vernadskite .... Andalusite
Vernadskite .... Antlerite
Viridine .... Andalusite
Vivianite-meta .... Metavivianite
Volchonskoite .... Volkonskoite
Voltine-meta .... Metavoltine
Vorobievite .... Beryl
Waldheimite .... Richterite
Wathlingite .... Kieserite
Weinschenkite .... Churchite-(Y)
Weinschenkite .... Magnesio-hastingsite
Wellsite .... Phillipsite
Wernerite .... Marialite
Wernerite .... Meionite
Westgrenite .... Bismutomicrolite
Wilhendersonite .... Willhendersonite
Wilkeite .... Apatite
Wilkeite .... Fluorellestadite
Wimsite .... Vimsite
Winchite .... Alumino-winchite
Winchite .... Ferri-winchite
Winchite .... Ferro-alumino-winchite
Winchite .... Ferro-ferri-winchite
Wittingite .... Neotocite
Woehlerite .... Wöhlerite
Woelsendorfite .... Wölsendorfite
Wolchonskoite .... Volkonskoite
Wood Tin .... Cassiterite
Woodfordite .... Ettringite
Wuelfingite .... Wülfingite
Wuestite .... Wüstite
Wurtzite Wurtzite-2H
Wyartite I .... Wyartite
Wyllieite-ferro .... Ferrowyllieite
Xanthophyllite .... Clintonite
Xilingoite .... Xilingolite
Xinganite .... Hingganite-(Y)
Yanzhongite .... Kotulskite
Yeelimite .... Ye'elimite
Yenshanite .... Vysotskite
Yokosukaite .... Nsutite
Yttro-orthite .... Allanite-(Y)
Yttrocerite .... Fluorite
Yttrohatchettolite .... Yttropyrochlore-(Y)
Zillerite .... Actinolite
Zillerthite .... Actinolite
Zinc blende .... Sphalerite
Zincblende .... Sphalerite
Zinckenite .... Zinkenite
Zincosite .... Zinkosite
Zippeite .... Cobalt-zippeite
Zippeite .... Magnesium-zippeite
Zippeite .... Sodium-zippeite
Zippeite-nickel .... Nickel-zippeite
Zirconolite Zirconolite-2M
Zirconolite .... Zirkelite
Zirkophyllite .... Zircophyllite
Zirlite .... Gibbsite